Springer Proceedings in Advanced Robotics 22

Series Editors

Bruno Siciliano
Dipartimento di Ingegneria Elettrica
e Tecnologie dell'Informazione
Università degli Studi di Napoli
Federico II
Napoli, Napoli
Italy

Oussama Khatib
Robotics Laboratory
Department of Computer Science
Stanford University
Stanford, CA
USA

Advisory Editors

Gianluca Antonelli, Department of Electrical and Information Engineering, University of Cassino and Southern Lazio, Cassino, Italy

Dieter Fox, Department of Computer Science and Engineering, University of Washington, Seattle, WA, USA

Kensuke Harada, Engineering Science, Osaka University Engineering Science, Toyonaka, Japan

M. Ani Hsieh, GRASP Laboratory, University of Pennsylvania, Philadelphia, PA, USA

Torsten Kröger, Karlsruhe Institute of Technology, Karlsruhe, Germany

Dana Kulic, University of Waterloo, Waterloo, ON, Canada

Jaeheung Park, Department of Transdisciplinary Studies, Seoul National University, Suwon, Korea (Republic of)

The Springer Proceedings in Advanced Robotics (SPAR) publishes new developments and advances in the fields of robotics research, rapidly and informally but with a high quality.

The intent is to cover all the technical contents, applications, and multidisciplinary aspects of robotics, embedded in the fields of Mechanical Engineering, Computer Science, Electrical Engineering, Mechatronics, Control, and Life Sciences, as well as the methodologies behind them.

The publications within the "Springer Proceedings in Advanced Robotics" are primarily proceedings and post-proceedings of important conferences, symposia and congresses. They cover significant recent developments in the field, both of a foundational and applicable character. Also considered for publication are edited monographs, contributed volumes and lecture notes of exceptionally high quality and interest.

An important characteristic feature of the series is the short publication time and world-wide distribution. This permits a rapid and broad dissemination of research results.

Indexed by SCOPUS, SCIMAGO, WTI Frankfurt eG, zbMATH.

All books published in the series are submitted for consideration in Web of Science.

More information about this series at https://link.springer.com/bookseries/15556

Fumitoshi Matsuno · Shun-ichi Azuma ·
Masahito Yamamoto
Editors

Distributed Autonomous Robotic Systems

15th International Symposium

 Springer

Editors
Fumitoshi Matsuno
Mechanical Engineering and Science
Kyoto University
Kyoto, Japan

Shun-ichi Azuma
Mechanical Systems Engineering
Nagoya University
Nagoya, Japan

Masahito Yamamoto
Computer Science and Information Technology
Hokkaido University
Sapporo, Hokkaido, Japan

ISSN 2511-1256   ISSN 2511-1264 (electronic)
Springer Proceedings in Advanced Robotics
ISBN 978-3-030-92789-9   ISBN 978-3-030-92790-5 (eBook)
https://doi.org/10.1007/978-3-030-92790-5

© The Editor(s) (if applicable) and The Author(s), under exclusive license to Springer Nature Switzerland AG 2022

This work is subject to copyright. All rights are solely and exclusively licensed by the Publisher, whether the whole or part of the material is concerned, specifically the rights of translation, reprinting, reuse of illustrations, recitation, broadcasting, reproduction on microfilms or in any other physical way, and transmission or information storage and retrieval, electronic adaptation, computer software, or by similar or dissimilar methodology now known or hereafter developed.

The use of general descriptive names, registered names, trademarks, service marks, etc. in this publication does not imply, even in the absence of a specific statement, that such names are exempt from the relevant protective laws and regulations and therefore free for general use.

The publisher, the authors and the editors are safe to assume that the advice and information in this book are believed to be true and accurate at the date of publication. Neither the publisher nor the authors or the editors give a warranty, expressed or implied, with respect to the material contained herein or for any errors or omissions that may have been made. The publisher remains neutral with regard to jurisdictional claims in published maps and institutional affiliations.

This Springer imprint is published by the registered company Springer Nature Switzerland AG
The registered company address is: Gewerbestrasse 11, 6330 Cham, Switzerland

Preface

DARS-SWARM2021 is a joint symposium of the 15th International Symposium on Distributed Autonomous Robotic Systems (DARS2021) and the 4th International Symposium on Swarm Behavior and Bio-Inspired Robotics (SWARM2021). This joint symposium was originally scheduled to take place in November 2020 in Kyoto, Japan. However, due to the COVID-19 pandemic, we decided to postpone it until June 2021, in the hope that we would be able to meet in person in Kyoto by that time. As the global pandemic was still ongoing, DARS-SWARM2021 was held June 1–4, 2021, as an online meeting.

The International Symposium on Distributed Autonomous Robotic Systems (DARS) has a history of 29 years. Distributed robotics is an interdisciplinary area that combines research in computer science, communication systems and engineering. Distributed robotic systems are expected to solve complex problems autonomously while operating in highly unstructured real-world environments. The International Symposium on Swarm Behavior and Bio-Inspired Robotics (SWARM) was started in 2015. Living beings have evolved through natural selection to move in swarms. A swarm can do many things that its individuals cannot do alone. Swarms in nature can not only adapt to their environments, but can also construct suitable habitats to their own advantage. As we move to the future, a constructive understanding of the intelligence of living things will be pivotal for progress in biology and engineering.

This joint DARS-SWARM2021 symposium aims to create a bridge between biologists and engineers interested in the distributed intelligence of living things and to establish a new academic field by integrating knowledge from both disciplines. For example, the analysis and control of human swarm behaviors can improve our understanding of viral transmission mechanisms and may lead to the development of solutions to mitigate for the effects of the COVID-19 pandemic.

The 33 papers in this book have been presented at DARS2021. These papers provide a very good overview of the state of the art in distributed robotic systems (DARS). They reflect current research themes in DARS with important contributions. We hope that this book helps to sustain the interest in DARS and triggers new research.

We sincerely hope that you, your families and colleagues remain safe and in good health through these challenging times.

Fumitoshi Matsuno
Masahito Yamamoto
Shun-ichi Azuma

Organization

Executive Committee

General Chair

Fumitoshi Matsuno — Kyoto University, Japan

Program Co-chairs

Shun-ichi Azuma — Nagoya University, Japan
Masahito Yamamoto — Hokkaido University, Japan

Vice Program Co-chairs

Hyo-Sung Ahn — Gwangju Institute of Science and Technology, Korea
Julien Bourgeois — Univ. Bourgogne Franche-Comté, France
Nikolaus Correll — University of Colorado Boulder, USA

Publication Chairs

Keitaro Naruse — The University of Aizu, Japan
Ikuo Suzuki — Kitami Institute of Technology, Japan

Finance Chair

Kazuyuki Ito — Hosei University, Japan

Publicity Chair

Toshiyuki Yasuda — University of Toyama, Japan

Registration Chairs

Tetsushi Kamegawa	Okayama University, Japan
Motoyasu Tanaka	The University of Electro-Communications, Japan

Local Arrangement Chairs

Shinya Aoi	Kyoto University, Japan
Kazunori Sakurama	Kyoto University, Japan

Secretary

Takahiro Endo	Kyoto University, Japan
Yuichi Ambe	Tohoku University, Japan
Ryo Ariizumi	Nagoya University, Japan
Keita Nakamura	University of Aizu, Japan
Ryusuke Fujisawa	Kyushu Institute of Technology, Japan
Jun Ogawa	Yamagata University, Japan
Tasumasa Tamura	Tokyo Institute of Technology, Japan
Kazuki Umemoto	Nagaoka University of Technology, Japan
Keisuke Yoneda	Kanazawa University, Japan

Advisory Committee

Hajime Asama	University of Tokyo, Japan
Young-Jo Cho	Electronics and Telecommunications Research Institute, Korea
Nak Young Chong	Japan Advanced Institute of Science and Technology, Japan
Tamio Arai	International Research Institute for Nuclear Decommissioning, Japan
Toshio Fukuda	Meijo University, Japan
Raja Chatila	Université Pierre and Marie Curie, France
Rüdiger Dillmann	Karlsruher Institut für Technologie, Germany
Maria Gini	University of Minnesota, USA
Alcherio Martinoli	The Swiss Federal Institute of Technology in Lausanne, Switzerland
Francesco Mondada	The Swiss Federal Institute of Technology in Lausanne, Switzerland
Gregory S. Chirikjian	National University of Singapore, Singapore
M. Ani Hsieh	University of Pennsylvania, USA
Roderich Gross	University of Sheffield, UK
Nikolaus Correll	University of Colorado Boulder, USA
Marcelo H. Ang	National University of Singapore, Singapore
Dong Sun	City University of Hong Kong, Hong Kong

Lynne E. Parker	University of Tennessee, USA
Vijay Kumar	University of Pennsylvania, USA
Fumitoshi Matsuno	Kyoto University, Kyoto, Japan
Julien Bourgeois	Univ. Bourgogne Franche-Comté, France
Haruhisa Kurokawa	National Institute of Advanced Industrial Science and Technology, Japan
Katia Sycara	Carnegie Mellon University, USA

Contents

Generating Goal Configurations for Scalable Shape Formation in Robotic Swarms 1
Hanlin Wang and Michael Rubenstein

Leading a Swarm with Signals 16
Nofar Menashe and Noa Agmon

Byzantine Fault Tolerant Consensus for Lifelong and Online Multi-robot Pickup and Delivery 31
Kegan Strawn and Nora Ayanian

Decentralized Multi-robot Planning in Dynamic 3D Workspaces 45
Arjav Desai and Nathan Michael

Optimized Direction Assignment in Roadmaps for Multi-AGV Systems Based on Transportation Flows 58
Valerio Digani and Lorenzo Sabattini

Datom: A Deformable Modular Robot for Building Self-reconfigurable Programmable Matter 70
Benoît Piranda and Julien Bourgeois

The Impact of Network Connectivity on Collective Learning 82
Michael Crosscombe and Jonathan Lawry

On the Communication Requirements of Decentralized Connectivity Control: A Field Experiment 95
Jacopo Panerati, Benjamin Ramtoula, David St-Onge, Yanjun Cao, Marcel Kaufmann, Aidan Cowley, Lorenzo Sabattini, and Giovanni Beltrame

Behavioral Simulations of Lattice Modular Robots with VisibleSim ... 108
Pierre Thalamy, Benoît Piranda, André Naz, and Julien Bourgeois

Evolving Robust Supervisors for Robot Swarms in Uncertain Complex Environments 120
Elliott Hogg, David Harvey, Sabine Hauert, and Arthur Richards

Distributed Cooperative Localization with Efficient Pairwise Range Measurements 134
Anwar Quraishi and Alcherio Martinoli

Robust Localization for Multi-robot Formations: An Experimental Evaluation of an Extended GM-PHD Filter 148
Michiaki Hirayama, Alicja Wasik, Mitsuhiro Kamezaki, and Alcherio Martinoli

Opportunistic Multi-robot Environmental Sampling via Decentralized Markov Decision Processes 163
Ayan Dutta, O. Patrick Kreidl, and Jason M. O'Kane

A PHD Filter Based Localization System for Robotic Swarms 176
R. A. Thivanka Perera, Chengzhi Yuan, and Paolo Stegagno

An Innate Motivation to Tidy Your Room: Online Onboard Evolution of Manipulation Behaviors in a Robot Swarm 190
Tanja Katharina Kaiser, Christine Lang, Florian Andreas Marwitz, Christian Charles, Sven Dreier, Julian Petzold, Max Ferdinand Hannawald, Marian Johannes Begemann, and Heiko Hamann

Multi-agent Reinforcement Learning and Individuality Analysis for Cooperative Transportation with Obstacle Removal 202
Takahiro Niwa, Kazuki Shibata, and Tomohiko Jimbo

Battery Variability Management for Swarms 214
Grace Diehl and Julie A. Adams

Spectral-Based Distributed Ergodic Coverage for Heterogeneous Multi-agent Search 227
Guillaume Sartoretti, Ananya Rao, and Howie Choset

Multi-agent Deception in Attack-Defense Stochastic Game 242
Xueting Li, Sha Yi, and Katia Sycara

Tractable Planning for Coordinated Story Capture: Sequential Stochastic Decoupling 256
Diptanil Chaudhuri, Hazhar Rahmani, Dylan A. Shell, and Jason M. O'Kane

Errors in Collective Robotic Construction 269
Jiahe Chen, Yifang Liu, Adam Pacheck, Hadas Kress-Gazit, Nils Napp, and Kirstin Petersen

Optimal Multi-robot Perimeter Defense Using Flow Networks 282
Austin K. Chen, Douglas G. Macharet, Daigo Shishika, George J. Pappas, and Vijay Kumar

Classification-Aware Path Planning of Network of Robots 294
Guangyi Liu, Arash Amini, Martin Takáč, and Nader Motee

Monitoring and Mapping of Crop Fields with UAV Swarms Based on Information Gain 306
Carlos Carbone, Dario Albani, Federico Magistri, Dimitri Ognibene, Cyrill Stachniss, Gert Kootstra, Daniele Nardi, and Vito Trianni

A Discrete Model of Collective Marching on Rings 320
Michael Amir, Noa Agmon, and Alfred M. Bruckstein

Map Learning via Adaptive Region-Based Sampling in Multi-robot Systems 335
Gianni A. Di Caro and Abdul Wahab Ziaullah Yousaf

Collective Transport via Sequential Caging 349
Vivek Shankar Vardharajan, Karthik Soma, and Giovanni Beltrame

ReactiveBuild: Environment-Adaptive Self-Assembly of Amorphous Structures 363
Petras Swissler and Michael Rubenstein

Processes for a Colony Solving the Best-of-N Problem Using a Bipartite Graph Representation 376
Puneet Jain and Michael A. Goodrich

Decentralized Navigation in 3D Space of a Robotic Swarm with Heterogeneous Abilities 389
Shota Tanaka, Takahiro Endo, and Fumitoshi Matsuno

Using Reinforcement Learning to Herd a Robotic Swarm to a Target Distribution 401
Zahi Kakish, Karthik Elamvazhuthi, and Spring Berman

Preservation of Giant Component Size After Robot Failure for Robustness of Multi-robot Network 415
Toru Murayama and Lorenzo Sabattini

Swarm Localization Through Cooperative Landmark Identification ... 429
Sarah Brent, Chengzhi Yuan, and Paolo Stegagno

Author Index 443

Generating Goal Configurations for Scalable Shape Formation in Robotic Swarms

Hanlin Wang[✉] and Michael Rubenstein

Northwestern University, Evanston, IL 60201, USA
h.w@u.northwestern.edu, rubenstein@northwestern.edu

Abstract. In this paper, we present an algorithm that automatically encodes a user-defined complex 2D shape to a set of cells on a grid each characterizing a robot currently in the swarm. The algorithm is validated via up to 100 simulated robots as well as up to 100 physical robots. The results show that the goal configurations generated by the algorithm for the swarms with any size are consistent with the input shapes, moreover, it allows the swarm to adapt to the swarm size change quickly and robustly. The supplementary materials for this paper can be found at: https://tinyurl.com/2huc42t6.

Keywords: Swarm systems · Shape formation

1 Introduction

Shape formation is an important and well studied problem in the robotic swarm systems. Here, the task is to move a group of robots to form a user-defined shape. In the past, the problem has received a lot of attentions due to its extensive real-world applications such as automated warehouse [1], entertainment applications [2], and more [3].

Many past efforts have concentrated on a vanilla version of the problem. In the vanilla shape formation problem, the swarm size is assumed to stay the same all the times, and the representation of the desired shape is often pre-computed and given to the swarm as an input. Many methods to represent the desired shapes has been presented in the past, including curves or regions explicitly described by a mathematical formula [4,5], potential fields [6], masked grid [7–9], and more [10,11]. The mathematical formula-based representations [4–6,12] can help to derive the formation control laws when the robot's kinematic or dynamic constraints need to be considered. However, when the desired shape is complex, it is time-consuming (sometimes even impossible) to encode the desired shape to a mathematical formula. Masked grid, also known as "binary image" in 2D case [7] or "binary volumes" in 3D case [8], is a grid where each cell is labeled with either a 1 or a 0, indicating whether the cell is in the shape or not, and the desired shape is described as the set of in-shape cells on the grid [7–9].

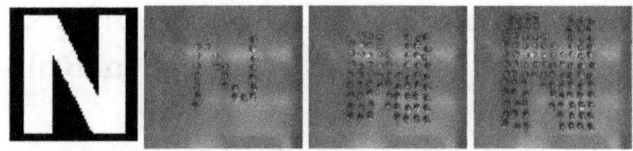

Fig. 1. From left to right: the target shape – "N"; the swarms with different sizes forming the configurations generated by the proposed algorithm

Masked grid is a convenient way to encode the complex shape, in addition, for the swarm of modular robots that are with discrete attachment locations [13,14], the masked grid is a natural way to describe the collective's configuration.

Beyond the vanilla shape formation, the other version of the problem is so called *scalable* (or *scale-independent*) shape formation [7,8]. In the *scalable* shape formation problem, robots can be removed from/added to the swarm in real-time. When the removal or addition of the robots occurs, there are two strategies for the swarm to adapt. One option is to keep the scale of the desired shape fixed, and change the density of the robots [15]. One drawback of this method is that: when the robot's physical size is finite, the size of swarm to display the shape will be limited, as one can fit only finite amount of robots in a unit of space. On the contrary, the other option is to keep the density of the robot fixed and change the size of the desired shape [7,8]. When using the masked grid to describe the target shape, there are two options to scale the goal configuration: change the number of robots fitted in each cell and fix the number of in-shape cells [7], or, change the number of in-shape cells and fit exactly one robot to each cell [8]. As shown in [7,8], both of these two methods can offer the swarm the capability of self-healing, making the system resilient to the removal and addition of robots. On the other hand, for the algorithm proposed in [7], when the swarm size changes, it takes the swarm a long time to adapt, as the swarm needs to wait "long enough" to sense the change of the swarm size. Moreover, the algorithm presented in the [8] only works for certain types of shapes, and the generated configurations can be perfectly formed only by the swarms with certain sizes.

In this paper, we present an algorithm that automatically encodes an input 2D shape (given by an binary image) to a masked square grid where each in-shape cell characterizes a robot current in the swarm. Given an input shape and the swarm size n, the algorithm will first use naive binary image scaling methods to generate two reference grids, in which one has slightly more than n in-shape cells and the other has slightly less than n in-shape cells, then use a second subroutine, called *interpolation*, to refine those two reference grids so as to obtain the final output – a masked grid with exactly n in-shape cells. The algorithm is validated via both the simulated and physical experiments, the results show that the goal configurations generated by the algorithm are consistent with the original input shapes, moreover, when the swarm size changes, it allows the swarm to adapt quickly and robustly.

2 Preliminaries

In this section, we will formally state the problem, and introduce the notations frequently used in the rest of the paper.

2.1 Goal Configuration Generation: Problem Statement

The proposed algorithm takes two inputs: an binary image describing the desired shape, and the size of the swarm to display the desired shape. Note that a binary image is essentially a 2D masked grid, therefore, for the sake of description, in the rest of the paper, we use the word "pixel" and the word "cell" interchangeably. The output of the algorithm is a masked grid such that: the number of in-shape cells must equal to the input swarm size.

When designing the algorithm, there are two factors to be considered: first, the generated goal configurations should be consistent with the input shape; second, in order to allow the swarm to quickly adapt to the removal and addition of the robots, the goal configurations generated for different swarm sizes should be similar to each other as well.

2.2 Notations

Let $\mathcal{G}_i$ be a 2D $m \times m$ masked square grid i, we use $c_i^{(x,y)} \in \{0, 1\}$ to denote the label of the cell in the x-th row and y-th column from the left-top corner, and we use $\mathcal{S}(\mathcal{G}_i) = \{(x, y) \mid c_i^{(x,y)} = 1\}$ to denote the set of the coordinates of all in-shape cells on $\mathcal{G}_i$. Given a set $\mathcal{A}$, we use $|\mathcal{A}|$ to denote its cardinality. For a pair of sets $\mathcal{A}$ and $\mathcal{B}$: $\mathcal{A} \cup \mathcal{B}$ denotes the union of set $\mathcal{A}$ and set $\mathcal{B}$; $\mathcal{A} \cap \mathcal{B}$ denotes the intersection of set $\mathcal{A}$ and set $\mathcal{B}$; $\mathcal{A} - \mathcal{B}$ denotes the set of all the elements that are in set $\mathcal{A}$ but not in set $\mathcal{B}$. For $n \geq 3$ sets $\mathcal{A}_0, \mathcal{A}_1, \ldots, \mathcal{A}_{n-1}$, their union is denoted as $\bigcup_{i=0}^{n} \mathcal{A}_i$.

3 Approach

We assume the desired shape is given to the swarm in the format of a 100×100 binary image, however this approach can be generalized to any size binary image. The proposed algorithm consists of two subroutines – *scaling* and *interpolation*. Given the desired shape $\mathcal{G}_{in}$ and the swarm size n, the algorithm will first use the *scaling* subroutine to find two reference masked grids $\mathcal{G}_o^l$ and $\mathcal{G}_o^h$ such that: $\mathcal{G}_o^h$ has slightly more in-shapes than n and $\mathcal{G}_o^l$ has slightly less in-shapes than n, then apply the *interpolation* subroutine to $\mathcal{G}_o^l$ and $\mathcal{G}_o^h$ so as to obtain an output that is with exactly n in-shape cells. A graphical illustration of the overall pipeline is shown in Fig. 2 and a detailed description of algorithm's overall pipeline is shown in Algorithm 1.

Algorithm 1: Pipeline for proposed algorithm

Input: Input shape $\mathcal{G}_{in}$, swarm size n
Output: Configuration for the swarm $\mathcal{G}_o$

1 $\mathcal{G}_o^l, \mathcal{G}_o^h \leftarrow scaling(\mathcal{G}_{in}, n)$
2 **if** $|\mathcal{S}(\mathcal{G}_o^l)|$ is n **then**
3 $\quad \mathcal{G}_o \leftarrow \mathcal{G}_o^l$
4 **else**
5 $\quad \mathcal{G}_o \leftarrow interpolation(\mathcal{G}_o^l, \mathcal{G}_o^h, n)$
6 **return** $\mathcal{G}_o$

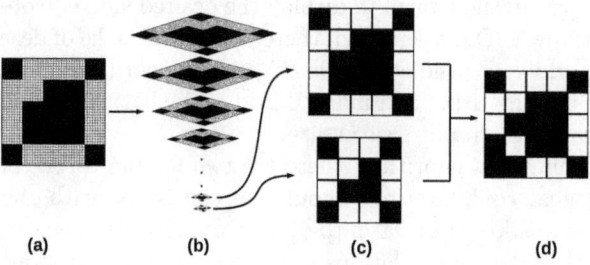

Fig. 2. The graphical illustration of the overall pipeline of the presented algorithm. From left to right: (a) The input shape, which is given in the format of a binary image. In this example, the task is to find a goal configuration for a swarm of 12 robots; (b) We apply *scaling* subroutine to the input masked grid so as to find two reference grids with approximately 12 in-shape cells; (c) Two reference masked grids with each pixel enlarged for the visualization purpose; (d) A configuration with 12 in-shape cells is constructed by the *interpolation* subroutine using those 2 reference grids in (c).

Algorithm 2: *scaling* subroutine

Input: Input shape $\mathcal{G}_{in}$, swarm size n
Output: Two reference masked grids $\mathcal{G}_o^l, \mathcal{G}_o^h$

1 $m \leftarrow 100$ // inialize m to the size of $\mathcal{G}_{in}$
2 $iter \leftarrow 1$
3 $last \leftarrow \mathcal{G}_{in}$ // variable to store the last scaled grid
4 **if** $|\mathcal{S}(\mathcal{G}_{in})|$ is n **then**
5 $\quad \mathcal{G}_o^l \leftarrow \mathcal{G}_{in}, \mathcal{G}_o^h \leftarrow \mathcal{G}_{in}$
6 **if** $|\mathcal{S}(\mathcal{G}_{in})| < n$ **then**
7 $\quad$ **while** 1 **do**
8 $\quad\quad m \leftarrow m+1$ // gradually increase scaled grid size
9 $\quad\quad cur \leftarrow$ scale the grid $\mathcal{G}_{in}$ to size $m \times m$
10 $\quad\quad$ **if** $|\mathcal{S}(cur)| \geq n$ **then**
11 $\quad\quad\quad \mathcal{G}_o^l \leftarrow last, \mathcal{G}_o^h \leftarrow cur$
12 $\quad\quad\quad$ **break**
13 $\quad\quad$ **else**
14 $\quad\quad\quad last \leftarrow cur$
15 $\quad\quad\quad iter \leftarrow iter + 1$
16 **if** $|\mathcal{S}(\mathcal{G}_{in})| > n$ **then**
17 $\quad$ **while** 1 **do**
18 $\quad\quad m \leftarrow m-1$ // gradually decrease scaled grid size
19 $\quad\quad$ **if** $m < 15$ **then** // switch to skeletonization
20 $\quad\quad\quad cur \leftarrow$ skeletonize the grid $last$
21 $\quad\quad\quad$ **if** $|\mathcal{S}(cur)| \leq n$ **then**
22 $\quad\quad\quad\quad \mathcal{G}_o^l \leftarrow cur, \mathcal{G}_o^h \leftarrow last$
23 $\quad\quad\quad\quad$ **break**
24 $\quad\quad\quad$ **else**
25 $\quad\quad\quad\quad$ Error: the input n is too small.
26 $\quad\quad cur \leftarrow$ scale the grid $\mathcal{G}_{in}$ to size $m \times m$
27 $\quad\quad$ **if** $|\mathcal{S}(cur)| \leq n$ **then**
28 $\quad\quad\quad \mathcal{G}_o^l \leftarrow cur, \mathcal{G}_o^h \leftarrow last$
29 $\quad\quad\quad$ **break**
30 $\quad\quad$ **else**
31 $\quad\quad\quad last \leftarrow cur$
32 $\quad\quad\quad iter \leftarrow iter + 1$
33 **return** $\mathcal{G}_o^l, \mathcal{G}_o^h$

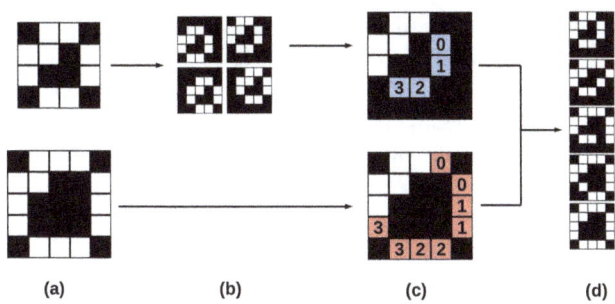

Fig. 3. The graphical illustration of the *interpolation* subroutine. From left to right: (a) Two reference grids $\mathcal{G}_o^l$ and $\mathcal{G}_o^h$ with 9 and 13 in-shape cells, respectively; (b) We align $\mathcal{G}_o^l$ to $\mathcal{G}_o^h$ (Algorithm 3, Line 1). There are 4 possible locations to place $\mathcal{G}_o^l$ on a 5×5 gird, and we choose the one in the right-up corner because the *difference score* between this grid and $\mathcal{G}_o^h$ is the lowest; (c) We calculate the set $\mathcal{D}_{l-h}$, which is the set of cells filled with blue color, and the set $\mathcal{D}_{h-l}$, which is the set of cells filled with red color, and then split these two sets into two sets of 4 subsets $\{d_{l-h}^0, \ldots, d_{l-h}^3\}$ and $\{d_{h-l}^0, \ldots, d_{h-l}^3\}$, the number on each cells indicates the subset that it belongs to (Algorithm 3, Line 2–24); (d) 5 configurations generated using $\mathcal{G}_o^l$ and $\mathcal{G}_o^h$ with different input swarm size n s. From top to bottom: the configuration generated for the swarm with a size of $9, 10, 11, 12, 13$, respectively (Algorithm 3, Line 25–27).

3.1 Scaling

In the *scaling* subroutine, we first use the *image scaling* to change the number of in-shape pixels. Image scaling is a well studied topic [16,17], here, the task is to create a new version of the image with a different width and/or height in pixels. Many strategies to scale a binary image have been proposed in the past, in this paper, we use the *nearest neighbor interpolation* [16] as our *image scaling* method.

Given an input shape $\mathcal{G}_{in}$ and swarm size n, there are three possible cases: If the number of in-shape cells on the input grid $|\mathcal{S}(\mathcal{G}_{in})|$ equals to n, the algorithm will return $\mathcal{G}_{in}$ directly (Algorithm 2, Line 4–5). If the $|\mathcal{S}(\mathcal{G}_{in})| < n$, the algorithm will first keep upscaling the $\mathcal{G}_{in}$ (Algorithm 2, Line 6–15) until a grid that has more than n in-shape cells is found (Algorithm 2, Line 10–12), then return the two scaled images obtained most lately, and exit this subroutine (Algorithm 2, Line 11–12). Similarly, if $|\mathcal{S}(\mathcal{G}_{in})| > n$, the algorithm will keep downscaling the $\mathcal{G}_{in}$ until finding a grid that contains less than n in-shape cells (Algorithm 2, Line 16–32).

One issue for using the *image scaling* to reduce the number of in-shape cells is that: When the size of the scaled image is too small ($\leq 15 \times 15$ according to our experiments), it often fails to preserve the main structure of the input shape. Therefore, to prevent the main structure of the shape in scaled image being distorted by over-downsampling, the size of the scaled image cannot be smaller than a threshold (Algorithm 2, Line 20). This limits the minimal number of in-shape cells in the outputs that can be generated.

Besides the *image scaling*, an alternative to reduce the pixels required to display a shape is the operation *skeletonization* [18]. The operation *skeletonization* generates a "thinner" version of the input shape that emphasizes shape's geometrical and topological properties. In the *scaling* subroutine, we use the operation *skeletonization* to extend the range of swarm sizes for which our method can work: if the image scaled with the minimal size still has more than n in-shape cells, we then apply the operation *skeletonization* to this scaled image so as to obtain the shape's skeleton, which is a masked grid with fewer in-shape cells (Algorithm 2, Line 20). It is possible that the number of in-shape cell on the skeleton is still more than n, if that happens, algorithm will raise an error to tell the user that the input swarm size n is too small for displaying the desired shape $\mathcal{G}_{in}$ (Algorithm 2, Line 25). See Algorithm 2 for the detailed pseudo code for the *scaling* subroutine.

3.2 Interpolation

The high-level idea behind the *interpolation* subroutine can be described as follows: Say we have a $l \times l$ binary image $\mathcal{G}_o^l$ and a $h \times h$ binary image $\mathcal{G}_o^h$ with a and b amount of in-shape cells, respectively. Assume $h > l$ and $b > a$, we want to generate a sequence of $h \times h$ binary images $\mathcal{G}_a, \ldots, \mathcal{G}_b$, in which each generated image $\mathcal{G}_i$ has exactly i amount of in-shape cells. To do so, we first place the input grid $\mathcal{G}_o^l$ on a empty $h \times h$ grid at a location such that the overlapping between the newly formed binary image and the $\mathcal{G}_o^h$ is maximized. Then, we calculate the difference between the newly formed $\mathcal{G}_o^l$ and $\mathcal{G}_o^h$, which can be characterized by the cells that are with different labels on those two grids. The sequence of the binary images $\mathcal{G}_a, \ldots, \mathcal{G}_b$ can be constructed by gradually toggling the labels of those difference cells on the grid $\mathcal{G}_o^l$. To be more specific, for the aligned $\mathcal{G}_o^l$ and $\mathcal{G}_o^h$, their difference can be characterized by two sets: $\mathcal{D}_{l-h} = \mathcal{S}(\mathcal{G}_o^l) - \mathcal{S}(\mathcal{G}_o^h)$, which is the set of cells that are in the shape on $\mathcal{G}_o^l$ but off the shape on $\mathcal{G}_o^h$, and $\mathcal{D}_{h-l} = \mathcal{S}(\mathcal{G}_o^h) - \mathcal{S}(\mathcal{G}_o^l)$, which is the set of cells that are in the shape on $\mathcal{G}_o^h$ but off the shape on $\mathcal{G}_o^l$. (Algorithm 3, Line 3–4). Next, let $k = |\mathcal{S}(\mathcal{G}_o^h)| - |\mathcal{S}(\mathcal{G}_o^l)|$ be the difference between the numbers of in-shape cells on $\mathcal{G}_o^l$ and $\mathcal{G}_o^h$ (Algorithm 3, Line 2), we split those two sets into two groups of k disjointed subsets $d_{l-h}^0, \ldots, d_{l-h}^{k-1}$ and $d_{h-l}^0, \ldots, d_{h-l}^{k-1}$ such that: $\bigcup_{i=0}^{k-1} d_{l-h}^i = \mathcal{D}_{l-h}$ and $\bigcup_{i=0}^{k-1} d_{h-l}^i = \mathcal{D}_{h-l}$. In addition, for a pair of subsets d_{h-l}^i and d_{l-h}^i, we enforce their sizes to be such that: $|d_{h-l}^i| = 1 + |d_{l-h}^i|$ (Algorithm 3, Line 5–24). With this constraint on each subset's cardinality, given an swarm size n, its corresponding configuration can be constructed as follows: first, set the labels the cells $\bigcup_{i=0}^{n-|\mathcal{S}(\mathcal{G}_o^l)|} d_{l-h}^i$ to 0 on $\mathcal{G}_o^l$, then, set the labels of the cells $\bigcup_{i=0}^{n-|\mathcal{S}(\mathcal{G}_o^l)|} d_{h-l}^i$ to 1 on $\mathcal{G}_o^l$ (Algorithm 3, Line 25–27). It is straight forward to examine that the masked grid constructed via the procedure above will have exactly n in-shape cells. The pseudo code for the *interpolation* subroutine is shown in Algorithm 3. A graphical illustration of the *interpolation* subroutine is shown in Fig. 3.

Assume that the reference grid $\mathcal{G}_o^l$ has a size of $l \times l$ and $\mathcal{G}_o^h$ has a size of $h \times h$, where $h \geq l$ according to Algorithm 2. The *interpolation* subroutine will

Algorithm 3: *interpolation* subroutine

Input: Reference grids $\mathcal{G}_o^l, \mathcal{G}_o^h$, swarm size n
Output: The generated masked grid $\mathcal{G}_o$

1 $\mathcal{G}_o^l \leftarrow$ align $\mathcal{G}_o^l$ to $\mathcal{G}_o^h$
2 $k \leftarrow |\mathcal{S}(\mathcal{G}_o^h)| - |\mathcal{S}(\mathcal{G}_o^l)|$ // calculate the difference of in-shape numbers on two reference grid
3 $\mathcal{D}_{l-h} \leftarrow \mathcal{S}(\mathcal{G}_o^l) - \mathcal{S}(\mathcal{G}_o^h)$ // the set of cells that are in the shape on $\mathcal{G}_o^l$ but off the shape on $\mathcal{G}_o^h$
4 $\mathcal{D}_{h-l} \leftarrow \mathcal{S}(\mathcal{G}_o^h) - \mathcal{S}(\mathcal{G}_o^l)$ // the set of cells that are in the shape on $\mathcal{G}_o^h$ but off the shape on $\mathcal{G}_o^l$
5 Initialize $sz^0, \ldots, sz^{k-1}$ to be all 0s   // the size of each subset used in interpolation
6 **for** $i \leftarrow 0, \ldots, k-1$ **do**   // determine each subset's size
7   **if** $i \leq |\mathcal{D}_{l-h}| \bmod k$ **then**
8      $sz^i \leftarrow \lfloor \frac{|\mathcal{D}_{l-h}|}{k} \rfloor + 1$
9   **else**
10     $sz^i \leftarrow \lfloor \frac{|\mathcal{D}_{l-h}|}{k} \rfloor$

11 $dt(\mathcal{G}_o^l) \leftarrow$ apply *distance transform* to $\mathcal{G}_o^l$ // calculate each in-shape cell's distance to the boundary
12 $\mathit{buf\_sub} \leftarrow$ use each cell's value in $dt(\mathcal{G}_o^l)$ as the key to sort $\mathcal{S}(\mathcal{G}_o^l)$ in ascending order
13 Initialize $d_{l-h}^0, \ldots, d_{l-h}^{k-1}$ to be all $\varnothing$s   // the subsets of cells to be turned off on $\mathcal{G}_o^l$
14 **for** $i \leftarrow 0, \ldots, k-1$ **do**   // assemble the subsets of cells to be turned off on $\mathcal{G}_o^l$
15   $d_{l-h}^i \leftarrow d_{l-h}^i \cup \{\text{the first } sz^i \text{ cells in } \mathit{buf\_sub}\}$
16   $\mathit{buf\_sub} \leftarrow \mathit{buf\_sub} - d_{l-h}^i$

17 Initialize $d_{h-l}^0, \ldots, d_{h-l}^{k-1}$ to be all $\varnothing$s   // the subsets of cells to be turned on on $\mathcal{G}_o^l$
18 $\mathit{buf\_add} \leftarrow \mathcal{D}_{h-l}$
19 **for** $i \leftarrow 0, \ldots, k-1$ **do**
20   $d_{h-l}^i \leftarrow d_{h-l}^i \cup \{sz^i \text{ amount of cells in } \mathit{buf\_add} \text{ that are closest to the cells in } d_{l-h}^i\}$
21   $\mathit{buf\_add} \leftarrow \mathit{buf\_add} - d_{h-l}^i$

22 **for** $i \leftarrow 0, \ldots, k-1$ **do**
23   $d_{h-l}^i \leftarrow d_{h-l}^i \cup \{\text{the first cell in } \mathit{buf\_add}\}$
24   $\mathit{buf\_add} \leftarrow \mathit{buf\_add} - d_{h-l}^i$

25 $\mathcal{G}_o \leftarrow$ a copy of $\mathcal{G}_o^l$ // make a copy of $\mathcal{G}_o^l$ and toggle the labels of cells on it so as to construct the output
26 set labels of all the cells in $\bigcup_{i=0}^{n-|\mathcal{G}_o^l|} d_{l-h}^i$ to 0 on $\mathcal{G}_o$
27 set labels of all the cells in $\bigcup_{i=0}^{n-|\mathcal{G}_o^l|} d_{h-l}^i$ to 1 on $\mathcal{G}_o$
28 **return** $\mathcal{G}_o$

---

first draw the $\mathcal{G}_o^l$ on an empty $h \times h$ grid (Algorithm 3, Line 1). Before drawing $\mathcal{G}_o^l$ on this empty $h \times h$ grid, we need to determine the location to place the $\mathcal{G}_o^l$ on this $h \times h$ grid, as there might be multiple choices since $h \geq l$. To do so, we first define a metric called *difference score* as follows:

Definition 1. *Given two masked grids $\mathcal{A}$ and $\mathcal{B}$, the difference score between $\mathcal{A}$ and $\mathcal{B}$ is given by $|\mathcal{S}(\mathcal{A}) - \mathcal{S}(\mathcal{B})| + |\mathcal{S}(\mathcal{B}) - \mathcal{S}(\mathcal{A})|$, i.e., the number of cells that are in shape on $\mathcal{A}$ but not in shape on $\mathcal{B}$ plus the number of cells that are on in shape on $\mathcal{B}$ but not in shape on $\mathcal{A}$.*

With this metric, the location to place the grid $\mathcal{G}_o^l$ on the new $h \times h$ grid can be determined as follows: we exhaustively search all possible translations, and choose the translation that gives the minimal *difference score* between the grid $\mathcal{G}_o^h$ and the newly generated $\mathcal{G}_o^l$ (Algorithm 3, Line 1).

After aligning $\mathcal{G}_o^l$ to $\mathcal{G}_o^h$, we first calculate the difference between $\mathcal{G}_o^l$ and $\mathcal{G}_o^h$ (Algorithm 3, Line 2-4), then pack the set $\mathcal{D}_{l-h}$ into k subsets (Algorithm 3, Line 5-16). sz^i denotes the size of the subset d_{l-h}^i of $\mathcal{D}_{l-h}^i$, we first calculate the size of each subset d^i_{l-h} (Algorithm 3, Line 5–10). After determining the size of each subset d^i_{l-h}, we then start to determine the contents of each subset d^i_{l-h}. When removing the cells from the shape, we want to remove the cells following the order such that: the cells closer to shape's boundary will be removed first. This order helps to avoid generating "holes" in the remaining shape. To address this design consideration, in the algorithm, for all the cells in $\mathcal{D}_{l-h}$, we first calculate each cell's Manhattan distance to the boundary using the operation *distance transform* [19] (Algorithm 3, Line 11), and then use each cell's distance to boundary as the key to sort all the cells in $\mathcal{D}_{l-h}$ in ascending order (Algorithm 3, Line 12). The sorted $\mathcal{D}_{l-h}$ is stored in the variable *buf_sub*. Next, we start to assemble each subset d^i_{l-h} according to the determined pack size sz^i (Algorithm 3, Line 14–16): for each subset d^i_{l-h}, we pack the first sz^i cells in *buf_sub* into it (Algorithm 3, Line 15), and then delete those sz^i cells from the *buf_sub* right after so as to avoid the case where the same cell shows up in two difference subsets (Algorithm 3, Line 16).

Next, we start to assemble the subsets $d^0_{h-l}, \ldots, d^{k-1}_{h-l}$ (Algorithm 3, Line 18–24). We first make a copy of $\mathcal{D}_{h-l}$ and store it to variable *buf_add* (Algorithm 3, Line 18). Note that when removing a subset d^i_{l-h} of cells from the shape, we will damage the structure of the shape. To reduce the effect of the removal of d^i_{l-h}, when packing each subset d^i_{h-l}, which are the sets to be added to the remaining shape, we want the cells in d^i_{h-l} to be as "close" to cells in d^i_{l-h} as possible (Algorithm 3, Line 20). To be more specific, given a subset d^i_{l-h} and the set *buf_add*, we treat each cell in d^i_{l-h} as a "job" and each cell currently in *buf_add* as a "worker", and the cost for each "worker" doing each "job" is given by the Manhattan distance between those two cells. We use the Hungarian algorithm [20] to assign exactly 1 "worker" to each "job" in each subset d^i_{h-l} such that the total cost is minimized, and these assigned "workers" will be packed into d^i_{h-l} (Algorithm 3, Line 20). Recall that as stated in the overall description of the *interpolation* subroutine, we have a constraint on the each pair of subsets' cardinalities that: $|d^i_{h-l}| = 1 + |d^i_{l-h}|$. On the other hand, it is straight forward to examine that after Algorithm 3, Line 19–21, each pair of subsets d^i_{h-l} and d^i_{l-h} have the same cardinality. In order to satisfy the cardinality constraint above, in Algorithm 3, Line 22–24, we add one extra cell to each of those subsets $d^0_{h-l}, \ldots, d^{k-1}_{h-l}$.

So far, both the subsets $d^0_{l-h}, \ldots, d^{k-1}_{l-h}$ and the subsets $d^0_{h-l}, \ldots, d^{k-1}_{h-l}$ have been assembled, we then construct the desired masked grid $\mathcal{G}_o$ following the procedure described at the beginning of the Sect. 3.2 (Algorithm 3, Line 25–27). See Fig. 3 for a graphical illustration of the *interpolation* subroutine.

In *interpolation* subroutine, one key element is the way to determine sz^i, which is the size of each subset d^i_{h-l} (Algorithm 3, Line 6–10). Given two masked grids $\mathcal{G}^l_o$ and $\mathcal{G}^h_o$, there might be multiple feasible combinations of each subset's size. One can consider a case where $|\mathcal{D}_{l-h}| = 2, |\mathcal{D}_{h-l}| = 4, k = 2$, one way to split $\mathcal{D}_{l-h}$ and $\mathcal{D}_{h-l}$ is: $|d^0_{l-h}| = 1, |d^1_{l-h}| = 1, |d^0_{h-l}| = 2, |d^1_{h-l}| = 2$, and the other way is: $|d^0_{l-h}| = 2, |d^1_{l-h}| = 0, |d^0_{h-l}| = 3, |d^1_{h-l}| = 1$. According to the overall pipeline of the algorithm (Algorithm 1), for the same pair of reference grids $\mathcal{G}_o^l$ and $\mathcal{G}_o^h$, they could be used to construct $|\mathcal{S}(\mathcal{G}_o^h)| - |\mathcal{S}(\mathcal{G}_o^l)| + 1$ different configurations with $n = |\mathcal{S}(\mathcal{G}_o^l)|, |\mathcal{S}(\mathcal{G}_o^l)|+1, \ldots, |\mathcal{S}(\mathcal{G}_o^h)|$ amount of in-shape cells, respectively. It is trivial to see that different ways to determine each subset's size will result in different *difference scores* among these generated masked grids. Recall that as stated in the Sect. 2.1, to allow the swarm to quickly adapt to the swarm size change, one of our design considerations is: the configurations generated for different swarm sizes should be similar to each other. Responding to this design consideration, given a pair of reference grids $\mathcal{G}_o^l$ and $\mathcal{G}_o^h$, for the configurations generated from them, a desirable way to determine each subset's size should make the *difference score* between any pair of masked grids with adjacent number of in-shape cells as small as possible. In the following, we show that: given two reference grids $\mathcal{G}_o^l$ and $\mathcal{G}_o^h$, the way that we determine each subset's size (Algorithm 3, Line 6–10) is actually the optimal way that can minimize the maximal *difference score* between any pair of generated masked grids whose difference of in-shape cell number is one.

Problem 1 *(Fair packing).* *Given a set $\mathcal{A}$ and an integer k, $\mathcal{P}_k(\mathcal{A})$ denotes a k-partition of the set $\mathcal{A}$, which is a set of k subsets $\{a_0, \ldots, a_{k-1}\}$ such that: (i) $\bigcup_{i=0}^{k-1} a_i = \mathcal{A}$, and (ii) $\forall i \neq j, a_i \cap a_j = \varnothing$. The task is to find a $\mathcal{P}_k(\mathcal{A})$ that minimizes the maximal cardinality among all those k subsets $a_i \in \mathcal{P}_k(\mathcal{A})$. That is, given a set $\mathcal{A}$ and an integer k, find a partition $\mathcal{P}_k^*(\mathcal{A})$ such that:*

$$\mathcal{P}_k^*(\mathcal{A}) = \mathrm{argmin} \max_{a_i \in \mathcal{P}_k^*(\mathcal{A})} |a_i|$$

To interpret Problem 1, one can consider a simple instance of it: Say we have 10 balls and we are tasked to put those 10 balls into 3 bins. We want to find a way to assign those 10 balls to those 3 bins such that the maximal number of balls among all 3 bins is minimized. Next, in the Lemma 1, we shows an sufficient condition for a solution to be optimal to Problem 1.

Lemma 1. *Given a set $\mathcal{A}$ and an integer k, let $\mathcal{P}_k'(\mathcal{A})$ be a k-partition of the set $\mathcal{A}$, if the $\mathcal{P}_k'(\mathcal{A})$ is made such that:*

$$\max_{a_i \in \mathcal{P}_k'(\mathcal{A})} |a_i| - \min_{a_i \in \mathcal{P}_k'(\mathcal{A})} |a_i| \leq 1 \tag{1}$$

Then $\mathcal{P}_k'(\mathcal{A})$ is an optimal solution to Problem 1.

Proof. See Sect. 1 in [21]. ∎

Theorem 1 *(Smooth transition).* *Given two reference masked grids $\mathcal{G}_o^l$ and $\mathcal{G}_o^h$ with a and b amount of in-shape cells, respectively, let $\mathcal{G}_a, \mathcal{G}_{a+1}, \ldots, \mathcal{G}_b$ be the masked grids generated for the swarms with a size of $a, a+1, \ldots, b$. Among all the ways to determine the size of each subset used in interpolation subroutine, Algorithm 3 Line 6–10 is the optimal way for minimizing the following objective:*

$$\max_{a \leq i \leq b-1} |\mathcal{S}(\mathcal{G}_i) - \mathcal{S}(\mathcal{G}_{i+1})| + |\mathcal{S}(\mathcal{G}_{i+1}) - \mathcal{S}(\mathcal{G}_i)| \tag{2}$$

Proof. See Sect. 2 in [21]. ∎

4 Performance Evaluation

In this section, we empirically study the performance of algorithm proposed in this paper. Given a goal shape, in Sect. 4.1, we first study the quality of configurations generated for the swarms with different sizes, then, in Sect. 4.2 and Sect. 4.3, the generated configurations were formed by a swarm of simulated robots as well as a swarm of physical swarms using the shape formation algorithm proposed in [9], and the results show that the proposed algorithm can indeed make the swarm adapt to the swarm size change quickly and robustly.

In the experiments, we use four complex shapes as our goal shapes: the "N", the "star", the "wrench", and the "circle". These four goal shapes are shown in the Fig. 4.

4.1 Experiments on Generated Configurations

First, given each goal shape, we use the proposed algorithm to generate the goal configurations with in-shape cell number ranging from around 20 to 1024. Recall that as stated in the Sect. 2.1, we have two design considerations: the similarity between each generated configuration and the goal shape, and the similarity between the configurations generated for difference swarm sizes. The videos of the configurations generated for different swarm sizes can be found in [21]. In addition, we introduce two metrics to qualitatively evaluate the similarity between each pair of the binary images whose difference of in-shape cell number is one: the *normalized difference score* (NDS), and the *normalized inter-shape distance* (NISD). The NDS is the ratio between those two configurations' *difference score* and the sum of two configurations' in-shape cell numbers. The NISD is defined as follows: given two masked grid $\mathcal{A}$ and $\mathcal{B}$ where $|\mathcal{S}(\mathcal{A})| \leq |\mathcal{S}(\mathcal{B})|$, we assign cells in $\mathcal{S}(\mathcal{B})$ to the cells in $\mathcal{S}(\mathcal{A})$ in a way such that: (i) for each in-shape cell on $\mathcal{A}$, we assign exactly one in-shape cell on $\mathcal{B}$ to it, in addition, (ii) each in-shape cell on $\mathcal{B}$ can be assigned to no more than one cell on $\mathcal{A}$. The cost of each pair of in-shape cell on $\mathcal{A}$ and its assigned in-shape cell from $\mathcal{B}$ is given by the Manhattan distance between them in cells. The NISD is the ratio between the minimal total cost that any feasible assignment can achieve and sum of two configurations' in-shape cell numbers. Intuitively, the NDS shows the "mismatch" between two configurations, and NIDS essentially characterizes the minimal average distance traveled by the swarm to transform from one configuration to the other. The plots showing these two metrics over difference in-shape cell numbers for each

Fig. 4. Goal shapes used in the experiments. From left to right: the shape "letter N", the shape "star", the shape "wrench", and the shape "circle".

Generating Goal Configurations for Scalable Shape Formation   11

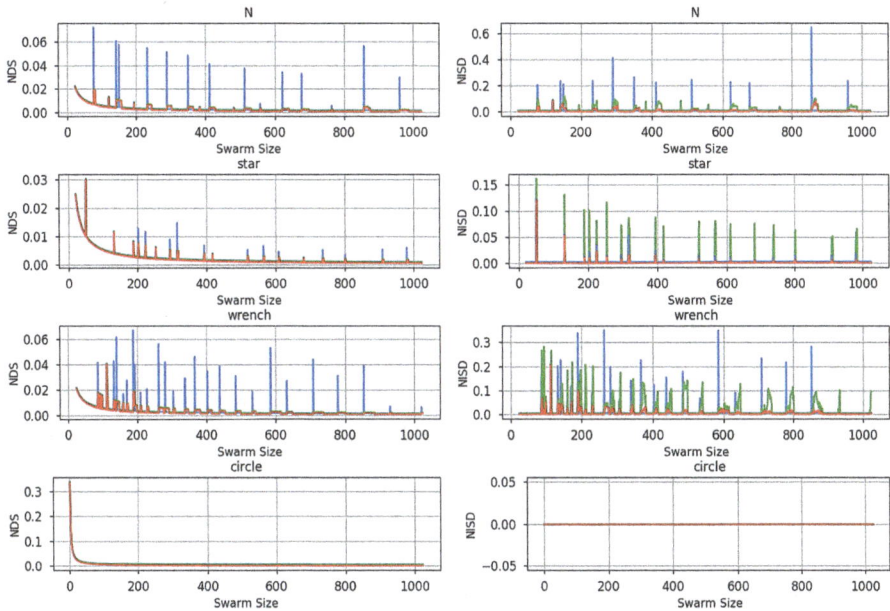

Fig. 5. Comparison between the proposed algorithm and two baselines. For each swarm size n, its corresponding data points on the plots are the NDS and NISD between the generated masked grids with n and $n+1$ in-shape cells, respectively. The red plots are the results for the proposed algorithm, the blue plots are the results for the baseline 1, and the green plots, which overlap with the red plots in all NDS plots, are the results for the baseline 2.

goal shape are shown in Fig. 5. In addition, we also compare the proposed algorithm with two baselines. Both of those two baselines use the same pipeline as our algorithm does. The difference between our algorithm and the baseline 1 is that: when executing the *interpolation* subroutine, instead of using Algorithm 3 Line 6–10 to determine each subset's size, the baseline 1 will aggressively set sz^0 to be $|\mathcal{D}_{l-h}|$ and set $sz^i \ldots sz^{k-1}$ to be 0. The difference between the our algorithm and the baseline 2 is that: when executing the *interpolation* subroutine, instead of using Algorithm 3 Line 11–24 to assemble each subset, the baseline 2 will naively fit the cells into each subset by the lexical order of each cell's coordinate. In the *interpolation* subroutine, there are two subproblems to be solved: how to determine each subset's size, and how to determine the content of each subset. These two baselines are essentially two naive solutions to those two subproblems.

As expected, in these plots, we can see that our algorithm outperforms two baselines with respect to both NDS and NISD for all four goal shapes, confirming that our way to determine the size and content of each subset in *interpolation* subroutine (Algorithm 3 Line 6–10, Line 11–24) can indeed make the transition between goal configurations generated for different swarm sizes more "smooth".

Note that our algorithm and baseline 2 uses the same way to determine each subset's size in *interpolation* subroutine, as a result, in all NDS plots, our algorithm (black) overlaps with baseline 2 (green). The other counter-intuitive observation here is that: the NISD for the shape "circle" is 0 for all the swarm sizes, this is because: using the pipeline presented in the paper, for any swarm size n, the "circle" generated for the swarm size n will always be "inside" the "circle" generated for the swarm size $n + 1$, therefore, the "circle's" NISD is by definition 0 for all the swarm sizes.

4.2 Experiments on Simulated Robotic Swarm

In the simulation, a swarm of up to 100 simulated *Coachbot* robots were tasked to use the shape formation algorithm proposed in [9] to form the configurations generated from our algorithm. The simulation consists two main components: a world engine written in C that simulates robot's on-board hardware resources, and a user-code loader written in python that executes the user's code. The world engine simulates the *Coachbot* robot's motion, sensing and communication in a very realistic way, that is, the specifications of all the simulated hardware, including the maximal speed of robot's wheel, sampling rate of robot's positioning sensor, throughput of the inter-robot communication channel, etc., are made to be consistent with the real robot. In addition, the user-code loader is designed in a way that: the code used to operate the simulated robot can be used to operate the actual *Coachbot* robot without any modification. In the simulation, the communication rate is 20 Hz, the maximal speed of robot's wheel is $0.1 m/s$, and each edge on the grid has a length of 0.3 m. The demonstration videos of the simulation can be found in [21].

In the first experiment, we study the how the addition of the robots will affect the swarm's behavior. In this experiment, the swarm size is initialized to be around 20. Every time when robots currently in the swarm complete forming the shape, we add one more robot in a random location near the swarm and broadcast the new swarm size to the swarm. The robots will update their goal configuration according to the new swarm size, and then start to form the new goal configuration. This process will be repeated until the swarm size gets to 100. For each goal shape, we repeat this experiment 50 times, and in each trial, we study two metrics: the response time (RT), which is time between the swarm size change and the swarm forming the shape at the new scale, and the normalized travel distance (NTD), which the average distance traveled by the swarm to form the shape at the new scale. The results from 50 trials are shown in Fig. 6.

In addition, besides the addition of the robots, we are also interested in effect of the removal of robots on the swarm's behavior. In the second experiment, the swarm size is initialized to 100, and similar to the first experiment, every time when the current swarm complete forming the shape, we randomly choose one robot currently in the swarm, remove it from the swarm, and broadcast the new swarm size to the robots. The robots will update their goal configuration according to the new swarm size, and then start to form the new goal configuration. This process will be repeated until the swarm size gets to minimal swarm size

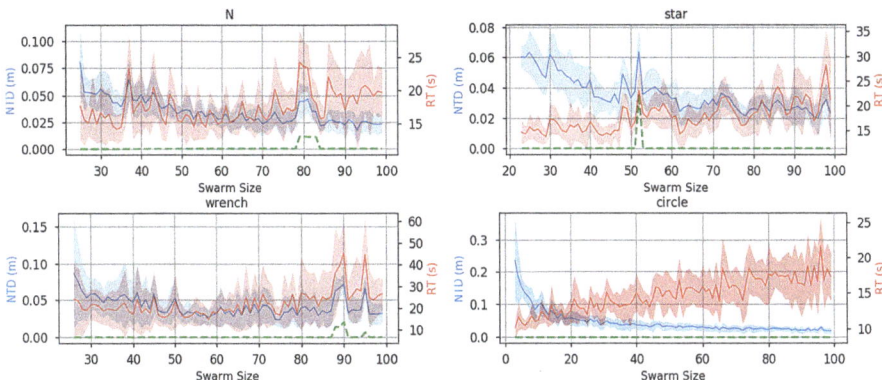

Fig. 6. The results from the addition experiment. Each solid line is the average result from 50 trials, and the colored shade areas show the confidence intervals for NTD and RT over swarm size at a confidence level of two σ. Each green dotted line is the shape's NISD obtained from the previous section.

required to display the shape. For each goal shape, we repeat this experiment 50 times, and in each trial, the metric RT and NTS are investigated, see Fig. 7 for the results from all 50 trials. As we can see in the plots, in both addition and subtraction experiments, every time when the swarm size change occurs, the swarm is able to adapt quickly, within 40 s to be more specific. Moreover, one can observe that for each shape, at some certain swarm sizes, the RT and NTD change sharply in both subtraction and addition experiments. For example, for the shape "wrench", the RT and NTD spike at the swarm size 52. To investigate the cause of these spikes, we compare the NISD obtained from previous section (green dotted line) with the NTD and RT obtained from the simulation. Unsurprisingly, the results show that the swarm sizes where the NTD and RT spike are consistent with the swarm sizes where the shape's NISD spikes, in other words, the swarm sizes where the RT and NTD spike are the swarm sizes where the generated goal configurations change greatly.

4.3 Experiments on Physical Robotic Swarm

Beyond the simulations, we also experiment on a swarm of up to 100 real *Coachbot* robots. In the experiment, the robots are tasked to form the shape "N". The swarm size starts to be 100, every time when the current swarm complete forming the shape, we remove a batch of robots from the swarm and then broadcast the new swarm size to the robots. The robots use the presented algorithm to update their goal configuration according to the new swarm size, and then start to form the new goal configuration. This process is repeated until the swarm size gets to 23. See Fig. 1 for the still images from the experiment, and the video for this experiment can be found in [21]. As we can see in the video, when the swarm size changes, the robots adapt quickly and robustly.

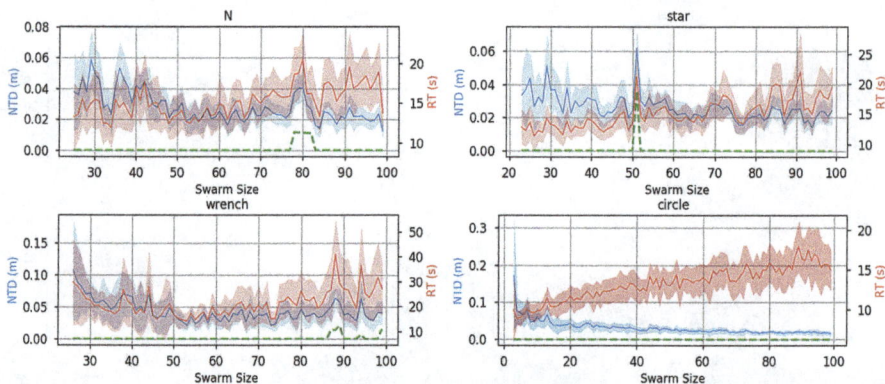

Fig. 7. The results from the substraction experiment. Each solid line is the average result from 50 trials, and the colored shade areas show the confidence intervals for NTD and RT over swarm size at a confidence level of two σ. Each green dotted line is the shape's NISD obtained from the previous section.

5 Conclusion

In this paper, we present an algorithm that encodes a 2D shape to a set of cells on a grid each characterizing a robot currently in the swarm. The performance of the algorithm is thoroughly evaluated via both the simulated swarm and physical swarms. The experiments show that the generated goal configuration for the swarm is consistent with the input shape, in addition, when the swarm size changes, it allows robots to adapt quickly and robustly.

References

1. Scher, G., Kress-Gazit, H.: Warehouse automation in a day: from model to implementation with provable guarantees. In: IEEE 16th International Conference on Automation Science and Engineering, pp. 280–287. IEEE (2020)
2. Le Goc, M., Kim, L.H., Parsaei, A., Fekete, J.-D., Dragicevic, P., Follmer, S.: Zooids: building blocks for swarm user interfaces. In: Proceedings of the 29th Annual Symposium on User Interface Software and Technology, pp. 97–109 (2016)
3. Trüg, S., Hoffmann, J., Nebel, B.: Applying automatic planning systems to airport ground-traffic control – a feasibility study. In: Biundo, S., Frühwirth, T., Palm, G. (eds.) KI 2004. LNCS (LNAI), vol. 3238, pp. 183–197. Springer, Heidelberg (2004). https://doi.org/10.1007/978-3-540-30221-6_15
4. Mong-ying, A.H., Kumar, V.: Pattern generation with multiple robots. In: 2006 Proceedings 2006 IEEE International Conference on Robotics and Automation, pp. 2442–2447. IEEE (2006)
5. Cheah, C.C., Hou, S.P., Slotine, J.J.E.: Region-based shape control for a swarm of robots. Automatica **45**(10), 2406–2411 (2009)
6. Hsieh, M.A., Kumar, V., Chaimowicz, L.: Decentralized controllers for shape generation with robotic swarms. Robotica **26**(5), 691–701 (2008). ISSN 0263-5747

7. Rubenstein, M., Shen, W.-M.: Scalable self-assembly and self-repair in a collective of robots. In: 2009 IEEE/RSJ International Conference on Intelligent Robots and Systems, pp. 1484–1489. IEEE (2009)
8. Stoy, K., Nagpal, R.: Self-repair through scale independent self-reconfiguration. In: 2004 IEEE/RSJ International Conference on Intelligent Robots and Systems (IROS), vol. 2, pp. 2062–2067. IEEE (2004)
9. Wang, H., Rubenstein, M.: Shape formation in homogeneous swarms using local task swapping. IEEE Trans. Robot. **36**, 597–612 (2020)
10. Rubenstein, M., Shen, W.-M.: A scalable and distributed approach for self-assembly and self-healing of a differentiated shape. In: 2008 IEEE/RSJ International Conference on Intelligent Robots and Systems, pp. 1397–1402. IEEE (2008)
11. Gauci, M., Nagpal, R., Rubenstein, M.: Programmable self-disassembly for shape formation in large-scale robot collectives. In: Groß, R., et al. (eds.) Distributed Autonomous Robotic Systems. SPAR, vol. 6, pp. 573–586. Springer, Cham (2018). https://doi.org/10.1007/978-3-319-73008-0_40
12. Alonso-Mora, J., Breitenmoser, A., Rufli, M., Siegwart, R., Beardsley, P.: Multi-robot system for artistic pattern formation. In: 2011 IEEE International Conference on Robotics and Automation, pp. 4512–4517. IEEE (2011)
13. Romanishin, J.W., Gilpin, K., Claici, S., Rus, D.: 3D M-blocks: self-reconfiguring robots capable of locomotion via pivoting in three dimensions. In: 2015 IEEE International Conference on Robotics and Automation, pp. 1925–1932. IEEE (2015)
14. Yim, M., et al.: Modular self-reconfigurable robot systems [grand challenges of robotics]. IEEE Robot. Autom. Mag. **14**(1), 43–52 (2007)
15. Cheng, J., Cheng, W., Nagpal, R.: Robust and self-repairing formation control for swarms of mobile agents. In: AAAI, vol. 5 (2005)
16. Olivier, R., Hanqiang, C.: Nearest neighbor value interpolation. Int. J. Adv. Comput. Sci. Appl. **3**(4), 25–30 (2012)
17. Vaquero, D., Turk, M., Pulli, K., Tico, M., Gelfand, N.: A survey of image retargeting techniques. In: Applications of Digital Image Processing XXXIII, vol. 7798, pp. 779814. International Society for Optics and Photonics (2010)
18. Zhang, Y.Y., Wang, P.S.-P.: A parallel thinning algorithm with two-subiteration that generates one-pixel-wide skeletons. In: Proceedings of 13th International Conference on Pattern Recognition, vol. 4, pp. 457–461. IEEE (1996)
19. Fisher, R.: Distance transform. https://tinyurl.com/22uacjz2
20. Kuhn, H.W.: The Hungarian method for the assignment problem. Naval Res. Logist. Q. **2**(1–2), 83–97 (1955)
21. Wang, H., et al.: Supplementary Materials. https://tinyurl.com/2huc42t6

Leading a Swarm with Signals

Nofar Menashe and Noa Agmon[✉]

Bar Ilan University, Ramat Gan, Israel
agmon@cs.biu.ac.il

Abstract. The prevalence of autonomous agents has raised the need for agents to cooperate without having the ability to coordinate their moves in advance, or communicate explicitly. This is referred to as *ad-hoc teamwork*. Prior work in this field has examined the possibility of leading a flock of simple, swarm-like, agents to a desired behavior that maximizes joint group utility, using informed agents that act within the flock. In this work we examine the problem of leading a flock of agents using signals. In this problem, the leading agents are equipped with a tool allowing them to send a simple signal to the flock, "calling" them to act in a desired way. However, the agents may misinterpret the signal with some probability, and head to the opposite direction. We examine the best behavior for a leading agent, that is, deciding when to signal, which depends on the signaling range of the leader, the probability of the signal misinterpretation, the sensing range of the flocking agents, and their current behavior. We extend the analysis to multiple leading agents, and show that their location within the flock also plays a role in the outcome. Finally, we examine the use of signals by leading a swarm of agents in the dispersion problem, where the team's goal is to spread in the environment and demonstrate the limitation of signals. Specifically, we show that signals may have no influence on the performance of the swarm, even if they are perfectly interpreted.

Keywords: Ad-hoc teamwork · Leading a swarm · Signals · Flocking · Dispersion

1 Introduction

The concept of swarms in robotics research refers to a group of simple agents with decentralized control. Since each agent is simple and has no global knowledge of the swarm and the environment and usually makes decisions based on its current limited view of the world, each swarm member is replaceable and the swarm can continue its mission even in the presence of dynamic changes to the world or to the group members. In contrary, the simplicity of each agent limits the ability of the swarm to perform complicated tasks that requires more sophisticated synchronization between the teammates. This disadvantage of swarms has motivated the need to add external influencing leaders (ad-hoc agents) [8] with more capabilities and larger view of the environment to improve or change the swarm behavior in order to optimize its performance. Therefore, in this research we examine the problem of leading swarm of agents by adding external, more knowledgeable agents to the swarm, referred to as *leading agents*.

Each agent in the swarm has limited and partial knowledge on the environment, the task and the group, and thus could benefit from sharing information to decrease uncertainty and improve the chances to meet their goal. However, establishing such communication between the agents is problematic because each agent is designed to work independently. Inspired by flock of birds, which also do not have direct ongoing communication model such as speech, we designed a communication model between the leading agent and the swarm based on a simple signal, same as birds using call notes to inform others of danger.

In an ideal environment, the leading agents signal the swarm how to behave, and each agent will receive the signal and act accordingly. However, real world environments holds communication challenges and there is a possibility the signal will be misinterpreted by some of the teammates and cause the opposite result. Therefore, our leader's model of decision making is based on the assumption that each agent might interpret a signal wrong with some probability.

In this work we introduce our signaling model in detail and examine how signals influence the performance of the swarm in the *flocking* problem, in which the leader aims at converging the flock into a desired heading. Initially examining the problem for one leader, we examine the decision the leader should make as to when to signal, which is based on the characteristics of the swarm members, their headings, the misinterpretation probability and the signal strength.

When there are multiple leaders available, the placement of the leaders within the flock becomes crucial for the ability of the flock to converge, as well as the convergence time. As the problem of optimal placement is NP-hard, we examine different heuristic methods and show the superiority of a simple greedy-based algorithm in various settings.

Finally, we examine the signaling protocol in the *swarm dispersion* problem, in which the leaders aim at dispersing the swarm as much as possible in a given environment. We show that the use of signals in this domain has little influence, even when the signals are perfectly interpreted. This leads us to the conclusion that the impact of signaling depends heavily on the task in hand, and it is not necessarily worthwhile to develop signaling capabilities categorically.

2 Related Work

Our work focuses on the problem of leading teams of agents in ad-hoc settings for two missions: flocking and dispersion. Stone et al. [1,19] introduced the problem of leading a team in ad-hoc settings to an optimal joint action while trying to minimize the system's cost. They designed a model based on game-theoretic tools to find an optimal set of actions in order to influence the agent's actions, assuming best-response agents. They used actions as a way of communication between the leading agent and the best response teammate(s), whereas in our work we use signals. Genter et al. [8] introduced the problem of leading a *flock* of agents in ad-hoc settings. They initially examined the flocking problem for stationary agents where the leading agents had influence only on the agents' orientation. A later work of Genter and Stone [9], considered the case of adding multiple influencing agents to alter the flock's behavior, focusing on determining the initial location of the leading agents for maximizing their influence on the flock. In addition, they examined how the leading agents should behave when joining a moving flock.

Previous work in ad-hoc teams also suggested using a teacher that gives the team the required information to perform their task and the other agents passively react according to this information. This teacher can be human, such as in [10], and provide the agents feedback for their action to use in reinforcement learning. The role of the teacher can also be assumed by another agent with more information about the environment that is able to advise the others, as presented in [20].

Grizou et al. [11] focused on teams of agents performing a task, and one ad-hoc agent (leader in our case). They assumed the ad-hoc agent does not know which task the team is trying to perform or its communication protocol. As part of their work they introduce partially observable version of the ad-hoc teamwork problem where teammates have limited view and each one has the ability to broadcast information to the team except for the ad-hoc agent. In our work, teammates are unable to communicate explicitly, except the leader which has the ability to broadcast a signal.

Communication in ad-hoc setting is much less trivial than communication between homogeneous teammates, since each agent has no knowledge about its peers. Overcoming this challenge can be done using new forms of collaboration without communication. Mirsky et al. [18] suggested that although the teammates in ad-hoc scenarios are different, they might have similar communication protocols, and thus, have the ability to communicate with each other. Other solutions to this challenge involve new forms of indirect communication, such as motion [6], radio signals or light waves [4].

Potop-Butucaru et al. [17] surveys communication models for most common multiagent problems (flocking, gathering, exploration etc.) that tolerate one or more faulty agents, i.e. models in which groups can achieve their original goal even when some of the agents are damaged. Grizou et al. [12] also focused on the idea of using signals as a form of interaction. In their work, they designed a model to label instructions (signals) in human-machine interaction. Similarly to our work, there are two types of signals: correct and incorrect, although in our problem domain the signaler (leader) sends always the same signal and the receiver can misinterpret it.

3 Leading with Signals in the Flocking Problem

In this section we formally define the problem of leading a group of *flocking* agents in ad-hoc settings using signals, to optimize group utility when performing the joint task. We first examine the problem with one leader, and extend the definition and analysis to multiple leaders in Sect. 3.4.

Given a group of n homogeneous agents $A = \{a_1, a_2, ..., a_n\}$ and one leader a_l. Each agent in A can collect information from the environment according to its sensing range. An agents' sensing radius is denoted by r_s, where if $r_s = \infty$ the agent can sense all the other agents in the environment. Denote the distance between two agents a_i and a_j at time t by $d_t(a_i, a_j)$. The group of agents in agent a_i's visibility range is defined as its *neighborhood*, $N_i(t)$, that is, $N_i(t) = \{a_j | d(a_i(t), a_j(t)) \leq r_s\}$. We initially examine the case of infinite visibility radius in which the neighborhood of an agent is equal to the whole team ($N_i = A \cup \{a_l\}$), and examine the limited visibility case in Sect. 3.2.

The leader can signal the agents, notifying them to update their orientation to its own. However, the signal may be misinterpreted leading to an opposite outcome. We assume the signal is correctly interpreted with probability p and misinterpreted with probability $1 - p$. Similar to the sensing radius, the leader can signal and the agents that will receive the signal are those inside its signaling radius r_{signal}. In case $r_{signal} = \infty$, the signals the leader send will be received by all the swarm agents.

At each time step t the leader moves one step (step size is constant for all agents and denoted as δ) in its direction vector and decides whether to signal the team or let them move according to their original behavior model. Leader's chosen option is represented by the binary parameter s ($s = True$ for signal and $s = False$ when the leader chooses not to signal).

In response to the leader's action, each agent can be in one of three states, defined by parameter st: (1) **NoSig** - the leader does not signal and so the agent acts according to is basic behavior model (2) **CorInt** - the leader signals and the agent interprets correctly the signal and (3) **MisInt** - the leader signals and the agent misinterprets the signal and moves to opposite direction from the desired one (we chose opposite direction since its the worst case scenario). Figure 1 shows selected direction vectors for agent A with leader L and neighbors $N1$ and $N2$ for all three possible states no signal (orange), correctly interpreted signal (green) and misinterpreted signal (red).

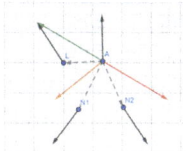

Fig. 1. Outcomes of the leader's possible actions

The leader decides how to operate by calculating the utility of each option (signal or not) and choosing the option that maximizes the *expected* group utility. We denote the group utility of action s at timestep t as $U^s(t)$. The error of an agent is inversely proportional to the utility and is defined as how far the agent is from its goal.

Therefore, the signaling criteria in case of a team of agents and one leader at timestep t will be demonstrated by the inequality: $U(s = True, t) - c - U(s = False, t) > 0$, where c is a parameter that represent the cost of signaling. As expected, and validated in experiments, as the cost increases the leader is less likely to signal, resulting in higher convergence time. Therefore we examine in this work the case in which the cost of signaling is negligible, to focus our attention on pure signaling decisions.

If the leading agent's signaling radius is infinite, the following holds (since the event that all signals are interpreted correctly occurs with probability p^n):

Theorem 1. *Given a group of n agents* $\{a_1, ..., a_n\} \in A$ *and one leader* a_l *with infinite signaling range* r_{signal}, *there exist a time step* $t_0 < \infty$ *in which all agents* $a_i \in A$ *interpret the signal correctly and thus, the agents will converge to the desired behavior, assuming the time between two steps in which the leader's decide to signal is bounded by* b.[1]

Therefore, theoretically, if there is no associated cost for signaling and the leader signals at each time step, then eventually all agents will converge, where convergence time depends on the probability p of correct signal interpretation. However, we do not want the leader to regularly signal but rather to decide whether or not to signal based on the signal efficiency and the signaling cost. The efficiency of a signal is defined by how well the signal succeeded in reducing the error of the group, i.e. how much the expected value of the error decreases on the following step.

Flocking Agents' Basic Behavior Model. Each teammate is represented by the forces acting on it. Based on a model presented by Couzin et al. [5], each of the flocking agents is influenced by two forces: orientation force (represented by the direction vector, the vector between previous and current locations) and attraction force (represented by the vector to the neighbor, the vector between agent's and neighbor's current location).

Following this model, in case the leader does not signal, i.e. $st = NoSig$, the agent calculates its next movement direction as the average direction between all of its neighbors' force vectors, and the agent's direction vector to preserve learned direction from previous steps. We denote the orientation force vector of agent a_j by v_j^o and attraction force vector of agent a_i to neighbor a_j by $v_{i,j}^a$ and calculate the force vector of neighbor a_j to a_i by their average: $v_{i,j}^d = \frac{1}{2}[v_j^o + v_{i,j}^a]$. Agent's self relative vector is calculated only by the orientation force vector, since attraction force vector $v_{i,i}^a$ is equal to zero.

Therefore, agent a_i's movement direction at timestep t equals the average force vector of its neighbors group N_i and its own direction at time step $t-1$:

$$v_i^o(st = NoSig, t) = \frac{1}{|N_i|+1} \sum_{a_j \in \{N_i \cup a_i\}} \frac{1}{2}[v_j^o(t-1) + v_{i,j}^a(t-1)] \quad (1)$$

Signal Response Model. In case the leader a_l signals, the agent calculates its next direction based only on the leader's orientation and the attraction vector toward the leader's location. The agent does not only ignore neighbors' direction during calculation but also its own previous direction. Formally, agent a_i's movement direction at timestep t, in case the leader signaled in the previous step and the agent interprets this signal correctly ($st = CorInt$), is equal to:

$$v_i^o(st = CorInt, t) = \frac{1}{2}[v_l^o + v_{i,l}^a(t-1)] \quad (2)$$

---

[1] Formal proofs and other additional material can be found in the supplamentary material in: www.cs.biu.ac.il/~agmon/DARS2021SignalSup.pdf.

If the leader signals and the signal is misinterpreted by agent a_i, its next movement direction will be equal to the opposite signaled direction:

$$v_i^o(st = MisInt, t) = -\frac{1}{2}[v_l^o + v_{i,l}^a(t-1)] \tag{3}$$

Group Utility. The leader decides at each step whether or not to signal the team based on the utility of each choice. Utility, in the flocking scenario, is defined as how close an agent's movement direction is to its desired direction. As mentioned before, signals might have a cost c which should also affect the decision of the leader. We calculate the angles between agent's direction to leader's orientation $\alpha_{i,l}^o = \angle(v_i^o, v_l^o)$ and attraction force vector $\alpha_{i,l}^a = \angle(v_{i,l}^a, v_i^o)$ and represent the utility as the average cosine of these two angles.

The decision of the leader is myopic, i.e. it calculates the utility of each option based only on the environment's state at the next step. We denote the direction of agent a_i at timestep t in case the leader does not signal as $v_i^o(st = NoSig, t)$. In case the leader does signal and the agent interprets the signal correctly $v_i^o(st = CorInt, t)$, and $v_i^o(MisInt, t)$ in case of misinterpreted signal. The angles are represented as $\alpha_{i,l}^o(st, t)$ and $\alpha_{i,l}^a(st, t)$ for orientation and attraction forces angles respectively. Therefore, utility of agent a_i at timestep t for state st is:

$$U_i(st, t) = \frac{1}{2}[cos(\alpha_{i,l}^o(st, t)) + cos(\alpha_{i,l}^a(st, t))] \tag{4}$$

In case the leader does not signal ($st = NoSig$) the group utility will be the sum of all agents' utilities.

$$U(s = False, t) = \sum_{i=1}^{n} U_i(st = NoSig, t) \tag{5}$$

The leader chooses whether to signal at each step based on the action with maximum utility.

$$U(s = True, t) - c > U(s = False, t) \tag{6}$$

Since signaling action has no deterministic outcome, we sum all possible outcomes of signaling to calculate the overall expected utility. If $h[i] = True$, agent i in option h interpreted the signal correctly and if $h[i] = False$ it misinterpret the signal. The probability of option h to be carried into effect is:

$$U(h) = \prod_{i=1}^{n} \left[\begin{cases} p & h[i] = True \\ 1-p & h[i] = False \end{cases} \right] \cdot \sum_{i}^{|N_l|} \begin{cases} U_{N_l[i]}(CorInt, t) & h[i] = True \\ U_{N_l[i]}(MisInt, t) & h[i] = False \end{cases} \tag{7}$$

Therefore, the utility of signaling, denoted by $U(s = True, t)$, considering all possible outcomes, all $h \in \{True, False\}^{|N_l|}$ (abbreviated $h_i \in [T, F]$) is equal to:

$$\sum_{h \in [T,F]^{|N_l|}} U(h) + \sum_{a_j \in A \setminus N_l} U_j(st = NoSig, t) \tag{8}$$

We define the error of agent a_i as the distance between its direction v_i^o and leader's v_l^o. After each step, we calculate the error for each teammate and check if its error does not exceed a predefined error tolerance rate ϵ. The swarm reaches its goal in case all teammates' errors are less than ϵ.

3.1 Basic Model Empirical Evaluation

All discussed behaviours were implemented as described in previous sections and tested in the MASON simulator [15]. We have examined how each parameter of the model influences the leader's decision to signal, agents' reaction and total team convergence time. First, we focused on two parameters: (1) the number of agents and (2) probability of agent to interpret leader's signal correctly. We tested our model by executing the simulation 50 times for each for populations of 1 to 20 agents and probabilities between 0.1 to 0.9 and calculated average number of steps until the entire team reaches the leader's direction (see Fig. 2).

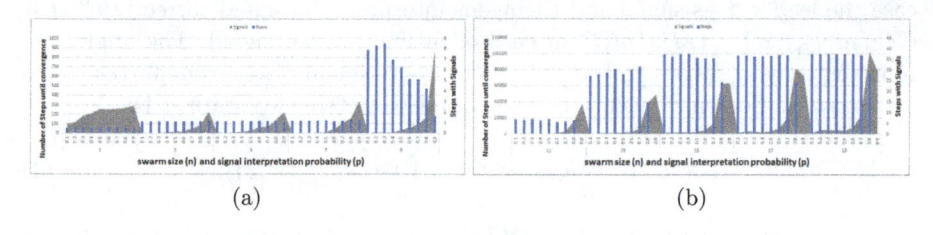

Fig. 2. Average steps until swarm convergence (limited by 100000) tested on different swarm sizes (1–20) and different probabilities for signal correct interpretation

Results shows that as the swarm size increases, the teammates converge slower to the leader's direction. It can also be inferred from the model description, since each agent calculates its next direction by averaging its neighbors' directions (when the leader does not signal). When the swarm size is small, there are less neighbors, and thus, the leader's direction has more influence on agent's next direction. From the signals graph (gray area) we can learn that for small groups, the number of steps does not necessarily decrease when the leader signals more often, because the leader has larger influence on each teammate even if it does not signal (there are less neighbors to affect the agent's direction). We can also see that signals have more impact on the swarm as swarm size increases. This is also related to the influence the leader has on each teammate when it does not signal. When the size of the group increases there are more neighbors to affect the agent's next direction and the team is more likely to converge to an average direction and not the desired one (the leader's).

3.2 Swarm with Limited Sensing Range

Real world robots, similar to humans and animals, observe the world using sensors with limited abilities. We therefore examine the case in which each agent can only view agents that are located inside its field of view, defined by circle with radius r_s around it. For this part of the work, we keep the leader's signal as a global signal with infinite strength.

This model was tested for different group sizes, probability for correctly interpreting signal p and sensing radius, where each set of parameters was executed 50 times with different random initial agents' positions and each execution was limited by 10000 steps. Results, presented in Fig. 3, show that increasing the sensing radius in most cases has positive impact over the convergence percentage of the team, the part of agents that were aligned to the leader at the end of the simulation (simulation may stop if all agents are aligned or after it reaches 10000 steps). As the number of agents grows, increasing the radius has less effect over the performance and might even harm the team's performance, since the difference between convergence percentage as the sensing radius increases when n = 4 is greater than the difference between those radii when n = 13. This is due to the fact that as the radius grows, each agent has more neighbors, and therefore the influence of the leader decreases when it is considered as a regular neighbor (if it does not signal).

Even though the leader's signaling radius is infinite, not all the agents converge to the leader's direction. The leader decides when to signal based on the utility of this action, if most of the agents are converged it might choose not to signal, so it will not lose the aligned agents (they can misinterpret the signal with probability p and move away).

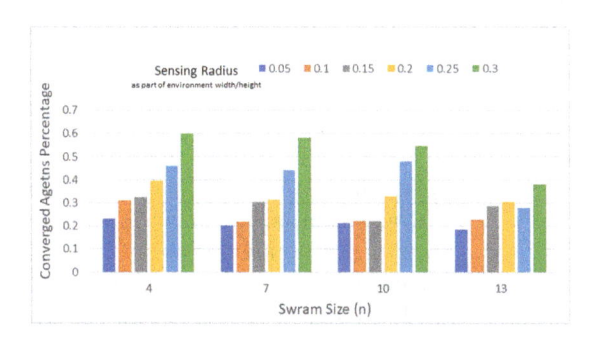

Fig. 3. Percentage of converged agents by the sensing radius of each agent for different swarm sizes

3.3 Limited Signal

Until now we assumed the leader's signal has infinite strength and can reach any agent regardless how far it is from the leader. In this section, we would like to apply the radius neighborhood model we presented in previous section also on the leader, i.e. if the leader signals only the agents in its signaling radius r_{signal}

will receive it and the rest will behave according to their basic behavior model (same as when there is no signal).

We examined the influence of different signaling radius with a high p value of 0.9, because with low probability the leader is not likely to signal. We have run the model with group size of 4,7,10 and 13 agents. Each experiment was executed 100 times, randomizing over initial agent locations.

In the results demonstrated in Fig. 4, we can see that for small groups, large signal radius achieved much better results than small radius. As the group size increases the improvement between small and large signal radius decreases. This is caused since there are more agents located randomly in the environment and therefore the probability the leader has agents in its signaling radius increases, more agents are influenced by the leader's signal and having more aligned agents helps those who are not aligned to converge faster.

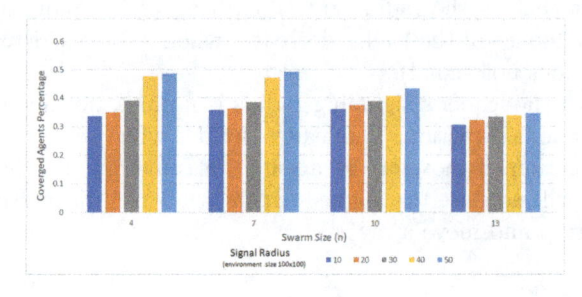

Fig. 4. Swarm converged agents percentage by signaling radius with probability $p = 0.9$ and for different swarm size

Limited signaling radius can cause isolated agents that have no neighbors (each agent sight radius is also limited) and the leader can not influence their behavior. Thus, these agents will continue to move in their original random direction, and if this direction is not accidentally directed towards the leader this agent will be lost after some steps. In order to overcome this issue of lost agents we need to use multiple leaders to keep maximal amount of influenced agents. We examine multiple leaders model in the next section.

3.4 Leader Position and Multiple Leaders

We have examined our model for different parameters such as probability to correctly interpret signal, sensing radius and signal propagation radius. If each leader has unlimited signaling range, adding leaders will not have an effect over the group, unless they do not signal and are considered as aligned neighbors. However, if the signaling range of the leader is limited, it influences less agents and thus by adding leaders more agents can be influenced. In this section, we will try to understand how using models to define leader's **initial position** can improve the results of the group.

Given k leaders with signaling radius r_{signal} and n agents with sensing radius r_s, we aim to find a position for each leader, maximizing the number of agents in the influence disks of all leaders. By a reduction from the geometric version of the maximum coverage problem (an $NP - hard$ problem [14]), the following holds:

Theorem 2. *The problem of finding k positions that maximizes the number of agents covered by a leader influence disk is NP-Hard.*

We therefore examine several heuristic approaches for placing the leaders, given the initial locations of the agents in the flock.

Previous Approaches. Previous research [7] examined how different models of positioning influencing agents affected the number of lost flocking agents. They compared different placement approaches for k influencing agents within a swarm of n agents, and concluded that a graph approach yields best results in terms of influenced agent. This approach is based on creating a connectivity graph between all agents (two neighbors are connected), calculating all points in the middle of two connected agents and points extremely close to each agent, and choosing from all points those who connect between maximum number of agents as influencing ones (directly or indirectly). While using the graph approach improved significantly the performance compared to a default random placement also in our setting, it does not consider the error of the group (how far each agent is from desired result). We therefore examined two new error based approaches for the leaders' placement: the *basic error* approach, and the *intersection error* approach.

Basic Error Approach. Since each agent influences its neighbors and we are looking for minimizing the overall error of the group, we have created a new positioning approach that considers these two aspects. We calculate for each agent the sum of its error and its neighbors' error, and the location of the leader is selected to be the one close to the agent with maximum sum of errors.

In order to increase the number of influenced agents, we prefer to select positions next to agents (with some small distance ϵ) that none of their neighbors are already chosen, i.e. if the first leader was chosen to be located next to agent a_x the location of the next leader will be chosen from $A\backslash\{N_x \cup a_x\}$ and so on, unless there are no agents that neither them nor their neighbors were already selected.

Intersection Error Approach. The next approach we tested is similar to the basic error approach except we consider groups of influenced agents by the intersection between their sensing radius instead of their neighbors. Therefore we are aiming at finding intersections that maximize the sum of the influenced agents' errors. The suggested greedy algorithm works as follows.

For each agent, we go over all the groups we have already created and check whether all of the group members are its neighbors and if they are, we add the agent to the group. In case the agent has a neighbor that was not found in any of the groups the agent is within, we create a new group for the agent and the neighbor. In case there are no existing groups or the agent does not have any neighbors, we create a new group that contains only the agent. After creating the intersection groups, we find all the middle point of each intersection between pair of agents and calculate for each point the number of affected agents and the sum of errors of all affected agents (there might be affected agents that are not in the group). Finally, we sort the groups in descending order by the sum of errors and the possible points P for each group first by number of affected agents and then by sum of errors of all affected agents and then choose iteratively the top position of each group until k positions are selected.

As proven in [16], when dealing with maximization over a monotone, non-negative and submodular function, greedy algorithm solution provides a $(1 - 1/e)$-Approximation to the optimal solution. Following Theorem 3, the Intersection Error approach provides an approximation ratio of $(1 - 1/e)$ to the optimal solution of determining the location of k agents that will minimize the sum of agents' error.

Theorem 3. *Given a selected positions group $L \subseteq P$ that covers group of agents $A_L \subseteq A$ we define $g(L) = \sum_{a \in A'} Error(a)$. Function g is monotone, non negative and submodular, i.e. for each $L_1, L_2 \subseteq P$ that cover agents $A_{L_1}, A_{L_2} \subseteq A$ respectively, $g(A_{L_1} \cap A_{L_2}) + g(A_{L_1} \cup A_{L_2}) \leq g(A_{L_1}) + g(A_{L_2})$.*

Genetic Algorithm Solution. We have examined the reliability of our greedy-based algorithms by comparing them to a solution that learns the best positions. We base our solution on the genetic algorithm presented by Atta et al. [2], which suggested a solution to a simplified version of our scenario, where there is set of finite optional locations to select from. We have implemented a similar version with adjustment to our specific problem domain, in which each position in the universe is possible. Parameters of the genetic algorithm implementation appear in the supplementary material.

Multiple Leaders - Empirical Comparison. We compared the performance of all suggested algorithms in terms of number of lost agents. The experiments varied over the number of agents, while fixing the probability of correctly interpreting the signal to 1. Each experiment was executed 80 times, randomizing over the initial locations of the agents.

The results, presented in Fig. 5, supported the theorem we proved above, showing that the genetic algorithm did not have significant advantage over our greedy solutions ($p-value > 0.05$), and the intersection model significantly outperformed all other greedy-based or random algorithms ($p-value < 0.02$).

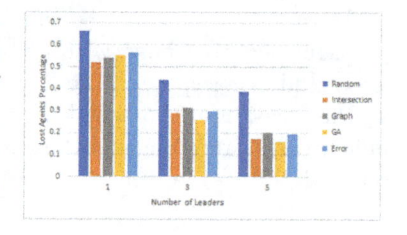

Fig. 5. Evaluation of the initial placement methods in terms of lost agents.

4 Signaling in the Dispersion Problem

In this section we briefly describe the results from introducing signaling to another swarm problem, *dispersion*, examining whether signaling can be beneficial there.

4.1 Dispersion Problem Definition

In the dispersion problem, a group of agents have to disperse in an environment, usually to explore unknown parts of it. We use a common potential field behavior as the **basic behavior model**, presented by Howard et al. [13]. In order to achieve area coverage we limit the environment settings and define boundaries such that an agent that reaches one of them changes its direction.

Behavior Model. Each agent moves in a random direction. Every dt steps it generates a new random direction and for the next dt steps it moves in this direction. If there are other agents in its vision area (within radius r_v), there are repulsive forces acting upon the agent. Based on Howard et al.'s work [13], the force F_x acting upon agent a_x, located in position P_x, is represented as the gradient of potential field U_x, defined as follows:

$$F_x = -\nabla U_x = -\frac{dU}{dP_x} \quad \text{and} \quad U_x = -\sum_{a_i \in N_x} \frac{1}{d(a_x, a_i)}$$

where $d(a_x, a_i) = \sqrt{(P_x - P_i)^2}$ is the euclidean distance between agent a_x and agent a_i and N_x is the neighborhood of agent a_x. Therefore, derivation of the potential field is defined as:

$$F_x = -\sum_{a_i \in N_x} \frac{1}{d(a_x, a_i)^2} \cdot \frac{d_{a_i \to a_x}}{d(a_x, a_i)}$$

where x is the location of agent a_x, x_i is the location of agent $a_i \in N_x$ and $d_{a_i \to a_x} = |x - x_i|$.

If the leader decides to signal and the signal is correctly interpreted, the direction of the agent a_x will be $d_x(st = CorInt) = -\frac{D}{|D|}$ (we normalize the direction vector). If the signal is misinterpreted, the direction of the agent will be $d_x(st = MisInt) = \frac{D}{|D|}$.

Evaluation Criteria. The quality of dispersion can be determined by many measurements. We chose to use two common measurements, commonly used in previous works (e.g., [3]): (1) average distance between each agent and its closest neighbor, (2) the area covered by the group as percentage of the size of the environment. The first one ensures that the agents are not too close to each other whereas the second one measures the quality of spreading in the environment. For both measurements we ignore the presence of leaders in the environment, and calculate their values based only on the group's agents. We would like to maximize both of them in order to achieve a good dispersion. For measuring the area coverage, we compute both the area visited by agents, and the area of the convex hull determined by the location of the agents.

4.2 Dispersion Model Results

We conducted many different sets of experiments, varying over the number of agents, probability of correctly interpreting the leader's signal, and different number of leaders. We report here one important experiment which resulted in the understanding that signaling in the dispersion problem might not be effective as in the flocking problem. In this experiment, we have used two types of leaders that travel according to the same algorithm, though they differ in one main characteristic: the first can signal perfectly ($p = 1$), and the second does not signal at all. Each experiment was executed 100 times. As seen in Fig. 6, there is no significant difference in the performance between these two types of leader's behaviors. We have also examined this with increasing signaling radius, between 10 (which is 0.05 percents of the environment size) and 55 (which is 0.28 percents of the environment size). Although the average distance to neighbors and size of convex hull increase as the signaling radius increases, the visited area remains the same. The reason is that the signaling agent can influence more agents in less time as its signal strength increases, but this does not necessarily translate into more visited areas, due to the basic random behavior of the agents.

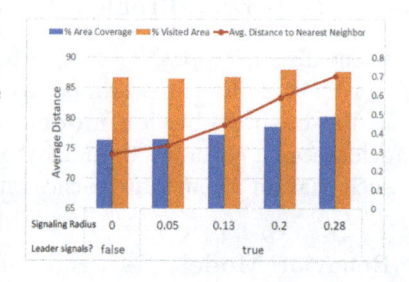

Fig. 6. Comparison of dispersion measurements with two types of leaders - signaling leader and only moving leader.

5 Conclusion and Future Work

In this paper we have introduced a simple signaling model for leading agents in ad-hoc settings, used to bring a swarm of simple agents to a desired optimal behavior. We have shown theoretically and empirically that in the flocking problem, signals can have a significant positive effect both in terms of convergence time, and percentage of converged agents. However, in the dispersion problem signaling has less of an effect in some cases.

In the future, we would like to further analyze the advantage and disadvantage of signaling in the dispersion and other swarm problems. In addition, we wish to examine other signaling models, such as costly signals, signals with other random effects when misinterpreted, and leading with heterogeneous signals.

Acknowledgments. This research was funded in part by ISF grant 2306/18.

References

1. Agmon, N., Stone, P.: Leading ad hoc agents in joint action settings with multiple teammates. In: Proceedings of the 11th International Conference on Autonomous Agents and Multiagent Systems, pp. 341–348 (2012)
2. Atta, S., Sinha Mahapatra, P.R., Mukhopadhyay, A.: Solving maximal covering location problem using genetic algorithm with local refinement. Soft. Comput. **22**(12), 3891–3906 (2017). https://doi.org/10.1007/s00500-017-2598-3
3. Batalin, M.A., Sukhatme, G.S.: Spreading out: a local approach to multi-robot coverage. In: Asama, H., Arai, T., Fukuda, T., Hasegawa, T. (eds.) Distributed Autonomous Robotic Systems 5, pp. 373–382. Springer, Tokyo (2002). https://doi.org/10.1007/978-4-431-65941-9_37
4. Becker, M., Blatt, F., Szczerbicka, H.: A concept of layered robust communication between robots in multi-agent search & rescue scenarios. In: 2014 IEEE/ACM 18th International Symposium on Distributed Simulation and Real Time Applications, pp. 175–180 (2014)
5. Couzin, I.D., Krause, J., James, R., Ruxton, G.D., Franks, N.R.: Collective memory and spatial sorting in animal groups. J. Theor. Biol. **218**(1), 1–11 (2002)
6. Das, B., Couceiro, M.S., Vargas, P.A.: MRoCS: a new multi-robot communication system based on passive action recognition. Robot. Auton. Syst. **82**, 46–60 (2016)
7. Genter, K.L. et al.: Fly with me: algorithms and methods for influencing a flock. Ph.D. thesis (2017)
8. Genter, K., Agmon, N., Stone, P.: Ad hoc teamwork for leading a flock. In: Proceedings of the 2013 International Conference on Autonomous Agents and Multi-agent Systems, pp. 531–538 (2013)
9. Genter, K., Stone, P.: Adding influencing agents to a flock. In: Proceedings of the 2016 International Conference on Autonomous Agents & Multiagent Systems, pp. 615–623 (2016)
10. Griffith, S., Subramanian, K., Scholz, J., Isbell, C.L., Thomaz, A.L.: Policy shaping: integrating human feedback with reinforcement learning. In: Advances in Neural Information Processing Systems, pp. 2625–2633 (2013)
11. Grizou, J., Barrett, S., Stone, P., Lopes, M.: Collaboration in ad hoc teamwork: ambiguous tasks, roles, and communication. In: AAMAS Adaptive Learning Agents (ALA) Workshop (2016)
12. Grizou, J., Iturrate, I., Montesano, L., Oudeyer, P.-Y., Lopes, M.: Interactive learning from unlabeled instructions. In: Proceedings of the Thirtieth Conference on Uncertainty in Artificial Intelligence (UAI) (2014)
13. Howard, A., Matarić, M.J., Sukhatme, G.S.: Mobile sensor network deployment using potential fields: a distributed, scalable solution to the area coverage problem. In: Asama, H., Arai, T., Fukuda, T., Hasegawa, T. (eds.) Distributed Autonomous Robotic Systems 5, pp. 299–308. Springer, Tokyo (2002). https://doi.org/10.1007/978-4-431-65941-9_30
14. Jin, K., Li, J., Wang, H., Zhang, B., Zhang, N.: Near-linear time approximation schemes for geometric maximum coverage. Theoret. Comput. Sci. **725**, 64–78 (2018)
15. Luke, S., Cioffi-Revilla, C., Panait, L., Sullivan, K.: MASON: a new multi-agent simulation toolkit. In: Proceedings of the 2004 Swarmfest Workshop, pp. 316–327 (2004)
16. Nemhauser, G.L., Wolsey, L.A., Fisher, M.L.: An analysis of approximations for maximizing submodular set functions-I. Math. Program. **14**(1), 265–294 (1978)

17. Potop-Butucaru, M., Raynal, M., Tixeuil, S.: Distributed computing with mobile robots: an introductory survey. In: 2011 14th International Conference on Network-Based Information Systems, pp. 318–324. IEEE (2011)
18. Reut, M., William, M., Andy, W., Harel, Y., Peter, S.: A penny for your thoughts: the value of communication in ad hoc teamwork. In: International Joint Conference on Artificial Intelligence (IJCAI) (2020)
19. Stone, P., Kaminka, G.A., Rosenschein, J.S.: Leading a best-response teammate in an ad hoc team. In: David, E., Gerding, E., Sarne, D., Shehory, O. (eds.) AMEC/TADA -2009. LNBIP, vol. 59, pp. 132–146. Springer, Heidelberg (2010). https://doi.org/10.1007/978-3-642-15117-0_10
20. Torrey, L., Taylor, M.: Teaching on a budget: agents advising agents in reinforcement learning. In: Proceedings of the 2013 International Conference on Autonomous Agents and Multi-agent Systems, pp. 1053–1060 (2013)

Byzantine Fault Tolerant Consensus for Lifelong and Online Multi-robot Pickup and Delivery

Kegan Strawn[✉] and Nora Ayanian

University of Southern California, Los Angeles, CA 90007, USA
{kegan.j.strawn,ayanian}@usc.edu

Abstract. Lifelong and online Multi-Agent Pickup and Delivery is a task and path planning problem in which tasks arrive over time. Real-world applications may require decentralized solutions that do not currently exist. This work proposes a decentralized and Byzantine fault tolerant algorithm building upon blockchain that is competitive against current distributed task and path planning algorithms. At every timestep agents can query the blockchain to receive their best available task pairing and propose a transaction that contains their planned path. This transaction is voted upon by the blockchain network nodes and is stored in the replicated state across all nodes or is rejected, forcing the agent to re-plan. We demonstrate our approach in simulation, showing that it gains the decentralized Byzantine fault tolerant consensus for planning, while remaining competitive against current solutions in its makespan and service time.

Keywords: Robotics · Blockchain · MAPD · Consensus

1 Introduction

Path planning for teams of robots is a fundamental problem for many applications of multi-robot systems. Often, teams of robots must reliably navigate to given locations without collisions. In Multi-Agent Pickup and Delivery (MAPD), given a set of tasks with known locations, agents must find an assignment of tasks to agents and plan the paths necessary to arrive at the locations of those tasks. We are interested in the lifelong and online version of MAPD (LO-MAPD), wherein tasks arrive over time for indefinite periods, in known environments. Examples include warehouse robots [33], video game characters [26], aircraft-towing vehicles [21], and general autonomous mobile robots [30].

In this work we focus on MAPD in a warehouse environment with multiple interested parties operating within the same system with no central authority. Here, multiple self-serving agent teams must collaborate to accomplish all tasks without complete trust in the other agents. Therefore, not only do centralized algorithms scale poorly for large numbers of agents, but are not a valid choice. Current distributed algorithms attempting to solve this, generally, produce longer paths for the overall system and make large assumptions on the behavior of agents. We assume all agents continuously seek to accomplish tasks, but make no further assumptions about the behavior of agents. For example, we do not assume that agents seek to perform the best tasks for the overall system or that agents do not attempt to plan paths with collisions. In this way, agents may be selfish or uncooperative and introduce Byzantine failures in planning that break current solutions.

These Byzantine failures are faults beyond simple failures (such as losing a message or agent) where there is imperfect information on whether there has been a failure. Existing work on swarm and multi-robot systems often consider themselves failure resistant due to the number of agents, but in many cases a single Byzantine failure is enough to stop the system from working [27]. Outside of the MAPD problem, blockchain based solutions have recently been proposed as a potential solution to provide Byzantine fault tolerance, a distributed and immutable storage, and anonymity for multi-robot systems [1,5,6,16,27].

Our algorithm builds upon a blockchain framework to provide Byzantine fault tolerance in planning for the decentralized warehouse MAPD problem. In our system and algorithm, each individual agent has a limited scope of information. They receive only the current timestep, the used location-timestep pairs planned so far to avoid collisions, the endpoint locations of agents where they stay indefinitely, and their best available task without any other attached agent or task information. This reduces the amount of state information sent between distributed agents and enables a larger system where multiple teams of agents operate without conflicts and without sharing identifiable data or trusting any single source for assignments. As we will show with our simulation results, our work gains this decentralized Byzantine fault tolerant consensus from building on top of the blockchain for task and path planning, while remaining competitive against current distributed solutions in its makespan and service time.

2 Background

2.1 MAPF and MAPD

MAPD is an extension and generalization of the popular Multi-Agent Path Finding (MAPF) problem. In the MAPF problem paths must be found for all agents from their starting locations to their destinations without colliding with other agents or obstacles in the environment. Much work exists on the MAPF problem, its variants, and benchmarks [19]. Similar to Target Assignment and Path Finding (TAPF) [18], MAPD extends MAPF to incorporate tasks that are composed of pickup and delivery locations agents must move to in order to accomplish them.

We address discrete LO-MAPD, wherein a stream of tasks arrive over time in an online setting on a general graph. The goal of the system is to accomplish each task while minimizing the cost. Depending on the application, costs could be: service time (the average number of timesteps needed to execute each task), the makespan (the last timestep used), and/or the runtime (the time it took to execute the planning of all tasks). To successfully solve a MAPD instance the algorithm must have a bounded service time and runtime. While not all MAPD instances are solvable, sufficient conditions exist that ensure solvability; we only work with such instances, called well-formed instances [29].

Many algorithms have been developed to solve the MAPF, TAPF, and MAPD family of problems. Examples include Conflict-Based Search and it's variants [3,8,10,18,25], Answer Set Programming [23], Enhanced Partial Expansion A* [9], suboptimal algorithms [11,28,31,32], and Windowed-Hierarchical Cooperative A* and similar approaches [15,24,26]. For LO-MAPD, Token Passing and Token Passing with Task Swaps are the state of the art distributed solutions [17,20]. However, the distributed approaches can suffer from deadlocks, make strong assumptions, and sacrifice performance for distribution.

2.2 Blockchain and Byzantine Faults

Our algorithm differs from existing solutions by taking advantage of blockchain technology. Created as part of Bitcoin [22], a blockchain framework is presented as a distributed ledger with peer-to-peer sharing of encrypted and linked data transactions with timestamps. Blockchain technology can provide an immutable state, decentralized consensus, fault tolerance, and a dynamic framework for the flexible control of a network of nodes [2]. A blockchain transaction can be developed to store different forms of data. In this work, we store different types of MAPD information as transactions on the blockchain. Each transaction has a string labeling the type of data (for example: a timestep, used timestep-location pairs, or task assignment) and the data itself in string format.

In the Byzantine Generals Problem, a set of generals with no trust in each other surround a castle and can only communicate via a messenger to agree upon when to attack. Lamport et al. [13] present this problem and prove a well-designed system can survive 1/3 of these Byzantine failures where the generals are unresponsive or unreliable. There exists a thorough body of work on Byzantine failures in collaborative networks in distributed systems literature [14]. In distributed systems, to be Byzantine fault tolerant (BFT), nodes in the network must agree regularly about the current state and have a 2/3 majority agreement, surviving 1/3 of all nodes differing in their judgement [7]. A blockchain can solve the Byzantine Generals Problem using cryptography and various validation models. Each blockchain framework defines the effort, investment, and resources a node must put into the system and the outcome of the vote in order to incentivize and maintain honest and positive behavior. It is important to note that the use of blockchain alone does not necessarily provide Byzantine fault tolerance, but that our work here builds upon blockchain to provide this additional fault tolerance in a specific multi-robot setting. In the context of robotics, the definition and categorization of Byzantine faults is not yet widely agreed upon. We focus on a subset of arbitrary behavior in adversarial agents' planning and communication, although Byzantine faults could take more advanced forms in real-world applications.

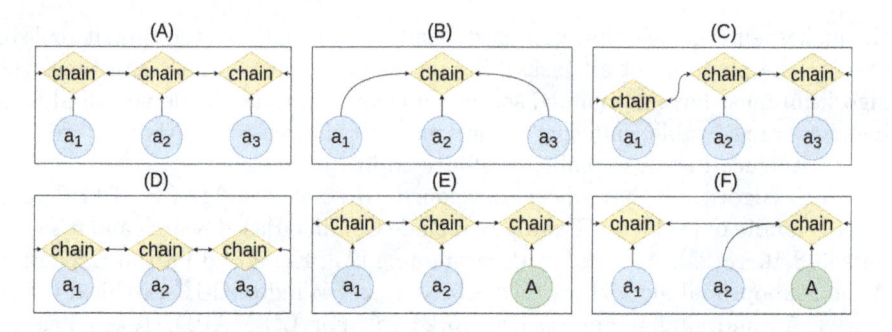

Fig. 1. Potential node-agent configurations. Yellow diamonds are Tendermint ABCI nodes running our BFTC Chain application. Blue circles a_1, a_2, a_3 are agents. A circles represent the simulation agent that sets the timestep and other simulation/environment variables on the blockchain.

2.3 Tendermint

Tendermint provides an Application Blockchain Interface (ABCI) for a peer-to-peer message passing protocol that checks programmer-defined functions to validate messages passed across the blockchain network. Each proposed transaction goes through Tendermint's voting process and provides a flexible interface to integrate the blockchain into our MAPD algorithm. The setup of the Tendermint network is flexible. Each MAPD agent in the system could run the Tendermint node as well as our algorithm, however, if an agent has insufficient computational resources it could communicate to the Tendermint node running on another computing system. In Fig. 1, some possible configurations are shown, such as all agents connected to separate nodes, all agents connected to the same node, one agent running the node on its local system, an agent A that can interact with the blockchain but does not accomplish tasks, and various other possible compositions for all nodes and agents.

We have designed a Tendermint ABCI application, called the Byzantine Fault Tolerant Consensus (BFTC) chain, that implements this interface and is similar to a key-value database to store used locations, edges, and endpoints as well as the current timestep and the set of available tasks. Tendermint runs this application on all nodes in the Tendermint network. The BFTC chain can be queried by MAPD agents running our algorithm in two ways. It can be sent either a Tendermint DeliverTX message that stores the transaction as part of the state if agreed upon by at least 2/3rds of all chain nodes as a valid transaction (such as no collisions), or a Tendermint Query message that returns identical replicated information about the state across all nodes that are part of the chain's network.

3 Methods

Our proposed BFTC system solves well-formed, LO-MAPD instances by building an application on top of the blockchain platform Tendermint [4] to reach

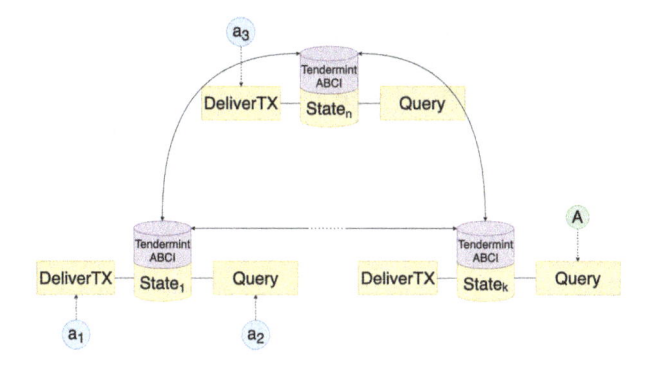

Fig. 2. A visual representation of our BFTC system. Agents send DeliverTX or Query messages to the nodes, who use the ABCI to utilize Byzantine fault tolerant Tendermint consensus before storing the transactions in the state storage on each node. All agents $a_1, \ldots, a_n$ communicate with nodes $1, \ldots, n$ that maintain identical states $state_1, state_2, \ldots, state_n$.

consensus on valid task assignments and valid paths in a decentralized network at every timestep. We assume agents must communicate on the network to receive and accomplish tasks, but not that all agents propose valid transactions. All valid agents follow our BFTC algorithm (Algorithm 1) simultaneously. All agents send messages to a Tendermint node on the network to receive a task and information to plan their paths, as well as propose these planned paths as transactions to collectively accomplish all tasks. In contrast, current state of the art algorithms pass full state knowledge around to each agent in a predefined order based on their ID number [20].

We assume an agent rests in the last location of its path and each agent plans its individual path using A*. Paths are sent as transactions, and, if valid, are stored in the distributed ledger that represents the world state with the order of transactions determined by the order of consensus rather than the order determined by a central system. The BFTC chain application has DeliverTX methods to handle requests for: adding a new timestep, adding a new task, posting an agent's location, assigning an agent to a task and storing the path to the task, storing an agent's path to a safe endpoint, and checking in for an agent at a timestep. The chain has Query methods to handle requests for: the current timestep, current delivery spots, best task for a specific agent's location (if any), used timestep-location pairs, edges, and endpoints, as well as if all agents have checked in for the current timestep. We use an agent A that communicates with the chain via a Tendermint node in order to increment each timestep once all agents have checked in for the previous timestep, adding any new tasks to the task set, and logging any accomplished task pickups or deliveries. This allows us to follow the same experimental model as in [20]. A visual representation of our BFTC system is shown in Fig. 2.

Every valid, non-Byzantine failing, node composing the decentralized BFTC chain simultaneously uses the Hungarian Method [12] in a decentralized fashion to find the best possible unassigned task-agent location pairings for the overall system, at every new timestep. To avoid collisions after assignments, our A* search and validation logic works similar to [20], planning collision-free paths using the time-location pairs and edges, called usedElements in Algorithm 1. Path costs only need to be found from a location to another endpoint, so heuristic values for A* are pre-computed. The chain's Hungarian method only uses this heuristic value map and does not compute the A* solution at every timestep for every task, saving computation resources and time. This new best possible task-agent pairing allows our decentralized BFTC algorithm to assign agents to tasks that might not be the closest to them, but minimizes the overall cost for the system. Thus, even without failures, our decentralized solution has the potential to outperform certain distributed algorithms; this is discussed further in Sect. 5.1.

Each agent begins its outer loop by checking if the state's timestep has been incremented. If it has been, an agent without a task will query the chain for the best possible task assignment for only itself. If it receives one it will use A* to plan a path to the pickup and then the delivery, sending this as a transaction containing a single path to the chain for assignment. If the chain replies with a successful assignment (meaning consensus was reached among Tendermint nodes), the agent will check in for this timestep. If the agent did not receive a best possible task, it will check to see if its current location will cause a collision. If it does, the agent will move out of this endpoint through exhaustive search from it's current location for a safe endpoint with the least cost path before checking in. Otherwise, it will stay in its current location and check in for the next timestep.

4 Benchmark Algorithms

Ma et al. present three algorithms for solving the LO-MAPD problem: two decoupled MAPD algorithms, Token Passing (TP) and Token Passing with Task Swaps (TPTS), and a centralized algorithm, CENTRAL [20]. The TP algorithm involves sending the token that stores all world state data to each agent that needs to choose a task and plan a path. TPTS modifies the TP algorithm to enable swapping tasks before the tasks have been picked up. In TP, all agents must wait until the agents before them have received the token and planned their paths/assignments. In TPTS all agents must wait until the agents before them have planned and all re-planning has taken place before they can be certain of their paths/assignments for that timestep. In decentralized systems this is an important negative attribute, as the system should not prioritize the wait time of certain agents without apparent reason.

Minimizing the sum of costs is NP-hard to solve optimally and NP-hard to approximate the makespan within a constant factor less than $4/3$ [34], so they compared their new algorithms with CENTRAL, a centralized strawman MAPD algorithm. CENTRAL uses the Hungarian Method [12] to assign tasks and ECBS [10] to plan conflict-free paths at every timestep in a simulated warehouse environment. CENTRAL provides a best possible benchmark to compare

Algorithm 1: BFTC (Byzantine Fault Tolerant Consensus)

for *each agent a_i in A* **do**
    $a_i.path = loc(a_i)$;
    /* Start all agent processes, calling IndividualAgent(a_i, chain) */
while *true* **do**
    Add all new tasks if any to the task set T;
    Put the new timestep on the chain;
    while *∃ a_i not checked-in in the chain for the timestep* **do**
        /* system waits, with timeout */
    AdvanceTimestep(chain);
    /* Move agents one step */

Function IndividualAgent(a_i, *chain*):
    while *true* **do**
        success ← False;
        if *a_i has no task* **then**
            τ ← QueryBestTaskAssignment($loc(a_i)$, chain);
            while *τ and not success* **do**
                usedElements← QueryUsedElements(chain);
                path ← AStarPath($loc(a_i)$, τ, usedElements);
                success ← DeliverPath(path, chain);
                if *success* **then**
                      assign a_i to τ;
                      $a_i.path$ = path;
                      break;

        if *a_i has no task* **then**
            if *no $\tau_j \in T$ with $g_j == loc(a_i)$* **then**
                while *not success* **do**
                      usedElements← QueryUsedElements(chain);
                      path ← AStarPath($loc(a_i)$, $loc(a_i)$, usedElements);
                      success ← DeliverPath(path, chain);
                $a_i.path$ = path;
            else
                while *not success* **do**
                      usedElements ← QueryUsedElements(chain);
                      safeEndpoint ← FindSafeEndpoint($loc(a_i)$));
                      path ← AStarPath($loc(a_i)$, safeEndpoint, usedElements);
                      success ← DeliverPath(path, chain);
                $a_i.path$ = path;

        DeliverCheckIn(a_i, chain);
        while *Not AdvancedTimestep(timestep, chain)* **do**
            timestep = GetTimestep(chain);

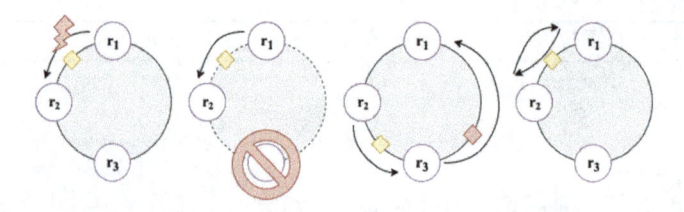

Fig. 3. TP and TPTS use the token passing system. This system can have multiple consensus failures if robots r_1, r_2, r_3, have limited trust in each other, if the network connection is faulty, or if agents can through error or intent reach false conclusions. From left to right are four examples of potential planning Byzantine failures in the MAPD problem: if passing the token fails, if an agent fails, if an incorrect message is passed, or if a cycle occurs when an agent incorrectly believes it has the better path.

the distributed algorithms against, not to beat. They proved that their algorithms solve a realistic subclass of well-formed MAPD instances and claim that TP can be extended to a fully distributed MAPD algorithm. Because of this distribution of computation it is their recommended choice for a real-time solution of MAPD instances with large numbers of agents and tasks. Their second algorithm TPTS requires more communication, and is presented as an in-between solution that trades computational complexity for improved solution quality. While presented as a distributed solution, these algorithms assume full trust between agents which restricts them from being decentralized, as shown in Fig. 3.

In our work we assume that all agents seek to accomplish tasks and that at most 1/3rd of the nodes support faulty Byzantine transactions. Our algorithm is similar to the algorithms presented in [20] and solves all well-formed MAPD instances, but in comparison reduces the assumptions made about the network of agents, uses our novel blockchain consensus framework rather than token passing, and allows swapping of tasks during rounds (not between them).

The TP and TPTS algorithms work well for general MAPD, but fail to work in the presence of multi-team collaboration and communication failures. Thus, by comparing against these algorithms we can see that our algorithm performs just as well, while also successfully working in the multi-team collaboration scenario with and without failures.

5 Experimental Evaluation

We ran our simulations on a 5.0 GHz Turbo Intel Core i9-9900K desktop computer with 16GB RAM. We implemented TP, TPTS, CENTRAL, and BFTC in Python to run on a simulated warehouse environment with obstacles as shown in Fig. 4. We implemented all algorithms in Python, with the exception of CENTRAL which uses an existing ECBS implementation in C++ [10].

We generated a single set of 500 randomly selected tasks to be used across all algorithms, task frequencies, and number of agents. Five different task frequencies were used to experiment with the rate of new tasks added to the set of live tasks over time: 0.5 (one task every other timestep), 1 (one task every timestep),

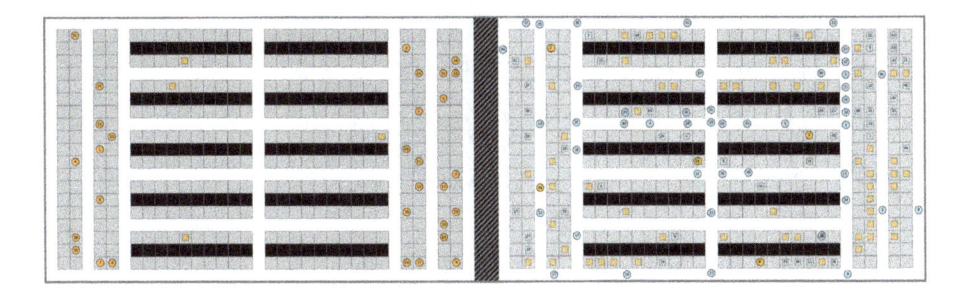

Fig. 4. This figure shows two separate captures of the output of our simulation: a 4-neighbor grid that represents the layout of a simulated warehouse environment. Shown on the left of the divider is the output of the simulation with 30 agents at their initial locations. The black cells are obstacles, the gray cells are endpoints, the orange squares are pickup locations of released tasks, and the orange colored circles are the initial locations of agents and their IDs. Shown on the right of the divider is a capture of the simulation output as it is running. The blue circles are agents that have picked up a task and are delivering them to their corresponding delivery spot ID number.

Table 1. Simulated warehouse comparison

Frequency	Agents	Makespan (timesteps)				Service time (timesteps)				Runtime per timestep (sec)			
		BFTC	TP	TPTS	CENTRAL	BFTC	TP	TPTS	CENTRAL	BFTC	TP	TPTS	CENTRAL
0.5	10	**1568**	1683	1610	1261	154.15	152.188	**142.312**	83.144	3.446	0.096	0.235	0.077
0.5	30	**1044**	1087	1054	1042	32.646	44.164	**31.668**	24.786	4.656	0.408	11.098	0.108
0.5	50	1050	1065	**1045**	1038	30.702	43.512	**29.638**	23.088	6.072	0.385	19.909	0.151
1	10	**1546**	1555	1570	1204	**198.928**	216.972	200.056	134.776	3.491	0.100	0.164	0.075
1	30	666	700	**645**	546	55.764	56.588	**53.61**	26.706	4.643	0.254	1.195	0.171
1	50	**576**	595	**576**	542	35.514	46.312	**33.516**	23.982	5.431	0.615	19.6729	0.220
2	10	1547	1608	**1544**	1186	**209.704**	230.224	213.468	152.41	3.494	0.096	0.359	0.076
2	30	668	654	**639**	470	81.66	86.202	**80.08**	48.932	4.667	0.238	1.181	0.168
2	50	469	495	**457**	318	54.25	60.446	**53.99**	28.094	5.618	0.359	4.207	0.333
5	10	**1549**	1581	1590	1171	221.628	227.556	**218.662**	166.592	3.486	0.086	0.152	0.800
5	30	659	692	**635**	445	**88.556**	96.856	92.79	60.284	4.668	0.209	1.168	0.177
5	50	**453**	506	464	396	**65.216**	72.56	66.748	40.216	5.789	0.316	4.577	0.272
10	10	**1551**	1654	1580	1174	222.162	230.576	**218.594**	165.62	3.473	0.092	0.173	0.083
10	30	673	677	**626**	487	**95.856**	98.196	97.652	61.578	4.592	0.245	1.229	0.185
10	50	**471**	505	480	302	**68.08**	72.992	69.904	42.91	5.572	0.372	4.487	0.363

2 (two tasks every timestep, 5 (five tasks every timestep), and 10 (ten tasks every timestep). For each task frequency we ran three different agent set sizes. Each agent set was generated ahead of time with random initial agent locations and each set was used for all algorithm and frequency combinations. Table 1 reports the makespan, service time, and runtime per timestep for experiments without Byzantine failures. Figure 6 shows the performance of the three algorithms when two instances of Byzantine failures are present.

5.1 Makespan and Service Time Comparison

The distributed MAPD algorithms in descending order of average makespan across all non-Byzantine failure experiments (best is last) is: TP (1004), TPTS

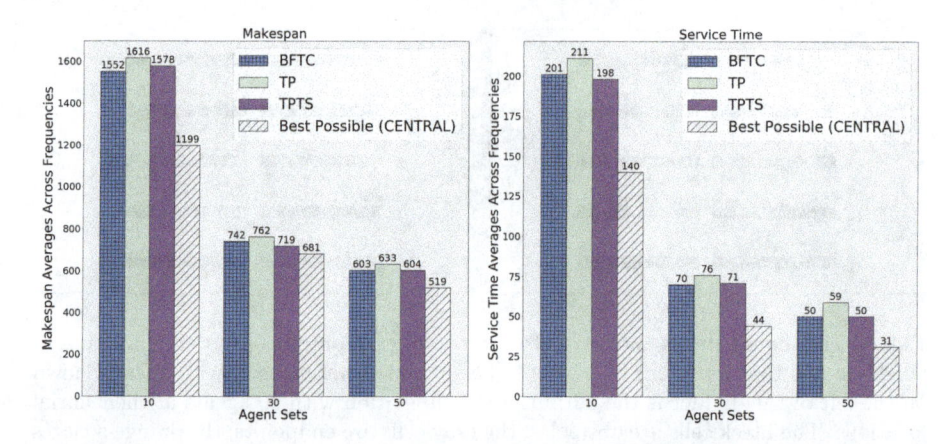

Fig. 5. Makespans and service times averaged across all task frequencies presented for all four algorithms for each set of agents with no failures present. From left to right: BFTC, TP, TPTS, and best possible (CENTRAL). Generally, makespans and service times decrease with more agents as all agent sets serve the same set of tasks.

(967), and BFTC (965). The distributed MAPD algorithms in descending order of average service time across all non-Byzantine failure experiments (best is last) is: TP (115), and both TPTS and BFTC tied with (107). TP lets each agent make the best greedy assignment for itself, producing less optimal makespans. TPTS allows swapping of tasks before the tasks are picked up, decreasing the makespan/service times by improving upon the greedy assignments each individual agent may have made. BFTC is able to take advantage of the replicated state across all nodes in the blockchain network to reply to each agent with the optimal assignment at that timestep, but does not let them swap. Letting them swap could further improve the makespan/service time of BFTC but does not scale well, as more computation and message passing would be necessary. CENTRAL demonstrates the best (sub-optimal) performance possible, outperforming any distributed algorithm, but will not work for systems that cannot have a central authority. Thus, BFTC is just as good, and in some cases better than, current state of the art solutions even in non-Byzantine cases (Fig. 5).

5.2 Runtime Comparison

The distributed MAPD algorithms in increasing order of average runtime per timestep across all non-Byzantine failure experiments is: TP, TPTS, and BFTC. The BFTC runtime is longer as it is the only algorithm using the network to send messages and takes on the runtime of the decentralized Tendermint network's consensus protocol. This increases the amount of network messages being sent and BFTC's runtime. In our experiments, each DeliveryTX transaction took 1–2 s to be committed or rejected and returned to the agent. At the start of every round, agent A adds any new tasks and increments the timestep on the chain through two DeliveryTX calls. Then, agent A must wait for all agents to

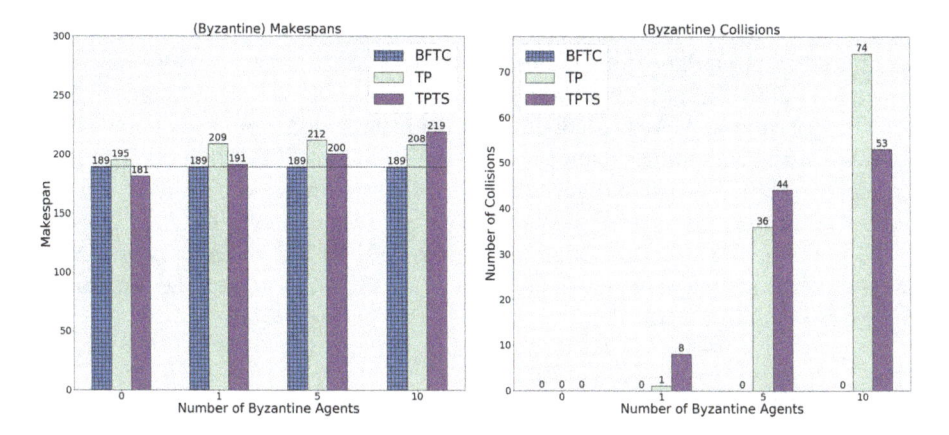

Fig. 6. Makespans and number of collisions presented for the three distributed algorithms as the number of Byzantine agents is increased. From left to right: BFTC, TP, and TPTS. BFTC has zero collisions across all Byzantine agent sets.

check in. All agents when planning and checking in make two DeliveryTX calls concurrently with each other. Since each DeliveryTX call takes 1–2 s BFTC's average runtime per timestep (across all task frequencies and agent sets) of 4.63 is dominated by the DeliveryTX message passing. In the BFTC algorithm agents are able to plan and receive confirmation of their plan in under 1–2 s concurrently with each other.

Using Table 1, we can compare how the algorithms perform as the number of agents increase. Previous work shows that the runtime of CENTRAL and TPTS increases at a much faster rate than TP, and thus CENTRAL and TPTS do not allow for real-time lifelong operation for large numbers of agents [20]. Conversely, while the runtime of BFTC does increase with additional agents, the runtime increases at a slower rate than TPTS and TP. For example, the change of average runtimes per timestep for each distributed algorithm from 10 to 30 agents in decreasing order (best is last) of rate of growth is: TPTS (1, 661%), TP (200%), and BFTC (36%). This shows that the runtime of BFTC grows at a slower rate with the addition of more agents than the other two distributed algorithms.

5.3 Byzantine Failure Comparison

To see how the distributed algorithms performed when Byzantine failures are present we ran two experiments with different types of Byzantine failures in the shared warehouse scenario. In both experiments 30 agents completed a set of 100 tasks with a frequency of 1, varying the number of Byzantine agents (0, 1, 5, 10). The first experiment had Byzantine agents that attempted to assign themselves to the furthest task available. The second experiment had Byzantine agents ignore the paths of other agents. In Fig. 6 we can see BFTC outperforms the other two distributed algorithms in all cases with Byzantine agents present. The distributed MAPD algorithms in decreasing order of average collisions across all Byzantine failure experiments (best is last) is: TP (37), TPTS (35), and BFTC (0). The distributed MAPD algorithms in decreasing order of average makespan across all Byzantine failure experiments (best is last) is: TP (206), TPTS (198), and BFTC (189). The other algorithms show increasing makespans as Byzantine agents are introduced into the system and result in many collisions that increase as the number of Byzantine agents are increased. BFTC remains at the same makespan it had before Byzantine agents were introduced into the simulation and never results in a collision.

6 Conclusion

In this paper, we present a system that solves the lifelong and online MAPD problem by building on top of a blockchain framework to maintain a distributed ledger of assigned tasks and planned paths that agents can query for limited information to individually plan their paths. Our Byzantine Fault Tolerant Consensus algorithm and system demonstrated better makespans and service times and scaled better in runtime as the number of agents increased than state of the art distributed algorithms. BFTC is also the only algorithm able to run in a completely decentralized network with no central authority and is the only algorithm that can survive Byzantine failures in planning and communication that resulted in collisions and increased makespans in the state of the art algorithms. The BFTC algorithm and system is an example of how blockchain can be used in multi-robot and swarm systems to enable consensus when there is no central authority or trust among robots and the deployment of collaborative, decentralized, multi-robot systems.

There are potential trade-offs to consider when using a blockchain-based system: for example, one must consider the latency or throughput of the network, validation models that fit some but not all types of problems, and the computational capabilities of the nodes/robots in the system. While out of the scope of this paper, we consider these not as roadblocks but avenues for related and future research to see what best fits a particular robotic application.

Acknowledgement. This work was partially funded by ARL DCIST CRA W911NF-17-2-0181.

References

1. Afanasyev, I., Kolotov, A., Rezin, R., Danilov, K., Kashevnik, A., Jotsov, V.: Blockchain solutions for multi-agent robotic systems: related work and open questions. In: Proceedings of the 24th Conference of Open Innovations Association. FRUCT'24, Helsinki (2019)
2. Atlam, H.F., Alenezi, A., Alassafi, M.O., Wills, G.: Blockchain with internet of things: benefits, challenges, and future directions. Int. J. Intell. Syst. Appl. **10**(6), 40–48 (2018)

3. Boyarski, E., et al.: ICBS: the improved conflict-based search algorithm for multi-agent pathfinding. In: Eighth Annual Symposium on Combinatorial Search. Citeseer (2015)
4. Buchman, E.: Tendermint: byzantine fault tolerance in the age of blockchains. Ph.D. thesis, The University of Guelph (2016)
5. Calvaresi, D., Dubovitskaya, A., Calbimonte, J.P., Taveter, K., Schumacher, M.: Multi-Agent systems and blockchain: results from a systematic literature review. In: Demazeau, Y., An, B., Bajo, J., Fernández-Caballero, A. (eds.) PAAMS 2018. LNCS (LNAI), vol. 10978, pp. 110–126. Springer, Cham (2018). https://doi.org/10.1007/978-3-319-94580-4_9
6. Castello, E., Hardjono, T., Pentland, A.: Editorial: proceedings of the first symposium on blockchain and robotics, MIT media lab, 5 Dec. 2018. Ledger **4** (2019). https://doi.org/10.5195/ledger.2019.179
7. Castro, M., Liskov, B.: Practical byzantine fault tolerance and proactive recovery. ACM Trans. Comput. Syst. **20**(4), 398–461 (2002). https://doi.org/10.1145/571637.571640
8. Cohen, L., Uras, T., Kumar, T.K.S., Xu, H., Ayanian, N., Koenig, S.: Improved solvers for bounded-suboptimal multi-agent path finding. In: Proceedings of the Twenty-Fifth International Joint Conference on Artificial Intelligence, IJCAI 2016, pp. 3067–3074. AAAI Press (2016)
9. Goldenberg, M., et al.: Enhanced partial expansion a*. J. Artif. Int. Res. **50**(1), 141–187 (2014)
10. Hönig, W., Kiesel, S., Tinka, A., Durham, J.W., Ayanian, N.: Conflict-based search with optimal task assignment. In: AAMAS (2018)
11. Khorshid, M., Holte, R., Sturtevant, N.R.: A polynomial-time algorithm for non-optimal multi-agent pathfinding. In: SOCS (2011)
12. Kuhn, H.W.: The Hungarian method for the assignment problem. Naval Res. Logist. Q. **2**, 83–97 (1955)
13. Lamport, L., Shostak, R., Pease, M.: The byzantine generals problem. ACM Trans. Program. Lang. Syst. **4**(3), 382–401 (1982). https://doi.org/10.1145/357172.357176
14. Laprie, J.-C.: The dependability approach to critical computing systems. In: Nichols, H., Simpson, D. (eds.) ESEC 1987. LNCS, vol. 289, pp. 231–243. Springer, Heidelberg (1987). https://doi.org/10.1007/BFb0022116
15. Li, J., Tinka, A., Kiesel, S., Durham, J.W., Kumar, T.K.S., Koenig, S.: Lifelong multi-agent path finding in large-scale warehouses. In: AAMAS (2020)
16. Lopes, V., Alexandre, L.A.: An overview of blockchain integration with robotics and artificial intelligence. Ledger **4** (2019). https://doi.org/10.5195/ledger.2019.171
17. Ma, H., Hönig, W., Kumar, T.K.S., Ayanian, N., Koenig, S.: Lifelong path planning with kinematic constraints for multi-agent pickup and delivery. In: Proceedings of the AAAI Conference on Artificial Intelligence, vol. 33, no. 01, pp. 7651–7658 (2019). https://doi.org/10.1609/aaai.v33i01.33017651
18. Ma, H., Koenig, S.: Optimal target assignment and path finding for teams of agents. In: Proceedings of the 2016 International Conference on Autonomous Agents and; Multiagent Systems, AAMAS 2016, pp. 1144–1152. International Foundation for Autonomous Agents and Multiagent Systems, Richland (2016)
19. Ma, H., Koenig, S.: Ai buzzwords explained: multi-agent path finding (MAPF). AI Matters **3**(3), 15–19 (2017). https://doi.org/10.1145/3137574.3137579

20. Ma, H., Li, J., Kumar, T.S., Koenig, S.: Lifelong multi-agent path finding for online pickup and delivery tasks. In: Proceedings of the 16th Conference on Autonomous Agents and MultiAgent Systems, AAMAS 2017, pp. 837–845. International Foundation for Autonomous Agents and Multiagent Systems, Richland (2017)
21. Morris, R., et al.: Planning, scheduling and monitoring for airport surface operations. In: AAAI Workshop: Planning for Hybrid Systems (2016)
22. Nakamoto, S.: Bitcoin: A peer-to-peer electronic cash system. Cryptography Mailing list at https://metzdowd.com, March 2009
23. Nguyen, V., Obermeier, P., Son, T.C., Schaub, T., Yeoh, W.: Generalized target assignment and path finding using answer set programming. In: Proceedings of the 26th International Joint Conference on Artificial Intelligence, IJCAI 2017, pp. 1216–1223. AAAI Press (2017)
24. Okumura, K., Machida, M., Défago, X., Tamura, Y.: Priority inheritance with backtracking for iterative multi-agent path finding. In: Proceedings of the Twenty-Eighth International Joint Conference on Artificial Intelligence, IJCAI-19, pp. 535–542. International Joint Conference on Artificial Intelligence Organization, July 2019. https://doi.org/10.24963/ijcai.2019/76
25. Sharon, G., Stern, R., Felner, A., Sturtevant, N.R.: Conflict-based search for optimal multi-agent pathfinding. Artif. Intell. **219**, 40–66 (2015)
26. Silver, D.: Cooperative pathfinding. In: Proceedings of the First AAAI Conference on Artificial Intelligence and Interactive Digital Entertainment, AIIDE 2005, pp. 117–122. AAAI Press (2005)
27. Strobel, V., Castelló Ferrer, E., Dorigo, M.: Managing byzantine robots via blockchain technology in a swarm robotics collective decision making scenario. In: Proceedings of the 17th International Conference on Autonomous Agents and MultiAgent Systems, AAMAS 2018, pp. 541–549. International Foundation for Autonomous Agents and Multiagent Systems, Richland (2018)
28. Surynek, P.: A novel approach to path planning for multiple robots in bi-connected graphs. In: 2009 IEEE International Conference on Robotics and Automation, pp. 3613–3619 (2009)
29. Čáp, M., Vokřínek, J., Kleiner, A.: Complete decentralized method for on-line multi-robot trajectory planning in well-formed infrastructures. In: Proceedings of the Twenty-Fifth International Conference on Automated Planning and Scheduling, ICAPS 2015, pp. 324–332. AAAI Press (2015)
30. Veloso, M., Biswas, J., Coltin, B., Rosenthal, S.: CoBots: robust symbiotic autonomous mobile service robots. In: Proceedings of the 24th International Conference on Artificial Intelligence, IJCAI 2015, pp. 4423–4429. AAAI Press (2015)
31. Wang, K.H., Botea, A.: MAPP: a scalable multi-agent path planning algorithm with tractability and completeness guarantees. J. Artif. Intell. Res. (JAIR) **42**, 55–90 (2011). https://doi.org/10.1613/jair.3370
32. Wilde, B.D., Mors, A., Witteveen, C.: Push and rotate: cooperative multi-agent path planning. In: AAMAS (2013)
33. Wurman, P., D'Andrea, R., Mountz, M.: Coordinating hundreds of cooperative, autonomous vehicles in warehouses. AI Mag. **29**, 9–20 (2008)
34. Yu, J., LaValle, S.: Structure and intractability of optimal multi-robot path planning on graphs. In: AAAI (2013)

Decentralized Multi-robot Planning in Dynamic 3D Workspaces

Arjav Desai[✉] and Nathan Michael

The Robotics Institute, Carnegie Mellon University, Pittsburgh, PA 15232, USA
{arjavdesai,nmichael}@cmu.edu

Abstract. We consider the problem of decentralized multi-robot kinodynamic motion planning in dynamic workspaces. The proposed approach leverages offline precomputation on an invariant planning representation (invariant geometric tree) for low latency online planning and replanning amidst unpredictably moving dynamic obstacles to generate kinodynamically feasible and collision-free time-parameterized polynomial trajectories. Simulation results with up to 10 robots in dynamic workspaces composed of varying obstacle densities (up to 30% by volume) and speeds (up to 2.5 m/s) suggest the use of the proposed methodology for real-time kinodynamic replanning in dynamic workspaces.

1 Introduction

Reliable and responsive multi-robot deployments in application domains such as search and rescue necessitate an online motion planning methodology to compute kinodynamically feasible and collision-free motion plans for each robot in the team. Additionally, the planning methodology must be robust to (1) unexpected changes in operator intent over the course of the deployment and (2) to the presence of dynamic obstacles (human operators, debris, displaced physical objects) sharing the workspace. In this work, we seek to develop a planning and coordination framework for multi-robot teams for generating kinodynamically feasible and safe motion plans for dynamic 3D workspaces.

This problem is challenging due to several reasons. *First*, kinodynamic planning in dynamic workspaces involves searching for a feasible sequence of states in a high-dimensional search space (challenge C1). The high dimensionality is attributed to the need to encode the higher-order kinematics and dynamics of the robot, as well as a time dimension to account for the spatiotemporal characteristics of the the dynamic obstacles in the workspace. *Second*, evolving mission conditions (e.g. online changes in goal locations) precludes the use of precomputed motion plans; therefore the multi-robot team must be able to replan online from potentially non-stationary initial states (challenge C2). This is challenging as robots cannot wait in place as in [11] in order to avoid conflicts. *Third*, long term trajectory prediction for stochastic dynamic obstacles is difficult due to

---

A. Desai and N. Michael—We gratefully acknowledge support from industry.

compounding effects of modeling inaccuracies as well as state and motion uncertainty [1]. A short prediction horizon is thus favoured which in turn necessitates planning approach which can replan at high-rates while maintaining kinodynamic feasibility and safety (challenge C3).

Related Works: Several works address motion planning in dynamic workspaces. Hierarchical Cooperative A* (HCA*) proposed by Silver et al. [10] searches for geometric paths in the full space-time search space under the guidance of a lower dimensional heuristic. While HCA* guarantees the optimality of the solution, it is essentially an offline algorithm. Vemula et al. [12] sacrifice optimality for efficiency and extend the notion of adaptive dimensionality for path planning in dynamic workspaces. The dimensionality of the search space is selectively increased in regions of conflict and ignored everywhere else. While this approach provides higher success rates and lower computation times than HCA*, it does not explicitly consider kinodynamic constraints and assumes complete knowledge of the trajectories of dynamic obstacles. Phillips et al. in [9] propose a search-based approach for kinodynamic motion planning in unpredictably dynamic environments. The authors exploit the observation; in a dynamic environment, a particular robot configuration is collision-free for only a few time steps (a safe interval); to reduce the search space dimensionality. The success of this approach is predicated on the stationarity of the initial robot states. This assumption renders the approach in [9] impractical for online replanning scenarios. Sampling-based approaches such as Multipartite RRT (MP-RRT) proposed by Zucker et al. [14] proposes replanning amidst dynamic obstacles by biasing the sampling distribution and reusing sub-trees over the course of the search. RRT^x proposed by Otte et al. [8] for single-robot replanning in dynamic environments address the limitations of MP-RRT; the search-tree in RRT^x is rooted at the goal which eliminates the additional collision-checks and tree rewiring operations due to the motion of the robot. These algorithms work well for geometric planning tasks, however, they are not well suited for real-time planning in high-dimensional state spaces for agile systems like quadrotors due to the computational cost associated with solving a two-point boundary value problem for *each* sampled configuration and evaluating the corresponding solution for feasibility and safety. Optimization-based approaches such as the one proposed by Zhu et al. [13] also consider motion planning in dynamic environments. However, due to high computational complexity, these are typically restricted to sparsely cluttered environments.

Contributions: This work addresses the aforementioned challenges for multi-robot motion planning in dynamic workspaces. Our approach leverages offline reachability analysis on an invariant local planning representation (initially proposed in [3]) and decentralized planning to counter the *curse of dimensionality* (addressing C1) The proposed planning representation allows for replanning from non-stationary initial states (addressing C2). Additionally the representation allows for search in the lower dimensional geometric space and guarantees a kinodynamically feasible and safe solution if a geometric solution is found, without additional refinement. Finally, we propose a fast collision checking approach in dynamic environments that allows for low latency replanning (addressing C3).

2 Problem Formulation

2.1 Notation and Assumptions

$\mathbb{R}$ and $\mathbb{N}$ denote the set of real and natural numbers respectively. $\mathcal{S}$, $\mathbf{M}$, $\mathbf{v}$, and s denote sets, matrices, vectors, and scalars respectively. $|\mathcal{S}|$ denotes the set cardinality. The assumptions used in this work are as follows. First, the robots employed are differentially flat and jerk-controlled [6] quadrotor systems. The team composition is assumed to be homogeneous and each robot is physically modelled as a ball of radius r denoted by $\mathcal{B}_r(\mathbf{x})$ where $\mathbf{x} \in \mathbb{R}^3$ denotes the cartesian coordinates of the centre of the ball. Second, the static portions of the 3-D workspace are known a priori. Third, the current states (position, velocity) and the bounding volumes of the dynamic obstacles are observable. The future trajectories are unknown and must be predicted. Fourth, robots can communicate without latency within a communication radius of r_{comm}.

2.2 Problem Statement

Consider a team of n robots with a p dimensional state space deployed in an uncertain dynamic workspace denoted by $\mathcal{W}$. The set of points corresponding to the known and static obstacles is given by $\mathcal{W}_s$ and the set of points occupied by the dynamic obstacles at time t are given by $\mathcal{W}_d^t$. At any time $t \in \mathbb{R}_{>0}$, the set of occupied points $\mathcal{W}_{\text{occ}}^t$ is the union set of $\mathcal{W}_s$ and $\mathcal{W}_d^t$ and the set of free points is given by $\mathcal{W}_{\text{free}}^t = \mathcal{W} \backslash \mathcal{W}_{\text{occ}}^t$. Let $\mathcal{I} \in \mathbb{R}^{n \times p}$ denote the set of initial robot states and $\mathcal{F} \in \mathbb{R}^{n \times p}$ denote the set of desired terminal states for n robots.

The objective of the motion planner is to generate a time-parameterized sequence of trajectories for each robot in the team that lead the robots from $\mathcal{I}$ terminate within an ϵ-ball, $\mathcal{B}_\epsilon$, of the terminal states $\mathcal{F}$ where $\epsilon \in \mathbb{R}^p$ are continuous up-to the second derivative of position, i.e., acceleration.

Let $\{\xi_i^{[t_0,t_1]}(t), \ldots, \xi_i^{[t_{k-1},t_k]}(t)\}$ denote the time-parameterized trajectories for robot i, and $\{\xi_j^{[t_0,t_1]}(t), \ldots, \xi_j^{[t_{k-1},t_k]}(t)\}$ denote the trajectories for robot j. At any time $t \in \mathbb{R}$ such that $t_{k-1} \leq t \leq t_k$, the corresponding states $\mathbf{x}_i^t \in \mathbb{R}^p$ and $\mathbf{x}_j^t \in \mathbb{R}^p$ satisfy the following constraints. First, each robot in the team must lie in the free space i.e. $\mathcal{B}_r(\text{pos}(\mathbf{x}_i^t)) \in \mathcal{W}_{\text{free}}^t$. Second, trajectories must satisfy the kinodyanmic constraints of the system. This corresponds to constraints on maximum acceleration and jerk. Third, robots must maintain clearances greater than $2r$ i.e. $||\mathbf{x}_i^t - \mathbf{x}_j^t|| > 2r$.

3 Approach

This section is organized as follows. In Sect. 3.1 we discuss the static and dynamic workspace representation. Section 3.2 describes the kinodynamic planner, the dynamic obstacle avoidance procedure, the replanning algorithm, and the decentralized planning setup.

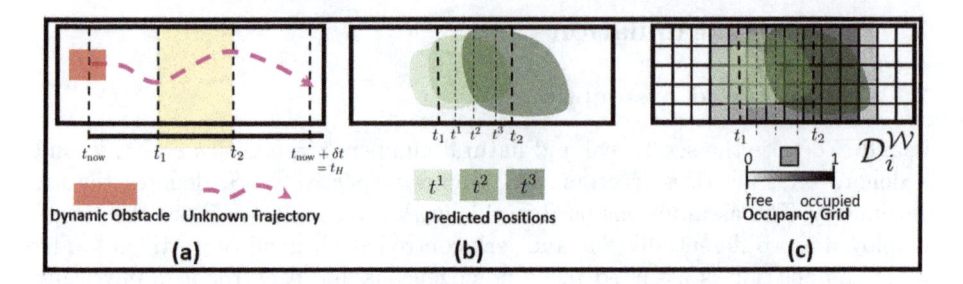

Fig. 1. Dynamic workspace representation. (a) The prediction horizon is split into k major intervals. (b) Each major interval is split into k' minor intervals and the position of the dynamic obstacles is predicted at each of the k' time instances. (c) An occupancy grid, $\mathcal{D}_i^{\mathcal{W}}$, with the collated occupancy values of k' minor intervals is generated for each major interval.

3.1 Workspace Representation

The dynamic workspace $\mathcal{W}_d^t$ is represented via two voxelgrids denoted by $\mathcal{S}$ and $\mathcal{D}$ that correspond to the static and dynamic components of the workspace.

Static Workspace Representation: Since the static obstacles are known *a priori*, the static voxelgrid $\mathcal{S}$ is computed offline. Additionally we precompute a sparse roadmap of the environment with respect to $\mathcal{S}$ using the SPARS2 algorithm [5]. The sparse roadmap captures the free space topology and provides approximate cost-to-go estimates to the kinodynamic planner (refer Sect. 3.2).

Dynamic Workspace Representation: The dynamic occupancy grid $\mathcal{D}^{\mathcal{W}}$ is a four dimensional grid (cartesian coordinates x, y, and z and time t) that encodes the spatiotemporal occupancy of the dynamic obstacles in the 3-D workspace over a time horizon t_H. Each cell in $\mathcal{D}$ is assigned a value, $c \in \mathbb{R}; 0 \leq c \leq 1$, which denotes the occupancy probability of that cell. Let t_{now} denote the current time at which the state of the dynamic obstacles is observed and t_H denote the time at the prediction horizon. The time-period $[t_{\text{now}}, t_H]$ is split into k major time-intervals and a 3D occupancy grid is maintained corresponding to each time-interval (Fig. 1). Each time-interval is further discretized into k' minor time instances and the state of the dynamic obstacles is predicted for each of these time instances (Fig. 1b). The predicted states of the dynamic obstacles are used to update the 3D occupancy grid corresponding to that interval (Fig. 1c). Thus, each 3D occupancy grid in $\mathcal{D}^{\mathcal{W}}$ represents the *collated* spatiotemporal occupancy of the dynamic obstacles in that time interval. When multiple occupancy probabilities are possible for a voxel (e.g. due to multiple dynamic obstacles), the highest occupancy probability is assigned to that voxel.

Dynamic Obstacle Modelling and Prediction: The motion of the dynamic obstacles is predicted via the Linear Velocity Polyhedron (LVP) model proposed in [7]. Let $\mathcal{H}$ denote a halfspace in $\mathbb{R}^m$, i.e., $\mathcal{H} = \{\mathbf{p} \mid \mathbf{a}^T\mathbf{p} \leq b, \mathbf{p} \in \mathbb{R}^m\}$. A convex polyhedron, $\mathcal{C}$, is defined as an intersection set of k halfspaces. Let the jth

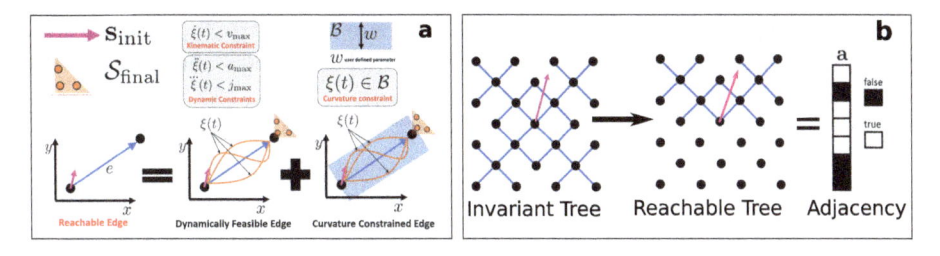

Fig. 2. Figure shows (a) reachable edges that are dynamically feasible and curvature constrained (b) cartesian projection of the reachable tree given the initial higher-order state at the root and its alternate representation via a boolean valued adjacency vector.

halfspace at time instance $t = 0$ be defined as $\mathcal{H}_j^{t_0} = \{\mathbf{p} \mid \mathbf{a}_{\{j,t_0\}}^T \mathbf{p} \leq b_{\{j,t_0\}}\}$. Further, let the linear velocity of the points in $\mathcal{C}$ be $\mathbf{v}_\mathcal{C}$. The jth halfspace of $\mathcal{C}$ at $t = t_1$ is given by $\mathbf{a}_j(t_1) = \mathbf{a}_{\{j,t_0\}}$, and $b_j(t_1) = b_{\{j,t_0\}} + \mathbf{a}_{\{j,t_0\}}^T \mathbf{v}_\mathcal{C} \Delta_t$. Here, Δ_t is the time difference $t_1 - t_0$. Equations (1–2) do not account for uncertainty in the predicted motion of the dynamic obstacles. To do so, the obstacle's geometry is inflated over time by a constant inflation factor $\mathbf{v}_e$. While the equation for $\mathbf{a}_j(t_1)$ remains the same, $b_j(t_1)$ is updated as follows:

$$b_j(t_1) = b_{\{j,t_0\}} + (\mathbf{a}_{\{j,t_0\}}^T \mathbf{v}_\mathcal{C} + \|\mathbf{a}_{\{j,t_0\}}\| \|\mathbf{v}_e\|) \Delta_t \qquad (1)$$

We note that the primary contribution of this work is development and analysis of a local planning representation capable to efficient motion-planning in uncertain dynamic workspaces and thus, the LVP predictor can be replaced with state-of-the-art prediction methodologies. We refer the reader [13] and the references therein for a detailed exposition of various prediction mechanisms.

3.2 Kinodynamic Planner

Local Planning Representation: We summarize the key properties of the planning representation initially proposed in our prior work [4]. The planning representation is an invariant geometric tree, T, constructed by propagating a geometric lattice in an obstacle-free workspace for multiple time steps. The tree is defined by an invariant vertex set $\mathcal{V}_T$ and a set of directed edges $\mathcal{E}_T$. The initial and terminal points of a tree edge $e \in \mathcal{E}_T$ are represented by $e(0)$ and $e(1)$ respectively. Kinodynamic constraints are encoded in the definition of edges.

- A *dynamically feasible edge* is one such that for some initial higher-order derivative $\mathbf{s}_{\text{init}}$ at $e(0)$, a non-empty set of higher-order derivatives exists $\mathcal{S}_{\text{final}}$ at $e(1)$ such that the fixed duration time-parameterized polynomial trajectories [6] connecting $\mathbf{s}_{\text{init}}$ to $\mathcal{S}_{\text{final}}$ along e do not violate the kinodynamic constraints of the platform (Fig. 2a).
- A *curvature constrained edge* is one such that for some initial higher-order derivative $\mathbf{s}_{\text{init}}$ at $e(0)$, there exists a non-empty set of higher-order derivatives

$\mathcal{S}_{\text{final}}$ at $e(1)$ such that the trajectories connecting $\mathbf{s}_{\text{init}}$ to $\mathcal{S}_{\text{final}}$ are entirely contained within a cuboidal bounding box, $\mathcal{B}$, of length $l(e)$ and a fixed width and height, oriented along the edge e (Fig. 2a).
- A *reachable edge* is dynamically-feasible *and* curvature constrained.

A *reachable tree* is then a tree that is composed of reachable edges. For any higher-order state at the root node $\mathbf{s}_{\text{init}}$, the corresponding reachable tree is then defined by the tuple $T = (\mathcal{V}, \mathbf{p}, \mathbf{a}, \mathbf{c})$, where $\mathcal{V}$ is the vertex set in $\mathbb{R}^3$, $\mathbf{p} \in \mathbb{N}^{|\mathcal{V}_T|}$ stores the parent ids of each vertex in the tree, $\mathbf{c} \in \mathbb{R}^{|\mathcal{V}_T|}$ is a cost vector that stores the the cost-to-come to the ith vertex, and $\mathbf{a} \in \{0,1\}^{|\mathcal{V}_T|}$ is a boolean valued adjacency (0 denotes unreachable) vector that encodes the set of reachable edges subject to the *underlying planning context* such as higher-order derivatives at the root vertex, or collisions with obstacles (Fig. 2b).

Proposition 1. Let $\mathcal{S}$ be a higher-order derivative set at the root vertex of an invariant k-step tree. Let $T_{\mathbf{s}_i} = (V, \mathbf{p}, \mathbf{a}_{\mathbf{s}_i}, \mathbf{c})$ represent the reachable k-step tree of $\mathbf{s}_i \in \mathcal{S}$. The reachable k-step tree of the entire set, $T_{\mathcal{S}}$, is given by $\mathbf{a}_{\mathcal{S}} = \sum_{\forall i, \text{ bitwise}} \mathbf{a}_{\mathbf{s}_i}$. We refer to this as the **merge-tree** operation. If $\mathbf{a}_{\mathcal{S}}(i) = 1$, the path from the root to the ith vertex is composed of reachable edges and there exists at least one $\mathbf{s}_i \in \mathcal{S}$ for which this path is reachable. The proof follows from the fact that each vertex of a tree can have only one parent. □

Let $\mathbf{a}_T^{\text{kino}}$ denote the kinodynamically feasible tree corresponding to higher-order derivatives at the tree root, $\mathbf{a}_T^{\text{static}}$, the collision-free tree w.r.t the static obstacles (edges in $\mathbf{a}_T^{\text{static}}$ may be kinodynamically infeasible) and let $\mathbf{a}_T^{\text{dynamic}}$ denote the collision-free tree with respect to the dynamic obstacles. The reachable tree $\mathbf{a}_T$ with kinodynamic and collision constraints is obtained by.

$$\mathbf{a}_T = \mathbf{a}_T^{\text{kino}} \odot \mathbf{a}_T^{\text{static}} \odot \mathbf{a}_T^{\text{dynamic}} \qquad (2)$$

Here, $\odot$ refers to the elementwise multiplication operation. We refer the reader to our prior work [4] for details on construction of $\mathbf{a}_T^{\text{kino}}$ and $\mathbf{a}_T^{\text{static}}$. Construction of $\mathbf{a}_T^{\text{dynamic}}$ is discussed in the subsection on dynamic obstacle avoidance.

Online Planning Procedure: The single-robot planner constructs trajectories from the start to the goal via a two step process that decouples geometric search (step 1) and higher-order derivative assignment to the intermediate vertices of the geometric path (step 2). The geometric search proceeds by incrementally translating the invariant tree towards the goal via:

- **Infer (Merge) Reachable Tree:** In each iteration, a reachable tree T is inferred given the underlying higher-order state at the root node (Fig. 3a). The $\mathbf{a}_T$ vector is updated such that the unreachable vertices and edges are pruned from the tree as in Eq. (2). After the first iteration, a *set* of higher-order states may feasibly exist at the root node. This is due to the association of multiple kinodynamically feasible trajectories with each geometric edge (Fig. 2a). The reachable tree corresponding to the entire set is generated via the merge-tree operation (Proposition 1).

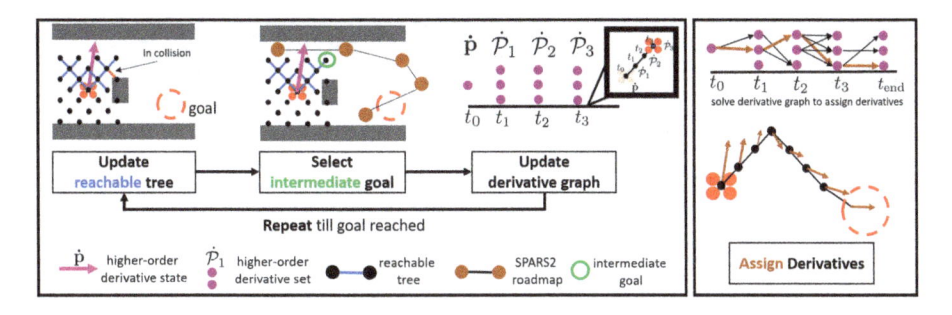

Fig. 3. Online planning procedure with invariant geometric trees

- **Select Intermediate Goal:** An intermediate goal is selected from the reachable vertices (using cost-to-go estimates from the SPARS roadmap) and the root of the tree is translated to the intermediate goal (Fig. 3b).
- **Update Derivative Graph:** A directed graph structure called the derivative graph $\mathcal{G}_{\text{der}}$ updated in each iteration with the time indexed distribution of higher-order derivatives that can *feasibly* exist at the intermediate vertices of the geometric path constructed *thus far* (Fig. 3c and inset).

This process continues till the root node of the tree reaches the goal after which, higher-order states are assigned to the intermediate geometric vertices by searching for the least cost (minimum jerk) path in the derivative graph Fig. 3d.

Dynamic Obstacle Avoidance: Here, we describe the procedure to identify the edges in the tree with a collision probability greater than p_{coll} and accordingly update the $\mathbf{a}_T^{\text{dynamic}}$ vector. While this can be done via explicit evaluation of edges for trees with a low cardinality, operating in dense multi-robot scenarios typically requires a high cardinality tree as higher cardinality corresponds to greater maneuverability. For efficient collision detection, we exploit the invariance of the planning representation and propose a two-stage offline-online procedure for obstacle avoidance (Fig. 4).

Offline Stage: In an obstacle free workspace, each edge of the tree T is discretized and a time index is associated with each discrete point. We only consider the time dimension up to t_H seconds (the prediction horizon). Similar to the construction of $\mathcal{W}^\mathcal{D}$, we split this time horizon into k intervals. For each interval, we construct a 3D voxelgrid where voxels marked occupied intersect with the tree edges spatially *and* in time. Thus, as in $\mathcal{D}^\mathcal{W}$, this gives rise to k three-dimensional voxelgrids (Fig. 4a). Let $\mathcal{D}^T$ denote the spatiotemporal tree voxelgrid. In addition to constructing $\mathcal{D}^T$, with each occupied voxel in $\mathcal{D}^T$, we associate the *minimal* set of edges that intersect with it i.e. if edges i, j, and k intersect with a voxel and if i is the parent of j and k, only edge i is associated with that voxel (Fig. 4b). This spatiotemporal voxelgrid representation and the voxel-to-edge mapping is precomputed and used during the online stage.

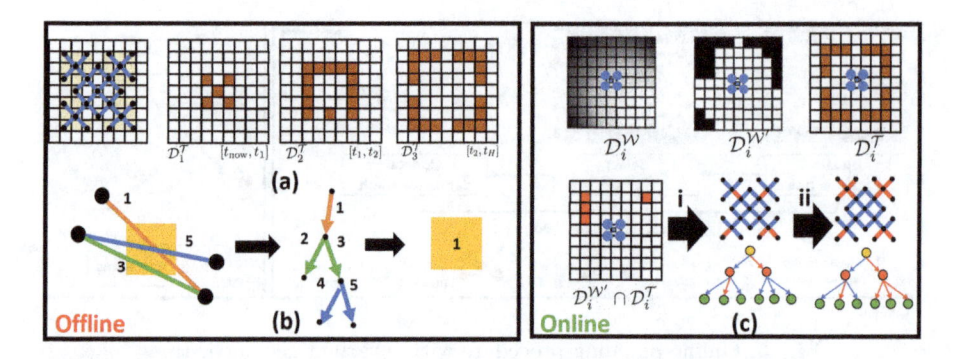

Fig. 4. Offline preprocessing of invariant geometric trees to construct (a) spatiotemporal voxel grids and (b) voxel-to-edge mapping. (c) During online operation, given the ith slice of the dynamic workspace grid $\mathcal{D}_i^\mathcal{W}$, the occupied voxels that also intersect with the ith slice of the spatiotemporal tree voxel grid $\mathcal{D}_i^\mathcal{T}$ are found. The minimal set of edges corresponding to the common voxels are queried from the voxel-to-edge mapping (process i) and Dijkstra's algorithm is run from the root to obtain the true set of reachable edges (process ii).

Online Stage: In the *online* stage (Fig. 4c), all voxels in $\mathcal{D}^\mathcal{W}$ with occupancy values greater than or equal to $1 - p_{\text{coll}}$ are marked as occupied i.e. the occupancy values of these voxels are set to 1. All other voxels in $\mathcal{D}^{\mathcal{W}'}$ with occupancy values less than $1 - p$ are marked as free. For each of the k voxelgrids in $\mathcal{D}^{\mathcal{W}'}$, we search for the set of voxels, $\mathcal{V}_{\text{common}}$, that are occupied by *both* the tree edges in $\mathcal{D}^\mathcal{T}$ and by the dynamic obstacles $\mathcal{D}^{\mathcal{W}'}$ given the current position of the robot and the state of $\mathcal{D}$. The edges that intersect with $\mathcal{V}_{\text{common}}$ i.e. $\text{occ}(\mathcal{D}^{\mathcal{W}'}) \cap \text{occ}(\mathcal{D}^\mathcal{T})$ are queried from the voxel-to-edge mapping and the corresponding indices in $\mathbf{a}_T^{\text{dynamic}}$ are marked unreachable. The set of unreachable vertices $\mathbf{a}_T^{\text{dynamic}}$ are obtained by running Dijkstra's algorithm from the root. All reachable edges in the resulting invariant tree have a probability of collision less than p_{coll}.

Replanning Procedure: Let t_{now} be the current time, ξ_i be the current trajectory of the ith robot in the team and $\mathcal{D}^\mathcal{W}$ be the most recent estimate of the dynamic voxelgrid. The trajectory ξ_i is evaluated for collisions with respect to $\mathcal{D}^\mathcal{W}$ in the time interval $[t_{\text{curr}}, t_H]$ where t_H is the time at the prediction horizon of $\mathcal{D}^\mathcal{W}$. If a collision is detected either with a dynamic obstacle or with the trajectory ξ_j of any robot in the team such that $i \neq j$, replanning is triggered. The replanning procedure typically requires a finite amount of time to execute. Thus, replanning from the current state of the robot $\xi_i(t_{\text{curr}})$ can introduce discontinuities in position as well as higher-order derivatives of the robot state that can lead to catastrophic failures. In order to avoid such a scenario, a state $\mathbf{x}_{\text{start}} \in \mathbb{R}^p$ that is $t_{\text{lookahead}}$ in the future is extracted from ξ_i and the robot replans from that state. If the actual execution time is greater than $t_{\text{lookahead}}$, the computed plan is rejected and a new replanning instance is initiated. While replanning, the single-robot planner with dynamic obstacle avoidance is executed. However, the dynamic obstacle avoidance is only considered until t_H seconds, after which, the dynamic obstacles are ignored. After successful replanning, each robot communicates its updated motion plans to its neighbors in $\mathcal{G}_{\text{comm}}$.

Decentralized Planning Architecture: Our decentralized planning architecture consists of a *commander* node (e.g. a human operator) that assigns labelled goal states, $\mathcal{F}$, to a subset of robots, $n' \leq n$, in the team. A *workspace observer* tracks the dynamic obstacles and broadcasts the timestamped obstacle states (position, velocity) to each robot in the team. On-board, each robot maintains and updates the dynamic voxelgrid $\mathcal{D}^{\mathcal{W}}$. All robots plan their trajectories *synchronously* and communicate the computed trajectories to their neighbors in the *communication graph* $\mathcal{G}_{\text{comm}}$ i.e. an undirected graph structure that encodes the time-varying communication topology of the multi-robot team. Two robots communicate if the distance between them is less than a predefined communication radius r_{comm}. At each time step, each robot avoids the trajectories computed by all the other robots in the previous time step thus guaranteeing safety.

4 Evaluation

4.1 Implementation Details and Experiment Design

The experimental evaluation was conducted using the Julia programming language on a Lenovo Thinkpad with an Intel 4-Core i7 CPU and 16 GB RAM. The proposed methodology is evaluated via three studies. All studies employ robots of radius 0.1 m with a 2.6 m/s velocity limit and a 6.8 m/s^2 acceleration limit.

Study 1: Comparison of the collision-checking method with lazy evaluation [2].

Study 2: Comparison of the local planning representation with motion-primitive trees commonly employed in state-of-the-art local planners [7].

Study 3: Decentralized planning architecture evaluation in dynamic 3D workspaces.

4.2 Results

Study 1: Evaluation of Collision-Avoidance Methodology: We compare the proposed dynamic collision-checking methodology with lazy evaluation [2]

Table 1. Mean and S.D. of the time required (in s) to search for collision-free sequence of edges in geometric trees of varying cardinalites and coverage volumes for 100 trials in random dynamic voxelgrids $\mathcal{D}$ with 0.1 m resolution. Multipliers are shown in blue.

Cardinality	$\|T_1\| = 70,491$			$\|T_2\| = 109,893$		
Coverage	$4 \times 4 \times 4$ m^3	$6 \times 6 \times 6$ m^3	$8 \times 8 \times 8$ m^3	$4 \times 4 \times 4$ m^3	$6 \times 6 \times 6$ m^3	$8 \times 8 \times 8$ m^3
Proposed	**0.007 ± 0.002**	**0.017 ± 0.008**	**0.047 ± 0.034**	**0.009 ± 0.010**	**0.019 ± 0.009**	**0.050 ± 0.06**
LazyEval (Forward)	0.022 ± 0.015 (**3.142x**)	0.042 ± 0.046 (**2.470x**)	0.079 ± 0.085 (**1.680x**)	0.032 ± 0.023 (**3.555x**)	0.048 ± 0.056 (**2.526x**)	0.102 ± 0.414 (**2.040x**)

Table 2. Comparison of the proposed planning representation (IGTree) with a state-of-the-art planning representation MPTree in nine control environments. Table reports the success rate (SR) and average response times (RT) for 900 trials. $|T|$ and $|U|$ denote the cardinality of IGTree and MPTree respectively and d denotes the depth.

Obst. density	A (4.6%)						B (7.5%)						C (9.0%)					
Obst. speed	Low (3.0 m/s)		Medium (4.0 m/s)		High (5.0 m/s)		Low (3.0 m/s)		Medium (4.0 m/s)		High (5.0 m/s)		Low (3.0 m/s)		Medium (4.0 m/s)		High (5.0 m/s)	
Metrics	SR %	RT (ms)	SR %	RT (ms)	SR %	RT (ms)	SR %	RT (ms)	SR %	RT (ms)	SR %	RT (ms)	SR %	RT (ms)	SR %	RT (ms)	SR %	RT (ms)
IGTree (our) $\|T\| = 14{,}763$	96.2	5.9	96.2	5.5	96.2	5.5	100.0	8.5	100.0	18.0	97.3	8.0	100.0	9.5	88.5	8.3	87.3	8.2
MPTree ($d=1$) $\|U\| = 5{,}329$	70.2	21.2	67.3	21.2	67.1	21.3	68.4	24.3	67.4	24.2	65.5	23.5	37.0	27.5	30.6	25.2	28.4	26.1
MPTree ($d=1$) $\|U\| = 14{,}641$	72.1	60.4	69.9	59.9	69.1	81.3	69.1	98.7	68.0	77.4	66.9	94.4	36.5	97.0	29.5	100.6	27.6	84.5
MPTree ($d=2$) $\|U\| = 14{,}763$	84.0	58.6	83.5	59.8	83.8	59.7	84.5	66.0	84.4	68.7	84.3	71.8	83.5	79.0	72.5	79.1	67.4	78.1

on six geometric trees of cardinalities $|T_1| = 70{,}491$ and $|T_2| = 109{,}893$ and spatial coverage volumes of $4 \times 4 \times 4\,\mathrm{m}^3$, $6 \times 6 \times 6\,\mathrm{m}^3$ and $8 \times 8 \times 8\,\mathrm{m}^3$. We conduct 100 trials for each tree and in each trial a random dynamic voxelgrid $\mathcal{D}^\mathcal{W}$ is generated. The objective of each trial is to select a sequence of edges in the tree that do not collide with the occupied voxels in $\mathcal{D}^\mathcal{W}$ and lead the robot towards a predefined goal, i.e., a feasible path selection problem. The lazy collision checking approach (abbrv. LazyEval) employs the Forward edge selector [2] in order to fully exploit the tree structure of the planning representation and avoid redundant edge evaluations. Table 1 reports the mean and standard deviation of the time required to find a collision-free path using both approaches. The proposed collision-checking approach achieves significantly lower computation times (upto 3.1 times for T_1 and 3.5 times for T_2) for all six geometric trees compared to LazyEval. The low computation times (approx. 50 Hz for T_2 with a coverage volume of $8 \times 8 \times 8$ m^3) suggest viability of the proposed approach for use in high-rate local planners typically used in dynamic workspaces. As the coverage volume increases, the time required by the proposed method approaches that of LazyEval but the key takeaway is that here is for volumes resembling local map dimensions, leveraging invariant structures is beneficial as offline preprocessing can significantly expedite online search.

Study 2: Evaluation of Local Planning Representation: We compare the proposed representation, i.e., invariant geometric trees (abbrev. IGTree) with control-input discretized motion-primitive trees (abbrev. MPTree) commonly employed by several state-of-the-art local planners [7]. The performance is evaluated by studying their behavior in nine 2D environments composed of varying obstacle densities and maximum obstacle speeds. In this study, 900 trials are conducted, and in each trial, the initial velocity of the robot $[v_x, v_y]^T$ is randomly selected from the range $v_x, v_y \in [-2.5, 2.5]$. The initial position of the robot is fixed at $[0.0, 0.0]^T$ and the projected or intermediate goal is fixed at $[4.0, 0.0]^T$. A trial is deemed successful if the local planner can compose kinodynamically-feasible and collision-free trajectories for a prediction horizon of 2.0 s. For this

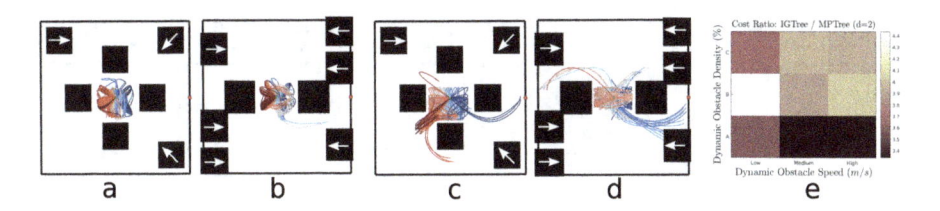

Fig. 5. Qualitative single-horizon solutions for `IGTree` (**a–b**) and `MPTree` (**c–d**) with $d = 2$ in 2D maps of varying obstacle densities. The black boxes indicate the obstacles and the white arrows indicate the direction of motion of the obstacles. (**e**) Heatmap with ratios of solution cost of `IGTree` compared to `MPTree`.

study, success rate (abbrev. SR), response time (abbrev. RT), and solution cost (total jerk for t_H seconds) are used as comparison metrics. From Table 2 we conclude that the proposed representation can safely tackle a wider range of dynamic local planning scenarios with a higher success rate and lower response times than the `MPTree` representation. This is because `IGTree` reasons over a *set* of higher-order states at each geometric vertex as opposed to committing to a singular state at each time-step as in the `MPTree` representation. In terms of the solution cost (Fig. 5-e), the `MPTree` representation outperforms the proposed representation as the trajectories obtained using our representation are constrained to remain within fixed geometric lattices thus incurring a higher control effort.

Study 3: Decentralized System Evaluation in 3D Workspaces: The control environments used in this study are $10 \times 10 \times 5$ m^3 workspaces consisting of two dynamic obstacle configurations—a *pillar* ($2 \times 2 \times 5$ m^3) that translates in 2-D and a *box* ($1 \times 1 \times 1$ m^3) that translates in 3-D. Forty environments with varying obstacle density and speeds are used. The obstacle density varies from 2.2% to 30.8% (by volume) and the obstacle speed varies from 1.0 m/s to 2.5 m/s. Ten planning experiments were conducted for team sizes of $n = 1$ to $n = 10$ each in the given control environments (Fig. 6-i). The resulting trajectories are evaluated for (1) robot-obstacle collisions (Fig. 6-ii), (2) inter-robot collisions (Fig. 6-iii), (3) kinodynamic feasibility (Fig. 6-iv, v), and response times. The clearance plots show that the robots maintain safe inter-robot clearance distances and on average, maintain a clearance distance of 2.5m from the dynamic obstacles. Further, based on the distribution of velocities and acceleration across all the computed motion plans, we conclude that the motion plans adhere to the specified kinodynamic constraints. Representative experiment videos are available at https://bit.ly/2VXn3in.

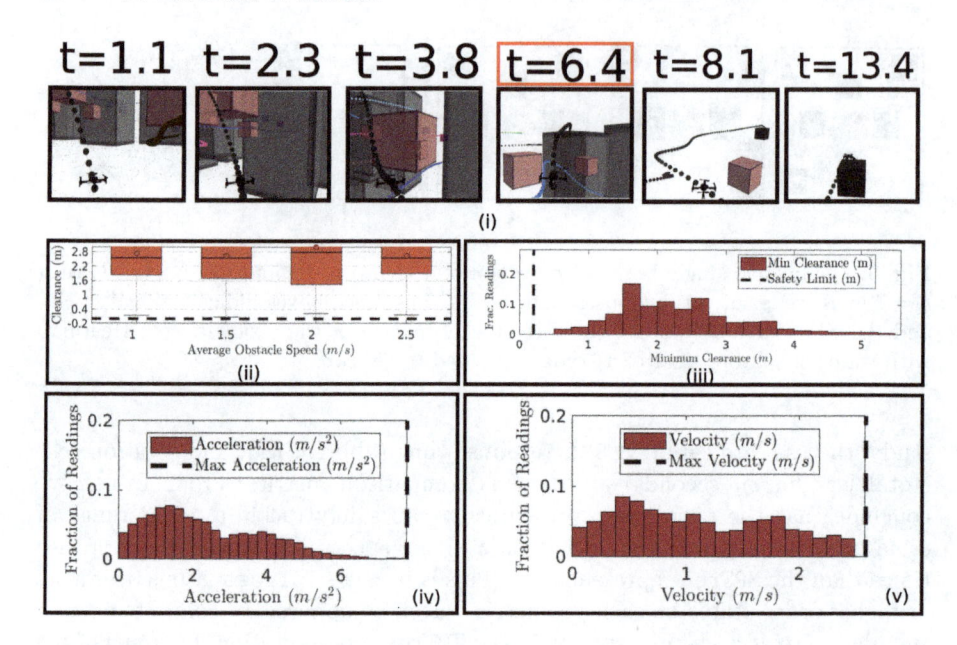

Fig. 6. (i) Snapshots of a single robot (dark green) at different time instances (in seconds) over the course of one entire planning experiment. The robot avoids the dynamic obstacles in the workspace and terminates in the goal region (dark green cube). At time 6.4 seconds, the robot replans a new trajectory from a non-stationary state to avoid collision with one of the dynamic obstacles. (ii) Figure shows the adjusted robot-obstacle clearance distances (observed clearance - safety limit) for different average obstacle speeds ranging from 0.5 m/s to 2.5 m/s (negative indicates collision). (iii)–(v) Bar plots shows distribution of inter-robot clearance distances, velocities, and accelerations across all experiments. Trajectories are dynamically feasible and safe.

5 Conclusion and Future Work

This work presents a decentralized planner and a planning representation for multi-robot navigation in uncertain 3D workspaces. The invariant nature of the kinodynamic planning representation allows for offline preprocessing thus enabling low-latency generation of kinodynamically feasible and collision-free trajectories online in dynamic 3D workspaces. As part of the future research, we intend to conceive a distributed heirarchical coordination framework composed of long-term deliberative and short-term reactive planning processes.

References

1. Aoude, G.S., Luders, B.D., Joseph, J.M., Roy, N., How, J.P.: Probabilistically safe motion planning to avoid dynamic obstacles with uncertain motion patterns. Auton. Rob. **35**(1), 51–76 (2013)
2. Dellin, C.M., Srinivasa, S.S.: A unifying formalism for shortest path problems with expensive edge evaluations via lazy best-first search over paths with edge selectors.

In: Twenty-Sixth International Conference on Automated Planning and Scheduling (2016)
3. Desai, A., Collins, M., Michael, N.: Efficient kinodynamic multi-robot replanning in known workspaces. In: 2019 International Conference on Robotics and Automation (ICRA), pp. 1021–1027. IEEE (2019)
4. Desai, A., Michael, N.: Online planning for quadrotor teams in 3-D workspaces via reachability analysis on invariant geometric trees. In: 2020 International Conference on Robotics and Automation (ICRA). IEEE (2020)
5. Dobson, A., Bekris, K.E.: Improving sparse roadmap spanners. In: 2013 IEEE International Conference on Robotics and Automation, pp. 4106–4111. IEEE (2013)
6. Hehn, M., D'Andrea, R.: Quadrocopter trajectory generation and control. IFAC Proc. Vol. **44**(1), 1485–1491 (2011)
7. Liu, S.: Motion planning for micro aerial vehicles (2018)
8. Otte, M., Frazzoli, E.: RRTx: asymptotically optimal single-query sampling-based motion planning with quick replanning. Int. J. Robot. Res. **35**(7), 797–822 (2016)
9. Phillips, M., Likhachev, M.: SIPP: safe interval path planning for dynamic environments. In: 2011 IEEE International Conference on Robotics and Automation, pp. 5628–5635. IEEE (2011)
10. Silver, D.: Cooperative pathfinding. AIIDE **1**, 117–122 (2005)
11. Turpin, M., Mohta, K., Michael, N., Kumar, V.: Goal assignment and trajectory planning for large teams of interchangeable robots. Auton. Robot. **37**(4), 401–415 (2014). https://doi.org/10.1007/s10514-014-9412-1
12. Vemula, A., Muelling, K., Oh, J.: Path planning in dynamic environments with adaptive dimensionality. In: Ninth Annual Symposium on Combinatorial Search (2016)
13. Zhu, H., Alonso-Mora, J.: Chance-constrained collision avoidance for MAVs in dynamic environments. IEEE Robot. Autom. Lett. **4**(2), 776–783 (2019)
14. Zucker, M., Kuffner, J., Branicky, M.: Multipartite RRTs for rapid replanning in dynamic environments. In: Proceedings 2007 IEEE International Conference on Robotics and Automation, pp. 1603–1609. IEEE (2007)

Optimized Direction Assignment in Roadmaps for Multi-AGV Systems Based on Transportation Flows

Valerio Digani[1(✉)] and Lorenzo Sabattini[2]

[1] System Logistics S.p.a., Fiorano Modenese, Italy
`valerio.digani@systemlogistics.com`
[2] Department of Sciences and Methods for Engineering (DISMI), University of Modena and Reggio Emilia, Modena, Italy
`lorenzo.sabattini@unimore.it`

Abstract. In this paper we propose a method for optimizing the design of a roadmap, used for motion coordination of groups of automated guided vehicles for industrial environments. Considering the desired flows among different locations in the environment, we model the problem as a multi-commodity concurrent flow problem, which allows us to assign the directions of the paths in an optimized manner. The proposed solution is validated by means of simulations, exploiting realistic layouts, and comparing the performance of the system with those achieved with a baseline roadmap.

Keywords: Traffic management · AGVs · Roadmap optimization

1 Introduction

Automated Guided Vehicles (AGVs) are mobile robots used in industrial applications, typically for logistics operations. More specifically, these systems are used for automatically transporting goods within a factory environment, replacing or complementing manually driven forklifts. Main advantages of AGV based solutions are related to increased performances and reduction in safety-related issues [16].

In order to obtain such advantages, a key factor is represented by traffic management. In fact, the overall transportation efficiency of the AGV system is mainly defined by the methodology that is used to assign tasks to the AGVs, and to control them for task completion. Since AGVs share a common environment, it is necessary to plan and control their motion in a coordinated manner. While several strategies exist in the literature for coordinated motion control of multi-AGV systems, in the vast majority of real-world industrial implementations, AGVs are constrained to move along a set of (virtual) paths, usually referred to as roadmap [21].

It is worth noting that the evolution of the traffic is heavily influenced by the way the paths that constitute the roadmap are designed. In fact, as shown in [10,17], globally optimizing the system performance requires to consider both the traffic management rules, and the design of the paths themselves.

Along these lines, several works appeared in the literature, that analyze how the design of the paths influences traffic, mostly considering road traffic, in urban and highway scenarios. The network of roads is modeled, in [15], as a graph, with the objective of controlling the traffic. In particular, the maximum flow approach is applied to optimize the definition of the speed limits, in order to ensure safety while maximizing the flow of vehicles. A similar method is applied in [1], where the maximum flow approach is exploited for analyzing the presence of bottlenecks in the road network, in order to understand what portions of the road network need to be improved, with the objective of optimizing the traffic flow. The maximum flow algorithm is also used in [22] to optimize road traffic, monitoring online the status of intersections, and hence avoiding traffic congestion.

When considering traffic problems for AGV systems in industrial applications, it is fundamental to consider the presence of multiple heterogeneous flows, generated by different sources, sharing the same environment. As such, the problem can be modeled as a multi-commodity concurrent flow problem [11]: the flow is generated by multiple sources (the commodities), and is propagated to a shared network (and is, then, concurrent on the same set of edges). Efficient solutions to these problems are proposed in [2,3].

The maximum concurrent multi-commodity linear flow problem is applied in [6] to road networks, explicitly considering node capacities: given a network, the solution to the problem finds what is the maximum achievable flow. In [14], the maximum flow algorithm is exploited to assess what is the maximum flow between any two nodes in a network, considering a dynamic network.

In this paper, we propose to model the roadmap for industrial AGV systems as a graph, and to exploit the solution to a multi-commodity concurrent flow problem for optimizing the design of the roadmap itself, in such a way that the performance of the AGV system is maximized.

Despite the importance of the roadmap on the system performance, its design is generally performed in a manual manner, by expert technicians, based on previous experience. As such, the achieved results are generally far from optimal. To solve this issue, a few methods have appeared in the literature to automatically design the roadmap, in an optimized manner. However, most of the methods that can be found in the literature are defined for motion planning of single robots, and are typically based on random sampling techniques [8,13,20]. A method for automatic creation of a roadmap for multiple AGVs was proposed in [4,9]. This method aims at maximizing the redundancy of the paths, with the objective of providing the AGVs with multiple options, thus leading to an easier (and more effective) coordinated motion coordination.

The contribution of this paper is in the definition of a novel method for optimizing the design of a roadmap, thus increasing traffic performance. In particular, considering a set of origin and destination configurations, and considering a set of roads interconnecting them, we propose (i) a new formulation based on a multi-commodity concurrent flow problem, and (ii) a method to assign each road with the best possible orientation, computed in order to optimize the desired flows originated by different sources in the plant, while minimizing traffic congestion.

2 Scenario and Problem Definition

2.1 Scenario

In this paper we consider a fleet of AGVs, used for goods transportation inside a factory environment. The factory environment contains a set of locations among which the AGVs need to move goods: such locations include production machines, warehouse positions, and delivery stations.

Several technological solutions are currently available to let AGVs localize themselves inside the environment, including magnetic systems, laser scanners detecting reflective markers, ultra wide band systems, vision and laser systems exploiting natural landmarks [19]. Despite the exploited localization method, due to the presence of infrastructure elements and obstacle, and in order to guarantee safety in environments shared with human operators, AGVs are generally constrained to move along a predefined set of paths, referred to as the *roadmap*.

2.2 Representation of the Roadmap

The roadmap is generally partitioned into segments, that represent portions of paths. Let then $\mathcal{R}$ be the roadmap, and let σ_i be the i-th segment of the roadmap. Then, we can write

$$\mathcal{R} = \{\sigma_1, \ldots, \sigma_S\} \quad (1)$$

where S is the total number of segments in $\mathcal{R}$.

Each segment is considered to be unidirectional, and to connect two points in the environment. Let v_i be the source node, and ψ_i be the sink node of segment σ_i, respectively. Based on its geometric characteristics, each segment σ_i is characterized by its capacity c_i, which is a positive number that represents the units of goods (e.g., pallets) that can travel along σ_i in the time unit (e.g., pallets per hour).

We can then represent the roadmap as a directed weighted graph $\mathscr{G} = \mathscr{G}(\mathcal{V}, \mathcal{E})$, where the vertex set $\mathcal{V}$ contains all the source and sink nodes, and each edge in the edge set $\mathcal{E}$ represents one segment of the roadmap, and is characterized by a weight corresponding to the capacity. With a slight abuse of notation, we will hereafter indicate with $\sigma_i = \sigma_i(v_i, \psi_i, c_i) \in \mathcal{E}$ as the edge corresponding to the i-th segment, with $v_i, \psi_i \in \mathcal{V}$ as the source and sink nodes, respectively, and $c_i > 0$ as the edge weight. We define $|\sigma_i| > 0$ as the length of the i-th segment.

We define also $\bar{\mathcal{E}}$ as the set of undirected edges obtained from $\mathcal{E}$: namely, each edge $\bar{\sigma}_i \in \bar{\mathcal{E}}$ has the same geometric characteristics, and the same capacity c_i as the corresponding edge $\sigma_i \in \mathcal{E}$, but no direction is assigned.

Furthermore, we define a path $\pi_j(v_h, v_k)$ as a sequence of edges that joins a sequence of vertices starting in v_h and ending in v_k. More specifically, if $\sigma_i, \sigma_{i+1} \in \pi_j$, then the sink node of σ_i corresponds to the source node of σ_{i+1}.

For each path π_j, we define $|\pi_j| > 0$ as its length, that represents the summation of the lengths of the segments corresponding to the edges in π_j, namely

$$|\pi_j| = \sum_{k \,|\, \sigma_k \in \pi_j} |\sigma_k| \qquad (2)$$

2.3 Representation of the Flows

In a factory environment, it is necessary to handle a certain quantity of different kinds of goods, which are produced in some nodes, and need to be delivered to other nodes.

This process can be modeled as a set of *flows*, within the roadmap. In particular, let $\mathcal{K}$ represent the set of flows. The k-th flow, namely $f_k \in \mathcal{K}$, represents the units of goods (e.g., pallets) of category k, that need to travel in the time unit (e.g., pallets per hour) from the source node $s_k \in \mathcal{R}$ to the sink node $t_k \in \mathcal{R}$ of the k-th flow.

Let $\Omega_k = \{\pi_j(s_k, t_k)\}$ be the set of paths that connect the source node s_k with the sink node t_k[1]. We define $\bar{\pi}(s_k, t_k)$ as the minimum length path, namely

$$\bar{\pi}(s_k, t_k) = \min_{\pi_j \in \Omega_k} |\pi_j| \qquad (3)$$

It is worth noting that, for complex and large roadmaps, the set Ω_k may contain paths that are significantly longer than the minimum length one. Since, in the analyzed scenario, long detours are commonly avoided by the path planner[2], we introduce a coefficient $\kappa > 1$ that quantifies the maximum considered deviation from the shortest path. Namely, we introduce the set $\Pi_k \subseteq \Omega_k$, defined as

$$\Pi_k = \{\pi_j(s_k, t_k) \in \Omega_k \text{ such that } |\pi_j| \leq \kappa \bar{\pi}(s_k, t_k)\} \qquad (4)$$

For each flow $f_k \in \mathcal{K}$, we define $d_k \geq 0$ as the desired flow, i.e., the desired number of units of goods of category k to be delivered, per time unit, from s_k to t_k. Since the roadmap, that is the set of paths exploited by the AGVs to perform the delivery operations, is a shared resource (i.e., multiple AGVs traveling at the same time, delivering goods of different categories), and since the number of AGVs is limited, the actual flow f_k is, in general, different from the desired one, namely d_k. In particular, we have $f_k \leq d_k$, as sketched in Fig. 1(a). Moreover, we define $f_k[\pi_j]$ the portion of flow f_k that travels through path π_j, as visually represented in Fig. 1(b).

---

[1] It is worth noting that, given a generic roadmap, multiple paths exist that connect each pair of nodes.

[2] It is worth noting that the most commonly traveled paths by the vehicles are generally the shortest paths. Alternative paths are chosen only in the presence of heavy traffic conditions or other unexpected events that would generate large waiting times. As an example, the reader is referred to [10].

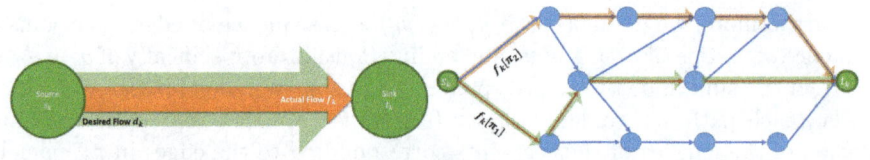

(a) Actual flow f_k and desired flow f_k from source node s_k to sink node t_k

(b) Multiple paths exist from the source node s_k and the sink node t_k. A portion of the overall flow f_k travels along each path.

Fig. 1. Representation of the flows

2.4 Problem Definition

Consider the following input parameters: (i) a given vertex set $\mathcal{V}$, containing all the nodes in the environment, (ii) a given set of undirected edges $\bar{\mathcal{E}}$ (where edges are characterized in terms of geometric characteristics, and of capacity), (iii) the set of desired flows $\{d_k\}$.

In this paper, we address the problem of defining the directed graph $\mathscr{G}(\mathcal{V}, \mathcal{E})$, building upon the given set of undirected edges $\bar{\mathcal{E}}$, in order to optimize the overall material flow, given the set of desired flows $\{d_k\}$. In other words, the objective is that of optimizing the choice of the directions in which each edge should be traveled by the AGVs, in such a way that the overall performance of the system is optimized.

Remark 1. The problem of geometrically designing the paths is out of the scope of this paper, and is hence not addressed here. For automated solutions to this issue, the reader is referred to, e.g., [4,9].

3 Proposed Method

In this section we propose a method for the solution of the problem introduced in Sect. 2. The proposed method consists in reformulating the problem considering the directed graph $\mathscr{G}$ representing the roadmap, and different *flows* along it. Each flow is represented as a *commodity*, and is associated with a *demand*, representing the desired amount of flow along $\mathscr{G}$. The problem of optimizing the assignment of flows to edges considering the demands of each commodity is referred to as *maximum concurrent flow problem*, in a *multi-commodity* setting. For additional details, and for examples of efficient algorithms for the solution of such problems, the reader is referred to [12].

Objective Function. Consider a set of heterogeneous flows $\mathcal{K}$, and let d_k be the desired value for each flow. Each flow $k \in \mathcal{K}$ represents a commodity, that is concurrent on the same set of edges, collected in the roadmap $\mathcal{R}$.

Let $f_k(v_1, v_2)$, with $v_1, v_2 \in \mathcal{V}$, be the portion of flow f_k that travels on the roadmap $\mathcal{R}$ from node v_1 to node v_2. We define then p_k as the *percentage of injected flow* related to flow f_k, as follows:

$$p_k = \sum_{\pi_j \in \Pi_k} \frac{f_k[\pi_j]}{d_k} \tag{5}$$

This quantity represents the ratio between the actual flow f_k that travels along the roadmap $\mathcal{R}$, and its demand (i.e., desired value) d_k. In fact, the numerator represent the summation of all the portions of f_k that travel along all the paths in Π_k.

Optimizing the definition of the roadmap $\mathcal{R}$ in terms of desired flows $\{d_k\}$ is achieved by maximizing the percentage of injected flow for each flow $f_k \in \mathcal{K}$. Hence, we propose the following max-min formulation:

$$\max \min_{k;\, f_k \in \mathcal{K}} p_k = \max \min_{k;\, f_k \in \mathcal{K}} \sum_{\pi_j \in \Pi_k} \frac{f_k[\pi_j]}{d_k} \tag{6}$$

In particular, we aim at maximizing the minimum percentage of injected flow, for each flow $f_k \in \mathcal{K}$.

Limited Capacity Constraints. It is worth remarking that each edge $\sigma_i \in \mathcal{E}$ is characterized by a limited capacity c_i: this constraint needs to be included in the optimization problem. In particular, considering the i-th edge $\sigma_i(v_i, \psi_i, c_i)$, we impose the following constraint:

$$\sum_{k;\, f_k \in \mathcal{K}} f_k(v_i, \psi_i) \leq c_i \quad \forall i;\, \sigma_i \in \mathcal{E} \tag{7}$$

Flow Constraints. Additional constraints are related to the definition of the flows in $\mathcal{K}$. In fact, each flow f_k is generated in the source node s_k, and terminated in the sink node t_k. This implies that flow is never accumulated nor destroyed in any edge. Hence, for each flow f_k, we can introduce the net flow across each edge in the map, $\sigma_i(v_i, \psi_i, c_i)$, as

$$\Delta_{k,i} = \sum_{v \in \mathcal{V}} f_k(v, v_i) - \sum_{v \in \mathcal{V}} f_k(\psi_i, v) \tag{8}$$

This quantity represents the difference between the portion of flow f_k entering and exiting edge σ_i. This quantity is:

- positive for the edges whose source nodes coincides with the source node of flow f_k,
- negative for the edges whose sink nodes coincides with the sink node of flow f_k,
- zero otherwise (i.e., the portion of flow f_k entering the edge corresponds to the portion exiting the edge itself).

We can then formalize the following constraint:

$$\Delta_{k,i} = \begin{cases} 0 & \forall k, i;\, \{f_k \in \mathcal{K}, \sigma_i \in \mathcal{E},\, v_i \neq s_k,\, w_i \neq t_k\} \\ \varphi_1,\, 0 \leq \varphi_1 \leq d_k & \forall k, i;\, \{f_k \in \mathcal{K}, \sigma_i \in \mathcal{E},\, v_i = s_k\} \\ \varphi_2,\, -d_k \leq \varphi_2 \leq 0 & \forall k, i;\, \{f_k \in \mathcal{K}, \sigma_i \in \mathcal{E},\, w_i = t_k\} \end{cases} \tag{9}$$

Connectivity Constraint. It is necessary to guarantee that the roadmap $\mathcal{R}$ is defined in such a way that a path exists that connects each pair of nodes $v_1, v_2 \in \mathcal{V}$: this is mandatory, to avoid dead-ends for the AGVs, or the existence of unreachable locations in the environment. In order to guarantee such property, we need to ensure that the graph $\mathscr{G}(\mathcal{V}, \mathcal{E})$ is strongly connected. As is well known from algebraic graph theory [5], this property is related to the eigenvalues of the Laplacian matrix of $\mathscr{G}$, for balanced graphs. However, is it worth noting that $\mathscr{G}$ is not balanced, for a generic roadmap $\mathcal{R}$.

Nevertheless, as shown in [18], it is always possible to find a set of edge weights $w \in \mathbb{R}^M$, where M is the number of edges in $\mathcal{E}$, that define a balanced version of $\mathscr{G}$, namely $\bar{\mathscr{G}}$. As shown in [18], $\bar{\mathscr{G}}$ can be found from $\mathscr{G}$, defining the edge weights as the solution of $\mathcal{I} w = \mathbb{O}$, where $\mathcal{I}$ is the incidence matrix of $\mathscr{G}$, and $\mathbb{O}$ is the null vector. The Laplacian matrix of $\bar{\mathscr{G}}$ can then be found as $\mathcal{L}(\bar{\mathscr{G}}) = \mathcal{I} \operatorname{diag}(w) \mathcal{I}^T$. It is worth remarking that $\bar{\mathscr{G}}$ is a balanced graph, defined on the same vertex set $\mathcal{V}$ as $\mathscr{G}$, and with the edges directed in the same direction as those in $\mathcal{E}$. Therefore, strong connectivity of $\bar{\mathscr{G}}$ implies strong connectivity of $\mathscr{G}$.

The algebraic connectivity of $\bar{\mathscr{G}}$ can then be exploited to formulate the connectivity constraint for the roadmap $\mathcal{R}$, as follows:

$$\Re\left\{\lambda_2^{\bar{\mathscr{G}}}\right\} > 0 \tag{10}$$

Overall Optimization Problem. Consider now the set of undirected edges $\bar{\mathcal{E}}$: the definition of the roadmap $\mathcal{R}$ consists in associating a direction δ_i to each undirected edge $\bar{\sigma}_i$. Define now the vector of directions as

$$\delta = [\delta_1, \ldots, \delta_S] \tag{11}$$

Considering the objective introduced in (6), and the constraints introduced in (7), (9), and (10), the roadmap definition problem can then be solved finding vector δ as the solution of the following optimization problem:

$$\max_{\delta} \min_{k;\, f_k \in \mathcal{K}} \sum_{\pi_j \in \Pi_k} \frac{f_k[\pi_j]}{d_k}$$

subject to: $\sum_{k;\, f_k \in \mathcal{K}} f_k(v_i, w_i) \leq c_i \quad \forall i;\, \sigma_i \in \mathcal{E}$

and $\Re\left\{\lambda_2^{\bar{\mathscr{G}}}\right\} > 0$

and $\Delta_{k,i} = \begin{cases} 0 & \forall k, i;\, \{f_k \in \mathcal{K}, \sigma_i \in \mathcal{E}, v_i \neq s_k, w_i \neq t_k\} \\ \varphi_1,\, 0 \leq \varphi_1 \leq d_k & \forall k, i;\, \{f_k \in \mathcal{K}, \sigma_i \in \mathcal{E}, v_i = s_k\} \\ \varphi_2,\, -d_k \leq \varphi_2 \leq 0 & \forall k, i;\, \{f_k \in \mathcal{K}, \sigma_i \in \mathcal{E}, w_i = t_k\} \end{cases}$

$$\tag{12}$$

It is worth noting that the the optimization variables collected in vector δ are binary values. Moreover, the constraints are nonlinear. Hence, (12) represents an integer nonlinear optimization problem, which is computationally very hard to solve.

However, following the procedure introduced in [12], it is possible to exploit the multi-commodity concurrent flow formulation to reduce the complexity of (12). In particular, the objective function of (12) can be rewritten as the solution of a constrained linear optimization problem (LP), in the multi-commodity concurrent flow formulation, as follows:

$$\max \lambda$$
$$\text{subject to: } \sum_{v \in \mathcal{V}} f_k(s_k, v) \geq d_k \lambda \qquad (13)$$
$$\text{and } \lambda \in \mathbb{R}, \lambda \geq 0 \in \mathbb{R}$$

The new optimization variable λ in (13) represents an auxiliary variable, that defines the amount of each flow (in a multi-commodity problem) actually on the network. In particular, for commodity k, the quantity λd_k represents the actual demand related to the k-th flow. It is worth noting that (13) represents a simple LP problem, which can be solved in a very efficient manner.

Putting (13) into the original optimization problem (12), we obtain the following new formulation:

$$\max \lambda$$
$$\text{subject to: } \sum_{v \in \mathcal{V}} f_k(s_k, v) \geq d_k \lambda$$
$$\text{and } \lambda \in \mathbb{R}, \lambda \geq 0 \in \mathbb{R}$$
$$\text{and } \sum_{k; f_k \in \mathcal{K}} f_k(v_i, w_i) \leq c_i \quad \forall i; \sigma_i \in \mathcal{E}$$
$$\text{and } \Re\left\{\lambda_2^{\mathscr{G}}\right\} > 0$$
$$\text{and } \Delta_{k,i} = \begin{cases} 0 & \forall k, i; \{f_k \in \mathcal{K}, \sigma_i \in \mathcal{E}, v_i \neq s_k, w_i \neq t_k\} \\ \varphi_1, 0 \leq \varphi_1 \leq d_k & \forall k, i; \{f_k \in \mathcal{K}, \sigma_i \in \mathcal{E}, v_i = s_k\} \\ \varphi_2, -d_k \leq \varphi_2 \leq 0 & \forall k, i; \{f_k \in \mathcal{K}, \sigma_i \in \mathcal{E}, w_i = t_k\} \end{cases}$$
$$(14)$$

Even though the inner problem (i.e., the LP problem given in (13)) can be solved easily with a linear programming solver in polynomial time [12], the outer assignment problem is still formulated as an integer problem. As noted, this optimization problem is generally NP-hard. However there are several algorithms that can generate a good approximate solution to this class of problems, for an effort that increases only slowly as a function of S, that is the size of vector δ. In our approach, a local search method is used. In particular, a *Tabu Search* approach is combined with a *Monte Carlo* method [7].

4 Evaluation

The performance of the proposed methodology was evaluated on real industrial plants, considering the two different scenarios depicted in Fig. 2.

(a) AS/RS (b) Block Storage

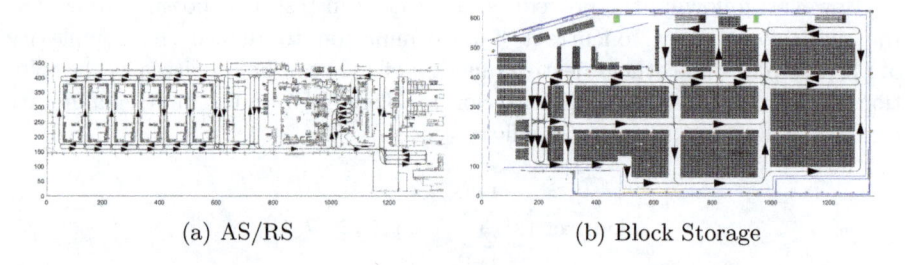

Fig. 2. Experimental scenarios used for the evaluation. Automatic roadmaps with optimized directions are shown.

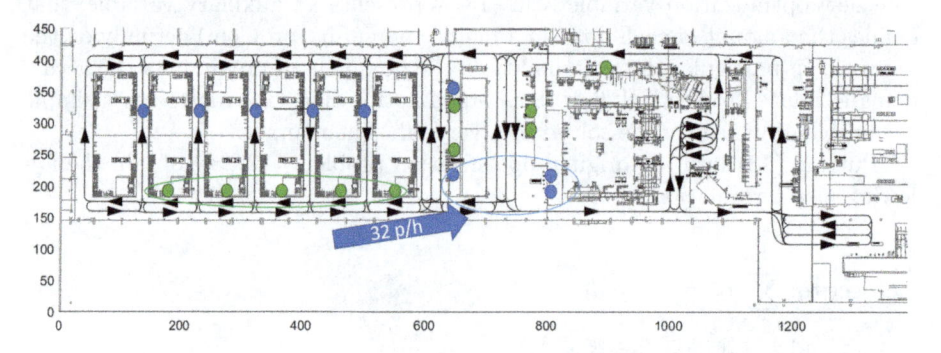

Fig. 3. Origin-destination flows. Green and blue dots represent, respectively, the source and the sink nodes of the different flows. As an example, the actual flow (in pallets per hour $[p/h]$) between the highlighted areas is indicated by the arrow.

The two scenarios have been chosen since they are well representative of different real-world operative conditions. In particular, Fig. 2(a) represents an automatic storage and retrieval system (**AS/RS**) scenario where goods are stocked by means of high density racks; Fig. 2(b) represents a typical **block storage** scenario, where the goods (pallets) are stocked directly on the floor. The two scenarios can be used to model most of the real industrial automatic warehouses. Due to lack of space, only one example about the real flows used in simulation is shown in Fig. 3.

For evaluation purposes, simulations have been performed on a discrete-event simulator developed by the industrial partner. The performance of the system were measured in terms of average number of transportation operations performed per time unit. The results achieved on the *optimized roadmap*, computed according to the method proposed in this paper, were compared to a *baseline roadmap*, where the directions of the edges are manually designed by expert technicians (based on their experience), that is currently utilized in the real industrial plants.

Optimized Direction Assignment   67

(a) AS/RS   (b) Block Storage

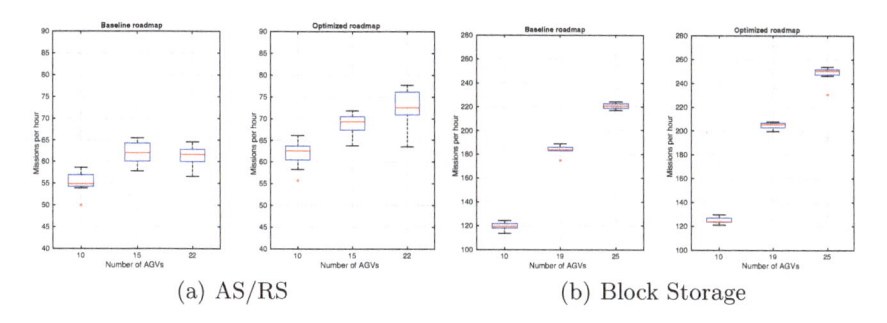

Fig. 4. Performance of the system (number of accomplished transportation missions per hour) on the two scenarios. Comparison between the *baseline roadmap* and the *optimized roadmap*. 10 trials per configuration.

(a) AS/RS   (b) Block Storage

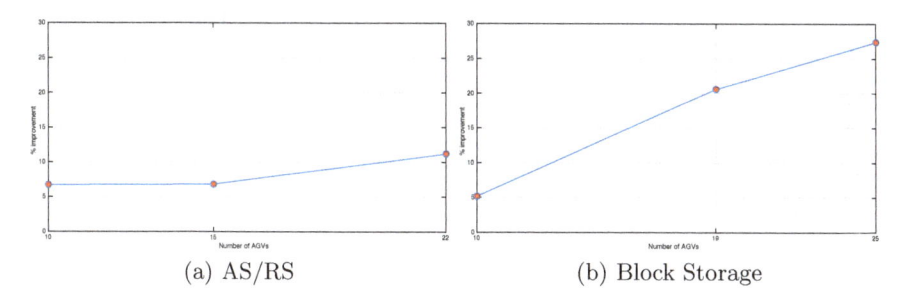

Fig. 5. Percentage performance increase, comparing the *baseline roadmap* with the *optimized roadmap*.

For each scenario, we performed the simulations using different number of AGVs[3], namely 10, 15, 22 for **AS/RS**, and 10, 19, 25 for **Block Storage**. For each configuration, 10 simulation trials were generated, with duration 3 h, constantly feeding the system with transportation tasks to be completed, defined by the warehouse management system according to the desired flows.

The results of the simulation, in terms of number of accomplished transportation missions per hour, are summarized in Fig. 4. These results clearly show that the *optimized roadmap* provides a significant increase in the performance, with respect to the *baseline roadmap*. Moreover, such increase in more significant as the number of AGVs increases. In order to numerically assess the performance increase, Fig. 5 depicts the percentage improvement in the performance, obtained with the *optimized roadmap*, with respect to the *baseline roadmap*. It is possible to notice that the performance increase is significant, with values up to 10% to 20% for large numbers of AGVs (depending on the specific scenario).

---

[3] Differences in the maximum number of AGVs used in the simulations are due to the different size of the scenarios.

5 Conclusions

In this paper we proposed a methodology for optimized roadmap design, considering, as a target application, groups of AGVs used for industrial logistics.

In particular, we propose a formulation based on the multi-commodity concurrent flow problem. Such formulation allows us to take into account the desired flow of material through the different locations of the environment, and to optimize the paths definition in such a way that the overall transportation performance of the systems is maximized.

Performance is evaluated in terms of average number of transportation missions completed per hour, comparing the proposed solution with a (manually designed) baseline roadmap. For evaluation purposes, a discrete-event simulator was exploited for simulating the behavior of the multi-AGV system, in realistic environments. Roadmaps designed with the proposed method provide consistently improved performance, with respect to baseline roadmaps: the performance increase appears particularly significant for large numbers of AGVs.

Future work will aim at improving the definition of the edge capacity, including a dynamic definition based on the current traffic level. Furthermore, we aim at investigating the use of queuing theory, for modeling dynamic effects in the flows. Moreover, we will investigate the introduction of an extra commodity, to model the flow of empty vehicles through the environment.

References

1. Abdullah, N., Hua, T.K.: Using ford-fulkerson algorithm and max flow-min cut theorem to minimize traffic congestion in Kota Kinabalu, Sabah. J. Inf. **2**(4), 18–34 (2017)
2. Babonneau, F.: Solving the multicommodity flow problem with the analytic center cutting plane method. Ph.D. thesis, University of Geneva (2006)
3. Bauguion, P.O., Ben-Ameur, W., Gourdin, E.: Efficient algorithms for the maximum concurrent flow problem. Networks **65**(1), 56–67 (2015)
4. Beinschob, P., Meyer, M., Reinke, C., Digani, V., Secchi, C., Sabattini, L.: Semi-automated map creation for fast deployment of AGV fleets in modern logistics. Robot. Auton. Syst. **87**, 281–295 (2017)
5. Bullo, F., Cortés, J., Martínez, S.: Distributed Control of Robotic Networks. Applied Mathematics Series. Princeton University Press (2009). http://coordinationbook.info
6. Chien, T.Q., Lau, N.D.: Algorithm finding maximum concurrent multicommodity linear flow with limited cost in extended traffic network with single regulating coefficient on two-side lines. Int. J. Comput. Netw. Commun. (IJCNC) **9**(2) (2017)
7. Chopin, N., Schafer, C.: Adaptive Monte Carlo on multivariate binary sampling spaces. Working Papers, Centre de Recherche en Economie et Statistique (2010)
8. Devaurs, D., Siméon, T., Cortés, J.: Optimal path planning in complex cost spaces with sampling-based algorithms. IEEE Trans. Autom. Sci. Eng. **13**(2), 415–424 (2015)
9. Digani, V., Sabattini, L., Secchi, C., Fantuzzi, C.: An automatic approach for the generation of the roadmap for multi-AGV systems in an industrial environment.

In: Proceedings of the IEEE/RSJ International Conference on Intelligent Robots and Systems (IROS), Chicago, IL, USA, September 2014
10. Digani, V., Sabattini, L., Secchi, C., Fantuzzi, C.: Ensemble coordination approach in multi-AGV systems applied to industrial warehouses. IEEE Trans. Autom. Sci. Eng. **12**(3), 922–934 (2015)
11. Even, S., Itai, A., Shamir, A.: On the complexity of time table and multi-commodity flow problems. In: 16th Annual Symposium on Foundations of Computer Science (SFCS 1975), pp. 184–193. IEEE (1975)
12. Garg, N., Koenemann, J.: Faster and simpler algorithms for multicommodity flow and other fractional packing problems. SIAM J. Comput. **37**(2), 630–652 (2007)
13. Geraerts, R., Overmars, M.H.: A comparative study of probabilistic roadmap planners. In: Boissonnat, J.-D., Burdick, J., Goldberg, K., Hutchinson, S. (eds.) Algorithmic Foundations of Robotics V. STAR, vol. 7, pp. 43–57. Springer, Heidelberg (2004). https://doi.org/10.1007/978-3-540-45058-0_4
14. Kaanodiya, K., Rizwanullah, M.: Minimize traffic congestion: an application of maximum flow in dynamic networks. J. Appl. Math. Stat. Inf. (JAMSI) **8**(1), 63–74 (2012)
15. Moore, E.J., Kichainukon, W., Phalavonk, U.: Maximum flow in road networks with speed-dependent capacities-application to bangkok traffic. Songklanakarin J. Sci. Technol. **35**(4) (2013)
16. Sabattini, L., et al.: The pan-robots project: advanced automated guided vehicle systems for industrial logistics. IEEE Robot. Autom. Mag. **25**(1), 55–64 (2018). https://doi.org/10.1109/MRA.2017.2700325
17. Sabattini, L., Digani, V., Secchi, C., Fantuzzi, C.: Hierarchical coordination strategy for multi-AGV systems based on dynamic geodesic environment partitioning. In: Proceedings of the IEEE/RSJ International Conference on Intelligent Robots and Systems (IROS), Daejeon, Korea, October 2016
18. Sabattini, L., Secchi, C., Chopra, N.: Decentralized estimation and control for preserving the strong connectivity of directed graphs. IEEE Trans. Cybern. **45**(10), 2273–2286 (2015)
19. Sabattini, L., et al.: Technological roadmap to boost the introduction of AGVs in industrial applications, pp. 203–208 (2013)
20. Tsardoulias, E., Iliakopoulou, A., Kargakos, A., Petrou, L.: A review of global path planning methods for occupancy grid maps regardless of obstacle density. J. Intell. Robot. Syst. **84**(1–4), 829–858 (2016)
21. Vis, I.F.: Survey of research in the design and control of automated guided vehicle systems. Eur. J. Oper. Res. **170**(3), 677–709 (2006)
22. Yamuna, M., Raza, A., Anand, K., Kumar, S., Kumar, S.: Traffic control system using maximum flow algorithm. Int. J. Curr. Eng. Technol. **6**(5) (2016)

Datom: A Deformable Modular Robot for Building Self-reconfigurable Programmable Matter

Benoît Piranda[✉] and Julien Bourgeois

FEMTO-ST Institute, Univ. Bourgogne Franche-Comté, CNRS, 1 Cours Leprince-Ringuet, 25200 Montbéliard, France
{benoit.piranda,julien.bourgeois}@femto-st.fr

Abstract. Moving a module in a modular robot is a very complex and error-prone process. Unlike in swarm, in the modular robots we are targeting, the moving module must keep the connection to, at least, one other module. In order to miniaturize each module to few millimeters, we have proposed a design which is using electrostatic actuator. However, this movement is composed of several attachment, detachment creating the movement and each small step can fail causing a module to break the connection. The idea developed in this paper consists in creating a new kind of deformable module allowing a movement which keeps the connection between the moving and the fixed modules. We detail the geometry and the practical constraints during the conception of this new module. We then validate the capability of motion for a module in an existing configuration. This implies the cooperation of some of the other modules placed along the path and we show in simulations that it exists a motion process to reach every free positions of the surface for a given configuration.

1 Introduction

The idea of designing hardware robotic modules able to be attached together has given birth to the field of modular robotics and when these modules can move by themselves they are named Modular Self-reconfigurable Robots (MSR) [9, 12] also named earlier as metamorphic robotic systems [2] or cellular robotic systems [3]. There are five families of MSR namely: lattice-based when modules are aligned on a 3D lattice, chain-type when the modules are permanently attached through an articulation, forming a chain or more rarely a tree, hybrid which is a mix between lattice-based and chain-type, mobile when each module can move autonomously and more recently continuous docking [11] where latching can be made in any point of the module. Since then, there have been many robots proposed and built by the community using different scales of modules and different latching and moving technologies. However, none of them have succeeded to reach a market.

Our objective is to build programmable matter [1] which is a matter than can change one or several of its physical properties, more likely its shape, according to an internal or an external action. Here, programmable matter will be constructed using a MSR, i.e. a matter composed of mm-scale robots, able to stick together and turn around each

© The Author(s), under exclusive license to Springer Nature Switzerland AG 2022
F. Matsuno et al. (Eds.): DARS 2021, SPAR 22, pp. 70–81, 2022.
https://doi.org/10.1007/978-3-030-92790-5_6

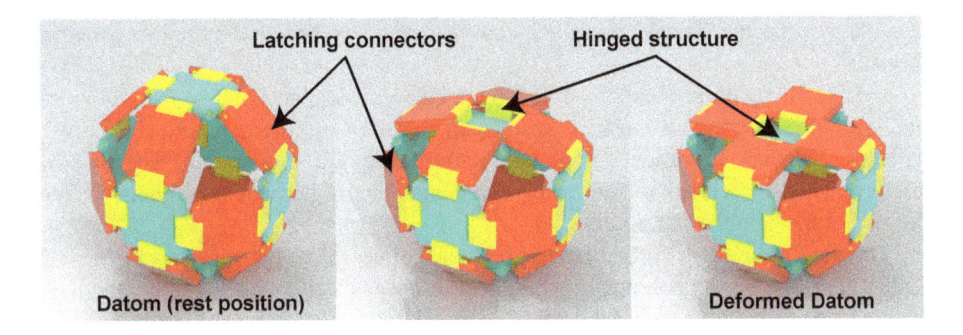

Fig. 1. Overview of the *Datom* in different deformation states.

other as it has been described in the Claytronics project [4]. The Programmable Matter Project[1] is a sequel of the Claytronics project and reuses most of its ideas and concepts. The requirements for each module are the following: mm-scale, being able to move in 3D, compute and communicate with their neighbors and the idea is to have thousands of them all linked together. Moving in 3D is the most complicated requirement as it needs a complex trade off between several parameters during the design phase. For example, moving requires power and, therefore, power storage, which adds weight to the module, the trade off being between having more power by adding more power storage and having a module as light as possible for easing the movement. We are currently building and testing a quasi-spherical module we designed [7]. This module rolls on another module using electrostatic electrodes. This way of moving creates uncertainty in the success of the movement as it is a complex sequence of repulsing/attaching/detaching actuations and we would like to study a movement where the moving module always stay latched to the pivot module.

The idea that drives this work is to design a motion process which never disconnects the moving and the fixed modules. We propose to define a deformable module named *Datom* (for Deformable Atom as a reference to the Claytronics Atom, Catom). Each module is strongly connected to neighbors in the Face-Centered Cubic (FCC) lattice with large latching connectors (drawn in red in Fig. 1). Two connected modules must deform their shapes to align future latched connectors while the previous connection is maintained as shown in Fig. 1. When new connectors are aligned they are strongly attached and the previous connection is released. Finally, the two modules return to their original shape.

2 Related Works

Crystalline Robot [8], developed by Rus et al. in 2001 is an interesting solution. These robots can move relatively to each other by expanding and contracting. A robot can move a neighbor by doubling its length along $\overrightarrow{x}$ and $\overrightarrow{y}$ axes. These robots are grouped in meta-modules of 4×4 units placed in a 2D square grid. Robot to robot attachment

---

[1] http://www.programmable-matter.com/.

is made by a mechanical system called "lock-and-key" located on the square connected faces.

In [10] Suh et al. propose the Telecube, a cubic robot able to move in a cubic lattice. Similarly to previous work, Telecube can shrink using internal motors to move a neighbor. Telecube are grouped in meta-modules made of $2 \times 2 \times 2$ units. The six arms are terminated by sensors to detect neighbors and electro-permanent magnets connect the arm of the neighboring module.

The *3D Catom* model presented in [7] is a robot geometry that can move in a FCC lattice in rolling on the border of its neighbors. It uses electrostatic actuators, both for latching on planar connectors and rolling around cylindrical parts separating connectors.

3 The *Datom* Model

We propose in the paper a design of a deformable robot based on the same lattice as the *3D Catom*. But the deformation must maintain some connectors in contact during the motion. Figure 2 shows the decomposition of the movements of a mobile *Datom B* (in yellow) moving around a fixed *Datom A* (in red) to go from the position shown in Fig. 2.a to the position shown in Fig. 2.e. We consider that connectors A_4 and B_5 are initially attached.

The first part of the motion consists in simultaneous deformations of both modules while keeping the connection between A_4 and B_5. At the middle of the deformation process (see Fig. 2.c), four connectors of B are aligned in front of four connectors of A: ($\{(A_2, B_3), (A_3, B_2), (A_4, B_5), (A_5, B_4)\}$).

At this stage, four different attachment can be applied in order to reach four different positions. To move to the final destination presented in the Figure, connectors A_3 and B_2 are then attached and the connection (A_4, B_5) is released. A mirrored deformation from the previous ones moves module B to its final position.

3.1 Theoretical Geometry of the Deformable Module

The shape of the module is deduced from the shape of the *3D Catom* proposed in [7]. From this initial geometry, we retain the position of the 12 square connectors, centered at P_i. These positions are imposed by the placement of modules in the FCC lattice.

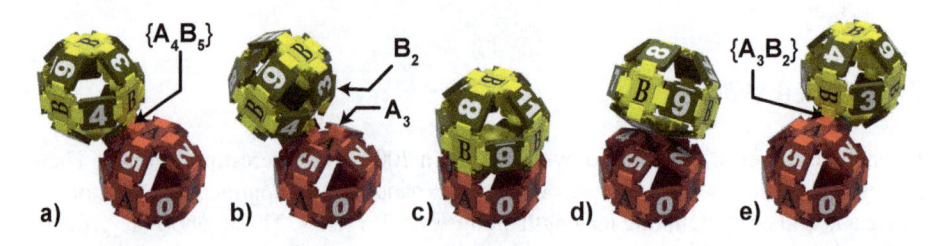

Fig. 2. Five steps of the motion of a *Datom B* around the fixed *Datom A* (the pivot). Darker elements are connectors, numbered from 0 to 11.

Datom: A Deformable Modular Robot for Building Programmable Matter 73

$$
\begin{array}{c|c|c|c}
P_0(r,0,0) & P_1(0,r,0) & P_6(-r,0,0) & P_7(0,-r,0) \\
P_2(\frac{r}{2},\frac{r}{2},\frac{r}{\sqrt{2}}) & P_3(-\frac{r}{2},\frac{r}{2},\frac{r}{\sqrt{2}}) & P_4(-\frac{r}{2},-\frac{r}{2},\frac{r}{\sqrt{2}}) & P_5(\frac{r}{2},-\frac{r}{2},\frac{r}{\sqrt{2}}) \\
P_8(-\frac{r}{2},-\frac{r}{2},-\frac{r}{\sqrt{2}}) & P_9(\frac{r}{2},-\frac{r}{2},-\frac{r}{\sqrt{2}}) & P_{10}(\frac{r}{2},\frac{r}{2},-\frac{r}{\sqrt{2}}) & P_{11}(-\frac{r}{2},\frac{r}{2},-\frac{r}{\sqrt{2}})
\end{array}
\tag{1}
$$

The size of the *Datom* is given by the distance between its two opposite connectors, this diameter is equal to $2 \times r$ (where r is the radius). The external square face of the red part presented in Fig. 1 are used to latch two *Datoms* together. Electrostatic latching actuators presented in [7] with *3D Catom*, need a maximum surface. Therefore, to ensure the best connection strength between two *Datoms*, we need to maximize the size c of the square connectors. In order to express the several dimensions of the proposed solution, we consider the plane of four co-planar connectors of a deformed *Datom*. In Fig. 3, connectors of length c are drawn in red and the piston actuator of length c also is drawn in blue. Mechanical links (drawn in green) of length e are placed between piston and connectors. Figure 3a shows 2 connectors C_0 and C_1 viewed from the top. In order to align them, we propose to turn them around the $\vec{z}$ axis at points P_0 and P_1 with an angle of $+45°$ for C_0 and $-45°$ for C_1 as shown in Fig. 3b. We can observe that the maximum width of a connector is limited by the 'diagonal' length ℓ which is equal to $3 \times c$, we can express also $\ell = \sqrt{2}(r + \frac{c}{2})$. We deduce the expression of c depending of the radius r:

$$
c = \frac{2 \times r}{3\sqrt{2}-1} \approx 0.61678 \times r \tag{2}
$$

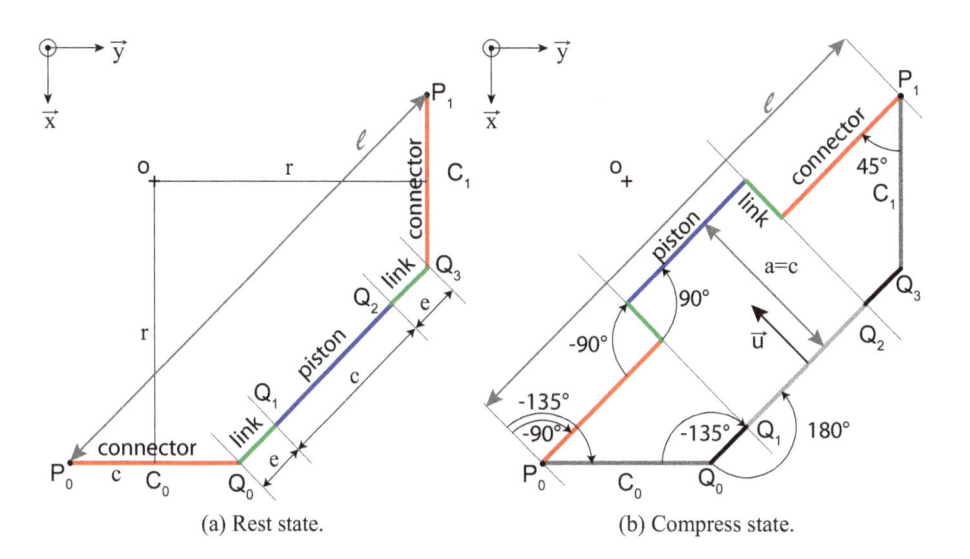

(a) Rest state.   (b) Compress state.

Fig. 3. Size and position of the parts around one of the piston of the deformable structure. The translation of the piston leads the rotation of the links and then the connectors.

Considering Fig. 3a, we can write a relation between r and c and e parameters:

$$r = \frac{c}{2} + \left(\frac{c}{2} + e\right)\sqrt{2} \tag{3}$$

That allows to express e in relation to the radius r:

$$e = r\left(\frac{2-\sqrt{2}}{3\sqrt{2}-1}\right) \approx 0.18065 \times r \tag{4}$$

3.2 Deformation Process of a Single *Datom*

Considering Fig. 3b, we can now express the amplitude (a) of of the piston during the deformation process.

$$a = \frac{\sqrt{2}}{2}c + e = \frac{\sqrt{2}}{2}c + \left(\frac{2-\sqrt{2}}{3\sqrt{2}-1} \times \frac{3\sqrt{2}-1}{2}c\right) = c \tag{5}$$

We obtain that the amplitude of motion of the piston is equal to the size of a connector.

The deformation to compress one side of the module is obtained by translating the corresponding piston along its $\vec{u}$ axis. It implies that the angle of joint between links and connectors (Q_0) goes from $-135°$ to $-90°$ and angle of joint between links and piston (Q_1) goes from $180°$ to $90°$. Finally the angle of joint between fixed links and connectors (P_0) goes from $-135°$ to $-90°$ as shown in Fig. 3b.

During this deformation, only one of the 6 pistons must move at a time in order to use the other elements as fixed supports at P_0 and P_1 joints.

3.3 Design of a Thick *Datom*

The theoretical shape of the *Datom* is not usable as is. To create a real functional module, we must consider that connectors have a not null thickness.

Let t be the thickness of the several mobiles parts of the module (connectors, links and pistons). In order to place the *Datom* in the FCC lattice, the main constraint is to keep the distance between two opposite connectors equal to $2 \times r$. We define $r' = r - \frac{1}{2}t$ as the new radius taking into account the connector thickness. Using this new radius, we can express c' and e' respectively the new size of connectors and links:

$$c' = \frac{2 \times r'}{3\sqrt{2}-1} \tag{6}$$

$$e' = \frac{2-\sqrt{2}}{3\sqrt{2}-1}r' \tag{7}$$

The geometry of the link part implies that the thickness t must be less than $e' = 0.19859 \times r$ (See Fig. 4).

We use rotation limits of each joint (between connector and link and also between link and piston) to shape blocking plots. These blocking plots help for the stability of the whole system. For example, Fig. 5 shows blocking plots for the joint between the connector and the link parts.

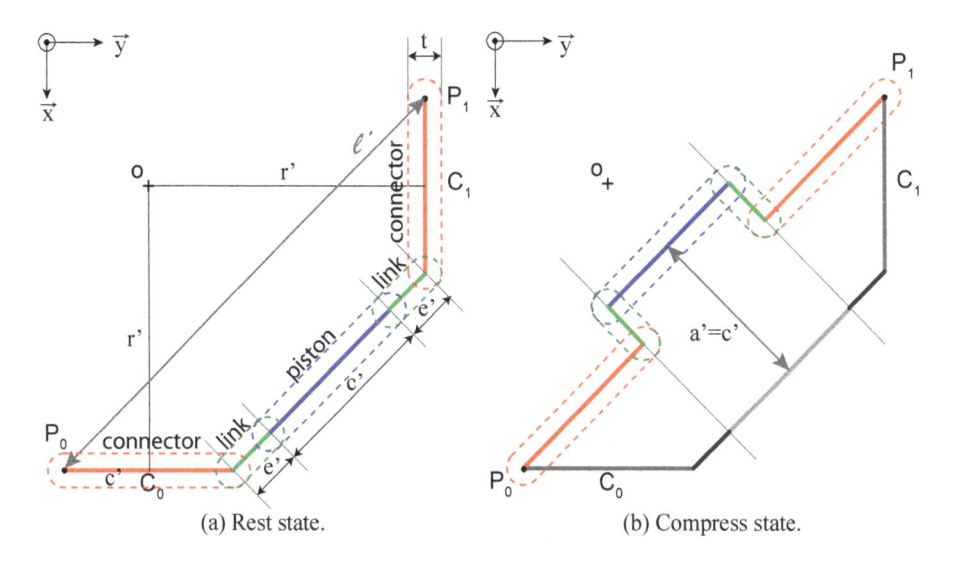

Fig. 4. Size and position of components taking into account of the thickness.

4 Motion Capabilities in an Ensemble

We now consider a configuration of several *Datoms* placed in a FCC lattice. The motion of a *Datom B* is possible around another connected *Datom A* (called the pivot) only. Considering two *Datoms* linked by a couple of connectors, it exists 2 available directions of motions, each one associated to a piston of the pivot. And we will show below that for each of these directions there is 3 possible motions ("*Turn left*", "*Go ahead*" and "*Turn right*"). Finally, each mobile *Datom* having 12 connectors, it implies that each *Datom* is able to make 72 different motions, but only a few of them are possible (or valid) at a time. Then we need a tool to detect the real list of valid motions depending on the neighborhood of the couple of *Datoms*.

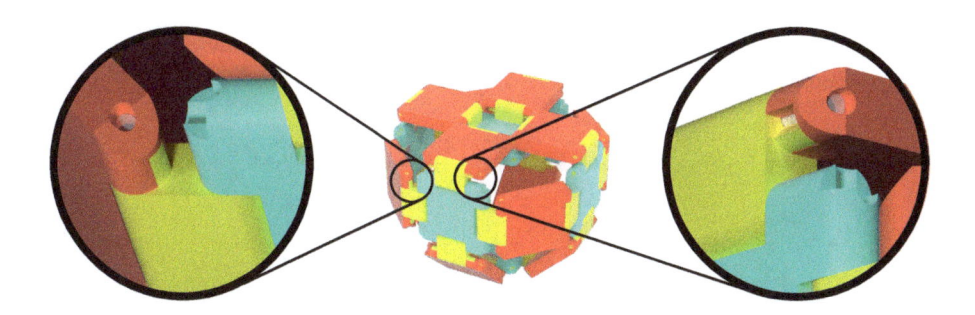

Fig. 5. Zooms on mechanical joints to show the angular blocking plots between the connector and the link parts.

The motion rules method proposed by Piranda et al. in [6] consists in defining a list of motions that are available for a considered module, taking into account several constraints in the neighboring cells of the lattice.

To simplify the explanation, we consider a couple of connectors of the *Datom B* and the pivot A that produces a motion in the plane $(A, \vec{x}, \vec{y})$, with the translation of the piston placed at the top of A. To simplify notations in the following, we use the same letter to name a free cell of the lattice and the *Datom* placed in the cell if it exists.

Definition 1. *A motion rule is a list of tuples (P,S) where P is a position in the grid relative to the pivot A and S is a status of the cell placed at position P. Status S can have one of the following values, or a combination of $\emptyset$ and one of these values:*

- *$\emptyset$, if the cell must be empty (no module at this position),*
- *a module name, if the cell must be filled by this Datom,*
- *def($\vec{X}$) if the cell must be filled by a deformed module, the deforming piston being oriented in the direction $\vec{X}$,*
- *def($\vec{X}, \vec{Y}$), if the cell must be filled by a Datom initially deformed along $\vec{X}$ axis and deformed along $\vec{Y}$ axis at the end of the motion.*

Theorem 1. *A motion rule is valid if all tuples of its list are validated by the current configuration. The Table 2 gives the list of tuples for each motion rules.*

Table 2 gives the list of tuples for the three possible motions of B around the pivot A using the deformation consisting in translating the piston of A defined by vector $\vec{U}$. The direction to the right of $\vec{BA}$ is defined by $\vec{R} = \vec{BA} \times \vec{U}$ and the front direction $\vec{F} = \vec{U} \times \vec{R}$ are expressed relatively to the positions of A and B (cf. Fig. 6d).

Every motion rule is defined relatively to a pivot A placed at the origin of the system and B at the top rear position of A. Using these relative directions we can express a motion rule for all the 72 different possible motions. We add the two contextual following rules $\{((0,0,0),A),(\vec{U}-\vec{F},B)\}$ that express that the position $(0,0,0)$ must be filled by A and the position of B is obtained adding the vector $(\vec{U}-\vec{F})$ to the position of A.

Figure 6 shows for each basic motion, an initial configuration that make it valid and every cells used by at least one motion rule tuple.

First, we consider the horizontal plane of cells containing the moving module B in yellow (level 0), which also contains H, I and J cells. Two of these cells are drawn in

Table 1. Position of *Datoms* concerned by the motion of B (relatively to B).

Cell	Position	Cell	Position	Cell	Position
A	$(1,1,-1)$	E	$(1,2,1)$	J	$(1,1,0)$
B	$(0,0,0)$	F	$(2,1,1)$	H	$(0,1,0)$
C	$(0,1,1)$	H	$(0,1,0)$	K	$(0,0,1)$
D	$(1,0,1)$	I	$(1,0,0)$		

(a) Turn left.	(b) Go ahead.	(c) Turn right.	(d) Motion rule axes.

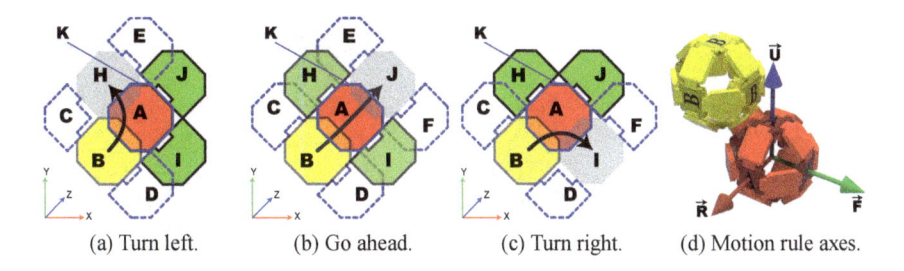

Fig. 6. The three possible motions of a module B linked to a pivot A to reach position G. For each motion we show the cells used by "motion rules" in the neighborhood of A. Last picture shows relative axes for a couple of *Datom* and a piston (here the top of A).

green and the third one corresponding to the goal cell is in grey. The pivot A (drawn in red) is placed in the underneath plane of cells (level −1). Also, we must take into account the cells placed on the plane above B: C, D, E, F and K placed on level 1.

Table 1 precises the coordinates of each *Datom* relatively to B, and Fig. 7a shows a 3D view of the "*Turn left*" initial configuration using the same color system.

Cells with dotted border of Fig. 6 may contain a module or be free but the goal cells must initially be free of module. At the top plane, the cell K placed over A must be free, while cells C, D, E and F may contain a module that must be deformed to free the path for B.

The three available motions are presented separately, Fig. 6a: "*Turn left*" with a goal placed at H, Fig. 6b: "*Go ahead*" with J as goal and Fig. 6c: "*Turn right*" allowing B to reach cell I. In the case of "*Turn left*" (resp. "*Turn right*") motion, if the cell C (resp. D) is filled, these corner *Datoms* must be deformed twice during the motion. The first deformation allows B to go on the top of A, then the deformation changes to allow B to reach its final position.

For example, in the first case (Fig. 6c), before moving module B to the goal cell I using A as a pivot, we must verify that the cell K on the plane on top of A is empty and then ask modules eventually placed at the C, D, F, H and J cells to deform themselves in order to free the path. Figure 7 shows some key steps of the "*Turn left*" motion of the module B in 3D and a video[2]: available on *YouTube* shows every details of the three different motions:

a) Initial configuration, *Datom B* plans to turn to left, it sends messages to the cells I, J, C, D and E to ask *Datoms* eventually filling these cells to free the path by deforming themselves (in the Figure, only I, J, D and E are represented).
b) *Datoms* I, J, D are now deformed, B can start the motion.
c) B is actuating synchronously with the pivot A to create the motion.
d) A and B are in the middle stage of the motion, they both change the connectors attachment. If there is a *Datom* in cell F, it changes its deformation to allow the final motion of B.
e) B reaches its final position, and asks C, D, E, F and H to release their deformation.
f) Final configuration.

---

[2] Motions of simulated and printed Datoms on video: https://youtu.be/EHuBtkRV4Jo.

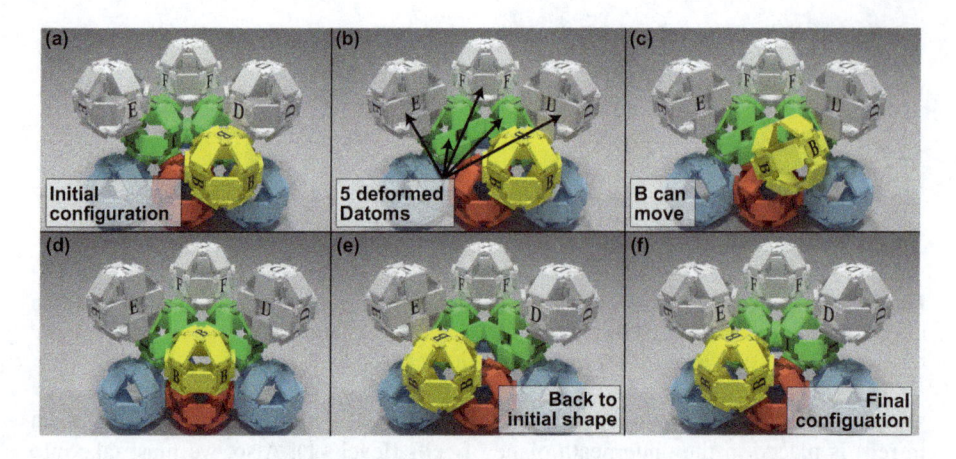

Fig. 7. Six steps of *turn left* displacements of the module B in a constrained configuration. a) Initial configuration, b) First the blocking modules must be deformed. c–e) Motion steps. f) Final configuration.

Table 2. Motion rules applied from the pivot

Rule	Tuples list by motion	Cell
All	$\{((0,0,0), A),$	A
	$(\vec{U} - \vec{F}, B),$	B
	$(2\vec{U}, \emptyset)\}$	K
Turn left	$\{(\vec{U} - \vec{R}, \emptyset),$	Goal
	$(\vec{U} + \vec{F}, \emptyset \vee \mathsf{def}(-\vec{F})),$	J
	$(\vec{U} + \vec{R}, \emptyset \vee \mathsf{def}(-\vec{R})),$	I
	$(2\vec{U} + \vec{R} - \vec{F}, \emptyset \vee \mathsf{def}(-\vec{R})),$	D
	$(2\vec{U} - \vec{R} - \vec{F}, \emptyset \vee \mathsf{def}(\vec{R}, \vec{F})),$	C
	$(2\vec{U} - \vec{R} + \vec{F}, \emptyset \vee \mathsf{def}(-\vec{F}))\}$	E
Turn right	$\{(\vec{U} + \vec{R}, \emptyset)$	Goal
	$(\vec{U} - \vec{R}, \emptyset \vee \mathsf{def}(\vec{R})),$	H
	$(\vec{U} + \vec{F}, \emptyset \vee \mathsf{def}(-\vec{F})),$	J
	$(2\vec{U} - \vec{R} - \vec{F}, \emptyset \vee \mathsf{def}(\vec{R})),$	C
	$(2\vec{U} + \vec{R} - \vec{F}, \emptyset \vee \mathsf{def}(-\vec{R}, \vec{F})),$	D
	$(2\vec{U} + \vec{R} + \vec{F}, \emptyset \vee \mathsf{def}(-\vec{F}))\}$	F
Go ahead	$\{(\vec{U} + \vec{F}), \emptyset$	Goal
	$(\vec{U} - \vec{R}, \emptyset \vee \mathsf{def}(\vec{R})),$	H
	$(\vec{U} + \vec{R}, \emptyset \vee \mathsf{def}(-\vec{R})),$	I
	$(2\vec{U} - \vec{R} - \vec{F}, \emptyset \vee \mathsf{def}(\vec{R})),$	C
	$(2\vec{U} + \vec{R} - \vec{F}, \emptyset \vee \mathsf{def}(-\vec{R})),$	D
	$(2\vec{U} + \vec{R} + \vec{F}, \emptyset \vee \mathsf{def}(-\vec{R})),$	F
	$(2\vec{U} - \vec{R} + \vec{F}, \emptyset \vee \mathsf{def}(\vec{R}))\}$	E

A particular case must be considered when C, D, E or H modules are only attached by one of the 4 connectors linked to the compressed piston. In this case, they must be removed first in order to make the motion of B possible.

5 Simulation

Simulations have been executed in *VisibleSim* [5], a modular robot simulator[3]. The goal of these experiments is to show that a *Datom* can reach every free positions at the surface of a configuration in successively applying the unitary motions presented above.

5.1 Algorithms

We implement a first algorithm that places a *Datom* at a goal cell G placed on the surface of a configuration. Then from this position, it computes every valid motions from this point testing successively the rules presented Table 2, the reached positions are memorized in every neighbor modules. According to Theorem 2, it exists a sequence of motions to go from each of these cells to G. More precisely, we implement a gradient algorithm that affects the distance 0 to the G cell, then the distance 1 is set to each cell which can reach the G cell after exactly one basic motion, and so on. It allows to define a gradient of distances to go from every reachable cells to G.

Theorem 2 (Bidirectional motions). *Each displacement is bidirectional, if a motion rule is valid to go from a cell X to a cell Y, it exists a valid motion rule to go from Y to X.*

Proof. If it exists a valid *"Go ahead"* motion rule to go from X to Y, as the motion constraints are symmetrical relatively to the up direction $\overrightarrow{U}$ of pivot A, the *"Go ahead"* motion rule will be valid for a motion from Y to X, using the same pivot A.

"Turn left" and *"Turn right"* motion rules are symmetrical relatively to the plane $(A, \overrightarrow{F}, \overrightarrow{U})$. If it exists a valid *"Turn left"* motion rule to go from X to Y, it exists a valid *"Turn right"* motion rule to go from Y to X, and reciprocally.

A second algorithm has been implemented to move a module B (with $ID = 1$) from one cell (a free cell of the border) to the goal position. The *Datom* B computes the list of reachable free cells from its current position. It then selects one of the cell which the minimum distance value. It sends a message to all modules that must be deformed to allow its motion and, after an acknowledgement applies the motion. And so on, until it reaches the goal cell which distance is 0.

5.2 Results

For this experiment, we construct a simulated configuration made of 141 *Datoms*. A $7 \times 7 \times 2$ box is covered by an obstacle making an arch whose hole is two *Datoms* high

---
[3] Video presentation of *VisibleSim*: https://youtu.be/N09KElCbUNk.

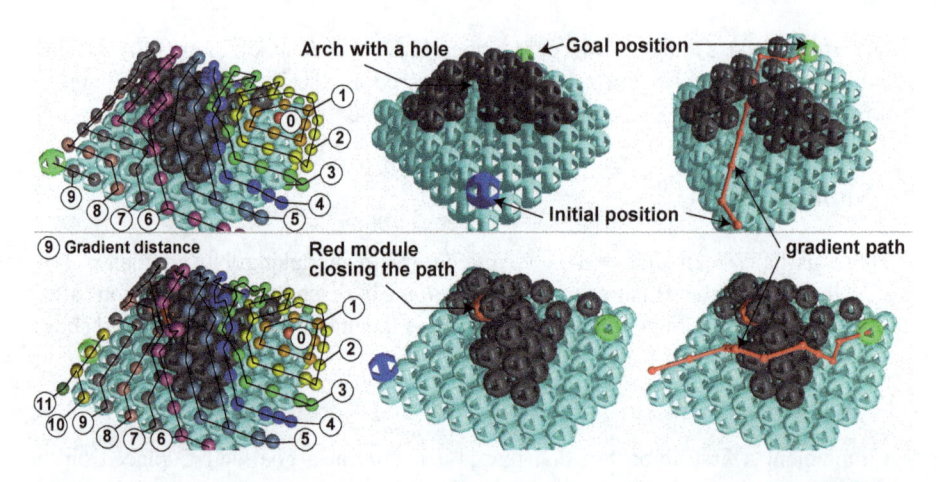

Fig. 8. Simulation results of the two algorithms (gradient and motion) on two similar configurations.

(cf. Fig. 8.a). And in a second time, we add a red *Datom* that reduces the size of the hole to one *Datom* high only (cf. Fig. 8.b).

For these two configurations, we calculate the distance from the position $G(6,5,2)$ in the lattice (the position of the green module) to all reachable cells. The distance of these cells is represented in the second screenshot (center) where the configuration is viewed from the top. Colored spheres are placed at the center of the cells reachable from the goal position (marked by the red sphere), the spheres' color represents the distance from the cell to the goal. We can observe that distances of cells at the left of the configuration are higher in the second case because the red *Datom* suppresses the shortcut passing through the arch.

The third screenshot presents the results of the second algorithm for the two cases. The red line shows the steps of the motion of the green module from the position $(0,0,2)$ marked by the blue *Datom* to the goal position (green *Datom*). In the top image, the *Datom* can pass under the arch while in the second image the path goes above the obstacle.

A video that shows the deformation of a 3D printed version of the *Datom* and some results obtained on the simulator is available on *YouTube*[4].

6 Conclusion

This work proposes a new model of deformable robot for programmable matter called a *Datom* which allows to realize safe motions in a FCC lattice. The size of the components relatively to the diameter of the robot and the angular limits between these pieces are precisely detailed for the realization of a real robot.

We study precisely how to implement the motion of a module in an ensemble to allow a module to step by step reach every free cell at the surface of a configuration.

---
[4] *YouTube* video: https://youtu.be/EHuBtkRV4Jo.

These motions are possible if many other modules collaborate and must synchronize their own deformation, in order to free the path for another one.

In order to build a mm-scale *Datom*, we envision the following steps. First, we will use our microtechnology center to produce the 3D printed structure of the *Datom*. This structure can fully be printed without further assembly as the hinges can be directly printed. The latching can use the same electrostatic electrodes developed for the *3D Catom* [7] and the deformation could be obtained using a soft bimorph actuator.

Acknowledgment. This work was partially supported by the ANR (ANR-16-CE33-0022-02), the French Investissements d'Avenir program, the ISITE-BFC project (ANR-15-IDEX-03), and the EIPHI Graduate School (contract ANR-17-EURE-0002).

References

1. Bourgeois, J., et al.: Programmable matter as a cyber-physical conjugation. In: IEEE International Conference on Systems, Man, and Cybernetics (SMC), pp. 002,942–002,947. Budapest, Hungary (2016)
2. Chirikjian, G.S.: Kinematics of a metamorphic robotic system. In: IEEE International Conference on Robotics and Automation (ICRA), pp. 449–455. IEEE (1994)
3. Fukuda, T., Kawauchi, Y., Buss, M.: Communication method of cellular robotics cebot as a selforganizing robotic system. In: IEEE/RSJ International Workshop on Intelligent Robots and Systems (IROS), pp. 291–296. IEEE (1989)
4. Goldstein, S.C., Mowry, T.C.: Claytronics: an instance of programmable matter. In: Wild and Crazy Ideas Session of ASPLOS. Boston, MA (2004)
5. Piranda, B.: VisibleSim: your simulator for programmable matter. In: Algorithmic Foundations of Programmable Matter (Dagstuhl Seminar 16271). Dagstuhl (2016)
6. Piranda, B., Bourgeois, J.: A distributed algorithm for reconfiguration of lattice-based modular self-reconfigurable robots. In: PDP 2016, 24th Euromicro International Conference on Parallel, Distributed, and Network-Based Processing, pp. 1–9. Heraklion Crete, Greece (2016)
7. Piranda, B., Bourgeois, J.: Designing a quasi-spherical module for a huge modular robot to create programmable matter. Auton. Robot. **42**(8), 1619–1633 (2018). https://doi.org/10.1007/s10514-018-9710-0
8. Rus, D., Vona, M.: Crystalline robots: self-reconfiguration with compressible unit modules. Auton. Robot. **10**(1), 107–124 (2001)
9. Støy, K., Brandt, D., Christensen, D.J., Brandt, D.: Self-reconfigurable Robots: An Introduction. MIT Press, Cambridge (2010)
10. Suh, J.W., Homans, S.B., Yim, M.: Telecubes: mechanical design of a module for self-reconfigurable robotics. In: ICRA, vol. 4, pp. 4095–4101 (2002)
11. Swissler, P., Rubenstein, M.: Fireant: a modular robot with full-body continuous docks. In: Proceedings of the 2018 IEEE International Conference on Robotics and Automation (2018)
12. Yim, M., et al.: Modular self-reconfigurable robot systems [grand challenges of robotics]. IEEE Robot. Autom. Mag. **14**(1), 43–52 (2007)

The Impact of Network Connectivity on Collective Learning

Michael Crosscombe[✉] and Jonathan Lawry

Department of Engineering Mathematics, University of Bristol, Bristol BS8 1UB, UK
{m.crosscombe,j.lawry}@bristol.ac.uk

Abstract. In decentralised autonomous systems it is the interactions between individual agents which govern the collective behaviours of the system. These local-level interactions are themselves often governed by an underlying network structure. These networks are particularly important for collective learning and decision-making whereby agents must gather evidence from their environment and propagate this information to other agents in the system. Models for collective behaviours may often rely upon the assumption of total connectivity between agents to provide effective information sharing within the system, but this assumption may be ill-advised. In this paper we investigate the impact that the underlying network has on performance in the context of collective learning. Through simulations we study small-world networks with varying levels of connectivity and randomness and conclude that totally-connected networks result in higher average error when compared to networks with less connectivity. Furthermore, we show that networks of high regularity outperform networks with increasing levels of random connectivity.

Keywords: Collective learning · Small-world networks · Multi-agent systems

1 Introduction

Reasoning about collective behaviours in autonomous systems can be difficult when the system-level behaviour emerges from local-level interactions, i.e., between individual agents. Due to the complexity of such systems we often rely on a series of modelling assumptions to effectively reason about their resulting dynamics. One such assumption is the "well-stirred system" assumption, which we restate to be the following: In a well-stirred system each agent is equally likely to encounter any other agent in the system and, therefore, each interaction is regarded as an independent event [20]. In other words, in a well-stirred system we treat agents as nodes in a totally-connected network or complete graph and the stochastic interactions of agents can hence be modelled by selecting edges at random from the network. While this assumption captures the notion that agents conducting random walks are likely to encounter one another at random, e.g., whilst exploring an environment, the impact of this assumption is not well-understood. Alternative network structures, such as small-world networks [29], have been studied in both artificial and biological systems [17,18] and provide a means of varying the connectivity of the network while preserving desirable properties such as short, consistent path-lengths.

In this paper we will study consensus formation in a collective learning setting, in which agents learn both from evidence gathered directly from the environment as well as indirectly from one another, through a process of belief fusion. In this context we shall investigate the impact of the underlying network structure on collective learning by studying small-world networks with varying degrees of connectivity and stochasticity. The rest of this paper is structured as follows: In Sect. 2 we provide an overview of the relevant literature on various formulations of collective learning as well as small-world networks. In Sect. 3 we describe a propositional approach to collective learning in which agents attempt to learn a full state description of the environment, followed by a brief overview of small-world networks in Sect. 4. Then, in Sect. 5 we detail simulation experiments in which we study small-world networks of varying degrees of connectivity and random rewiring. Finally, in Sect. 6 we close with some discussions and conclusions.

2 Related Work

Consensus formation within a large group of individuals has long been studied in the form of 'opinion pooling' since the works of Stone [25] and DeGroot [7]. These early works considered the (usually linear) weighted fusion of beliefs between 'experts' initialised with prior beliefs and without considering the influence of external evidence. Instead of learning a description of the world, models of opinion dynamics are typically concerned with a single proposition of contention and beliefs are then denoted by a single real value that represents an agent's certainty in the truth or falsity of the proposition [14]. Agents simply conduct weighted combinations of their beliefs under interaction limitations imposed by concepts such as bounded confidence [6,13]. Three-valued representations of beliefs have also been studied in the context of opinion dynamics in [1], which thresholds the underlying real value into three states, and in the context of consensus formation in [22], which proposes a three-valued fusion operator that assigns the third truth state to resolve conflict between two strictly opposing opinions.

Collective learning is the combination of two distinct processes: belief fusion between agents; and the updating of beliefs based on evidence gathered from the environment. The interaction of these two processes in the context of social epistemology has been explored in [9], in which the argument is made that communication between agents not only acts to propagate information more effectively through the system but also provides an error-correcting effect when the evidence being gathered may be erroneous. Further studies of this effect in multi-agent systems and swarm robotics can be found in [4,8,16]. In a robotics setting this evidence may take the form of signals received by the robots' onboard sensors. For example, in [27] experiments are conducted in which Kilobots use their ambient light sensors to determine the quality (i.e., light intensity) of a particular location [23].

Due to recent developments in multi-robot systems and swarm robotics, the increasing viability of their deployment outside of the lab has led to a surge of interest in the development and understanding of collective behaviours [3,24]. Many of the current solutions are based in nature, e.g., the house-hunting behaviours studied in ants and honeybees [11] are instances of the best-of-n problem [21,26]. It is in this context that [20] first proposed the adoption of the well-stirred system assumption which is to be found, implicitly or explicitly, in many recent models for collective learning [5,15,16]. However, there has been increasing effort to understand the impact of the underlying network topology on collective behaviours in multi-agent systems. In [19] the authors compare the effects that different network topologies, including small-world networks [29] and random networks [10], have on the consensus formation process. In the context of consensus formation, Baronchelli [2] investigates convergence dynamics for various network topologies. In the broader context of distributed systems, Hamann [12] identifies the common properties that determine system performance across several domains, from parallel computing to swarm robots to network science. Specifically, one such property is the system's underlying network topology which dictates the amount of information being shared across the system. Hamann finds that performance degrades both when there is too little or too much information being propagated through the system, i.e., when connectivity is too low or too high, respectively.

3 A Propositional Model for Collective Learning

Consider a collective learning problem whereby a population of agents are attempting to reach a consensus about the true state of their environment. Let us suppose that such an environment can be described by a set of n propositional variables $\mathcal{P} = \{p_1, \ldots, p_n\}$ and that an agent's belief about the environment (i.e., a possible world) is an assignment of truth values to each of the propositional variables denoted by $b : \mathcal{P} \to \{0, \frac{1}{2}, 1\}^n$. Here we adopt a three-valued propositional logic with the truth values 0 and 1 corresponding to false and true, respectively, while the third truth value $\frac{1}{2}$ denotes *unknown*. This additional truth value allows for agents to express uncertainty in their beliefs about the world. For notational convenience, let us represent an agent's belief B by the n-tuple $\langle B(p_1), \ldots, B(p_n) \rangle$. Then an agent's belief may express uncertainty about the world by the assignment of the truth value $\frac{1}{2}$ to any of the propositional variables in $\mathcal{P}$. For example, for $n = 2$ propositional variables the belief $B = \langle 1, 0 \rangle$ expresses an agent's belief that p_1 is true while p_2 is false. We can say that this belief expresses absolute certainty in the state of the world. Alternatively, the belief $B' = \langle \frac{1}{2}, \frac{1}{2} \rangle$ expresses an agent's uncertainty about both of the propositions p_1 and p_2, thus indicating the agent's total lack of certainty regarding the state of the world.

We propose to combine the beliefs of agents in a pairwise manner as follows: the fusion operator $\odot$ is a binary operator defined on $\{0, \frac{1}{2}, 1\}$ as given in Table 1. This operator is applied element-wise to all of the propositional variables p_i in

Table 1. The fusion operator $\odot$ applied to beliefs B and B'.

$B(p_i)$	$\odot$	0	$\frac{1}{2}$	1
	0	0	0	$\frac{1}{2}$
	$\frac{1}{2}$	0	$\frac{1}{2}$	1
	1	$\frac{1}{2}$	1	1

with column header $B'(p_i)$.

$\mathcal{P}$ so that, given two beliefs B and B' corresponding to the beliefs of two agents, the fused belief is given by

$$B \odot B' = \langle B(p_1) \odot B'(p_1), \ldots, B(p_n) \odot B'(p_n) \rangle. \tag{1}$$

A pairwise consensus is thus formed by both agents adopting the fused belief $B \odot B'$.

Evidential updating is the process by which an agent selects a proposition (e.g., a feature of its environment) to investigate and, upon receiving evidence, updates its belief to reflect this evidence. Firstly, to decide which proposition to investigate an agent selects a single proposition at random from the set of propositions about which they are uncertain, i.e., where $B(p_i) = \frac{1}{2}$. Having selected a proposition p_i to investigate, an agent then receives evidence with probability r or learns nothing with probability $1 - r$, where r is an evidence rate quantifying the sparsity of evidence in the environment. Evidence takes the form of an assertion about the truth value of the chosen proposition p_i as follows: $E = \langle \frac{1}{2}, \ldots, S^*(p_i), \ldots, \frac{1}{2} \rangle$ where $S^* : \mathcal{P} \to \{0, 1\}^n$ denotes the true state of the world. Secondly, upon gathering evidence E, the agent then updates its belief B to $B|E$ using the same fusion operator given in Table 1 such that

$$\begin{aligned} B|E &= \langle B(p_1) \odot E(p_1), \ldots, B(p_i) \odot E(p_i), \ldots, B(p_n) \odot E(p_n) \rangle \\ &= B \odot E. \end{aligned} \tag{2}$$

Notice that we are also using the fusion operator $\odot$ to update beliefs based on evidence and that updating in this manner does not therefore alter the truth values for the propositions $p_j \in \mathcal{P}$ where $p_j \neq p_i$ because $E(p_j) = \frac{1}{2}$. An agent repeats this process of gathering evidence until the set of propositions about which it is uncertain is empty, or rather that it holds a belief of total certainty, at which point it chooses to stop looking for evidence. Also notice that while evidential updating in this manner can only lead to agents becoming more certain in their beliefs, the process of agents combining their beliefs via the fusion operator $\odot$ can also lead to agents becoming more uncertain when the fusing agents disagree about the truth value of a given proposition. For example, supposing that $B_1(p_i) = 1$ and $B_2(p_i) = 0$, then upon the agents fusing their beliefs such that $B_1 \odot B_2(p_i) = \frac{1}{2}$, both agents will attempt to seek additional evidence about proposition p_i, either directly from the environment or indirectly via fusion with other agents, having become uncertain about the truth value of p_i.

Let us now assume that the evidence gathering process may be noisy (e.g., due to sensor noise or a noisy environment). Evidence shall then take the following form:

$$E(p_i) = \begin{cases} S^*(p_i) & : \text{ with probability } 1-\epsilon \\ 1 - S^*(p_i) & : \text{ with probability } \epsilon \end{cases} \qquad (3)$$

where $\epsilon \in [0, 0.5]$ is a noise parameter denoting the probability that the evidence is erroneous.

To measure the performance of a given population in the context of a collective learning problem we introduce a measure of the average error of the population.

Definition 1. *Average error*
The average error of a population of m agents is the normalised difference between each agent's belief B and the true state of the world S^* averaged across the population as follows:

$$\frac{1}{m}\frac{1}{n}\sum_{j=1}^{m}\sum_{i=1}^{n}|B_j(p_i) - S^*(p_i)|.$$

As mentioned previously, we often adopt the well-stirred system assumption which states that an interaction between any two agents is an independent event and is therefore equally likely for any pair of agents in the population [20]. In the network agents are represented by nodes and the existence of edges between them represents the ability of the agents to communicate directly with one another. In networks with less-than-total connectivity the lack of an edge between two agents means that they cannot communicate directly, although information may still be shared via other agents in the population through the process of belief fusion. In the following section we introduce small-world networks to study the impact that the underlying network structure has on the collective learning process and to challenge the well-stirred system assumption.

4 Small-World Networks

The network topology of many real-world systems often lies somewhere between total regularity and total randomness, as discussed by Watts and Strogatz in their seminal work introducing small-world networks [28,29]. Indeed the concept of a small-world network – in which nodes are connected to their k nearest neighbouring nodes – more accurately reflects the kinds of networks that emerge in both natural and artificial self-organising systems, with examples in social networks [18] and neural networks [17]. Many social systems, both biological and engineered, exhibit these small-world dynamics due to their spatial properties. For example in swarm robotics, where each embodied agent is typically composed of low-cost hardware and consequently possesses a limited radius within which it can communicate [23], the distance between agents determines the connectivity

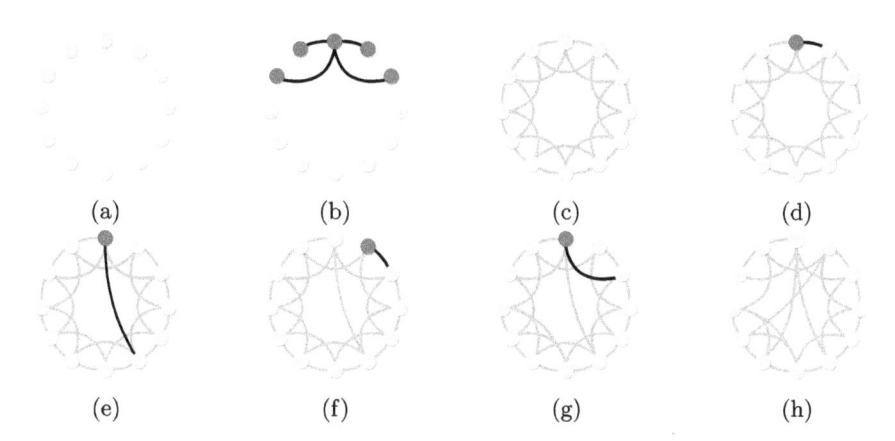

Fig. 1. The generation of a Watts-Strogatz small-world graph illustrated for $m = 12, k = 4$ and $\rho = 0.2$.

of the underlying communication network which will often resemble a small-world network.

Small-world networks are parameterised by two variables k and ρ, where k denotes the number of nearest neighbours to which each node in the graph is connected and ρ denotes the probability of rewiring an existing edge to a different agent. A small-world graph, as illustrated in Fig. 1, is generated in the following way: (a) begin with n vertices ordered in a ring; (b) for each vertex in the graph, connect it to its k nearest neighbours until; (c) a regular small-world graph is formed. (d) For each vertex (moving clockwise around the ring), select the edge connected to its clockwise-nearest neighbour; (e) with probability ρ reconnect that edge to another vertex selected uniformly at random, unless doing so produces a duplicate edge. (f) Continue this process clockwise for all vertices. (g) Repeat this process for their second-nearest clockwise neighbour until; (h) each original edge in the graph has been considered once. Notice that a small-world network with $k = (m-1)$ is equivalent to a totally-connected network in which each agent is connected to every other agent.

In the following section we describe simulation experiments applying the model in Sect. 3 to small-world networks with varying degrees of connectivity and regularity.

5 Agent-Based Simulations

We study a collective learning scenario in which the environment can be described by $n = 100$ propositions. Without loss of generality we define the true state of the world S^* to be $\langle 1, 1, 1, \ldots, 1 \rangle$. For each experiment we initialise a population of $k = 100$ agents holding totally ignorant beliefs, i.e., at time step $t = 0$ each agent holds the belief $B(p_i) = \frac{1}{2}$ for $i = 1, \ldots, 100$. By Definition 1 the average error of such a belief, representing complete uncertainty, is 0.5 and

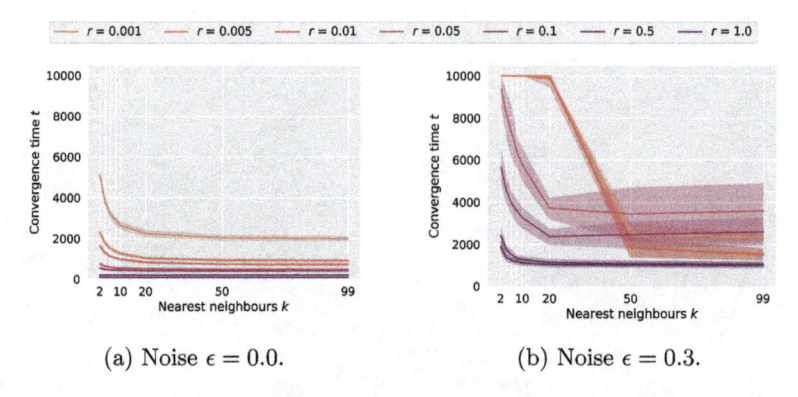

Fig. 2. Convergence time for a population of 100 agents across nearest neighbours k. Each line depicts a different evidence rate $r \in [0.01, 1]$.

therefore each population shall begin with an average error of 0.5. Furthermore, should a population converge on the true state of the world S^* then the average error shall be 0. A population-wide evidence rate $r \in (0, 1]$ determines the frequency with which agents successfully obtain evidence from the environment. In other words, during each time step every agent has an equal probability r of updating their belief based on evidence. As described previously, each piece of evidence pertains to a single proposition about which the investigating agent is uncertain, i.e., where $B(p_i) = \frac{1}{2}$. This evidence is also likely to be noisy with $\epsilon \in [0, 0.5]$ denoting the probability that a piece of evidence is incorrect. Notice that for $\epsilon = 0.5$ the evidence becomes random with an equal probability of learning that the investigated proposition is either true or false. Finally, in addition to evidential updating, at each time step one edge in the graph is selected at random and the connected pair of agents combine their beliefs using the fusion operator defined in Table 1.

For a given set of parameter values we average the results over 100 independent runs and each run takes a maximum of 10,000 time steps, or until convergence occurs. Here we define convergence as the beliefs of the population remaining unchanged for 100 interactions, where an interaction is updating either based on evidence or on the beliefs of other agents, i.e., via fusion. For line plots, the shaded regions represent 10th and 90th percentiles.

5.1 Convergence Results for Regular Small-World Networks

In this section we show convergence results for regular small-world networks without random rewiring (i.e., where $\rho = 0$) as depicted in Fig. 1c. Figure 2 shows the average convergence time across different networks with nearest neighbours k for different levels of noise ϵ. Each solid line represents a different evidence rate r between 0.001 and 1. For $\epsilon = 0$ in Fig. 2a we depict a noise-free scenario in which evidence is always accurate. Broadly, we see that time to convergence decreases as the network connectivity k increases and that a totally-connected network

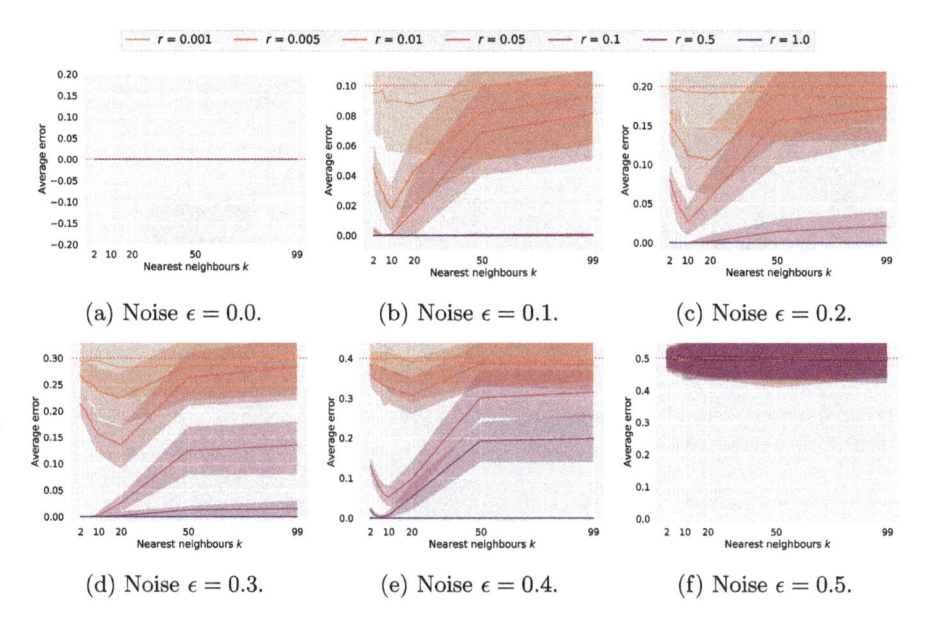

The Impact of Network Connectivity on Collective Learning 89

Fig. 3. Average error of a population of 100 agents for nearest neighbours k. Each solid line depicts a different evidence rate $r \in [0.01, 1]$, while the red dotted line depicts the noise level ϵ.

results in the fastest convergence times for all evidence rates r. Additionally, we see that the average convergence time decreases as the evidence rate r increases. This is to be expected as the greater the frequency with which the population receives evidence, the faster the agents learn about their environment. In Fig. 2b we see that for a moderately noisy environment with $\epsilon = 0.3$ and lower evidence rates $r \leq 0.01$ the population no longer converges in under 10,000 time steps for networks of connectivity $k < 50$. For networks with greater connectivity, i.e., $k \geq 50$, the population once again reaches a steady state in under 3,000 time steps on average. For higher evidence rates $r \geq 0.05$ we see again that convergence time decreases as both the evidence rate r and network connectivity k increases. We also see that for these higher evidence rates, increasing network connectivity k to beyond 20 nearest neighbours does not lead to further reductions in convergence time. While Fig. 2 demonstrates the ability of the model to reach a steady state under certain conditions, it does not demonstrate the learning accuracy of our model. To this end, we now present results showing the average error of the population according to Definition 1.

In Fig. 3 we show the average error of the population at steady state against the number of nearest neighbours k in the network. Figure 3a shows that when the system is free from noise (i.e., when $\epsilon = 0$) the model always converges on the true state of the world with 0 average error at steady state. This is true for all values of k and all evidence rates r including $r = 0.001$ which, for a population of 100 agents, corresponds to the population receiving a single piece of evidence

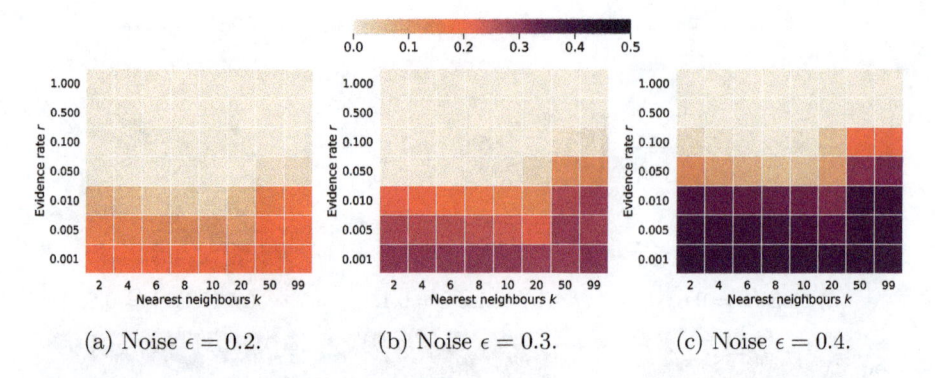

(a) Noise $\epsilon = 0.2$. (b) Noise $\epsilon = 0.3$. (c) Noise $\epsilon = 0.4$.

Fig. 4. Average error of a population of 100 agents for both evidence rates $r \in [0.01, 1]$ and nearest neighbours $k = 2, \ldots, 99$.

once every 10 time steps on average. However, in a scenario in which evidence is always accurate, these results are to be expected. Another edge case is when $\epsilon = 0.5$, resulting in agents always receiving random evidence. Unsurprisingly, as shown in Fig. 3f, the population always converges on an average error of around ϵ due to its inability to receive informative evidence for any particular proposition.

Figures 3b to 3e show the average error of the population for increasing levels of noise ϵ from 0.1 to 0.4. Compared with the noise-free scenario, the connectivity of the network has a clear impact on the learning accuracy of our model when agents encounter noisy evidence. For all noise levels ϵ in this range we see that the model performs well with respect to accuracy as the population consistently achieves an average error at or below ϵ for all evidence rates r. While the population does not always learn the true state of the world, the average error of the population can be significantly reduced below ϵ by adopting a less connected network. Indeed, for a totally-connected network, i.e., where $k = 99$, the average error is very often greater than networks with lower connectivity. In the extreme cases, with evidence rates $r \geq 0.5$ the population always learns the true state of the world for $\epsilon < 0.5$.

Alternatively, in Fig. 4 we show heatmaps of the average error at steady state for evidence rate r and the number of nearest neighbours k. Focussing on a range of ϵ from 0.2 to 0.4 we see again that small-world networks with less connectivity outperform networks with greater connectivity, including totally-connected networks. Broadly, given a noisy scenario and an environment with sparse evidence, there is a network with connectivity k that outperforms networks of both lesser and greater connectivity. For environments with higher evidence rates r, a less connected small-world network improves the accuracy of collective learning.

(a) $k = 2$. (b) $k = 6$. (c) $k = 10$.

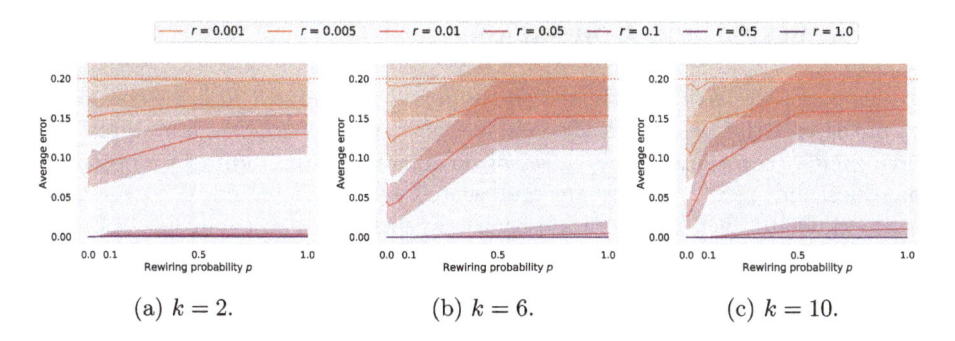

Fig. 5. Average error of a population of 100 agents against rewiring probability $\rho \in [0, 1]$. Each solid line depicts a different evidence rate $r \in [0.01, 1]$ with noise $\epsilon = 0.2$, while the red dotted line depicts the noise level ϵ.

5.2 Convergence Results for Small-World Networks with Random Rewiring

Having studied the primary parameter k of small-world networks in Sect. 5.1, we now study the secondary parameter associated with small-world networks: the rewiring probability ρ. This randomising parameter reduces the regularity of small-world networks by rewiring connections between neighbouring nodes to agents of greater separation in the network. The purpose of this parameter is to introduce additional connections (or 'paths') in the network that, for small values of k, are likely to improve information propagation in the network.

Figure 5 shows the average error of the population at steady state with moderate noise $\epsilon = 0.2$ and for different rewiring probabilities ρ between 0 and 1. In Fig. 5a we see that for $k = 2$, corresponding to a ring network in which agents are connected to their two nearest neighbours only, the average error of the population increases with ρ for r from 0.005 to 0.05 while remaining stable for very low and very high evidence rates, i.e., for $r = 0.001$ and $r \geq 0.1$. For example, with $r = 0.01$, when the network is totally regular with $\rho = 0$, the average error is 0.081, while for $\rho = 1$, a network with totally random connectivity, the average error increases by 60% to 0.130. The same effect is broadly observed in Figs. 5b and 5c for the same range of evidence rates, except that random rewiring increases the average error of the population to a greater extent in small-world networks with greater connectivity. For example, in Fig. 5c with $k = 10$ and an evidence rate $r = 0.01$, when the network is totally regular (i.e., when $\rho = 0$) the average error of the population is 0.027. When $\rho = 0.1$, the average error increases to 0.085 while for total randomness (i.e., when $\rho = 1$) the average error is 0.162 which is a 600% increase in average error compared with the $\rho = 0$ case.

As the network connectivity is altered from total regularity to total randomness, i.e., from $\rho = 0$ to 1, respectively, the population consistently performs worse when attempting to learn the true state of the world, with the average error of the population increasing as ρ increases. It is clear, therefore, that irregularity in the network negatively impacts performance in a collective learning setting.

6 Discussion and Conclusion

In this paper we have investigated the importance of considering the connectivity of the underlying network for collective learning in autonomous systems. The environment is represented by a set of descriptors in the form of propositional variables and agents' beliefs are represented by an allocation of three-valued truth values to each of the propositions. Agents adopt a combined process of evidential updating, learning directly from the environment, and belief fusion, combining their beliefs with other agents to form a pairwise consensus while correcting for inconsistencies that have arisen from noisy evidence. In this context, we have studied how the structure of the underlying network impacts the dynamics of a system of 100 agents learning the state of the world for a range of scenarios with varying levels of evidence sparsity and noise.

We have shown that a less-connected small-world network leads to greater accuracy on a collective learning task when compared with a totally-connected network. Through simulation studies our results show that, when the evidence in the environment is both sparse and noisy, then a network with moderate connectivity k outperforms networks with lower or higher connectivity for some combination of evidence rate r and noise level ϵ. Broadly, the optimal level of connectivity is always lower than total connectivity when the underlying network retains high regularity, i.e., $\rho \approx 0$. As the network connectivity becomes increasingly random, i.e., as $\rho \to 1$, the accuracy of the system decreases.

Acknowledgements. This work was funded and delivered in partnership between Thales Group, University of Bristol and with the support of the UK Engineering and Physical Sciences Research Council, ref. EP/R004757/1 entitled "Thales-Bristol Partnership in Hybrid Autonomous Systems Engineering (T-B PHASE)."

References

1. Balenzuela, P., Pinasco, J.P., Semeshenko, V.: The undecided have the key: interaction-driven opinion dynamics in a three state model. PLoS ONE **10**(10), 1–21 (2015)
2. Baronchelli, A.: The emergence of consenus: a primer. R. Soc. Open Sci. **5**(2), 172,189 (2018)
3. Brambilla, M., Ferrante, E., Birattari, M., Dorigo, M.: Swarm robotics: a review from the swarm engineering perspective. Swarm Intell. **7**(1), 1–41 (2013)
4. Crosscombe, M., Lawry, J.: A model of multi-agent consensus for vague and uncertain beliefs. Adapt. Behav. **24**(4), 249–260 (2016)
5. Crosscombe, M., Lawry, J., Hauert, S., Homer, M.: Robust distributed decision-making in robot swarms: exploiting a third truth state. In: 2017 IEEE/RSJ International Conference on Intelligent Robots and Systems (IROS), pp. 4326–4332 (2017)

6. Dabarera, R., Wickramarathne, T.L., Premaratne, K., Murthi, M.N.: Achieving consensus under bounded confidence in multi-agent distributed decision-making. In: 2019 22th International Conference on Information Fusion (FUSION), pp. 1–8 (2019)
7. DeGroot, M.H.: Reaching a consensus. J. Am. Stat. Assoc. **69**(345), 118–121 (1974)
8. Douven, I.: Optimizing group learning: an evolutionary computing approach. Artif. Intell. **275**, 235–251 (2019)
9. Douven, I., Kelp, C.: Truth approximation, social epistemology, and opinion dynamics. Erkenntnis **75**(2), 271 (2011)
10. Erdös, P., Rényi, A.: On random graphs. Publ. Math. (Debrecen) **6**, 290–297 (1959)
11. Franks, N., Pratt, S., Mallon, E., Britton, N., Sumpter, D.: Information flow, opinion-polling and collective intelligence in house-hunting social insects. Philos. Trans. B Biol. Sci. **357**(1427), 1567–1583 (2002)
12. Hamann, H.: Superlinear scalability in parallel computing and multi-robot systems: shared resources, collaboration, and network topology. In: Berekovic, M., Buchty, R., Hamann, H., Koch, D., Pionteck, T. (eds.) ARCS 2018. LNCS, vol. 10793, pp. 31–42. Springer, Cham (2018). https://doi.org/10.1007/978-3-319-77610-1_3
13. Hegselmann, R., Krause, U.: Opinion dynamics driven by various ways of averaging. Comput. Econ. **25**(4), 381–405 (2005)
14. Hegselmann, R., Krause, U., et al.: Opinion dynamics and bounded confidence models, analysis, and simulation. J. Artif. Soc. Soc. Simul. **5**(3) (2002)
15. Lawry, J., Crosscombe, M., Harvey, D.: Epistemic sets applied to best-of-n problems. In: Kern-Isberner, G., Ognjanović, Z. (eds.) ECSQARU 2019. LNCS (LNAI), vol. 11726, pp. 301–312. Springer, Cham (2019). https://doi.org/10.1007/978-3-030-29765-7_25
16. Lee, C., Lawry, J., Winfield, A.: Negative updating combined with opinion pooling in the best-of-n problem in swarm robotics. In: Dorigo, M., Birattari, M., Blum, C., Christensen, A.L., Reina, A., Trianni, V. (eds.) ANTS 2018. LNCS, vol. 11172, pp. 97–108. Springer, Cham (2018). https://doi.org/10.1007/978-3-030-00533-7_8
17. Masuda, N., Aihara, K.: Global and local synchrony of coupled neurons in small-world networks. Biol. Cybern. **90**, 302–9 (2004)
18. Newman, M.E.J., Watts, D.J., Strogatz, S.H.: Random graph models of social networks. Proc. Natl. Acad. Sci. **99**(suppl 1), 2566–2572 (2002)
19. Olfati-Saber, R., Fax, J.A., Murray, R.M.: Consensus and cooperation in networked multi-agent systems. Proc. IEEE **95**(1), 215–233 (2007)
20. Parker, C.A.C., Zhang, H.: Cooperative decision-making in decentralized multiple-robot systems: the best-of-n problem. IEEE/ASME Trans. Mechatron. **14**(2), 240–251 (2009)
21. Parker, C.A.C., Zhang, H.: Biologically inspired collective comparisons by robotic swarms. Int. J. Robot. Res. **30**(5), 524–535 (2011)
22. Perron, E., Vasudevan, D., Vojnović, M.: Using three states for binary consensus on complete graphs. In: Proceedings - IEEE INFOCOM, pp. 2527–2535 (2009)
23. Rubenstein, M., Ahler, C., Nagpal, R.: Kilobot: a low cost scalable robot system for collective behaviors. In: 2012 IEEE International Conference on Robotics and Automation, pp. 3293–3298 (2012)
24. Schranz, M., Umlauft, M., Sende, M., Elmenreich, W.: Swarm robotic behaviors and current applications. Front. Robot. AI **7**, 36 (2020)
25. Stone, M.: The opinion pool. Ann. Math. Stat. **32**(4), 1339–1342 (1961)
26. Valentini, G., Ferrante, E., Dorigo, M.: The best-of-n problem in robot swarms: formalization, state of the art, and novel perspectives. Front. Robot. AI **4**, 9 (2017)

27. Valentini, G., Hamann, H., Dorigo, M.: Efficient decision-making in a self-organizing robot swarm: on the speed versus accuracy trade-off. In: Proceedings of the 2015 International Conference on Autonomous Agents and Multiagent Systems, AAMAS 2015, pp. 1305–1314. International Foundation for Autonomous Agents and Multiagent Systems, Richland, SC (2015)
28. Watts, D.J.: Small Worlds: The Dynamics of Networks between Order and Randomness. Princeton University Press, USA (2003)
29. Watts, D.J., Strogatz, S.H.: Collective dynamics of small-world networks. Nature **393**(June), 440–442 (1998)

On the Communication Requirements of Decentralized Connectivity Control
A Field Experiment

Jacopo Panerati[1(✉)], Benjamin Ramtoula[1], David St-Onge[2], Yanjun Cao[1], Marcel Kaufmann[1], Aidan Cowley[3], Lorenzo Sabattini[4], and Giovanni Beltrame[1]

[1] Department of Software and Computer Engineering, Polytechnique Montréal, Montréal, QC, Canada
jacopo.panerati@polymtl.ca
[2] Department of Mechanical Engineering, École de technologie supérieure, Montréal, QC, Canada
[3] European Astronaut Centre, European Space Agency, Cologne, Germany
[4] Department of Sciences and Methods for Engineering, Università degli Studi di Modena e Reggio Emilia, Reggio Emilia, Italy

Abstract. Redundancy and parallelism make decentralized multi-robot systems appealing solutions for the exploration of extreme environments. However, effective cooperation can require team-wide connectivity and a carefully designed communication. Several recently proposed decentralized connectivity maintenance approaches exploit elegant algebraic results drawn from spectral graph theory. Yet, these proposals are rarely taken beyond simulations or laboratory implementations. The contribution of this work is two-fold: (i) we describe the full-stack implementation—from hardware to software—of a decentralized control law for robust connectivity maintenance; and (ii) we assess, in the field, our robots' ability to correctly exchange the information required to execute it.

1 Introduction

Multi-robot systems can be used to tackle complex problems that benefit from physical parallelism and the inherent fault-tolerance provided by redundancy—surveillance, disaster recovery, and planetary exploration being a few notable examples. Decentralized control strategies further improve the reliability of these systems by partially relaxing communication bandwidth requirements and eliminating the risks posed by single points of failure.

For many multi-robot applications, an essential requirement for effective cooperation is the enforcement of global connectivity. That is, the ability for every robot to find a communication path to any other robot in the team. When only limited-range communication is available, global connectivity can require intermediate robots to also act as relays. Assessing and controlling the global connectivity of a communication graph (where robots are nodes and radios create links) in a decentralized fashion is not trivial [2]. Several recent approaches [11,15] exploit the spectral graph theory result stating that the second smallest eigenvalue of the Laplacian matrix L of the communication graph (often referred to as λ_2, λ, or algebraic connectivity), is non-zero *if and only if* the underlying communication graph is connected [4]. These proposals, however, are typically limited to simulations [15] or controlled laboratory experiments [11].

In this work, we provide two contributions to the research on decentralized assessment and control of algebraic connectivity (and, in general, multi-robot connectivity maintenance). First, we present how to implement a decentralized connectivity control law [5] in a team of quadcopters—from the computing and communication hardware level, to the robotic middleware and control software. Second, we report on the communication performance of field experiments conducted using flying three quadcopters endowed with our hardware/software stack.

2 Related Work

Fiedler wrote about the properties of the second smallest eigenvalue λ_2—also called Fiedler eigenvalue—of the unweighted Laplacian matrix of a graph in a seminal paper [4] where he derives, from the Perron–Frobenius theorem, that λ_2 "is zero if and only if the graph is not connected". More recent research discusses how to compute λ_2 in decentralized fashion in ad-hoc networks. The work of Sahai *et al.* [13] exploits wave propagation and fast Fourier transforms. Bertrand and Moonen [2] propose a method based on the power iteration algorithm.

As networked multi-robot systems research [1] proliferated over the last decade, many suggested to include algebraic connectivity in control laws aimed at preserving the global connectivity of robotic teams. Ji and Egerstedt [6] proposed—and evaluated in simulation—multiple feedback control laws ensuring connectivity for the rendezvous and formation control problems based on the weighted Laplacian matrix. Robuffo Giordano *et al.* [11] introduced a decentralized control law based on a potential function of algebraic connectivity. Their work was tested with four quadrotors in a laboratory setting (using Wi-Fi for communication and a commercial mo-cap solution for localization). Even so, the authors observed discrepancies "due to the presence of noise and small communication delays, and in general to all of those non-idealities and disturbances affecting real conditions" [11]. Sabattini *et al.* [12] evaluated their decentralized connectivity maintenance control law using four E-Puck robots. Solana *et al.* [15] used λ_2 for path planning in cluttered environments, yet only in simulation.

When aiming at field deployment in extreme areas (such as caves, planetary surfaces, and regions hit by natural disasters), however, one has to make sure that a control law's performance is robust against hardware and communication failures. Approaches only controlling the Fiedler eigenvalue might be unsuccessful as they can be blind to certain pathological configurations with highly vulnerable nodes. A combined control law—to simultaneously improve algebraic connectivity and robustness of a network—was proposed and evaluated in simulation by Ghedini *et al.* [5]. We brought this approach to a real-world implementation using eight K-Team Khepera IV robots and tested against faulty communication—albeit only through emulation—in [9]. Finer tuning of its hyper-parameterization and coverage approach were discussed in [8] and [14], respectively. The work in this article advances previous research by investigating the challenges of transferring these approaches beyond the reality gap, in field robotics.

3 Control Law

We consider the control law proposed in [5]. This law is intended to both (i) preserve connectivity and (ii) strengthen the communication topology against the failure of individual robots. This control law can be implemented in a fully decentralized fashion under the loose assumption of exploiting a situated communication model. That is, robots possess range and bearing information about their 1-hop neighbors (see Fig. 1). Considering robots modeled as m-dimensional single integrators[1], and defining $p_i \in \mathbb{R}^m$ as the position of the i-th robot, the control law is defined as the linear combination of connectivity, robustness, and (in this implementation) coverage contributions which, for robot i, can be written as:

$$\dot{p}_i = u_i = \sigma u_i^c + \psi u_i^r + \zeta u_i^{LJ} \tag{1}$$

The computation of $u^c, u^r, u^{LJ} \in \mathbb{R}^m$ is detailed in the following subsections. Offline and online schemes for the selection of hyper-parameters $\sigma, \psi, \zeta \in \mathbb{R}$ were presented in [8,9] and not further discussed here.

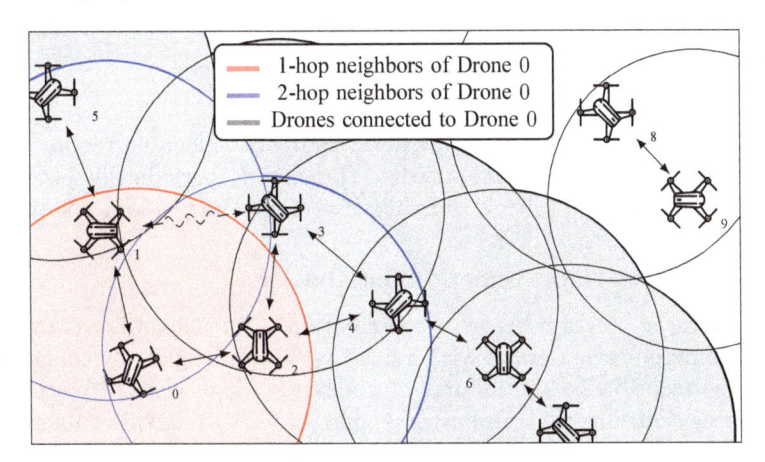

Fig. 1. In a multi-robot system with limited-range communication capabilities, we define as direct (or 1-hop) neighbors of a robot those robots that are within such range. We can then iteratively apply this notion to define 2-hop neighborhoods (Flaticon.com).

---

[1] Even though this is a very simple model, by endowing a robot with a sufficiently good Cartesian trajectory tracking controller, it can be used to represent the kinematic behavior of several types of ground and flying mobile robots [7].

3.1 Connectivity Maintenance Contribution

The first component on the right side of (1), u_i^c, is the one intended to maintain global connectivity, i.e., to prevent splits in the communication graph of the multi-robot system. Indeed, this is done through the control of λ_2. Algebraic connectivity is positive only when the graph is connected and also upper bounds the sparsest cut in the network. Decentralized computation of λ_2 in ad-hoc networks was demonstrated, among others, by [2] and [3]. Both of these approaches rely on the power iteration (PI) algorithm: they compute the largest eigenvalue (and associated eigenvector $\mathbf{x}$) of a matrix M using the update rule:

$$\mathbf{x}^{l+1} = M \mathbf{x}^l \qquad (2)$$

Over communication graphs, the update in (2) can be computed in a decentralized fashion for any shift operator (i.e., any matrix with the same sparsity pattern of the graph). The adjacency A and Laplacian L matrices are two such operators. For L the decentralized update rule becomes

$$x_k^{l+1} = L_{kk} \cdot x_k^l + \sum_{j \mid j \neq k \wedge L_{kj} \neq 0} L_{jk} \cdot x_j^l$$

where x_k^l is the k-th robot's estimate of the k-th entry of the eigenvector $\mathbf{x}$, at the l-th iteration, and L_{kj} is the element (k,j) of the Laplacian matrix L. Then, using an energy function $V(\lambda_2)$ that is non-negative, non-increasing with respect to λ_2, and that goes to infinity for λ_2 approaching zero (such as the one proposed in [12]), one can compute the connectivity contribution to (2) as follows

$$u_i^c = -\frac{\partial V(\lambda_2)}{\partial p_i} = -\frac{\partial V(\lambda_2)}{\partial \lambda_2} \frac{\partial \lambda_2}{\partial p_i} \qquad (3)$$

The main caveat is that a PI approach requires a "mean correction step" to avoid numerical instability. In practice, this entails periodically broadcasting information about each robot's estimate of vector $\mathbf{x}$ entry across the team.

3.2 Robustness Improvement Contribution

Motivation for adding a robustness contribution u_i^r to control law (1) was given in [5]. A communication graph with a positive λ_2 can be globally connected but still very susceptible to the failures of nodes with high centrality scores (e.g., betweenness centrality) [5]. Robustness aims at mitigating this vulnerability—critical for field experiments—quantified through the heuristic $\nu_i^k = \frac{|Path_i(k)|}{|\Pi_i|}$ where $|\Pi_i|$ is the number of 1- and 2-hops neighbors (see Fig. 1) of i, and $|Path_i(k)|$ is the number of nodes that are exactly 2-hops away from node i and relying on $\leq k$ 2-hops paths to communicate with i. Having defined $q_i^k \in \mathbb{R}^3$ as the barycentre of the robots in $Path_i(k)$, we compute the control input as:

$$u_i^r = \xi_r(\nu_i^k) \frac{q_i^k - p_i}{\|q_i^k - p_i\|} \qquad (4)$$

where $\xi_r(\cdot)$ evaluates as 0 or 1 depending on whether V_i^k surpasses threshold r or not [5]. The decentralized computation of u_i^r requires the robots to know about their 2-hop neighbors, i.e., to be able to exchange information about all their direct neighbors to all other members of this same neighborhood.

3.3 Coverage Improvement Contribution

The role of coverage contribution u_i^{LJ} in (1) is to homogeneously spread robots over an area of interest as well as to provide simple collision avoidance by introducing repulsive forces between nearby robots that grow quickly as robots get closer. The Lennard-Jones potential is a simple, well-known inter-molecular interaction model whose control contribution can be computed by deriving its expression and accounting for multiple neighbors as follows:

$$u_i^{LJ} = \sum_{n \in \mathcal{N}(i)} -\iota \left(\left(\frac{a \cdot \delta^a}{(p_n - p_i)^{a+1}} \right)^a - 2 \cdot \left(\frac{b \cdot \delta}{(p_n - p_i)^{b+1}} \right)^b \right) \qquad (5)$$

where a, b, δ, and ι are the potential's parameters and $\mathcal{N}(i)$ is the direct neighborhood of i. The decentralized computation of u_i^{LJ} only requires the 1-hop neighbors' positions.

4 Field Experiments

The disconnect between theoretical research and field robotics is often referred to as the reality gap. The field deployment and experiments described below are the major contributions of this paper. First, we developed the computing hardware and software framework to support the control law presented in Sect. 3 in a team of quadcopters. In particular, our software implementation focuses on the message passing required by the decentralized algorithms behind the three control contributions (3)–(5). The required middleware—in the form of ROS nodes to interface with the flight controller and the XBee sub-1 GHz RF modules—was also developed by the MIST Laboratory. Field experiments were conducted in Lanzarote, Spain during PANGAEA-X [16][2].

PANGAEA is the yearly geology training campaign organized by the European Space Agency for its astronauts. PANGAEA-X is an extension of this campaign giving the opportunity to universities and researchers to deploy and test their technologies in "scenarios that mimic human and robotic operations away from our planet". Because of its stringent fault-tolerance requirements and communication delays, space exploration beyond low Earth orbit is one of the applications that could benefit from decentralized multi-robot systems.

4.1 Robotic and Computing Hardware

Our robotic platform is the Spiri, a small quadrotor designed by Pleiades Robotics and intended for research and development. The Spiri is approximately $40 \times 40 \times 15$ centimetres and weighs 1.5 kg. Its flight controller is the

---
[2] http://blogs.esa.int/caves/2018/12/04/a-swarm-of-drones/.

PixRacer R14 which interfaces to three additional modules: a compass and GPS/GLONASS receiver, a range finder (to measure height) and a 2.4 GHz RF module to interact with its remote controller. The companion on-board computer is an NVIDIA Jetson TX2 board with 8 GB of LPDDR4 RAM, a hex-core ARMv8 CPU, and a 256-core Pascal GPU. As an operating system (OS), we use a stripped-down version of the 64-bit release of Ubuntu 16.04.6 LTS Xenial Xerus, installed through NVIDIA's JetPack SDK. A separate laptop, also running a Debian-based OS, acts as our ground station and interacts with the Spiris' Jetson TX2 boards through 5 GHz 802.11n Wi-Fi (before flight) and a Digi XBee PRO900/SX868 sub-1 GHz RF module (during flight). The ground station initiates take-off and acts as a safeguard, offering backup control to the drone team. These RF modules are also used on each Spiri for robot-to-robot communication.

4.2 Middleware and Software Implementation

For the software implementation of the decentralized control law in Sect. 3—and the corresponding communication strategy described below—we used the swarm-specific scripting language Buzz[3] by Pinciroli and Beltrame [10]. Buzz includes primitives supporting the implementation of typical swarm robotics operations such as polling from and broadcasting to all direct neighbors. The language has a simple syntax and was designed to allow researchers to create concise and composable programs. These can be executed in teams of (possibly heterogeneous) robots thanks to a portable, C-based virtual machine (VM). The VM allows to run Buzz scripts on multiple platforms such as the Khepera IV, the Matrice 100, and the Spiri. The Jetson TX2 computers onboard each Spiri run ROS Kinetic Kame and the MAVROS node to needed communicate with the flight controller. We then add two custom ROS nodes[4, 5]: ROSBuzz and XBeeMav. The former is a node encapsulating the Buzz VM to interface it with the PixRacer flight controller and other ROS nodes. ROSBuzz also supports RVO collision avoidance. XBeeMav is a node interfacing ROSBuzz with the XBee RF module for serializing Buzz messages into MAVlink standard payloads. Having this infrastructure in place, we want to study the feasibility of implementing (1) in a team of quadcopters. In particular, we want to evaluate the performance of the information exchanges needed for the decentralized computation of each one of the control contributions u^c, u^r, and u^{LJ}.

4.3 Inter-robot Communication with Buzz

The connectivity improvement contribution u^c (Subsect. 3.1) requires the estimation of λ_2. Executing the decentralized PI update, as explained in [2], needs a mean correction step. To make this possible, all robots are required to re-broadcast information so that it can be spread over multiple communication

---

[3] https://github.com/MISTLab/Buzz.
[4] https://github.com/MISTLab/ROSBuzz.
[5] https://github.com/MISTLab/XbeeMav.

hops. In Buzz, this can be done with a `broadcast` call within a `listen` call. This entails having information traveling possibly as many hops as the diameter of the communication graph. The mean correction step only needs to be performed periodically, for numerical stability. The coverage control contribution u^{LJ} (Subsect. 3.3) is the simplest to compute as it only requires information about the positions of 1-hop neighbors. This information in natively available within the runtime of Buzz (in a global `neighbor` structure). In this case, messaging does not have to be dealt with explicitly because it is managed by the virtual machine. Finally, the robustness improvement contribution u^r (Subsect. 3.2) is computed from the position information of 1- and 2-hop neighbors. As Buzz makes 1-hop information readily available, to diffuse 2-hop information, robots only need to further broadcast it once and listen to the corresponding messages from direct neighbors.

5 Results and Discussion

Our experiments were conducted using three Spiri quadcopters christened Mars, Pluto, and Valmiki. The flight area was set on the island of Lanzarote approximately 5 Km north-east of PANGAEA's main site in a 300 × 300 m open field around coordinates 29.067°N, 13.662°W. After two preliminary flights, all three drones were flown for about 350 s (roughly 50% of their ideal maximum flight time using 1600 mAh battery packs) under manual control while, at the same time, running the infrastructure and Buzz implementation described in Sect. 4. These experiments were meant to selectively stress-test the communication by forcing the drones to reach—large and small—inter-robot distances from which they would not have interacted, had they been solely controlled by (1). The data collection process was aimed at verifying that our field setup could achieve the communication performance required to compute all three contributions of the law in (1). Figure 2 presents the drones' trajectories, coordinates and inter-robot distances.

5.1 Timing Performance

Both Ubuntu and ROS are best-effort rather than real-time operating systems. Hence, a first step in assessing the relevance of our experimental results required to verify the synchronization between by the operations of ROS, the Buzz VM, and the actual passing of time. Figure 3 compares the evolution of the latitude and longitude logs—within Buzz, ROS, and with respect to the elapsed time—for two drones (Pluto and Valmiki). We observe that Buzz deviates by 1% or less from its ideal frequency 10 Hz. Thus, our implementation, albeit not strictly real-time, provides a timely best effort execution. In the plots of this section, we use Buzz iterations as the abscissae.

5.2 Connectivity

Figure 4 presents the results associated to the message passing required to compute u^c. The three charts in the left column present, for each one of the robots,

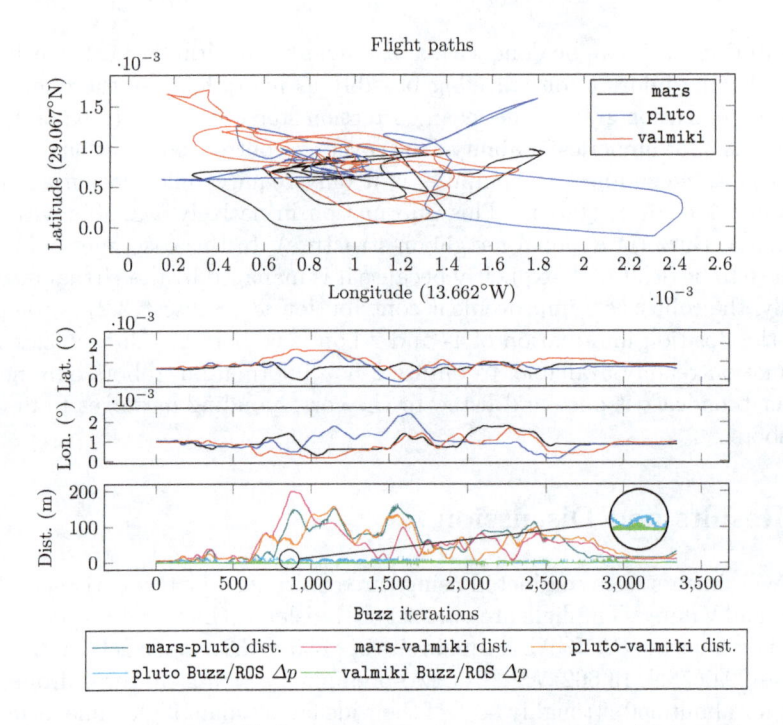

Fig. 2. From top to bottom: (i) the quadcopters' trajectories; their (ii) latitude and (iii) longitude; (iv) the inter-robot distances and the discrepancies in position Δp between the information stored in Buzz's logs and `rosbag` due to imperfect synchronization.

the number of received messages originating from different robots per every line of a textual log (these logs have ∼5000 entries as they can be written more than once in a single Buzz iteration, if multiple messages were queued). In an idealized, synchronous world, the number of such messages would steadily be 2. In practice, we observe that the plots constantly oscillate between 1 and 2. Yet, they are never 0, suggesting that the exchanges never broke down (at least, not until the end of the experiments, when robots were turned off).

Table 1. Buzz iterations (ratios) missing any of the 2-hop information messages. Correlations are computed until the 3000-th iteration, from the data in Fig. 5.

	Buzz iterations with 1 `robustness` message		A-B correlation	Buzz iterations without `robustness` messages
	From A	From B		
mars	0.240	0.266	−0.115	0.088
pluto	0.265	0.255	+0.129	0.051
valmiki	0.236	0.308	+0.131	0.052

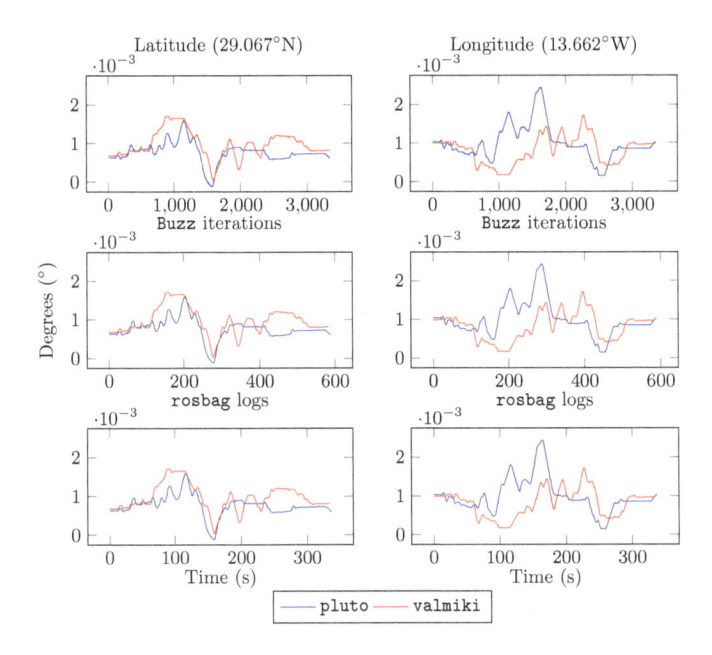

Fig. 3. Comparison of the evolution of latitude and longitude (from the experiment in Fig. 2) of Pluto and Valmiki against the progression of the Buzz VM, the `rosbag` log, and the absolute elapsed time when using a best-effort operating system.

The charts in the right column of Fig. 4 present the evolution of the Buzz iteration of origin of each of these messages. For each robot, the two lines (teal and magenta) in the three plots refer to different senders (the two neighbors). We can observe that, as time goes by, the received information stays current, i.e., originates in more recent Buzz iterations. Once again, in an ideal world, these trends would be perfectly linear and monotone, with constant positive slopes. In reality, we notice the presence of non-linear trends and very small oscillations (whose detail is magnified) caused by the recursive way in which we relay messages, making it possible for slightly older information to bounce over multiple hops and to reach a robot after the most up-to-date one. Thus, rapidly changing topologies will lead to inexact mean corrections for (2).

5.3 Robustness

The decentralized computation of the robustness improvement input u^r in (4) requires the relative positions of both 1- and 2-hop neighbors. The communication performance of its implementation is presented in Fig. 5 for all three drones (top six plots) versus the evolution the inter-robot distances (bottom plot). Table 1 summarizes, for each robot, the percentages of Buzz iterations in which either one or both messages coming from direct neighbors were not received, as well as the correlations between the omission of these messages. We can see in Fig. 5 that, for all three robots, the number of direct neighbors varies

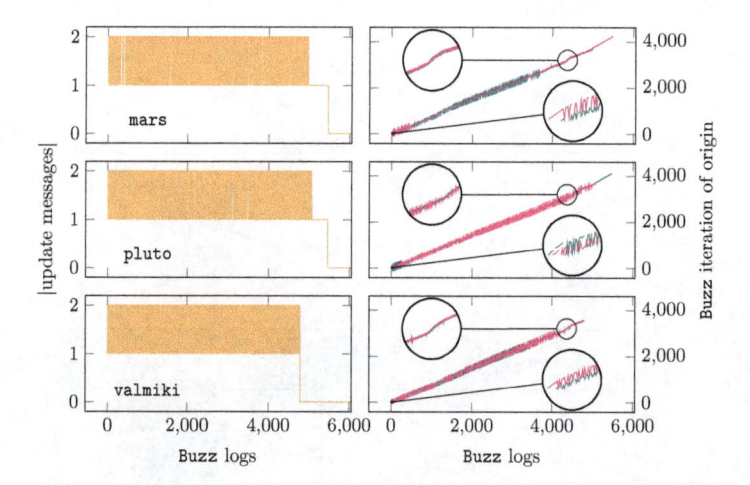

Fig. 4. Performance results of the message passing required for the decentralized computation of the connectivity maintenance contribution u^c (Subsect. 3.1) of the control law in (1). The left column shows the number of messages received by each robot while the right column displays their recentness (the magenta and teal lines representing the two different neighbors of origin).

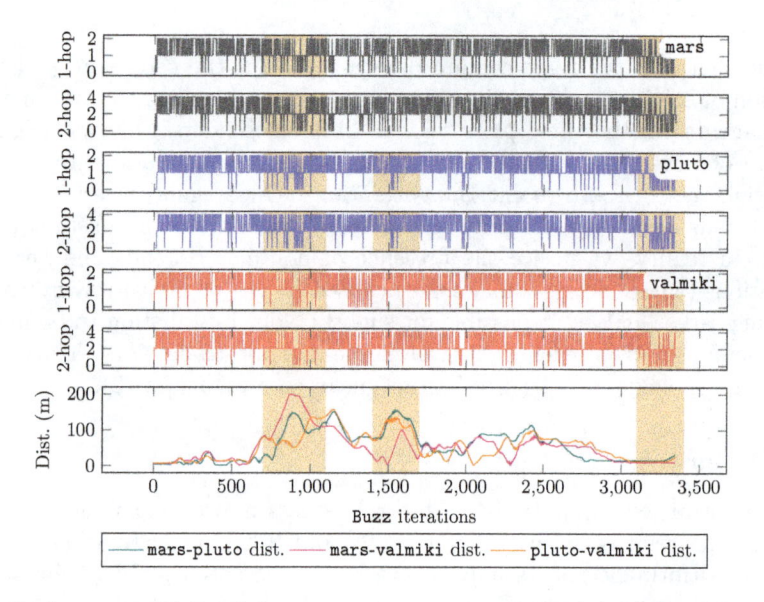

Fig. 5. Performance results of the message passing required for the decentralized computation of the robustness improvement contribution u^r (Subsect. 3.2). The number of 1- and 2-hop neighbors (including themselves) known to each robot are plotted against the inter-robot distances.

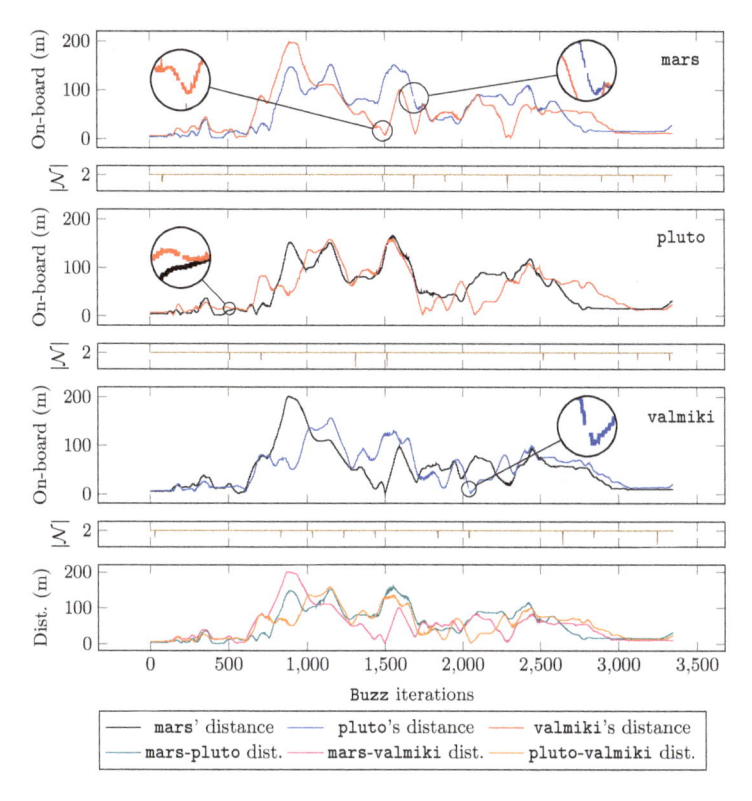

Fig. 6. Performance results of the message passing required for the decentralized computation of the coverage improvement contribution u^{LJ} (Subsect. 3.3) of the control law in (1). The estimated inter-robot distances onboard each robot are compared with the ground truth (the bottom plot). The brown lines show the number of entries stored within Buzz's `neighbor` structure.

and so does the number of indirect (2-hop) neighbors. More frequent drops in 1- and 2-hop neighbors in Fig. 5 coincide with periods of greater inter-robot distances and the very end of our experiments, after the robots have landed. (This latter phenomenon is likely explained by the ground plane obstructing the radio antennas.) The very low correlations between the lack of messages from 1-hop neighbors in Table 1 also suggest that these drops are more likely ascribed to external, independent causes (e.g., inter-robot distances) rather than intrinsic ones (e.g., a computational bottleneck).

5.4 Coverage

As we explained in Subsect. 4.2, the coverage improvement contribution u^{LJ} in (5) is the simplest to compute in a decentralized fashion as it only requires information about the relative positions of all direct neighbors of a drone. Figure 6 shows how this information evolves over time on-board each robot. We do so by plotting each robot's on-board, presumed inter-robot distances against the GPS-given ground truth—the bottom chart. We observe an almost perfect match: the robots only sporadically lose track of their neighbors for fractions of seconds (the zoomed-in bubbles), meaning that they can reliably compute u^{LJ}.

6 Conclusions

In this paper, we tackled the reality gaps associated to decentralized, robust, global connectivity control laws in a multi-robot system using three quadcopters communicating with sub-1 GHz RF modules. Prior to this work, most of the research in the area had only focused on numerical simulations and indoor experiments. Our first contribution was the creation of the hardware and software stack implementing the control law proposed in [5]. Then, we brought this stack to a team of quadcopters and performed field tests (in the context of ESA's PANAGEA-X training campaign) to assess the performance of our implementation, especially with respect to information exchanges. Our results indicate that the information required to compute all three components of the decentralized control law in Eq. 1 can be transmitted across multiple robots even when flying hundreds of meters apart. Yet, these tests also show that the reality gap—with respect to assumptions on communication made by previous simulation [5] and laboratory [9] studies—is still remarkable as, oftentimes, only part of the total information is available to each robot. The takeaway message is that theoretical research in multi-robot systems should not shy away from the nitty-gritty of implementation and field experiments as, behind their inconvenience, might lie the more practical insights.

References

1. Banfi, J., Basilico, N., Amigoni, F.: Multirobot reconnection on graphs: problem, complexity, and algorithms. IEEE Trans. Rob. **34**(5), 1299–1314 (2018). https://doi.org/10.1109/TRO.2018.2830418
2. Bertrand, A., Moonen, M.: Distributed computation of the Fiedler vector with application to topology inference in ad hoc networks. Signal Process. **93**(5), 1106–1117 (2013). https://doi.org/10.1016/j.sigpro.2012.12.002
3. Di Lorenzo, P., Barbarossa, S.: Distributed estimation and control of algebraic connectivity over random graphs. IEEE Trans. Signal Process. **62**(21), 5615–5628 (2014). https://doi.org/10.1109/TSP.2014.2355778
4. Fiedler, M.: Algebraic connectivity of graphs. Czechoslovak Math. J. **23**(2), 298–305 (1973). http://eudml.org/doc/12723
5. Ghedini, C., Ribeiro, C., Sabattini, L.: Toward fault-tolerant multi-robot networks. Networks **70**(4), 388–400 (2017). https://doi.org/10.1002/net.21784
6. Ji, M., Egerstedt, M.: Distributed coordination control of multiagent systems while preserving connectedness. IEEE Trans. Rob. **23**(4), 693–703 (2007). https://doi.org/10.1109/TRO.2007.900638
7. Lee, D., Franchi, A., Son, H.I., Ha, C., Bülthoff, H.H., Giordano, P.R.: Semiautonomous haptic teleoperation control architecture of multiple unmanned aerial vehicles. IEEE/ASME Trans. Mechatron. **18**(4), 1334–1345 (2013)

8. Minelli, M., Kaufmann, M., Panerati, J., Ghedini, C., Beltrame, G., Sabattini, L.: Stop, think, and roll: online gain optimization for resilient multi-robot topologies. In: Correll, N., Schwager, M., Otte, M. (eds.) Distributed Autonomous Robotic Systems. SPAR, vol. 9, pp. 357–370. Springer, Cham (2019). https://doi.org/10.1007/978-3-030-05816-6_25
9. Panerati, J., et al.: Robust connectivity maintenance for fallible robots. Auton. Robot. **43**(3), 769–787 (2018). https://doi.org/10.1007/s10514-018-9812-8
10. Pinciroli, C., Beltrame, G.: Swarm-oriented programming of distributed robot networks. Computer **49**(12), 32–41 (2016)
11. Robuffo Giordano, P., Franchi, A., Secchi, C., Bülthoff, H.H.: A passivity-based decentralized strategy for generalized connectivity maintenance. Int. J. Robot. Res. **32**(3), 299–323 (2013)
12. Sabattini, L., Chopra, N., Secchi, C.: Decentralized connectivity maintenance for cooperative control of mobile robotic systems. Int. J. Robot. Res. **32**(12), 1411–1423 (2013)
13. Sahai, T., Speranzon, A., Banaszuk, A.: Hearing the clusters of a graph: a distributed algorithm. Automatica **48**(1), 15–24 (2012)
14. Siligardi, L., et al.: Robust area coverage with connectivity maintenance. In: 2019 IEEE International Conference on Robotics and Automation (ICRA), pp. 2202–2208 (2019). https://doi.org/10.1109/ICRA.2019.8793555
15. Solana, Y., Furci, M., Cortés, J., Franchi, A.: Multi-robot path planning with maintenance of generalized connectivity. In: 2017 International Symposium on Multi-Robot and Multi-Agent Systems (MRS), pp. 63–70 (2017)
16. St-Onge, D., et al.: Planetary exploration with robot teams. IEEE Robot. Autom. Mag. (2019, in press). https://doi.org/10.1109/MRA.2019.2940413

Behavioral Simulations of Lattice Modular Robots with VisibleSim

Pierre Thalamy, Benoît Piranda[✉], André Naz, and Julien Bourgeois

Univ. Bourgogne Franche-Comté (UBFC), University of Franche-Comté (UFC),
FEMTO-ST Institute, UMR CNRS 6174, 1 Cours Leprince-Ringuet,
25200 Montbéliard, France
{benoit.piranda,julien.bourgeois}@femto-st.fr

Abstract. Robotics research needs complex hardware and software that is why simulation is often view as an alternative for testing. Large scale self-reconfiguring modular robotic systems needs a scalable simulation environment which cannot be physics-based.

This paper presents *VisibleSim*, an open-source behavioral simulator for lattice-based modular robots that uses discrete-event simulation to simulate ensembles of up to millions of modules. We describe the principles behind the simulator and introduce its features and usage from a user standpoint. *VisibleSim* is built with extensibility, versatility, and flexibility in mind, can be used as a powerful visualization tool, and already has a proven track record with several modular robotic architectures.

1 Introduction

Simulation can be used for multiple purposes in robotics research, one of them being, researching software solutions for managing complex systems that are not yet producible. This is particularly relevant for self-reconfigurable modular robotic systems [1] which assemble individual modules latched to one another. The software challenge is to coordinate all the modules to achieve a common goal like in self-reconfiguration [23].

In this paper, we present *VisibleSim*[1], a framework for creating *behavioral* simulators for distributed lattice-based modular robotic systems in regular 3D environments. *VisibleSim* can be used for studying the behavior and programmability of such distributed systems, but it does not comprise physics simulation.

Each module executes the same program as all other modules, generating communications and/or events that are handled deterministically by *VisibleSim*'s discrete-event scheduler. Beyond reproducibility, this allows for scaling up in the number of simulated modules which can be greater than 32 millions. A video presentation of the simulator and its aforementioned features is available on Youtube[2].

---

[1] https://projects.femto-st.fr/programmable-matter/visiblesim
[2] Video presentation of *VisibleSim*: https://youtu.be/N09KElCbUNk.

2 Related Works

There are many general simulation frameworks dedicated to robots [11]. Some of them have been used for many different kinds of robotics projects like Player [8], associated at Stage or at Gazebo [10]. Webots [13] is another reference of a commercial open-source simulator, now a leader in this field. ARGos [17] simulator for swarm robotics, which can simulate large and heterogeneous multi-robots systems. These simulators are not particularly well-suited for the specificities of modular self-reconfigurable systems.

On the one hand, regarding hardware-specific modular robot simulators, many platforms for simulating self-reconfiguration for lattice modular robots are unnamed simulators with core features implemented using Java3D [22,25].

On the other hand, several generic modular robot simulators have been developed, mostly physics-based and targeting chain or hybrid modular robots like Rebots [4], including models for Roombots [21], Smores [5] and Superbot [20]. Sim [24], USSR [3] including models for ATRON, Odin [12], and M-TRAN [9].

Nevertheless, there are still a few generic simulators that are designed for lattice modular robots, the type of robots targeted by *VisibleSim*. For example, SRSim [7] used with Sliding-Cube [7], Crystalline [19] and Superbot [6]. Finally, DPRSim [2] has been shown to efficiently simulate ensembles with up 20 millions of Catom modules, both in their 2D and 3D forms, by maximally leveraging the potential for multi-threading of the simulation and using computing clusters.

The common feature among all the aforementioned simulators except perhaps SRSim, is that they are all physics-based, which is useful when developing robotic designs, evolving controllers, and interacting with complex environments, but might be superfluous and prohibitively costly when researching distributed robotic control from a more fundamental, or *behavioral*, point of view. This is the kind of simulator that *VisibleSim* is thus intended to be, a framework for performing all kinds of behavioral simulations on lattice-based modular robotic systems with low environmental interactions. Furthermore, it is most similar to USSR and SRSim in its usage, being a framework for developing simulator instances rather than an actual monolithic executable software where all simulation parameters are interpreted.

3 Simulator Overview

VisibleSim is designed for researchers that have computer programming experience as it consists in a **C++ framework** for building lattice modular robot simulators controlled by distributed programming. Several sample modular robot simulators are provided with the software. *VisibleSim* takes the form of an open source project under AGPLv3 license and is available on Github[3].

In *VisibleSim* lingo, the distributed program that is executed on each module during the simulation is named a *BlockCode*. It is effectively the controller of the modules and where users will describe the behavior of the robot in response to

---

[3] https://github.com/ProgrammableMatterProject/VisibleSim.

all kinds of events whether external (interactions with the world, reception of a message, etc.), or internal (interruption or timer, initialization, end of a motion, etc.).

Unlike other simulators where each robot is fitted with a number of sensor and actuator components, this distinction is not materialized in *VisibleSim*. Modules from any type of robots are however fitted with a constant number of *interfaces*, depending on the geometry of their lattice, and which can both be used for sensing connected modules (by examining whether an interface is connected) and communicating with them. In the current state of the simulator communication between modules is only natively allowed in a peer-to-peer manner between connected neighbors.

Previous work on modular robots can be classified based on the shape of the robots and the type of grid in which they are placed. Each grid has a specific number of positions adjacent to each of its cells, which determines the number of neighbors a module in that grid can communicate with (2D square or hexagonal lattices, Face Centered Cubic lattice, etc.).

VisibleSim offers different classes of modular robots across these different lattices, as shown in Fig. 1.

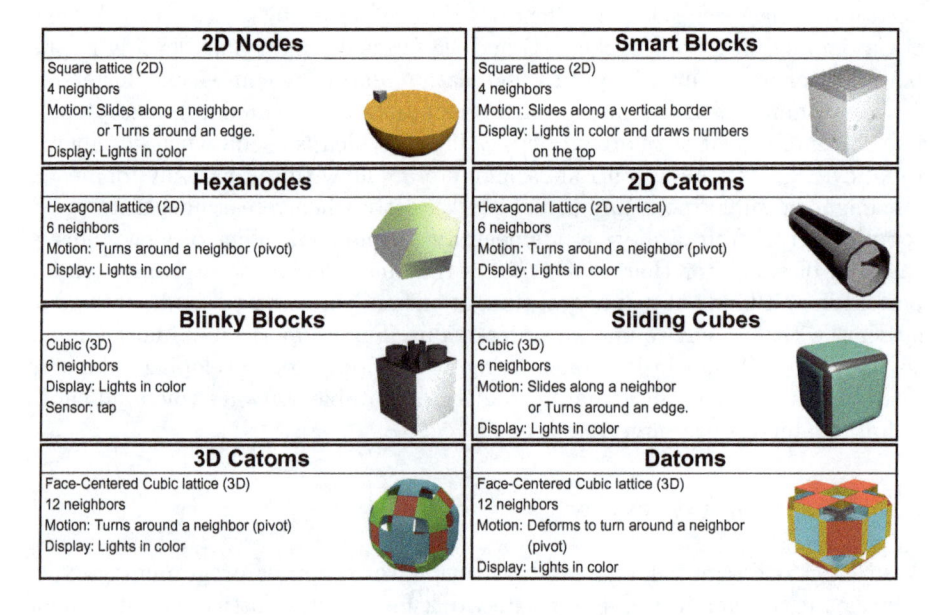

2D Nodes	Smart Blocks
Square lattice (2D) 4 neighbors Motion: Slides along a neighbor or Turns around an edge. Display: Lights in color	Square lattice (2D) 4 neighbors Motion: Slides along a vertical border Display: Lights in color and draws numbers on the top
Hexanodes	**2D Catoms**
Hexagonal lattice (2D) 6 neighbors Motion: Turns around a neighbor (pivot) Display: Lights in color	Hexagonal lattice (2D vertical) 6 neighbors Motion: Turns around a neighbor (pivot) Display: Lights in color
Blinky Blocks	**Sliding Cubes**
Cubic (3D) 6 neighbors Display: Lights in color Sensor: tap	Cubic (3D) 6 neighbors Motion: Slides along a neighbor or Turns around an edge. Display: Lights in color
3D Catoms	**Datoms**
Face-Centered Cubic lattice (3D) 12 neighbors Motion: Turns around a neighbor (pivot) Display: Lights in color	Face-Centered Cubic lattice (3D) 12 neighbors Motion: Deforms to turn around a neighbor (pivot) Display: Lights in color

Fig. 1. Several shapes of robots proposed in *VisibleSim*.

What characterizes a modular robot in *VisibleSim* is therefore: the geometry and visual aspect of its modules; the lattice in which they belong (hence their number of possible neighbors), and a specific mode of motion. Additional components and state visualization features such as a display, speakers, or tap sensors can however be added.

4 Programming Environment and Features

4.1 User Application Demonstration

This section presents an example of a *SlidingCube* modular robot application, where a message is broadcast distributively through the robot from a leader module (identified by its identifier) to instruct modules to perform a random motion. Though this application has no practical purpose, it demonstrates concisely the structure of a user application as well as elements of its motion and communication API.

Furthermore, a visual *BlockCode* generator is available online[4], which takes a target robotic architecture and a list of messages as input and returns a code template for that setup.

Listing 1. Sample BlockCode: Broadcast of a message across the robot from a master module and moves upon reception

```
#include "myBlockCode.h"

void MyBlockCode::startup() {
    addMessageEventFunc(BROADCAST_MSG, bind(&myBlockCode::onBroadcastRcvd,
                        this, std::placeholders::_1, std::placeholders::_2));
    if (module->blockId == 1) { // module #1 is the master
        this->broadcastReceived = true;
        sendMessageToAllNeighbors(new Message(BROADCAST_MSG));
    } else {
        this->broadcastReceived = false;
    }
}

void MyBlockCode::onBroadcastRcvd(shared_ptr<Message> msg,
                                  P2PNetworkInterface* sender) {
    if (not this->broadcastReceived) {
        this->broadcastReceived = true;
        // Propagate broadcast (ignoring sender interface)
        sendMessageToAllNeighbors(new Message(BROADCAST_MSG), sender);
        //  and move to first available location
        list<Cell3DPosition> dests = getPossibleDestinations();
        if (not list.empty()) initiateMotionTo(dests.front());
    }
}
```

Listing 2. Sample main file: initiates and cleans up the simulation

```
#include <iostream>
#include "robots/slidingCubes/slidingCubesSimulator.h"
#include "robots/slidingCubes/slidingCubesBlockCode.h"
#include "myBlockCode.h"

int main(int argc, char **argv) {
    // Create simulation world and modules
    //   attach a MyBlockCode instance to each module
    createSimulator(argc, argv, MyBlockCode::buildNewBlockCode);
    // Previous call returns only once scheduler has ended
    deleteSimulator();
    return 0;
}
```

---

[4] https://services-stgi.pu-pm.univ-fcomte.fr/visiblesim/generator.php.

Listing 3. Sample XML configuration file: describes the simulated world; the modules within it; and other simulation parameters

```xml
<?xml version="1.0" standalone="no" ?>
<world gridSize="20,20,20" windowSize="1920,1080">
    <blockList defaultColor="128,128,128" ids="RANDOM">
        <!-- Describe individual modules -->
        <block position="3,4,2" color="127,255,43" />
        <!-- or use Constructive Solid Geometry (CSG) -->
        <csg content="union() { cube([10, 5, 5]); cube([5, 10, 5]); }"/>
    </blockList>
    <targetList> <!-- Goal shape for reconfiguration -->
        <target format="csg">
            <csg content="sphere(10)"/>
        </target>
    </targetList>
</world>
```

4.2 User Interactions

In fixed-increment time progression mode, *VisibleSim* supports pausing and resuming of the simulation (programmatically or using the keyboard), which can be used to inspect the simulated world at any given time. This is especially useful since *VisibleSim* has a built-in console that provides useful information about a number of core (messages sent or received, motions, etc.) or custom (any user-implemented event or debugging trace) events. This includes the time of the event and any other useful information that is necessary for retracing what in the chain of events that led to the current state of the simulation. Not only can the traces concerning all the modules in the system be shown at once from within the simulation window, but individual threads of events relative to a specific module can be shown by selecting the module from the GUI (Fig. 2).

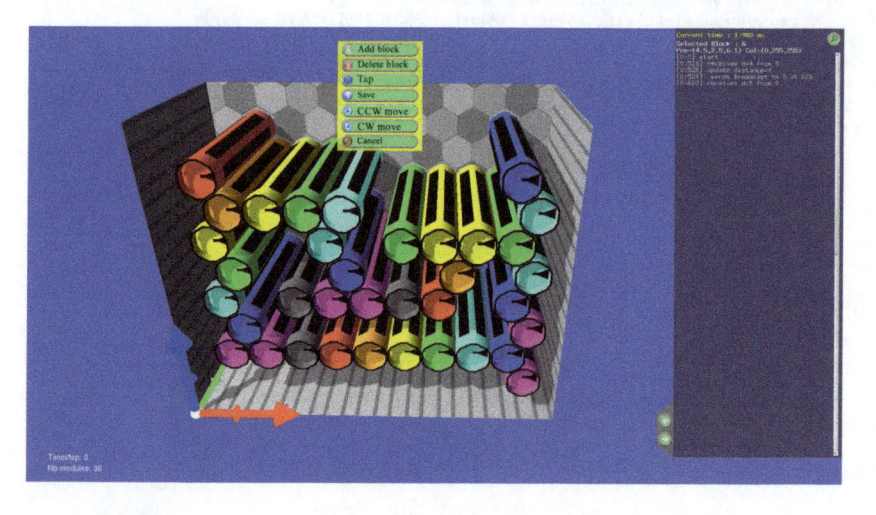

Fig. 2. Screenshot of a *VisibleSim* simulation of Catom 2D modules with the console and interaction menu open for the selected module.

Furthermore, left-clicking a module opens a pop-up for interacting with the simulated world and the module itself. These interactions are the addition and removal of neighbors on the interfaces of a module, motion commands, or a physical event such as an accelerometer tap. Finally, the current world configuration can be exported, making *VisibleSim* both a simulation software and a sandbox for building robotic configurations.

4.3 Customization Hooks

VisibleSim proposes a number of customization hooks that are called at various points of the simulation and that can be used to implement custom behaviors for a given *BlockCode* application. Some of these functions provide greater flexibility to the user, others simply facilitate debugging:

- Parsing custom configuration file elements pertaining to the world or individual modules.
- Parsing custom command line arguments exclusive to this specific *BlockCode* application.
- Responding to custom keyboard events generated by the user during simulation and specific to that application.
- Drawing custom graphical elements in the OpenGL world every time it is updated.
- Drawing custom text onto the OpenGL window to keep some essential information always visible.
- A custom function that gets called on a module whenever a *VisibleSim* assertion has been triggered for that module, and that can provide critical information on its current state.

4.4 Export Tools

An important aspect of *VisibleSim* is that it is *more than a simulation tool*. Its other main purpose is to produce impactful visual results for academic research, at a low cost for the user. It does so by allowing researchers to easily export various content from their simulated world: screenshots and videos of a simulation, 3D animation data for stunning videos in Blender or similar software, Stereolithography (STL) data for the 3D printing of a configuration, etc. This thus makes *VisibleSim* a *visualization tool* as well as a simulator, supporting research at multiple levels.

Apart from the standard simulation workflow of *VisibleSim*, where a simulation is run in real time either in terminal-mode or with a graphical output, *VisibleSim* offers to export a simulation to a file on disk, so that the simulation can be later be visualized on a *replayer* in an interactive fashion. This *replayer* allows the user to slowly and repeatedly analyse events generated during the simulation, record a part of it as a video or an animation, or compare results. This feature is particularly helpful for inspecting a problematic simulation when debugging, or for viewing the graphical output of a terminal-mode simulation that would have taken too long or too much memory to be computed graphically.

5 Usage and Evaluation

In this section, we highlight a number of different modular robots and applications that have been successfully simulated using *VisibleSim* in published research. Our aim is to highlight different ways *VisibleSim* can be used and has been used. We also show that the simulation of existing hardware system can show a high level of fidelity to hardware experiments. Finally, we bring to light the current capabilities of *VisibleSim* in terms of scalability.

5.1 Use Cases

VisibleSim has become over the years the dedicated simulation and experimentation tool of the *Programmable Matter Project*. As such, not only has it been used for simulating diverse modular robotic models, but also for a wide variety of applications (distributed time synchronization, self-reconfiguration, center election, etc.). Figure 3 thus illustrates the versatility and reliability of the *VisibleSim* framework by highlighting select research work that has relied on it in recent years.

5.2 Simulation Fidelity

In addition to faithfully reproducing the functional behavior of algorithms, *VisibleSim* also accurately simulates timing. Communication and clock models can be customized and passed to *VisibleSim* in order to fit best with the simulated modular robotic platform.

After having modeled the communication system of the *Blinky Blocks* in *VisibleSim* [14], we measured the execution time of the ABC-CenterV1 algorithm [14,15]—an algorithm for electing an approximate-center module in modular robots—on hardware *Blinky Blocks* and in simulations. Table 1 shows that the simulated execution times (average and standard-deviation) on *VisibleSim* closely match the execution time obtained experimentally on hardware *Blinky Blocks*, for small and larger configurations, and for sparse (e.g., lines), less-sparse (e.g., squares), compact (e.g., cubes) and mixed-density configurations with compact components linked by a critical path (e.g.,the dumbbell-like shape).

We have also modeled the *Blinky Blocks* hardware clocks in *VisibleSim* and evaluated the synchronization precision of the Modular Robot Time Protocol (MRTP) [16]—a protocol for providing global time synchronization across a modular robotic system—both with hardware modules and simulations. Experiments were conducted on a doubled L-shaped system composed of 10 *Blinky Blocks* over an hour, with a synchronization period of 5 s. Synchronization error distribution looks Gaussian both in simulation and hardware experiment results [14]. In the hardware *Blinky Blocks* system (resp. in *VisibleSim*), MRTP has an average precision of 0.06 ms (resp. -0.11 ms) and a standard-deviation of 1.62 ms (resp. 1.40 ms).

Results obtained using *VisibleSim* show a very high fidelity to the hardware results, which indicates that *VisibleSim* is able to perform an accurate timing simulation of the algorithms.

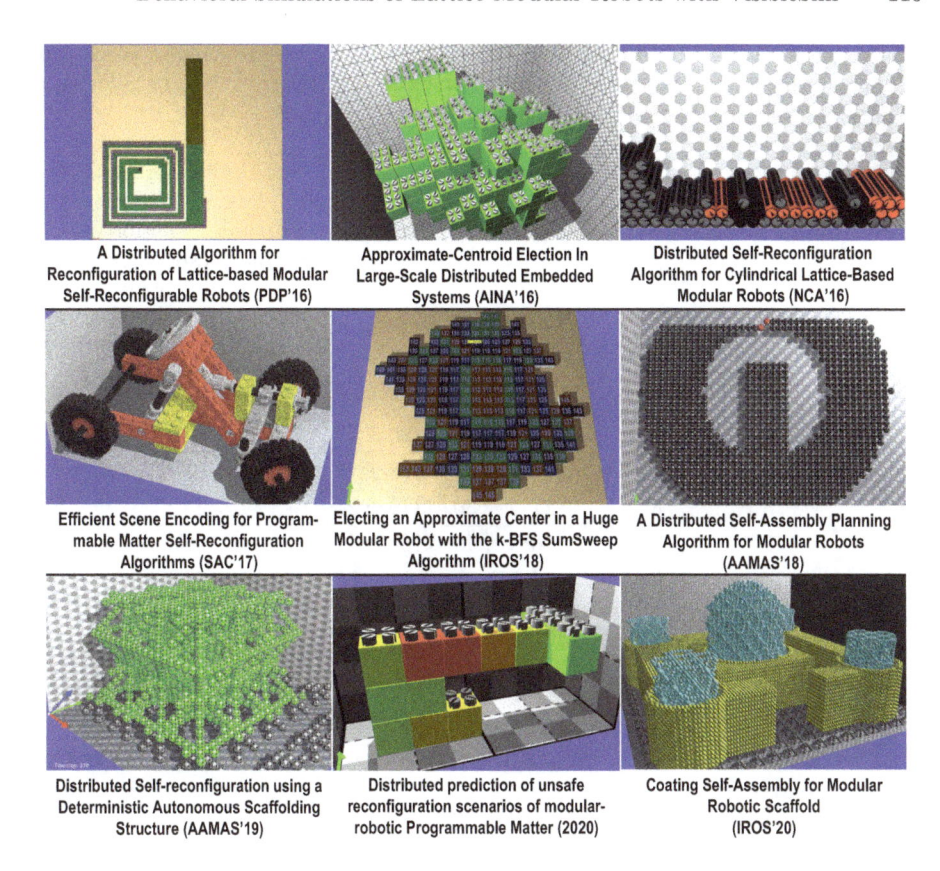

Fig. 3. Several results from previous work based on *VisibleSim* across several module types and tasks.

5.3 Scalability

In order to demonstrate the scalability of the *VisibleSim* simulation framework, we have designed a stress test experiment which consists in simulating a sort of *brownian motion* of as many modules as possible, within a square grid. The underlying *BlockCode* program is quite straightforward:

- At the start, a single leader module *activates* and sends an *activation* message to all its neighbors.
- Upon reception of an *activation* message, modules turn into the *activated* state.
- *Activated* modules then alternate between a 0.5 s wait, and a random motion lasting 1 s.
- When a motion ends for a module, it sends an *activation* message to its new neighbors, if any, before starting the next wait/move cycle.
- The simulation ends when all modules are in the *activated* state.

Table 1. Average execution time of ABC-CenterV1 on hardware Blinky Blocks and in simulations. Statistics on the execution time were computed over 25 runs for every configuration.

Shape	Size (module)	Diameter (hop)	Average execution time ± standard deviation (ms)		Absolute error of the average execution time simulator vs hardware (ms) (relative error)
			Hardware	Simulator	
Line	5	4	234 ± 1	244 ± 3	10 (4.27%)
	10	9	545 ± 5	544 ± 5	1 (0.18%)
	50	49	2873 ± 23	2885 ± 17	12 (0.42%)
Square	9	4	598 ± 45	588 ± 14	10 (1.67%)
	25	8	1117 ± 30	1119 ± 27	2 (0.18%)
	49	12	1684 ± 48	1686 ± 44	2 (0.12%)
Cube	27	6	1229 ± 56	1214 ± 31	15 (1.22%)
	64	9	1927 ± 51	1941 ± 33	14 (0.73%)
Dumbbell	59	15	1262 ± 56	1252 ± 57	10 (0,79%)

This simple distributed program will thus propagate agitation across an entire modular robotic system, generating a massive number of messages, motions, and wait events in the process. The aim is therefore to stress the *VisibleSim* scheduler as much as possible and show that a graphical simulation is still possible with a massive robotic ensemble. Executions of this stress test program can be seen in the video mentioned in footnote 2.

We run the program with a large set of square configurations and for each of them we compute the number of messages and the number of displacements that are necessary to activate the entire robotic ensemble. For each size of configuration, the initial set of modules is made by growing a tree of modules from a regular list of seeds, ending when branches reach a cell that is already in a filled state.

Figure 4 shows the number of messages and displacements as a function of the number of robots in the several configurations. As shown in the figure, we are able to simulate more than 32 million robots communicating and moving through the grid, which is to the best of our knowledge a new record in the field of modular robotic simulation. The first experiments, dealing with up to 3 million robots have been made on a laptop with 32 GB of RAM, and all subsequent simulations have been made on a server with 3 TB of RAM.

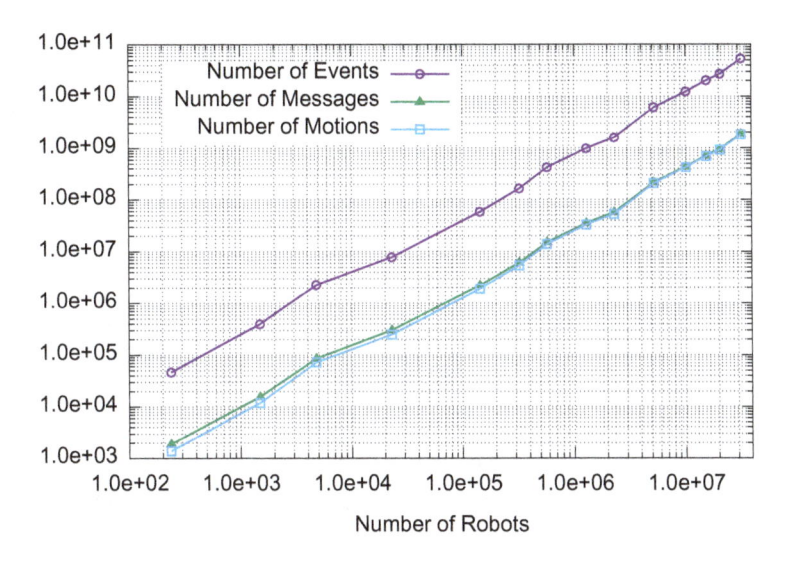

Fig. 4. Number of motions and messages simulated during the stress test experiment (Log-Log plot).

6 Conclusion and Future Work

In this paper, we have introduced *VisibleSim*, a C++ framework for simulating large-scale lattice-based distributed modular robotic ensembles. It differs from other modular robot simulators in its philosophy as a behavior-focused simulator, and its corresponding discrete-event-based style of scheduling. Various modular robotic designs supported by *VisibleSim* have been introduced, along with how to add new architectures by instantiating the OOP simulator framework, and implementing user applications. We have shown that it doubles as a powerful visualization software for effectively communicating research results, and that the simulator is flexible and easy to customize. Finally, we have outlined the versatility, reliability, and scalability of *VisibleSim*, by showing diverse usages of the software in published research, outlining the accuracy of simulations, and performing graphical simulations with more than a million individual modules. We therefore argue that *VisibleSim* can benefit any present of future research on the algorithmic foundation of modular robotic systems, especially since it is freely available as open source software. *VisibleSim* is an ongoing project and there are a number of features that are currently under investigation, detailed below. In its current implementation, all the scheduling tasks are performed on a single thread. While it guarantees an accurate simulation, this also limits the scalability of the software. We are thus enabling multi-thread scheduling for the simulator, which raises a number of challenges for the preservation of the integrity of the simulation flow. Moreover, with distributed algorithms being notoriously difficult to develop and debug, we are seeking to implement DPRSim-style debugging [18] to provide critical support to application development.

Acknowledgment. This work was partially supported by the ANR (ANR-16-CE33-0022-02), the French Investissements d'Avenir program, the ISITE-BFC project (ANR-15-IDEX-03), and the EIPHI Graduate School (contract ANR-17-EURE-0002).

References

1. Ahmadzadeh, H., Masehian, E., Asadpour, M.: Modular robotic systems: characteristics and applications. J. Intell. Robot. Syst. **81**(3), 317–357 (2016). https://doi.org/10.1007/s10846-015-0237-8
2. Ashley-Rollman, M.P., Pillai, P., Goodstein, M.L.: Simulating multi-million-robot ensembles. In: 2011 IEEE International Conference on Robotics and Automation, pp. 1006–1013. IEEE, Shanghai (2011). https://doi.org/10.1109/ICRA.2011.5979807. http://ieeexplore.ieee.org/document/5979807/
3. Christensen, D., Brandt, D., Stoy, K., Schultz, U.: A unified simulator for Self-Reconfigurable Robots. In: 2008 IEEE/RSJ International Conference on Intelligent Robots and Systems, pp. 870–876. IEEE, Nice (2008). https://doi.org/10.1109/IROS.2008.4650757. http://ieeexplore.ieee.org/document/4650757/
4. Collins, T., Shen, W.M.: ReBots: a drag-and-drop high-performance simulator for modular and self-reconfigurable robots. Technical reports 714, University of Southern California, Information Sciences Institute (2016)
5. Davey, J., Kwok, N., Yim, M.: Emulating self-reconfigurable robots - design of the SMORES system. In: 2012 IEEE/RSJ International Conference on Intelligent Robots and Systems, pp. 4464–4469. IEEE, Vilamoura-Algarve (2012). https://doi.org/10.1109/IROS.2012.6385845. http://ieeexplore.ieee.org/document/6385845/
6. Fitch, R., Butler, Z.: Million module march: scalable locomotion for large self-reconfiguring robots. Int. J. Robot. Res. **27**(3-4), 331–343 (2008). https://doi.org/10.1177/0278364907085097
7. Fitch, R., Butler, Z., Rus, D.: Reconfiguration planning for heterogeneous self-reconfiguring robots. In: 2003 IEEE/RSJ International Conference on Intelligent Robots and Systems, (IROS 2003), Proceedings, pp. 2460–2467 (2003). https://doi.org/10.1109/IROS.2003.1249239
8. Gerkey, B.P., Vaughan, R.T., Howard, A.: The player/stage project: tools for multi-robot and distributed sensor systems. In: In Proceedings of the 11th International Conference on Advanced Robotics, pp. 317–323 (2003)
9. Kamimura, A., Yoshida, E., Murata, S., Tomita, K., Kokaji, S.: A Self-reconfigurable modular robot (MTRAN) - hardware and motion generation software. In: 5th International Symposium on Distributed Autonomous Robotic Systems, p. 10 (2002)
10. Koenig, N., Howard, A.: Design and use paradigms for gazebo, an open-source multi-robot simulator. In: 2004 IEEE/RSJ International Conference on Intelligent Robots and Systems (IROS) (IEEE Cat. No.04CH37566), vol. 3, pp. 2149–2154. IEEE, Sendai (2004). https://doi.org/10.1109/IROS.2004.1389727. http://ieeexplore.ieee.org/document/1389727/
11. Kramer, J., Scheutz, M.: Development environments for autonomous mobile robots: a survey. Auton. Robots **22**(2), 101–132 (2007). https://doi.org/10.1007/s10514-006-9013-8. http://link.springer.com/10.1007/s10514-006-9013-8
12. Lyder, A., Garcia, R., Stoy, K.: Mechanical design of odin, an extendable heterogeneous deformable modular robot. In: 2008 IEEE/RSJ International Conference on Intelligent Robots and Systems, pp. 883–888. IEEE, Nice (2008). https://doi.org/10.1109/IROS.2008.4650888. http://ieeexplore.ieee.org/document/4650888/

13. Michel, O.: Webots: professional mobile robot simulation. J. Adv. Robot. Syst. 1(1), 39–42 (2004). http://www.ars-journal.com/International-Journal-of-Advanced-Robotic-Systems/Volume-1/39-42.pdf
14. Naz, A.: Distributed algorithms for large-scale robotic ensembles: centrality, synchronization and self-reconfiguration. Ph.D thesis, FEMTO-ST Institute, Univ. Bourgogne Franche-Comté, CNRS (2017)
15. Naz, A., Piranda, B., Goldstein, S.C., Bourgeois, J.: ABC-Center: approximate-center election in modular robots. In: 2015 IEEE/RSJ International Conference on Intelligent Robots and Systems (IROS), pp. 2951–2957. IEEE, Hamburg (2015). https://doi.org/10.1109/IROS.2015.7353784. http://ieeexplore.ieee.org/document/7353784/
16. Naz, A., Piranda, B., Goldstein, S.C., Bourgeois, J.: A time synchronization protocol for modular robots. In: 2016 24th Euromicro International Conference on Parallel, Distributed, and Network-Based Processing (PDP), pp. 109–118. IEEE, Heraklion (2016). https://doi.org/10.1109/PDP.2016.73. http://ieeexplore.ieee.org/document/7445320/
17. Pinciroli, C., et al.: ARGoS: a modular, parallel, multi-engine simulator for multi-robot systems. Swarm Intell. 6(4), 271–295 (2012). https://doi.org/10.1007/s11721-012-0072-5. http://link.springer.com/10.1007/s11721-012-0072-5
18. Rister, B.D., Campbell, J., Pillai, P., Mowry, T.C.: Integrated debugging of large modular robot ensembles. In: Proceedings 2007 IEEE International Conference on Robotics and Automation, pp. 2227–2234. IEEE, Rome (2007). https://doi.org/10.1109/ROBOT.2007.363651. http://ieeexplore.ieee.org/document/4209415/. ISSN 1050-4729
19. Rus, D., Vona, M.: Crystalline robots: self-reconfiguration with compressible unit modules. Auton. Robots 10(1), 107–124 (2001). https://doi.org/10.1023/A:1026504804984
20. Salemi, B., Moll, M., Shen, W.m.: SUPERBOT: a deployable, multi-functional, and modular self-reconfigurable robotic system. In: 2006 IEEE/RSJ International Conference on Intelligent Robots and Systems, pp. 3636–3641. IEEE, Beijing (2006). https://doi.org/10.1109/IROS.2006.281719. http://ieeexplore.ieee.org/document/4058969/
21. Spröwitz, A., et al.: Roombots-towards decentralized reconfiguration with self-reconfiguring modular robotic metamodules. In: 2010 IEEE/RSJ International Conference on Intelligent Robots and Systems (IROS) (2010). https://doi.org/10.1109/IROS.2010.5649504
22. Støy, K., Nagpal, R.: Self-reconfiguration using directed growth. In: Distributed Autonomous Robotic Systems, vol. 6, pp. 3–12 (2007). https://doi.org/10.1007/978-4-431-35873-2_1
23. Thalamy, P., Piranda, B., Bourgeois, J.: A survey of autonomous self-reconfiguration methods for robot-based programmable matter. Robot. Auton. Syst. 120, 103, 242 (2019). https://doi.org/10.1016/j.robot.2019.07.012. https://linkinghub.elsevier.com/retrieve/pii/S0921889019301459
24. Vonásek, V., Saska, M., Košnar, K., Přeučil, L.: Global motion planning for modular robots with local motion primitives. In: 2013 IEEE International Conference on Robotics and Automation (ICRA), pp. 2465–2470. IEEE (2013)
25. Yim, M., Zhang, Y., Lamping, J., Mao, E.: Distributed control for 3D metamorphosis. Auton. Robots 10(1), 41–56 (2001). https://doi.org/10.1023/A:1026544419097

Evolving Robust Supervisors for Robot Swarms in Uncertain Complex Environments

Elliott Hogg[1](✉), David Harvey[2], Sabine Hauert[1], and Arthur Richards[1]

[1] Bristol Robotics Laboratory, University of Bristol, Bristol, UK
elliott.hogg@bristol.ac.uk
[2] Thales UK, Reading, UK

Abstract. Whilst swarms have potential in a range of applications, in practical real-world situations, we need easy ways to supervise and change the behaviour of swarms to promote robust performance. In this paper, we design artificial supervision of swarms to enable an agent to interact with a swarm of robots and command it to efficiently search complex partially known environments. This is implemented through artificial evolution of human readable behaviour trees which represent supervisory strategies. In search and rescue (SAR) problems, considering uncertainty is crucial to achieve reliable performance. Therefore, we task supervisors to explore two complex environments subject to varying blockages which greatly hinder accessibility. We demonstrate the improved performance achieved with the evolved supervisors and produce robust search solutions which adapt to the uncertain conditions.

Keywords: Swarm robotics · Artificial evolution · Behaviour trees · Search and Rescue

1 Introduction

Swarm robotics studies the application of large numbers of agents which follow simple local rules to generate complex emergent behaviours, often inspired by swarms found in nature [1]. Swarms have been applied to problems from collective motion to decentralized consensus formation [2]. Whilst swarms show great promise, in practical real-world situations we need easy ways for supervisors to change the behaviour of the swarm in an intuitive and understandable way.

Supervision of swarms has previously been investigated within the area of human swarm interaction (HSI) that includes a human operator as part of the swarm system to perform supervision [3]. Performance can be improved by monitoring the swarm's state and using different methods of interaction to fix suboptimal behaviours. Much of the work in this area has focused on methods to allow humans to interact with swarms to influence their behaviour. Many investigations employ the idea of switching between different swarm algorithms to take advantage of different emergent behaviours [4]. Other methods include interacting directly with the swarm's environment to change low level goals and manually take over control of individual agents [5]. Each control method has shown the ability to positively affect the swarm's performance in varying scenarios. Other effects on human-swarm system performance include cognitive load on the human operator [6], their knowledge of swarm dynamics [7], and rate of interaction [8]. Our previous work has investigated methods to automate this process through the evolution of behaviour trees that can monitor and change the behaviour of the swarm. This produces human readable solutions, and enables systematic exploration of supervisory strategies [9].

In this paper, we explore the evolution of an artificial supervisor to control a swarm to search partially known, complex realistic environments. We aim to produce supervisory strategies which are robust to variations in the environment state and maintain high performance in comparison to solutions which specialize to a single environment state. The paper is structured as follows. Section 2 highlights similar areas of work and examples of exploration under uncertainty. Section 3 details the simulated scenario and the methodology used to apply artificial evolution to produce swarm supervision strategies. Section 4 then presents our findings and conclusions.

2 Related Work

Search and rescue (SAR) problems have been studied widely in swarm robotics as they are well suited to these problems through the use of large numbers of simple robots to cover large mission areas when compared to a single autonomous robot. This is demonstrated by Arnold *et al.* who show high performance with simple reactive behaviours [10]. The ability to search with limited sensing has also been demonstrated in real-world experiments with a swarm of drones exploring an indoor environment and only on-board sensing [11]. In addition, Hauert *et al.* evolve novel approaches to sweeping environments using agents with no global positioning information [12].

Examples of exploration in uncertain environments have been explored in varying levels of complexity within the field of swarm robotics. Yang *et al.* combine an ant colony search algorithm with pheromone mapping to efficiently cover an uncertain environment by limiting the amount of overlapping paths of agents [13]. Similarly, Pan *et al.* improve upon a particle-swarm optimization algorithm when searching environments in the presence of noise using optimal computing budget allocation [14]. Whilst in both cases performance is improved these problems are investigated in simple environments. Dirafzoon and Lobaton investigate the mapping of unknown environments using cockroach inspired swarm behaviours [15]. This follows similar concepts in HSI by adjusting parameters of the swarm behaviour during simulation to improve performance. Whilst the algorithms presented perform highly, this subject is discussed briefly and the environments that are investigated do not represent complex real-world environments.

3 Methodology

The following section will detail how artificial evolution has been applied to optimise swarm supervision strategies for a SAR scenario in uncertain environments. In this work we use behaviour trees (BT) to encode these supervisory strategies. We apply artificial evolution to optimize the structure of these trees and produce non-obvious high performing strategies.

3.1 Simulation

We investigate the exploration of indoor environments using a bespoke 2-D simulator. Swarm agents travel at a constant speed of 0.5 m/s and obstacle avoidance is achieved using potential fields. Agents have no perception of their surroundings beyond the avoidance of obstacles and can sense their distance to other agents. We measure the coverage of an environment based on the detection of objectives distributed over the environment which the swarm must find. An objective is detected when the euclidean distance between an agent and objective is less than 2.5 m and can only be found once.

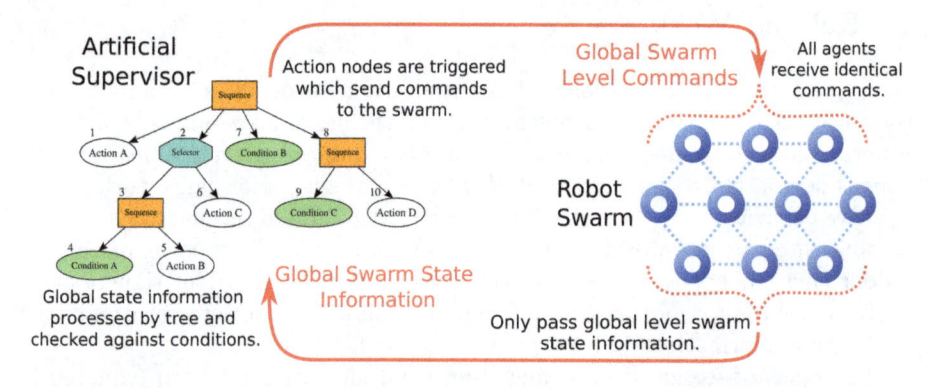

Fig. 1. Interface between the artificial supervisor and the swarm. The supervisor receives global state information from the swarm which is used to trigger different decisions. The supervisor sends global commands back to the swarm to change its behaviour.

Whilst swarm agents follow their own set of local rules, the non-embodied artificial supervisor which is separate to the swarm, observes the swarm's state at a global level and sends global level commands back to the swarm to facilitate supervision. Figure 1 presents the interaction between supervisor and swarm. The supervisor cannot perceive the shape of the environment and views the swarm through global identity-free swarm metrics without need for direct agent control. This is detailed further in Sect. 3.3.

Supervision of Robot Swarms in Uncertain Complex Environments   123

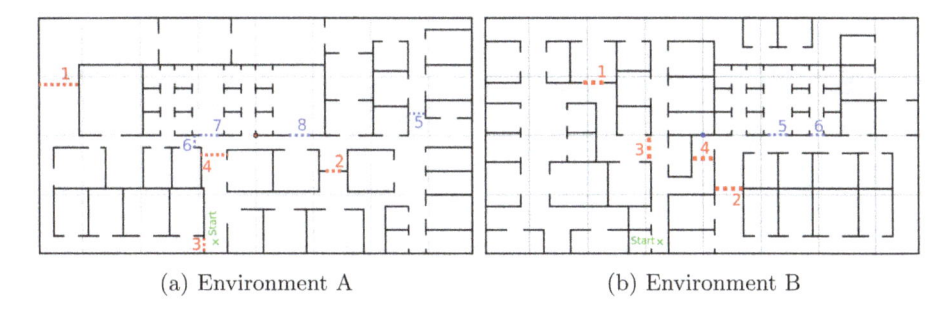

(a) Environment A

(b) Environment B

Fig. 2. Two example environments to perform a coverage task. Shown in each are the starting positions of the swarm and numbered blockage points. The red blockages highlight those which have the most effect on navigation which are chosen for testing states. Blue blockages are excluded.

3.2 Exploring Uncertain Environment States

In this paper we investigate the problem of searching known environments with varying levels of unknown blockages that hinder navigation. To do this, we first consider where blockages might occur. In this investigation we focus on the exploration of indoor environments and study two cases shown in Fig. 2. Environment A, is based upon the real Bristol Robotics Laboratory floor plan and environment B is a variation of A. Both environments measure 150 m × 80 m. In each environment the swarm is deployed from the indicated starting point and has to search the environment to find objectives which are evenly distributed in increments of 2.5 m for a total of 1800 objectives.

We explore blockages of pathways which alter the connectivity of the environment and make navigation more difficult. We first identify the worst case scenario which is the largest set of possible blockages such that the environment is not disconnected. In this case the environment can still be fully traversed, but the highest number of blockages are present. These set of blockages are highlighted in Fig. 2. With the blockages labelled as shown in Fig. 2, we measure how the addition of each individual blockage changes diffusion through the environment. Through this process we can highlight how critical blockages could change our requirements for supervisory search strategies and the need for evolution. We deploy a swarm of 1000 random walkers which are launched from the starting position and measure the probability of reaching different regions of the environment. The walkers motion is tuned to give the best level of coverage. At each position in the environment we record the proportion of time that the space is occupied over a duration of 1500 s. This produces a heat map over the environment representing the probability of reaching each position.

To measure the overall difficulty of navigation, we calculate the proportion of positions with less than 1% probability of being occupied under each blockage. With this information we identify 4 critical blockages in each environment which reduce diffusion the most as indicated by the red blockages in Fig. 2. The

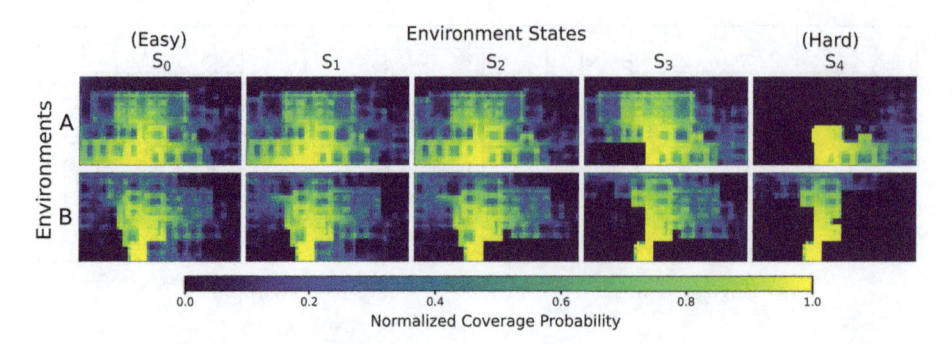

Fig. 3. Normalized probability maps of different environment states subject to blockages when deploying a swarm of 1000 random walkers. Each point in the map indicates the normalized probability that the position was covered by an agent, highlighting hard to reach blue regions.

blue blockages showed lower affect and were excluded. Given the selection of these blockages, we define a set of increasingly difficult environment states to investigate.

$$s_0 = \emptyset, s_1 = \{b_1\}, s_2 = \{b_1, b_2\}, s_3 = \{b_1, b_2, b_3\}, s_4 = \{b_1, b_2, b_3, b_4\}$$

Each environment state s_0 to s_4 refers to different possible configurations of environments A and B subject to different sets of blockages. With each of these states, we observe that the proportion of the environment which is left largely uncovered increases from state s_0 to s_4 as shown in Fig. 3. In s_1, the addition of blockages has only slight affect on diffusion whereas s_4 greatly affects navigation. Whilst we could investigate randomly changing blockages, by targeting the most significant blockages and worst-case scenario, we reduce the need to evolve over a large set of states and more efficiently target the problem. With a set of increasingly difficult environment states to explore, we next detail the implementation of artificial supervision of swarms to robustly search these set of environment states.

3.3 Swarm Supervisor

We represent the swarm supervisor in the form of a behaviour tree (BT). In order to produce control strategies we define a set of actions and conditions that allow the supervisor to interact with the swarm and can be constructed in the form of a BT. For greater detail on this design process, refer to our previous work [9].

Methods of Interaction. The supervisor can command the swarm to execute a desired swarm-level behaviour. This command is broadcast to all agents in the swarm and changes the algorithm which they execute. In addition, the supervisor can tune aspects of different behaviours by varying parameters of the local rules. In this investigation we include the following set of search based swarm behaviours.

- *Dispersion*: When clustered together, agents are repelled by an exponential force from one another causing them to disperse and cover a larger area (Fig. 4a). When significantly spread out, agents will travel with random motion [16,17]. The supervisor can also vary a parameter, R, which scales the strength of dispersion.
- *Directed fields*: The swarm will travel in a specified direction whilst repelling nearby agents using the same rules as the dispersion behaviour (Fig. 4b). We enable eight varying forms of this behaviour to direct the swarm north, south, east, and west. As well as, north west, north east, south west, and south east. These behaviours give much greater control to direct the swarm to particular regions. Control over spread is also given by parameter R.
- *Random walk*: We implement a simple random walk behaviour where agents travel at a constant speed and adjust their headings each time step based on a uniform probability distribution (Fig. 4c). The supervisor can control the degree of random motion with parameter K.
- *Rotational random walks*: By skewing the probability that agents choose to turn in a certain direction, we generate behaviours where agents move in random looping trajectories (Fig. 4d). We implement two forms of this behaviour, clockwise and ant-clockwise rotation. The supervisor can also can control the turning rate with parameter J.

Conditions. In order for the swarm supervisor to decide upon a certain action, knowledge of the swarm state is required. In this scenario the supervisor can observe the median position, spread of the swarm, and coverage achieved during simulation. The *median position* of the swarm is given in both the x direction, μ_x, and y direction, μ_y. *Spread*, σ, is defined as the average distance from agent to agent as defined in Eq. 1, where n is the total number of agents and each agent ordinate is defined as x_n and y_n. Coverage, γ, is the proportion of detected objectives to the total number of objectives.

This high level representation means the supervisor does not use knowledge of each agent state to enable control. In addition, the supervisor has no knowledge of the structure of the environment and must learn this through the evolutionary process.

$$\sigma = \frac{1}{n(n-1)} \sum_{k=1}^{n} \sum_{i=1:i \neq k}^{n} \sqrt{(x_i - x_k)^2 - (y_i - y_k)^2)} \qquad (1)$$

Condition nodes are constructed using these real-time metrics when compared to defined thresholds shown in Table 1. This enables the supervisor to perform simple checks to see if a metric is greater or less than a set value. The thresholds that can be selected for the median are bounded within the size of the environment, and similarly, the spread is limited to a high level of dispersion within the bounds of the environment.

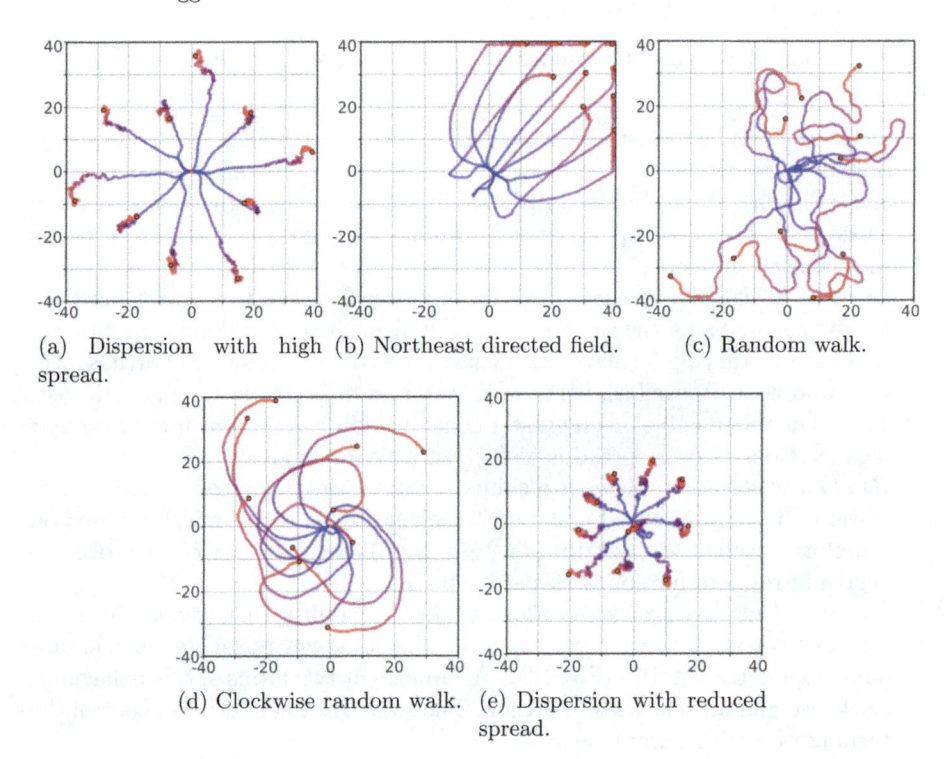

(a) Dispersion with high spread.
(b) Northeast directed field.
(c) Random walk.
(d) Clockwise random walk.
(e) Dispersion with reduced spread.

Fig. 4. Examples of available swarm behaviours used for navigation. Each figure depicts the swarm motion over 150 time steps with agent positions shown with shifting colour from blue to red with the progression of time.

3.4 Evolving the Swarm Supervisor

We apply genetic programming (GP) to evolve BTs and optimize their structure in order to produce high performing supervisory strategies. GP has previously been applied to BTs in other applications [18]. Table 1 presents the available nodes used to construct trees and highlights the limits that the conditional statements must satisfy.

Evolutionary Algorithm. We evaluate the fitness of strategies by the proportion of objectives that are detected. The reward for finding each objective also decays over time in order to promote fast exploration. Each objective has a unique decay constant based on the probability maps discussed in Sect. 3.2. The probability of finding each objective scales the rate of decay, ρ, by Eq. 2 where δ is the probability of finding an objective, t is the current time, and T is the total time duration. Therefore, objectives with a low probability of being found decay slower, maintaining a higher reward. Each environment state under investigation has a unique probability map such that the rate of decay varies

Table 1. The limits defined by the GP algorithm for the types of nodes that can be selected to produce BTs.

Node type	Selection Choices
Operator	Selector / Sequence (Between 2-7 children)
Action node	Emergent behaviours: Dispersion / North / South / East / West / North East / North West / South East / South West / Random walk / Clockwise random walk / Anti-clockwise random walk Param set: $R \in [1, 10, 20, 30, 40, 50, 60]$ $J \in [0.01, 0.02, 0.03, 0.04, 0.05, 0.06, 0.07, 0.08, 0.09]$ $K \in [0.005, 0.01, 0.015, 0.02, 0.025, 0.03, 0.035, 0.04]$
Condition node	$\mu_x>$-38, $\mu_x>$-34, $\mu_x>$-30, ... increment by 4 ..., $\mu_x>$30, $\mu_x>$34, $\mu_x>$38 $\mu_x<$-38, $\mu_x<$-34, $\mu_x<$-30, ... increment by 4 ..., $\mu_x<$30, $\mu_x<$34, $\mu_x<$38 $\mu_y>$-38, $\mu_y>$-34, $\mu_y>$-30, ... increment by 4 ..., $\mu_x>$30, $\mu_y>$34, $\mu_y>$38 $\mu_y<$-38, $\mu_y<$-34, $\mu_y<$-30, ... increment by 4 ..., $\mu_x<$30, $\mu_y<$34, $\mu_y<$38 $\sigma>$2, $\sigma>$6, $\sigma>$10, ... increment by 4 ..., $\sigma>$26, $\sigma>$30, $\sigma>$34 $\sigma<$2, $\sigma<$6, $\sigma<$10, ... increment by 4 ..., $\sigma<$26, $\sigma<$30, $\sigma<$34 $\gamma>$0.1, $\gamma>$0.2, $\gamma>$0.3, ... increment by 0.1 ..., $\gamma>$0.7, $\gamma>$0.8, $\gamma>$0.9 $\gamma<$0.1, $\gamma<$0.2, $\gamma<$0.3, ... increment by 0.1 ..., $\gamma<$0.7, $\gamma<$0.8, $\gamma<$0.9

depending on the difficulty of each state. This helps scale the fitness reward in accordance with the difficulty of each environment state.

$$\rho = \exp\left(-3\frac{\delta t}{T}\right) \quad (2)$$

The fitness function used to evaluate solutions is defined by Eq. 3. Each individual has n attempts to search the environment which is set to 4. The score per run, β, is the total reward achieved based on the detection of objectives in the environment over time and, α, is the total number of objectives. The size of evolved BTs are also limited to a maximum number of nodes τ. This applies pressure on the evolution to find concise and efficient solutions. When training a supervisor over different environment states, the fitness is averaged across each of those states. Evolution aims to maximize fitness.

$$Fitness = \begin{cases} \frac{1}{n\alpha}\sum_{k=1}^{n}\beta & \text{if } \tau <= 100 \\ 0 & \text{if } \tau > 100 \end{cases} \quad (3)$$

We implement a standard evolutionary island model using three islands with identical conditions [19]. We ran evolutionary runs over 300 generations with a population of 40 individuals on each island. At each generation, we used tournament selection with groups of 3 individuals followed by single point crossover, single point mutation, and sub-tree growth. Elitism is used to save the best 6 individuals from each generation. For each search, we set a time limit of 1500 s. This time period is sufficiently long to fully explore the environments.

4 Results

4.1 Behaviour Benchmarks

Before evaluating the performance of the evolved supervisory strategies, we test the performance of each individual swarm behaviour when deployed without supervision in each environment state (Fig. 5a and 5b) to highlight where evolved supervision is needed.

For environment A, we see that the random behaviours perform well, with the clockwise walk performing the highest, whilst random walk and dispersion also perform well. When deploying this behaviour we see that it is very useful for cycling in and out of rooms, following the edge of the interior, before then exiting to continue through the environment (Fig. 6a). For environment B, we see that again dispersion performs well, whereas, the rotational walks perform much lower than in environment A. In this case, there are several places where agents can become stuck looping over the same area when passing through certain pathways (Fig. 6b). The use of the rotation behaviours appear very effective in certain types of environments where many paths are open, however in environment B, these types of motions are not suitable in bent corridors. The benchmarks highlight that we cannot rely on these singular behaviours to solve these problems and highlight that behaviours become susceptible to changes in environment states, causing significant drops in performance.

4.2 Evolved Supervisory Control

The aim of this investigation is to produce solutions that can robustly search environments under the presence of unknown blockages. We first evolve supervision of the swarm in only state s_0 where no blockages are present and the supervisor is not exposed to changes in the environment state. We then run further configurations where each supervisor is exposed to at least two possible environment states during evolution. These different training configurations are presented in Figs. 5c and 5d. Each row shows the performance of an evolved supervisor which has been subject to a unique set of training states during evolution. We see the increase in performance in relation to Figs. 5a and 5b demonstrates the value of artificial supervision over the deployment of an unsupervised swarm. In addition, we observe the ability to produce generalizable solutions where supervisors achieve high performance across each environment state. In the following sections we examine these solutions and how they form robust strategies.

4.3 Qualitative Analysis of Evolved Supervision

In both environments, supervisors interacted with the swarm by selecting specific sequences of swarm behaviours with fine tuned control of spread and degree of random motion. We found that supervisors opted to use combinations of the

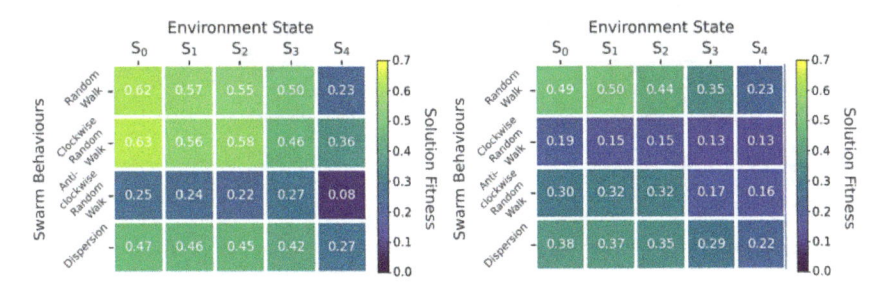

Supervision of Robot Swarms in Uncertain Complex Environments 129

(a) Swarm behaviours without supervision in environment A.

(b) Swarm behaviours without supervision in environment B.

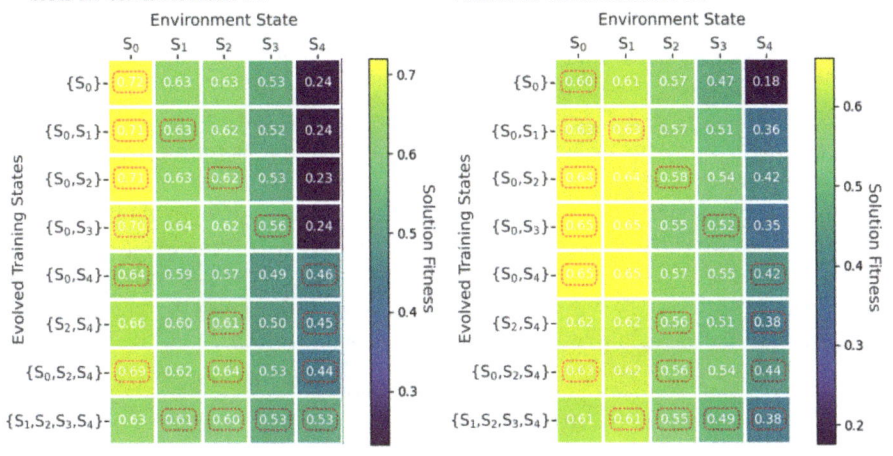

(c) Evolved solutions in environment A.

(d) Evolved solutions in environment B.

Fig. 5. Fitness of the evolved supervisory strategies under each possible environment state. Each data point is the average performance over 100 trials. Each row represents the best evolved supervisor from each set of training states. The red boxes indicate the performance in the states of which the supervisor was trained. Supervisors trained over multiple environment states produce more robust performance.

random walk, rotational walks, and the directed field behaviours. We saw very few uses of dispersion despite performing well in the benchmarks.

If we first examine the supervisor trained in only state s_0 in environment A, we see that the supervisor scores highly when searching in state s_0 as expected. The supervisor initially uses the random walk behaviour to spread out in all directions before switching to the clockwise walk. The random walk performs well initially, however agents become trapped in rooms more easily after a short period of time. The supervisor identifies this and switches to the clockwise random walk which is better suited to enter and exit rooms. The supervisor finishes by commanding the swarm to head east, forming a final sweep of the environment near the end of the search period. This approach was able to achieve up to 94% coverage of the environment. This shows the benefit of the supervisor to

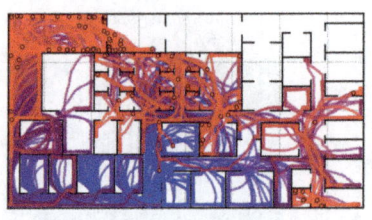

130   E. Hogg et al.

(a) Clockwise random walk coverage in environment A

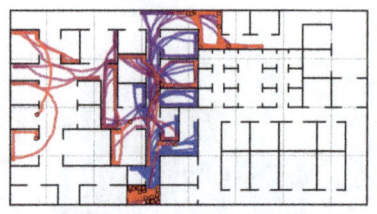

(b) Clockwise random walk coverage in environment B

Fig. 6. Coverage of a swarm of 100 agents when performing only a clockwise random walk. Agent trajectories are plotted with colour shifting from blue to red over time. Whilst very effective in environment A, the behaviour breaks easily under the right conditions where agents become trapped in environment B.

identify the advantages of each type of behaviour and combine them into an effective search strategy.

Whilst this supervisor performs well in state s_0, when evaluated in the blocked states we see that performance drops quickly with very low performance in s_4 as shown in Fig. 5c. Under the additional blockages the supervisor attempts to use the random walk which causes the majority of agents to become stuck near the starting position resulting in minimal exploration.

4.4 Robust Search Strategies

In environment A, when training over s_0 and one additional state up to s_3, there is shown to be little variation in overall performance and still poor performance in s_4. This suggests that the addition of blockages s_1 to s_3 do not have significant effect on how we should change supervision.

However, when the supervisor is evolved over s_4, we see that performance in this state is greatly improved. Every supervisor trained including state 4 takes a different approach to the previous supervisors as illustrated in Fig. 7 and shown in simulation (video link). Rather than first using random behaviours, 3 of the solutions first direct the swarm east away from the additional blockage stopping access to the center of the environment (Fig. 7a). This directs the swarm across to the bottom right of the environment to avoid the blockages. In one case the north movement is used with high spread which inadvertently pushes agents east then up the right side of the environment. Under state s_4, using the random behaviours initially performs poorly where the use of the clockwise random walk causes agents to become trapped within a loop near the starting point. This is the reason why solutions on row 1 to 4 which aren't trained over s_4 perform poorly. After being directed around the blockages in s_4, the supervisors switch to the clockwise random walk which becomes effective at passing upwards and around the blockages (Fig. 7b) and finishing with a normal random walk (Fig. 7c). In addition, when these solutions are deployed in states s_0 to s_3 with fewer

(a) Initially directs the swarm east away from blockages. Time-step 275.

(b) Switch to clockwise random walk to cycle upwards. Time-step 626.

(c) Finish by switching to random walk for greater coverage. Time-step 1420.

(d) The supervisory behaviour tree evolved over states s_1 to S_4.

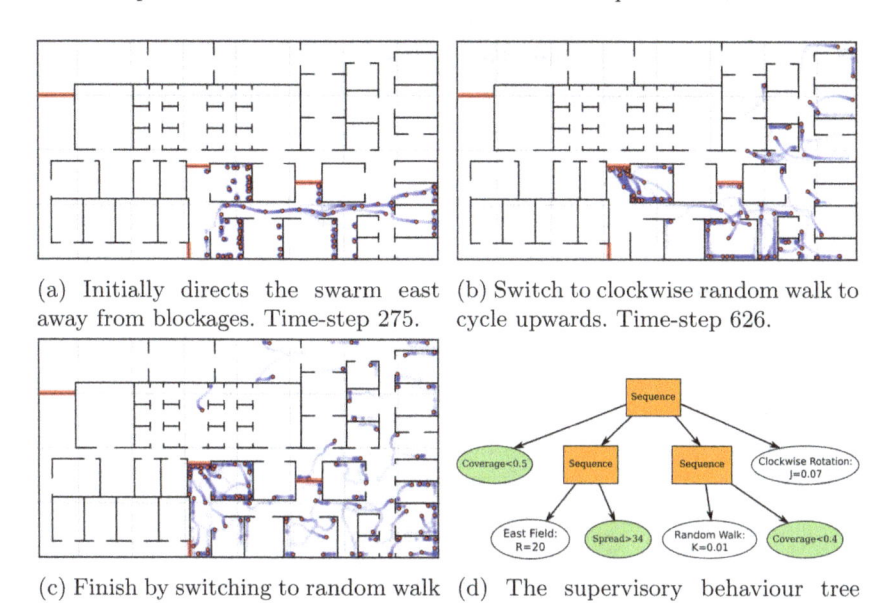

Fig. 7. The behaviour of the supervisor trained over states s_1 to s_4 when deployed in environment A. The stages of this solution are captured when searching the environment in state s_4 showing the swarm being diverted around blockages highlighted in red.

blockages, this approach is still effective and the supervisor switches faster to the clockwise random walk behaviour, identifying the lack of blockages.

4.5 Performance Variation Between Environments

In environment B, we observe different trends. As previously highlighted, we see a lack of robustness for the supervisor trained only in s_0. This is because the supervisor learns to use the random clockwise walk initially which performs poorly under the presence of additional blockages, causing agents to become trapped in certain areas. However, when trained over one additional environment state with s_1, we see improved performance even in s_4 despite not being trained over that state. This suggests that in this case, it's not crucial to expose the supervisor to the most difficult state s_4.

We also notice that the supervisor trained on s_0 and s_3 performs highly in each environment state but also achieves the highest scores in s_0 and s_1, outperforming the supervisor which specialises to s_0. These findings suggest in this case that it's always better to give the supervisor a small amount of exposure to other environment states to promote higher performance in all states without having to compromise performance in s_0. In both environments we observe that given exposure to additional states of the environment, the supervisor is capable of adapting its strategy to maintain good performance in the worst case scenario. Through this exposure, the supervisors identify where the most significant blockages occur and actively directs the swarm away from these areas of the environment, taking the safest route and ensuring robust performance.

5 Conclusions and Future Work

In order to achieve real-world deployment of swarms, we need to consider how to effectively monitor and influence their behaviour. The implementation of supervisory control of swarms enables us to promote high performing, robust solutions to complex problems with only minimal global level interaction. In this paper we showed that through the application of artificial swarm supervision, we were able to take advantage of different swarm behaviours to effectively search complex environments. We identified that when training a supervisor without the presence of uncertainty, the emergent strategies were not robust to variations in the environment. By evolving over the worst case scenario where critical blockages were present, we observed that the supervisor learned to direct the swarm away from these blockage points and adapt its approach and use different behaviours. We also identified that evolving over only slight variations in the environment made supervision significantly more robust to the most critical blockages which it had not been exposed to. All of these supervisory solutions were achieved using concise sequences of behaviours, relying on the autonomy of the swarm and learning how best to utilise the available behaviours.

Future work will explore how we can evolve robust supervision with the worst case scenario in greater detail and how this compares to evolving over randomly chosen states. In addition, we will investigate a broader spectrum of environments to understand to what extent our findings generalize.

Acknowledgements. This work was funded and delivered in partnership between the Thales Group and the University of Bristol, and with the support of the UK Engineering and Physical Sciences Research Council Grant Award EP/R004757/1 entitled "Thales-Bristol Partnership in Hybrid Autonomous Systems Engineering (T-B PHASE)".

References

1. Meng, X.B., Gao, X.Z., Lu, L., Liu, Y., Zhang, H.: A new bio-inspired optimisation algorithm: bird swarm algorithm. J. Exp. Theor. Artif. Intell. **28**(4), 673–687 (2016)
2. Valentini, G., Ferrante, E., Dorigo, M.: The best-of-n problem in robot swarms: formalization, state of the art, and novel perspectives. Front. Robot. AI **4**, 9 (2017). https://doi.org/10.3389/frobt.2017.00009
3. Kolling, A., Walker, P., Chakraborty, N., Sycara, K., Lewis, M.: Human interaction with robot swarms: a survey. IEEE Trans. Hum. Mach. Syst. **46**(1), 9–26 (2016)
4. Kolling, A., Nunnally, S., Lewis, M.: Towards human control of robot swarms. In: HRI 2012 - Proceedings of the 7th Annual ACM/IEEE International Conference on Human-Robot Interaction, pp. 89–96 (2012)
5. Walker, P., Amraii, S.A., Lewis, M., Chakraborty, N., Sycara, K.: Control of swarms with multiple leader agents. In: Conference Proceedings - IEEE International Conference on Systems, Man and Cybernetics, pp. 3567–3572 (2014)

6. Kolling, A., Sycara, K., Nunnally, S., Lewis, M.: Human swarm interaction: an experimental study of two types of interaction with foraging swarms. J. Hum. Robot Interact. **2**(6), 104–129 (2013)
7. Kapellmann-Zafra, G., Salomons, N., Kolling, A., Groß, R.: Human-robot swarm interaction with limited situational awareness. In: Dorigo, M., et al. (eds.) ANTS 2016. LNCS, vol. 9882, pp. 125–136. Springer, Cham (2016). https://doi.org/10.1007/978-3-319-44427-7_11
8. Walker, P., Nunnally, S., Lewis, M., Chakraborty, N., Sycara, K.: Levels of automation for human influence of robot swarms. In: Proceedings of the Human Factors and Ergonomics Society, pp. 429–433 (2013)
9. Hogg, E., Hauert, S., Harvey, D., Richards, A.: Evolving behaviour trees for supervisory control of robot swarms. Artif. Life Robot. **25**(4), 569–577 (2020). https://doi.org/10.1007/s10015-020-00650-2
10. Arnold, R.D., Yamaguchi, H., Tanaka, T.: Search and rescue with autonomous flying robots through behavior-based cooperative intelligence. J. Int. Humanit. Action **3**, 1–18 (2018). https://doi.org/10.1186/s41018-018-0045-4
11. Stirling, T., Roberts, J., Zufferey, J.C., Floreano, D.: Indoor navigation with a swarm of flying robots. In: Proceedings - IEEE International Conference on Robotics and Automation, pp. 4641–4647 (2012)
12. Hauert, S., Zufferey, J.C., Floreano, D.: Evolved swarming without positioning information: an application in aerial communication relay. Auton. Robot. **26**(1), 21–32 (2009)
13. Yang, F., Ji, X., Yang, C., Li, J., Li, B.: Cooperative Search of UAV Swarm Based on Improved Ant Colony Algorithm in Uncertain Environment
14. Pan, H., Wang, L., Liu, B.: Particle swarm optimization for function optimization in noisy environment. Appl. Math. Comput. **181**(2), 908–919 (2006)
15. Dirafzoon, A., Lobaton, E.: Topological mapping of unknown environments using an unlocalized robotic swarm. In: IEEE International Conference on Intelligent Robots and Systems, pp. 5545–5551 (2013)
16. Hsiang, T.-R., Arkin, E.M., Bender, M.A., Fekete, S.P., Mitchell, J.S.B.: Algorithms for rapidly dispersing robot swarms in unknown environments. In: Boissonnat, J.-D., Burdick, J., Goldberg, K., Hutchinson, S. (eds.) Algorithmic Foundations of Robotics V. STAR, vol. 7, pp. 77–93. Springer, Heidelberg (2004). https://doi.org/10.1007/978-3-540-45058-0_6
17. McLurkin, J., Smith, J.: Distributed algorithms for dispersion in indoor environments using a swarm of autonomous mobile robots. Distrib. Auton. Robot. Syst. **6**, 399–408 (2008)
18. Jones, S., Studley, M., Hauert, S., Winfield, A.: Evolving behaviour trees for swarm robotics. In: Groß, R., et al. (eds.) Distributed Autonomous Robotic Systems. SPAR, vol. 6, pp. 487–501. Springer, Cham (2018). https://doi.org/10.1007/978-3-319-73008-0_34
19. Squillero, G., Tonda, A.: Divergence of character and premature convergence: a survey of methodologies for promoting diversity in evolutionary optimization. Inf. Sci. **329**, 782–799 (2016)

Distributed Cooperative Localization with Efficient Pairwise Range Measurements

Anwar Quraishi[✉] and Alcherio Martinoli

Distributed Intelligent Systems and Algorithms Laboratory, School of Architecture, Civil and Environmental Engineering, École Polytechnique Fédérale de Lausanne, Lausanne, Switzerland
anwar.quraishi@epfl.ch
https://disal.epfl.ch/auvdistributedsensing

Abstract. We present a method based on covariance intersection for cooperative localization with pairwise range-only relative measurements. Our method was designed for underwater robots equipped with an acoustic communication and ranging system. Range measurements are not sufficient to compute a complete relative 3D position. Therefore, covariance intersection is performed in a transformed space along their relative estimated positions, while preserving cross-correlations between other state variables. Given the characteristics of the acoustic channel, only one robot can transmit data or a ranging request at a time, hence the pairwise limitation. We also present a heuristic for choosing a peer robot for a range measurement by maximizing mutual information. Our method places no further restrictions on the order, timing or scheduling of relative measurements. We evaluated our method for accuracy and consistency, and present results from simulations as well as outdoor experiments.

1 Introduction

A number of robot actions, such as path planning and spatial information gathering depend on accurate localization. Robots operating on land or in the air can often exploit external positioning references such as Global Navigation Satellite Systems (GNSSs), cameras and range finders. However, access to such external references is limited in many scenarios, such as indoor places, caves and underwater environments. In Cooperative Localization (CL), a team of robots operating together shares and fuses information and relative measurements to improve their individual localization accuracy. This allows new, more accurate information about position acquired by one robot to be propagated to other robots. We have previously demonstrated acoustic navigation for underwater robots with static and moving surface beacons [1]. However, by using CL and taking turns to resurface periodically for GNSS reception, underwater robots can function without relying on external beacons, as illustrated in Fig. 1.

An important characteristic of CL is that information sharing makes position estimates of all robots correlated [2]. Position estimates of other robots coupled

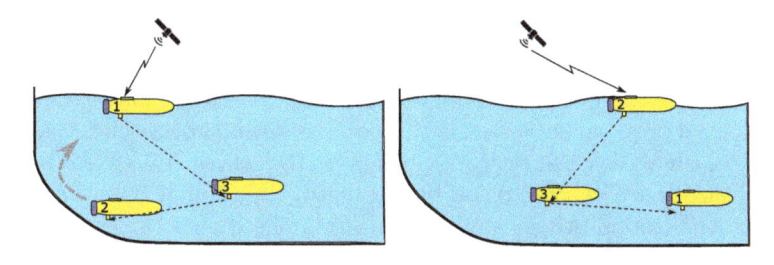

Fig. 1. Underwater robots periodically resurface for GNSS position reception, and then use CL to share the improved position estimate with other robots.

with relative measurements are key ingredients for CL. However, measurement models in localization frameworks used typically, such as Kalman filtering, often assume that the measurement information is uncorrelated with robot state estimates [3]. Therefore, relative measurement updates in CL need to be handled carefully to avoid inconsistent and overconfident position estimates. A further problem of peer-selection arises in CL with pairwise relative measurements. The ideal choice of a peer is that which provides a relative measurement resulting in the maximum information gain.

CL has received considerable attention in the realm of multi-robot systems. The simplest approach for CL is to gather all robot observations and relative measurements and process them at a central location [4]. Roumeliotis et al. in [5] showed that a centralized Kalman filter for CL can be decomposed into smaller, communicating filters which are distributed among the robots. Later, Luft et al. in [6] limited communication exclusively between the pair of robots that obtained a relative measurement. This reduced the communication cost for N robots from $\mathcal{O}(N)$ to $\mathcal{O}(1)$. However, both approaches require inter-robot cross-correlation terms to be communicated along with robot position estimates. This is necessary but adds communication cost.

Treating correlated information as independent, as in [7], can make new position estimates overconfident [8]. In the work of Bahr et al. [2], each robot maintains a set of several state estimates, and keeps track of their dependencies with other peers through careful book-keeping. A robot can use only those estimates of another peer that are not correlated (directly or via other peers) with its own estimate. However, the memory, computation and communication requirements of this approach grow exponentially with the number of robots.

In [9] and [10], the authors use an approach based on distributed Maximum Likelihood Estimation, where each robot optimizes its own state given the relative measurements. The problem of inter-robot correlations does not arise in this approach. While there is no formal proof of convergence, it performs well in practice.

Another technique for addressing the problem of inter-robot correlation is Covariance Intersection (CI) [8,11–13], which treats estimates as if they were maximally correlated. They do not require robots to communicate cross-correlation estimates, saving communication bandwidth. The price to pay is that they are pessimistic in that they overestimate uncertainty. This overestimation is addressed in a hybrid approach, called Split-CI [14,15]. It splits the covariances into dependent and independent components. However, it requires communication of both, the independent and the dependent covariance matrices.

The second problem addressed in this paper is optimal peer-selection, which is similar to optimal sensor or beacon selection. In [16], three best (fixed) ultrasonic ranging beacons are selected based on Geometric Dilution of Precision (GDOP) for indoor localization. An entropy-minimization-based sensor selection approach for fixed target tracking is presented in [17]. [18] presents a selection criteria based on mutual information for general Bayesian filtering problems.

In this paper, we present a CI-based fully distributed cooperative localization algorithm for range-only relative measurements. We have developed our method keeping in mind limitations of underwater acoustic ranging and communication. To that end, CI offers important advantages. It does not need inter-robot correlations to be computed and communicated. While this is inefficient in that the uncertainty in robot positions is *overestimated*, it allows for a completely distributed implementation of CL. Additionally, CI is provably consistent.

When a relative measurement comprises of a full relative pose along with the position estimate of another robot, it is straightforward to perform CI. This is not the case with range measurements, which are one dimensional, whereas robot positions can be two- or three-dimensional. Therefore, a range measurement update can have a direct influence on robot position only along the relative position vector between two robots. We perform CI in a transformed space aligned with an estimate of this vector to update the robot position and the corresponding position covariance. It is important to note that range measurements can indirectly influence all state variables via the cross-correlations between them. Our method accounts for and preserves the cross-correlation between the state variables. While our method adds computational cost, the cost of internal computation is much lower than that of acoustic communication.

We also derive a peer-selection heuristic for choosing the best peer for performing a pairwise range measurement. In the trivial case when uncertainty in positions of other robots is not known, the best choice is a peer robot that is along (or closest to) the direction of highest uncertainty. However, the knowledge of the said uncertainty exists because robots broadcast their position estimates during a range measurement. We use the mutual information between current position estimate of the robot and a potential range measurement to derive a mathematical expression for scoring the 'usefulness' of peer robots.

In summary, our work consists of two main contributions. (1) We derive a linear transform of the robot state in which to perform CI with range-only relative measurements, while preserving the cross-covariances in the robot state, and (2) we derive a mathematical expression to rank peer robots in the order of the amount of information a relative update would provide.

2 Methodology

We consider a team of N underwater robots navigating in a three dimensional space. Robots choose a peer with which to perform a range measurement and

(a) Peer selection (b) Query (c) CI (d) New estimate

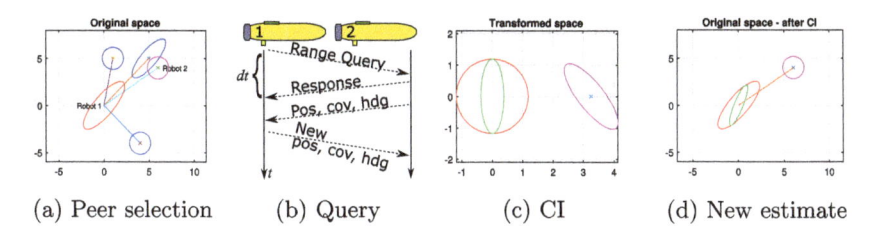

Fig. 2. (a) Robot 1 chooses Robot 2 for a range measurement. (b) R2 responds to the range query of R1 and also sends its position estimate. (c) CI is performed in a transformed space. (d) Updated estimate of R1 (in green).

transmit a ranging query. On receiving a response, CI is performed using our *projected covariance intersection* method, while preserving all the cross-correlation terms in the state. The sequence of steps are illustrated in Fig. 2, and explained in the rest of this section.

2.1 State Description

The states of the robots are assumed to be Gaussian random variables and are expressed using the mean-covariance parametrization,

$$s_i^t \sim \mathcal{N}(X_i^t, \Sigma_i^t), \tag{1}$$

where $i \in \{1, \ldots, N\}$. Our algorithm is agnostic to the formulation of the state variable, except that it requires the three-dimensional position represented in a fixed frame to be a part of the state.

$$X_i^t = [x_i, y_i, z_i, \ldots] \tag{2}$$

2.2 Motion and Individual Measurements

Motion and private measurement updates are purely internal to a robot do not require any communication between them. They are processed individually by robots using any kind of sensor fusion framework such as a Kalman filter.

2.3 Range-Based Covariance Intersection

Communication between two robots is required only when they perform a range measurement. No other robots are required to be involved in the communication. However, we assume all communication is broadcast, so other robots can listen to this communication.

Consider a robot, i with position and position covariance $\vec{p}_i, R_i$. It chooses to query another robot j and obtains a range measurement r_{ij} with standard deviation σ_r. Robot j also transmits its position estimate $\vec{p}_j, R_j$. We define the range vector as

$$\vec{r}_{ij} = r_{ij} \frac{\vec{p}_j - \vec{p}_i}{\|\vec{p}_j - \vec{p}_i\|}, \tag{3}$$

Note that only magnitude of $\vec{r}_{ij}$, i.e., r_{ij}, is measured, the actual vector is obtained from estimated quantities.

A transform $\mathcal{F}$ is applied to the state of robot i so that the dimension along $\vec{r}_{ij}$ is decoupled from the rest of the elements of the state.

$$q = \mathcal{F}(s_i - X_i), \tag{4}$$
$$\Sigma'_q = \mathcal{F}\Sigma_i\mathcal{F}^{\mathrm{T}}, \tag{5}$$

where q is the transformed state vector. Such a transform can be obtained by setting

$$\mathcal{F} = VW^{-\frac{1}{2}}T^{\mathrm{T}}, \tag{6}$$

where T and W are obtained from the eigenvalue decomposition of the state covariance matrix Σ_i. V is obtained from the Gram-Schmidt orthogonalization [19] starting with the vector $\vec{r}_{ij}$. The key feature of this transform is that q is zero-mean and the covariance matrix is identity, i.e., all the elements are uncorrelated.

$$q \sim \mathcal{N}(0_{M\times 1}, I_{M\times M}). \tag{7}$$

This method has similarly been used in [20] for truncating Gaussian PDFs given a hard constraint. The correlations between elements of X_i are not lost but encoded in T and W. The matrix V in the transform rotates the space in a way that the first dimension in the transformed space is along the range vector. CI is then easily performed in one dimension, and only applies to the first element of q, q_x.

$$q_x \sim \mathcal{N}(0,1). \tag{8}$$

Next, the transformation is applied to the other relevant quantities, namely $\vec{r}_{ij}$, σ_r, $\vec{p}_j$ and R_j. Note that $\mathcal{F}$ applies rotation as well as scaling.

$$\vec{r}' = \mathcal{F}\vec{r}_{ij}, \tag{9}$$
$$\sigma'_r = \|\mathcal{F}\hat{r}\|\sigma_r, \tag{10}$$
$$\vec{p}'_j = \mathcal{F}\vec{p}_j, \tag{11}$$
$$R'_j = \mathcal{F}R_j, \tag{12}$$

where $\hat{r}$ is the unit range vector. To perform CI, we are interested in the conditional probability distribution of $\vec{p}'_j$ along the transformed range vector $\vec{r}'$ as below. This is trivial to obtain for a Gaussian distribution.

$$P_j(p'_j|y'=0, z'=0) \sim \mathcal{N}(x'_j, \sigma'^2_j). \tag{13}$$

In the transformed space, using the range measurement and position of robot j, an estimate of the position of robot i is calculated as

$$\hat{x}'_i = x'_j - \vec{r}', \tag{14}$$
$$\hat{\sigma}'^2_i = \sigma'^2_j + \sigma'^2_r. \tag{15}$$

Thus, we have

$$\hat{q}_x \sim \mathcal{N}(\hat{x}'_i, \hat{\sigma}'^2_i). \tag{16}$$

CI is now performed between q_x and $\hat{q}_x$ (see Eqs. 8, 16) to obtain an updated estimate of the position of robot i in the transformed space.

$$\left[\sigma_{x'}'^2\right]^{-1} = \omega\left[1\right]^{-1} + (1-\omega)\left[\hat{\sigma}_i'^2\right]^{-1}, \tag{17}$$

$$\mu_{x'} = \left[\sigma_{x'}'^2\right] \cdot \left[0 \cdot \omega\left[1\right]^{-1} + \hat{x}_1' \cdot (1-\omega)\left[\hat{\sigma}_i'^2\right]^{-1}\right] \tag{18}$$

where ω is chosen to minimize $\sigma_{x'}'^2$. After this update, the resulting state and covariance of robot 1 in the transformed space are

$$q = [\mu_{x'}, 0, 0, \ldots], \tag{19}$$

$$\Sigma_q' = \mathrm{diag}\left(\sigma_{x'}^2, 1, 1, \ldots\right). \tag{20}$$

This is illustrated in Fig. 2c. Finally, the new estimate of the state of robot 1 in the original space is computed by performing an inverse transform.

$$\hat{X}_{i|j} = TW^{\frac{1}{2}}V^{\mathrm{T}}q + X_1, \tag{21}$$

$$\hat{\Sigma}_{i|j} = TW^{\frac{1}{2}}V^{\mathrm{T}}\Sigma_q' VW^{\frac{1}{2}}T^{\mathrm{T}}, \tag{22}$$

which follows from the inverse of the transform, $\mathcal{F}^{-1} = TW^{\frac{1}{2}}V^{\mathrm{T}}$, given that T and V are orthonormal matrices (see Fig. 2d).

2.4 Peer Selection

Consider a robot i that needs to choose another robot $j \in [1, n] \setminus i$ and perform a pairwise range update (we performed experiments with $n = 3$ and $n = 4$). We seek to choose a robot j that results in the lowest posterior uncertainty, as shown in Fig. 2a. Doing so would require current position estimates of all robots, which is not feasible in view of communication constraints. However, since all communication is broadcast, all robots receive position estimates as well as heading of other robots during a range measurement (see Fig. 2b and Sect. 2.5). This information, coupled with a constant velocity model is used to compute the current estimates for all other robots.

For the following analysis, we assume that the state consists of only the 3D positions. We have

$$s_k \sim \mathcal{N}(p_k, R_k). \tag{23}$$

Let z_{ij} be the random variable describing the potential range measurement between i and j (including the position of a peer robot j). We would like to choose a j for which the mutual information $I(s_i; z_{ij})$ is maximized. Formally, we seek to solve the problem

$$j^* = \arg\max_{j \in [1,n] \setminus i} I(s_i; z_{ij}). \tag{24}$$

We know that this mutual information between two random variables can be written in terms of their differential entropy as

$$I(s_i; s_j) = h(s_i) - h(s_i | z_{ij}), \tag{25}$$

where $h(s)$ is the differential entropy of random variable s [21]. When s has a Gaussian distribution, the $h(s)$ can be evaluated as

$$h(s) = \frac{1}{2}\log(2\pi e)^n |\Sigma_s| \text{ bits}, \qquad (26)$$

where $|\Sigma_s|$ is the determinant of the covariance matrix Σ_s. In Eq. 25, $P(s_i)$ is fixed, and only the posterior density $P(s_i|z_j)$ depends on the choice of j. Therefore, we can combine Eqs. 25, 26 to reformulate the problem in Eq. 24 as

$$j^* = \arg\min_{j \in [1,n] \setminus i} |\hat{\Sigma}_{i|j}|, \qquad (27)$$

where $\hat{\Sigma}_{i|j}$ is the covariance of the posterior probability density. This was previously computed in Eq. 22. On the right side of the equation, note that T and V are orthonormal matrices, hence their determinant is 1. The posterior covariance in the transformed space Σ'_q is shown in Eq. 20 to be a diagonal matrix such that its determinant will evaluate to $\sigma^2_{x'}$. Therefore, we have

$$|\hat{\Sigma}_{i|j}| = |W||\Sigma'_q| = |W|\sigma^2_{x'}. \qquad (28)$$

Considering that that W is independent of j, and following from Eqs. 15 and 17, the problem can be further simplified to

$$j^* = \arg\min_{j \in [1,n] \setminus i} \sigma'^2_j + \sigma'^2_{r,j}. \qquad (29)$$

We recall here that the first term is the conditional distribution of the position of robot j, and the second term is the variance of the range measurement, both in the transformed space. Note that it can be easily deduced that if the first term is ignored, the minimum is obtained for robot j which is along the direction of highest uncertainty of robot i.

However, broadcasts from other robots coupled with a constant velocity model provide a current estimate for the first term, σ'^2_j, as explained earlier. This serves as a heuristic for solving the optimization problem in Eq. 29.

2.5 Range Queries and Communication

The limitation of pairwise measurements is because of constraints of most underwater acoustic transceivers. To avoid interference, only one robot can transmit a signal at a time. However, all robots can listen to any transmitted signal, if they are within communication range. We use a Time Division Multiple Access (TDMA) scheme, where each robot is assigned pre-determined time slots by rotation. Practical approaches for implementing such a scheme in underwater robots have been discussed in [1].

During its assigned slot, a robot initiates a range query and the subsequent exchange shown in Fig. 2b. Range is computed by measuring two-way-travel-time. Finally, the robot performs CI and broadcasts its updated position estimate. The duration of individual time slots depends on the time it takes to perform the exchange in Fig. 2b, which is a characteristic of the acoustic transceiver hardware used.

3 Experimental Setup

The proposed method was implemented and tested in a group of three to four robots. We evaluated our algorithms in terms of localization accuracy and estimation consistency, both in simulation as well as with outdoor experiments. The kinematics in simulation were roughly calibrated to that of real robots. In both, simulation and real experiments, the robots were programmed to follow a pre-planned trajectory. The state estimator uses proprioceptive sensors and a motion model of the robot for inertial navigation, combined with range measurements. To emulate periodic surfacing events in a team of underwater robots, two of the three or four robots in the group were allowed periodic access to GNSS positioning information, which would be passed on to other robots through the cooperative localization framework. Otherwise, GNSS positions were used only for following the trajectory and as ground truth.

Experiments are performed with two strategies of choosing a peer for range measurement. In the first one, called *cyclic*, each robot queries other robots for a range measurement turn-by-turn, in a cycle. The second one, called best-selection or *bsel*, uses the proposed peer-selection approach.

Range measurement updates are performed using the proposed CI approach, as well as an Extended Kalman Filter (EKF) approach. The EKF approach uses the standard update equations, ignoring the correlation between inter-robot position estimates. We do not implement any centralized EKF for joint estimation of positions of all robots. Each robot runs an independent state estimator.

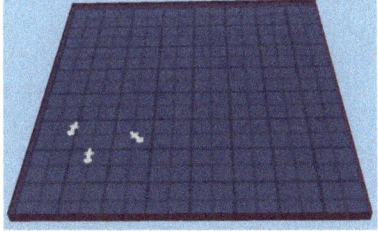

(a) ASV with acoustic modem   (b) Webots simulation

Fig. 3. (a) The ASV with the acoustic modem. When in operation, the modem, attached to a rod, is suspended in water from the rear of the robot. (b) A screenshot of the Webots simulation environment with three AUVs.

3.1 Simulation

Experiments were performed in simulation with three to four robots using Webots [22], a high-fidelity robotics simulation software. A picture of the simulation environment is shown in Fig. 3b. An acoustic modem is also simulated with the same propagation delay, ranging accuracy and bandwidth as the real acoustic modem. Simulations are performed with two different sets of trajectories, one with four robots and another with three robots. We restrict access to GNSS to two robots at 0.1 Hz or 0.05 Hz (once in 10 s or 20 s). The time-slot length for acoustic transmission is set to 5 s, which means there is one range measurement in the whole system every 5 s. We perform five experimental runs for each combination of trajectory, GNSS access, peer-selection method (cyclic or bsel) and range update method (EKF or CI).

3.2 Outdoor Experiments

Outdoor experiments were performed using three Autonomous Surface Vehicles (ASVs). The ASVs, pictured in Fig. 3a, are equipped with the same sensing and computing hardware as the Vertex Autonomous Underwater Vehicles [23] from Hydromea SA. Additionally, they have continuous GNSS reception, which serves as ground truth (but the state estimator has limited access to it). All robots are equipped with a Beringia Microlink acoustic transceiver for communication as well as range measurements. The maximum data transfer speed is 10 Bytes/s, and the ranging accuracy is about 2.5 m. A range measurement, including related data exchange (Fig. 2b), takes about 3.5 s. Experiments were performed in Lake Geneva.

The robots follow a pre-planned trajectory, which is occasionally disturbed by strong waves. Robot 1 has no access to GNSS for position estimation. Robots 2 and 3 are allowed GNSS updates once in 20 s (0.05 Hz). Acoustic range is recorded every 1.5 s between each of the three pairs of robots. For the purpose of this experiment, shorter acoustic time-slots were used without any guard times for avoiding echoes and interference, and only range measurements were recorded. Position information was recorded via radio.

The recorded data is re-processed offline with different range update methods (CI, EKF) and peer-selection strategies (cyclic, bsel). Depending on the peer-selection strategy, and subject to a more realistic time-slot length of 5 s, only a subset of the recorded range measurements are used. This gave us accuracy and consistency metrics for both range measurement update methods as well as peer-selection strategies.

4 Results

4.1 Evaluation Metrics

We use the Root Mean Squared Error (RMSE) as a measure of accuracy. We compute it as the Euclidean distance between the true and estimated positions averaged over all time steps since the beginning of the experiment.

$$\text{RMSE(T)} = \frac{1}{T} \sum_{t=1}^{T} \|x^t - \hat{x}^t\|. \tag{30}$$

(a) T1, 4 robots

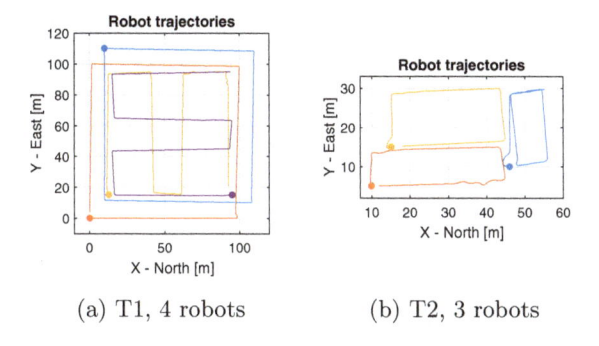

(b) T2, 3 robots

Fig. 4. Two sets of trajectories, T1 and T2, in simulation. The starting point of the trajectory for each robot is shown with a dot.

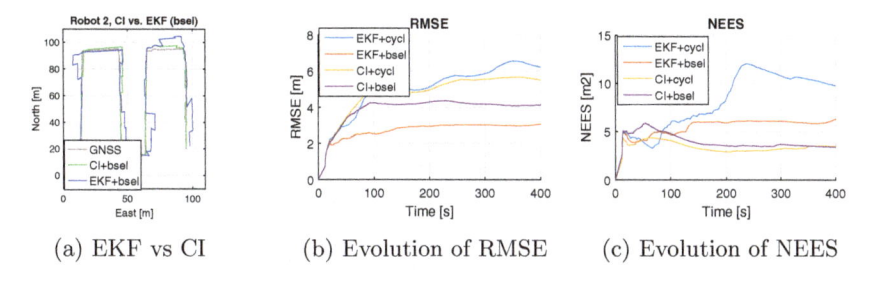

(a) EKF vs CI

(b) Evolution of RMSE

(c) Evolution of NEES

Fig. 5. For scenario 1: (a) Example estimated trajectory of one of the robots using CI and EKF. (b) Comparison of RMSE over time for the two peer-selection strategies. Our peer-selection method results in higher accuracy. (c) Comparison of NEES for CI and EKF. EKF has higher estimation error owing to ignoring inter-robot correlations.

RMSE is higher when the estimated trajectory is farther off from the true trajectory.

For measuring consistency, we use the Normalized Estimation Error Squared (NEES), averaged over all time steps since the beginning of the experiment.

$$\text{NEES}(T) = \frac{1}{T} \sum_{t=1}^{T} (x^t - \hat{x}^t)^{\text{T}} (\Sigma^t)^{-1} (x^t - \hat{x}^t). \tag{31}$$

Higher values of NEES indicate higher inconsistency between the estimated position covariance and the actual error in estimated position. This metric was introduced by Shalom et al. in [24].

4.2 Simulations

An example of a trajectory of one of the robots estimated with CI and EKF is shown in Fig. 5a. A comparison of evolution of RMSE is shown in Fig. 5b. Our peer-selection strategy, bsel, results in an improved accuracy of estimated

Table 1. RMSE and NEES for various scenarios and estimation methods.

Scenarios	CI RMSE-Cyclic NEES-Cyclic	CI RMSE-bsel NEES-bsel	EKF RMSE-Cyclic NEES-Cyclic	EKF RMSE-bsel NEES-bsel
1. T1,GNSS 0.10 Hz,2 Rob/4	5.51 3.45	4.14 3.38	6.22 9.76	3.06 6.27
2. T1,GNSS 0.05 Hz,2 Rob/4	6.19 5.43	4.57 4.02	6.77 11.89	3.85 6.10
3. T2,GNSS 0.10 Hz,2 Rob/3	2.23 2.71	2.14 3.69	2.34 4.58	2.05 4.31
4. T2,GNSS 0.05 Hz,2 Rob/3	2.60 2.74	2.33 3.41	2.88 5.93	1.94 4.17
5. Real,GNSS 0.05 Hz,2 Rob/3	3.28 5.03	2.17 4.56	3.23 6.33	2.44 7.22

position compared to a cyclic strategy. This is because robots are able to predict which peer can provide most useful information based on the heuristic in Eq. 29.

Evolution of NEES for scenario 1 is compared in Fig. 5c. EKF initially performs similar to CI. However, as uncertainties increase over time, they must be accounted for. Therefore, after a few range updates, NEES for EKF increases in comparison with that of CI. The results of various experimental scenarios, averaged over all runs and across all robots are tabulated in Table 1.

4.3 Outdoor Experiments

The estimated trajectory for one of the three robots along with ground-truth trajectories of the other two peers is shown in Fig. 6. The purely inertial estimate (without any GNSS or acoustic updates) is also shown for comparison. The evolution of RMSE and NEES is shown in Fig. 7a, 7b. The accuracy metrics are shown in the last row of Table 1.

The results demonstrate that the trajectory estimated with CI is more consistent compared to EKF. This is shown by a lower value of NEES, implying that the estimated uncertainty is in better agreement with the actual estimation error. Correct estimation of uncertainty is important because it is used to adaptively weight the influence of incoming measurements (e.g., via the Kalman gain in EKF). Therefore, erroneous uncertainty estimates are likely to cause higher trajectory RMSE eventually. The results also show that regardless of the range update method (CI or EKF), the proposed peer-selection strategy resulted in a lower RMSE. This shows the improvement in accuracy brought by an educated choice of peer for a relative measurement. The experimental results with real robots are in agreement with those obtained from simulations.

4.4 Computation and Data Overhead

The proposed CI approach is more expensive in computation compared to EKF. A range update with EKF would require a 2×2 matrix inversion (assuming

Robot Trajectories - lake experiments

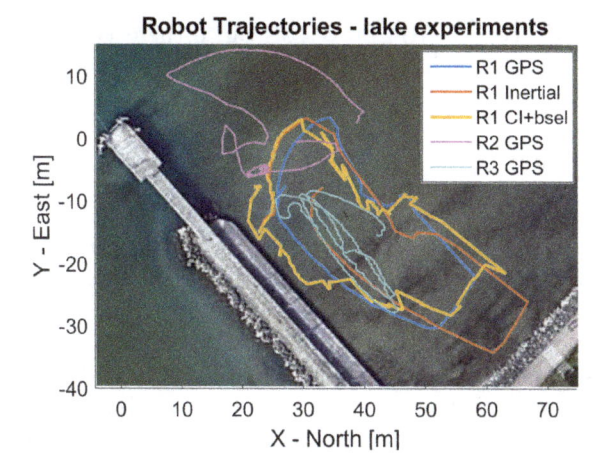

Fig. 6. Robot trajectories from the real experiment. For Robot 1, trajectory estimated with CI and the proposed peer-selection strategy is shown. The true (GNSS) trajectory and the inertial estimate (without range measurements) are also shown for comparison. For robots 2 and 3, only the true (GNSS) trajectory is shown.

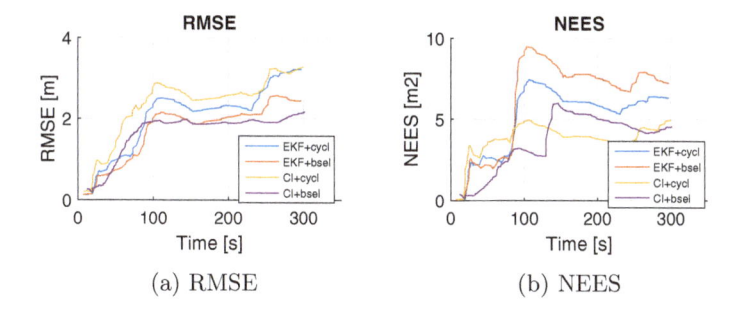

(a) RMSE

(b) NEES

Fig. 7. RMSE and NEES for the real experiment. The plots show that CI results in lower estimation error (NEES), which demonstrates more accurate estimation of uncertainty. They also show that using the proposed peer-selection strategy results in more accurate trajectory estimation.

position estimation in 2D, the vertical dimension is provided by a depth sensor), regardless of the size of the state variable. The proposed transformation requires eigenvalue decomposition, inversion of a diagonal matrix and several matrix multiplications, the sizes of which depend on the size of the state. This adds an overhead in computation time (EKF: 0.1 ms, CI: 1.5 ms, approx.).

For the peer-selection heuristic, the only additional information added to the communication is the robot heading. All the other information is also needed for range update. Computing this metric also requires many of the operations used for CI-based range update (eigenvalue decomposition, etc.). The computation time was found to be about 1.0 ms per peer on our setup. Our setup consists of C++ code compiled with -O2 optimization flag, running on a Raspberry

Pi Zero with single core 1 GHz processor. Since there is one range update in approximately 5 s, this combined computation delay is affordable.

5 Conclusion

We presented a range-based covariance intersection method for cooperative localization. CI provides a number of advantages for cooperative localization with underwater robots, which have severe communication constraints. CI does not require communication of inter-robot correlation terms. It can be run in a fully distributed fashion, and uses information only from the two robots involved in a relative measurement. It allows robots to exploit information from their peers without any adverse effects such as overconfident estimates. We showed that CI-based cooperative localization results in a better estimation of uncertainty.

We also derived a peer-selection heuristic for performing range measurements based on an information theoretic approach. We showed that our peer-selection strategy improves localization accuracy in comparison to sequentially querying peers for a range measurement.

A number of improvements to the system are possible. During real-world operation, range queries or responses made by a robot may get lost or corrupted. At the moment, we do not try to resend a query or query another peer, resulting in no updates being performed during some time slots. A recovery strategy may reduce this 'dead time' with no updates. The peer-selection heuristic is based on estimating position of peers using a constant velocity model. When a peer changes direction, this estimate becomes invalid. A better estimate can be made if planned trajectories of peer robots are communicated in advance.

Acknowledgement. This work was partially funded by the Swiss National Science Foundation under grant CRSII2_160726/1.

References

1. Quraishi, A., Bahr, A., Schill, F., Martinoli, A.: A flexible navigation support system for a team of underwater robots. In: International Symposium on Multi-Robot and Multi-Agent Systems, pp. 70–75 (2019)
2. Bahr, A., Leonard, J.J., Fallon, M.F.: Cooperative localization for autonomous underwater vehicles. Int. J. Robot. Res. **28**(6), 714–728 (2009)
3. Thrun, S., Burgard, W., Fox, D.: Probabilistic Robotics. MIT Press, Cambridge (2005)
4. Borenstein, J.: Experimental results from internal odometry error correction with the OmniMate mobile robot. IEEE Trans. Robot. Autom. **14**(6), 963–969 (1998)
5. Roumeliotis, S.I., Bekey, G.A.: Distributed multirobot localization. IEEE Trans. Robot. Autom. **18**(5), 781–795 (2002)
6. Luft, L., Schubert, T., Roumeliotis, S.I., Burgard, W.: Recursive decentralized localization for multi-robot systems with asynchronous pairwise communication. Int. J. Robot. Res. **37**(10), 1152–1167 (2018)

7. Panzieri, S., Pascucci, F., Setola, R.: Multirobot localisation using interlaced extended Kalman filter. In: IEEE/RSJ International Conference on Intelligent Robots and Systems, pp. 2816–2821 (2006)
8. Julier, S.J., Uhlmann, J.K.: A non-divergent estimation algorithm in the presence of unknown correlations. In: Proceedings of the 1997 American Control Conference (Cat. No.97CH36041), vol. 4, pp. 2369–2373 (1997)
9. Howard, A, Matarić, M.J., Sukhatme, G.S.: Localization for mobile robot teams using maximum likelihood estimation. In: IEEE/RSJ International Conference on Intelligent Robots and Systems, vol. 1, pp. 434–439 (2002)
10. Howard, A., Matarić, M.J., Sukhatme, G.S.: Localization for mobile robot teams: a distributed MLE approach. In: Siciliano, B., Dario, P. (eds.) Experimental Robotics VIII. Springer Tracts in Advanced Robotics, vol. 5, pp. 146–155. Springer, Berlin Heidelberg (2003). https://doi.org/10.1007/3-540-36268-1_12
11. Klingner, J., Ahmed, N., Correll, N.: Fault-tolerant covariance intersection for localizing robot swarms. Robot. Auton. Syst. **122**, 103–306 (2019)
12. Carrillo-Arce, L.C., Nerurkar, E.D., Gordillo, J.L., Roumeliotis, S.I.: Decentralized multi-robot cooperative localization using covariance intersection. In: IEEE/RSJ International Conference on Intelligent Robots and Systems, pp. 1412–1417, November 2013
13. Vasic, M., Mansolino, D., Martinoli, A.: A system implementation and evaluation of a cooperative fusion and tracking algorithm based on a Gaussian Mixture PHD filter. In: 2016 IEEE/RSJ International Conference on Intelligent Robots and Systems, pp. 4172–4179 (2016)
14. Li, H., Nashashibi, F., Yang, M.: Split covariance intersection filter: theory and its application to vehicle localization. IEEE Trans. Intell. Transp. Syst. **14**(4), 1860–1871 (2013)
15. Li, H., Nashashibi, F.: Cooperative multi-vehicle localization using split covariance intersection filter. IEEE Intell. Transp. Syst. Mag. **5**(2), 33–44 (2013)
16. Park, J., Lee, J.: Beacon selection and calibration for the efficient localization of a mobile robot, vol. 32, no. 1, pp. 115–131. Cambridge University Press (2014)
17. Wang, H., Yao, K., Pottie, G., Estrin, D.: Entropy-based sensor selection heuristic for target localization. In: Proceedings of the 3rd International Symposium on Information Processing in Sensor Networks, pp. 36–45 (2004)
18. Ertin, E., Fisher, J.W., Potter, L.C.: Maximum mutual information principle for dynamic sensor query problems. In: Zhao, F., Guibas, L. (eds.) IPSN 2003. LNCS, vol. 2634, pp. 405–416. Springer, Heidelberg (2003). https://doi.org/10.1007/3-540-36978-3_27
19. Pursell, L., Trimble, S.Y.: Gram-Schmidt orthogonalization by gauss elimination. Am. Math. Mon. **98**(6), 544–549 (1991)
20. Simon, D., Simon, D.L.: Constrained Kalman filtering via density function truncation for turbofan engine health estimation. Int. J. Syst. Sci. **41**(2), 159–171 (2010)
21. Cover, T.M., Thomas, J.A.: Elements of Information Theory. Wiley, Hoboken (1991)
22. Michel, O.: Cyberbotics LTD. Webots TM: professional mobile robot simulation. Int. J. Adv. Rob. Syst. **1**(1), 39–42 (2004)
23. Schill, F., Bahr, A., Martinoli, A.: Vertex: a new distributed underwater robotic platform for environmental monitoring. In: Groß, R., et al. (eds.) Distributed Autonomous Robotic Systems. SPAR, vol. 6, pp. 679–693. Springer, Cham (2018). https://doi.org/10.1007/978-3-319-73008-0_47
24. Bar-Shalom, Y., Li, X.R., Kirubarajan, T.: Estimation with Applications to Tracking and Navigation: Theory Algorithms and Software. Wiley, Hoboken (2004)

Robust Localization for Multi-robot Formations: An Experimental Evaluation of an Extended GM-PHD Filter

Michiaki Hirayama[1(✉)], Alicja Wasik[2], Mitsuhiro Kamezaki[1], and Alcherio Martinoli[2]

[1] Department of Modern Mechanical Engineering, Graduate School of Creative Science and Engineering, Waseda University, Tokyo, Japan
m_hirayama@sugano.mech.waseda.ac.jp, kame-mitsu@aoni.waseda.jp
[2] Distributed Intelligent Systems and Algorithms Laboratory, École Polytechnique Fédérale de Lausanne, Lausanne, Switzerland
{alicja.wasik,alcherio.martinoli}@epfl.ch

Abstract. This paper presents a thorough experimental evaluation of an extended Gaussian Mixture Probability Hypothesis Density filter which is able to provide state estimates for the maintenance of a multi-robot formation, even when the communication fails and the tracking data are insufficient for maintaining a stable formation. The filter incorporates, firstly, absolute poses exchanged by the robots, and secondly, the geometry of the desired formation. By combining communicated data, information about the formation, and sensory detections, the resulting algorithm preserves accuracy in the state estimates despite frequent occurrences of long-duration sensing occlusions, and provides the necessary state information when the communication is sporadic or suffers from short-term outage. Differently from our previous contributions, in which the tracking strategy has only been tested in simulation, in this paper we present the results of experiments with a real multi-robot system. The results confirm that the algorithm enables robust formation maintenance in cluttered environments, under conditions affected by sporadic communication and high measurement uncertainty.

Keywords: Multi-robot tracking · Formation control · Cooperative localization · Probability hypothesis density filter

1 Introduction

In recent years, we observe a slow increase in the number of applications where the advantages of Multi-Robot Systems (MRSs) are recognized and leveraged to achieve improved performances when compared to single-robot solutions [1]. For example, in industrial applications, multiple mobile manipulators carry objects that cannot be transported by a single robot [2]. In building inspection, multiple aerial vehicles inspect areas that are difficult to be reached by humans [3], eventually providing the same documentation gathered by experts. Robots can even assume different roles in their teams: for instance in [3], one of the robots inspects the conditions of the building while others provide illumination. Furthermore, socially assistive robotics is one of the most attractive applications for MRSs. Through appropriate task allocation strategies, robots provide services to users within a multi-region environment simultaneously [4].

Cooperation among multiple robots, however, introduces another level of complexity in the system. Methods such as formation control not only require powerful algorithms, especially for deployments in structured indoor environments [5], but also necessitate the existence of a reliable localization infrastructure. In particular, formation control requires each robot in a team to have continuous access to the state information of its neighbors, which is typically achieved through a wireless communication link. However, communicated messages can be delayed or even lost [6], while loss of communication even for a short period of time can lead to formation breaking. Issues with reliability of communication have been widely recognized in the context of cooperative positioning systems [6]. Different solutions have been sought, including a reduction of the broadcast data by filtering out unnecessary information [7], purposeful packet delays within controlled time slots [8], and bounds on the extent of the communication graph [9]. In general, it is considered a good practice to take into account limitations in the communication bandwidth when dealing with MRSs [10].

In our recent work [11], we introduced an approach to provide a reliable localization system for ID-based formation control methods – algorithms, where the robot position in the formation (also referred to as the role), depends on its unique identification number (ID). For this class of algorithms, robots must be capable of distinguishing each other, which can pose additional challenges when a tracking solution is sought. In particular, tracking of multiple homogeneous robots based on Laser Range Finders (LRFs) does not provide information for directly distinguishing the robots. To this end, we combine ID-less detections with ID-based communicated data, when available, in a multi-target tracking filter. Additionally, we leverage information related to the formation geometry to improve the estimates of the robot poses. The two aforementioned information sources are incorporated in an extension of the Gaussian Mixture Probability Hypothesis (GM-PHD) filter [12], called Formation Information GM-PHD (FI-GM-PHD) filter [11]. As the resulting estimates are anonymous, a role assignment procedure finds their optimal allocation to the roles in the formation. While the previously designed FI-GM-PHD filter has shown promising performances in high-fidelity calibrated simulations, the method was never evaluated on a real robotic platform. Although state-of-the-art simulators can yield results comparable to reality [13], the simulation-to-reality gap still exists as a consequence of the difficulty to capture accurate distributions of sensing and actuation noise [14], lack of incorporation of some subtle factors that affect the performance of the real robots, such as fluctuation of the on-board sensors during locomotion [15], and inadequate models for physical contact [16]. As a consequence, high-fidelity simulations cannot be considered a replacement for real world evaluations [16].

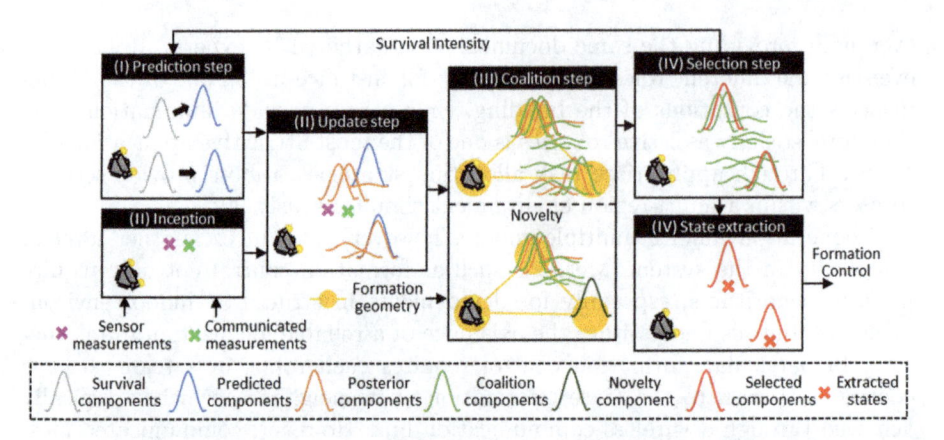

Fig. 1. Schematic diagram of the FI-GM-PHD filter.

For example, in the context of cooperative target tracking, the authors in [17] reported that the tracking performance with real robots outperforms the simulation results, while in [18] the opposite is observed, despite a careful calibration with real robot data. As it can be expected, within the field of MRSs, the simulation-to-reality gap is widened not only because of the presence of several robots with their associated single-robot simulation inaccuracies, but also because the effects of communication imperfections on the overall navigation performance are rarely assessed.

In this paper, we perform an extensive set of experiments with a real robotic platform. In particular, we carefully estimate the sensor-dependent detection error, the environment-dependent self-localization error, and the quality of communication to understand their effects on the performance of the FI-GM-PHD filter. Based on these experimental findings, we update the detection model of the filter to take into account the sensor-dependent distribution shown by empirical data. The results confirm the robustness of the FI-GM-PHD filter to measurement uncertainties, distorted formations due to navigation in cluttered environments, and challenging communication settings.

The rest of this paper is organized as follows. Section 2 provides background on the FI-GM-PHD filter and explains how the estimates are integrated into a formation control algorithm. In Sect. 3, we provide details of the calibration procedure and the experimental setup. Section 4 presents the results, followed by conclusions in Sect. 5.

2 Background

The FI-GM-PHD filter is an extension of the GM-PHD filter [12]. The standard GM-PHD filter has four steps, namely *prediction*, where the previous intensity evolves according to the motion model, *update*, where the intensity is updated with the acquired measurements, *selection*, which reduces the solution space, and *state extraction*. In addition, in our algorithm we perform an integrated *update-and-inception step*, in which we incorporate data communicated from the other robots, and a *coalition step*, in which we specify the formation geometry.

We provide a full description of the FI-GM-PHD filter in [11], while in this paper we briefly overview the main concepts as follows. The schematic diagram of the overall algorithm is shown in Fig. 1, in which each block graphically represents the steps described in Sect. 2.1. In Fig. 1, two survival components (gray) from the last step serves as input to the Prediction step. (I) Prediction: propagates the components (blue) according to the motion model. (II) Update-and-Inception: creates posterior components (orange) according to the detection probability and measurements including sensing (purple cross) and communication (green cross). (III) Coalition: combines the components from the update-and-inception step with the coalition components (green) derived from the formation geometry (yellow). If coalition components do not have corresponding components of the posterior intensity, a novelty component (dark green) is created. (IV) Selection and state extraction: prunes and merges components (red), then extracts the estimated states (red cross) from the components above a certain weight.

2.1 The FI-GM-PHD Filter

The multi-target state is approximated by an *intensity* – a Gaussian mixture in the form:

$$v_k(\mathbf{x}) = \sum_{i=1}^{J_k} w_k^{(i)} \mathcal{N}(\mathbf{x}; m_k^{(i)}, P_k^{(i)}) \qquad (1)$$

at time k, where $\mathcal{N}(\cdot; m, P)$ denotes a Gaussian density with mean m, covariance P and weight w, and $\mathbf{x}$ is a target state.

I. Prediction. The predicted intensity is a Gaussian mixture in the form:

$$v_{k|k-1}(\mathbf{x}) = v_{S,k|k-1}(\mathbf{x}) \qquad (2)$$

where $v_{S,k|k-1}(\mathbf{x})$ is referred to as the survival intensity. The components of the survival intensity are computed from the previous intensity components according to a linear Gaussian motion model.

II. Update-and-Inception. The posterior intensity is composed of two terms:

$$v_k(\mathbf{x}) = (1 - p_{D,k}(\mathbf{x}))v_{k|k-1}(\mathbf{x}) + \sum_{\mathbf{z} \in Z_k} v_{D,k}(\mathbf{x}; \mathbf{z}) \qquad (3)$$

The first term discounts the predicted components $v_{k|k-1}$ according to the state-dependent probability of detection $p_{D,k}(\mathbf{x})$. The second term, $v_{D,k}(\mathbf{x}; \mathbf{z})$, generates a new set of components for each measurement in $\mathbf{z}$.

When available, the communicated state information is incorporated as an additional set of measurements:

$$Z_k := Z_k \cup \sum_{j=1}^{\Delta_k} \mathbf{z}_k^{(j)} \qquad (4)$$

where Δ_k is the number of robots communicating their state information.

III. Coalition. First, the specification of the desired formation geometry is encoded in the form of a Gaussian mixture, referred to as the coalition intensity. At the locations of where the other robots in formation should be with respect to the tracking robot i, we generate Gaussian components with the means $m_{\zeta,k}^{(j)}$, where $j \neq i$ is the jth role in the formation.

The coalition step combines the intensities obtained during the update-and-inception step with the coalition intensity. All the components forming the posterior intensity are compared against the components of the coalition to find the matching that optimizes some criteria. In our implementation, we minimize the distance between the component means, while choosing components of the posterior intensity with significant weights:

$$o_k^{(j,l)} = \exp(||m_k^{(l)} - m_{\zeta,k}^{(j)}||) + (w_k^{(l)} + \epsilon)^{-1} \tag{5}$$

where j is the jth component of the coalition intensity, and l is the lth component of the posterior intensity $v_k(\mathbf{x})$. We first evaluate the best candidates for good matching by sorting the components of the posterior intensity according to the measure o_k.

New components are generated for each matching pair. To limit a number of new components, each coalition component j is assigned a budget $\Phi_{\zeta,k}^{(j)}$, which decreases with every posterior component that has been found close-by, with the amount of expended budget inversely proportional to the distance between. In other words, the budget of the coalition component decreases significantly with every posterior components found close to it. Once the budget of the component j is depleted, the matching procedure for that component is completed.

Finally, the coalition and the posterior components are coalesced to form a new Gaussian component:

$$\begin{aligned}
\overline{m}_k^{(n)} &= \Phi_k^{(l)} m_k^{(l)} + (1 - \Phi_k^{(l)}) m_{\zeta,k}^{(j)} \\
\overline{P}_k^{(n)} &= (\Phi_k^{(l)} + \epsilon)^{-1} P_k^{(l)} \\
\overline{w}_k^{(n)} &= \Phi_k^{(l)} w_k^{(l)}
\end{aligned} \tag{6}$$

where $\Phi_k^{(l)}$ is the matching score.

The budget $\Phi_{\zeta,k}^{(j)}$ left at the end of iteration indicates that one of the coalition components did not have a corresponding component in the posterior. In that case, the coalition component, from now on referred to as the *novelty*, is propagated to the final intensity. Since *novelty* can suffer when the formation is far from the desired topology, its weight is proportional to the overall matching error.

IV. Selection and State Extraction. Components with weights weaker than a certain threshold are pruned. Furthermore, components naturally clustered together are merged into a single component by first selecting a component with the highest weight and then merging with it all components within a prescribed distance. In the state extraction step, the means of the components that have weights greater than a predefined threshold are selected as state estimates.

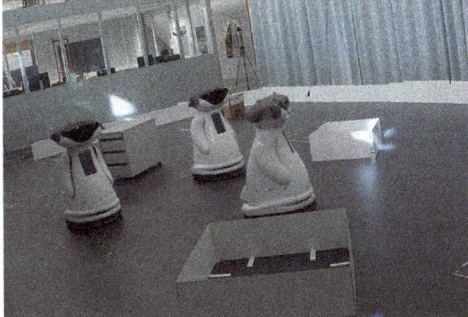

Fig. 2. (Left) The MBot robot. (Right) the experimental arena.

2.2 Graph-Based Formation Control

The formation is comprised of Δ holonomic robots, including one leader and $\Delta - 1$ followers. The leader moves on a predefined trajectory while followers maintain the desired formation as follows:

$$\dot{x}_i = \frac{1}{|\sum_j \mathcal{L}_{ij}|} \sum_{i \sim j} [-\mathcal{L}_{ij}(r_{ij}(t) \cos(\gamma_{ij}(t)) - b_{ij}^x(t))]$$
$$\dot{y}_i = \frac{1}{|\sum_j \mathcal{L}_{ij}|} \sum_{i \sim j} [-\mathcal{L}_{ij}(r_{ij}(t) \sin(\gamma_{ij}(t)) - b_{ij}^y(t))] \quad (7)$$

where $\mathcal{L}$ is a non-stationary Laplacian, r_{ij} and γ_{ij} are the Euclidean relative range and the bearing between the robots i and j. The bias $b_{ij} \in \mathbb{R}^2$ defines the desired inter-robot distance between i and j. The details of the algorithm can be found in [5].

2.3 Role Assignment

The role assignment procedure finds a permutation that assigns the ID-less estimates to the target positions in the formation. With respect to the detecting robot i, each estimate j, obtained from the state extraction step, is coupled with bias b_{il} that corresponds to the "l^{th}" place in the formation. An optimal assignment is found by computing the smallest cost between the estimates and the projected formation positions, brought to a common reference frame.

3 Experimental Campaign

Experiments are performed with three MBot robots – omnidirectional robots with a height of 0.98 m and a footprint of 0.65 m in diameter, shown in Fig. 2. The robots are equipped with two LRFs that provide 360° field of view and a 4 m sensing range. The robots are connected through a wireless network and self-localize using the AMCL [19] package offered in ROS. Ground truth positioning data is provided by a Motion Capture System (MCS) with millimeter accuracy. The experimental arena is approximately 8×10 m^2.

Table 1. The self-localization error e_L and the measurement error e_M of our setup, determined empirically through dedicated experiments.

$e_L[m]$		$e_M[m]$	
mean	std	mean	std
0.18	0.051	0.33	0.24

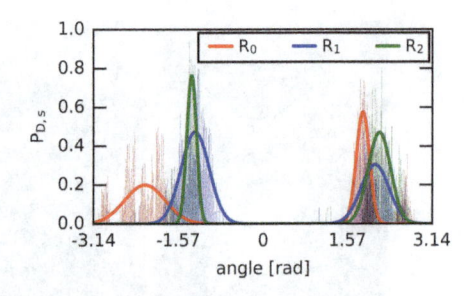

Fig. 3. The models of sensor-dependent missed detection probability for each robot (R_0, R_1, R_2).

3.1 Implementation

As our objective is to evaluate the robustness of the original FI-GM-PHD filter in reality, we keep the parametrization used in Section 7.1 of [11], with the notable exception of calibrating the sensor model according to the empirical data.

The state of the target is composed of its position and velocity $\mathbf{x} = [x, y, \dot{x}, \dot{y}]$. The measurement is the position $\mathbf{Z} = [z^x, z^y]^T$. Tracking is performed in the global frame; all methods are run onboard. The formation heading is fixed to the positive y-axis of the global coordinate frame. The message losses are simulated by modulating the communicated information input reported in Eq. 4 with a probability of $(1 - p_{md})$, where p_{md} stands for the message drop probability. For further details and comprehensive parametrization, please refer to [11].

3.2 Self-localization and Measurement Errors

The performance of our methods is affected by two sources of stochasticity. First, the self-localization error e_L, is included in the formation projections and in the positioning information communicated by the robots. Second, the measurement error e_M, is independent of e_L and affects the sensory data. Before evaluating the filter algorithm in its integrity, we have carried out a series of tests (i.e. T_I and T_{II}) to assess the extent to which the errors above may affect the performance of our system.

The self-localization error e_L is the difference between the self-localization estimates and ground truth data obtained through the MCS. To calculate e_L, in Test T_I, we let a robot moving around in the arena for 960 s and average all the data obtained each 100 ms.

The measurement error e_M is the difference between the estimated position of the detected robot and the actual position, also acquired 10 Hz through the MCS. In our system the error is higher in dynamic situations, where both the detecting and the detected robots are moving [20]. Therefore, to determine e_M, in Test T_{II}, we move two robots independently, keeping them within sensing reach, with the range and the bearing between the two varying throughout an experiment that lasts for 960 s.

The results are summarized in Table 1. The self-localization error corresponds to about half of the robot radius, therefore influencing in a limited way the tracking performance. The measurement error is instead close to the robot radius, thus having a larger effect on the tracking performance.

3.3 Tailoring the Probability of Detection

In our original model reported in Section 5.2 of [11], we already integrated effects related to a limited field of view and occlusions. Therefore, in the original model the probability of detection p_D was dependent on the robot pose and this allowed a robot to reduce the risk of losing track of another robot. Despite taking into account such sensing realism in the original filter algorithm, such considerations turned out to be insufficient for handling the effects present in reality. Indeed, to cope with additional sensing heterogenities across robots and real world uncertainties, in this paper we had to additionally model the sensor-dependent probability of missing a detection and incorporate it in p_D.

Our models are based on the empirical data collected in Test T_{II}. The models characterize specific sensors, therefore, in contrast to their simulated counterpart, they are different for each robot. The data and the models fitted to it are shown in Fig. 3. The spikes indicate the portion of the lost detections $p_{D,s}$ for a given angle. To the resulting distributions we fit Gaussian models using the curve_fit method from SciPy optimize [21]. Recall that the MBot robots are equipped with two LRFs. The sensors, each of them providing 240° field of view, are located at the front and at the back of the robot, while on the sides their ranges overlap. The overlapping, however, is skewed, resulting in higher probability of missing detection around the angles $-\pi/2$ and $\pi/2$. Moreover, it is worth noticing that the distribution of $p_{D,s}$ characterizing the robot R_0 is significantly different from the two other robots, especially for the angles between $-\pi/2$ and $-\pi$, probably due to a slightly different tilting of its LRFs.

3.4 Evaluation Metrics

For the evaluation of the multi-target tracking performance, we use the Optimal SubPattern Assignment (OSPA) metric [22]. OSPA is comprised of two components: the first accounting for the cardinality error in the target number, and the second for the positioning error. Therefore, the lower the OSPA metric, the higher the tracking performance. OSPA is tailored with two weighting parameters, p and c, the former related to the position accuracy and the latter to the cardinality. In our experiments, OSPA is computed between the ground truth positions and the estimated positions of all targets, with $c = 1.0$ and $p = 2.0$.

The second metric considered evaluates the formation control performance. The formation error e_F is the average difference between the desired distances and the actual distances between the robots in the formation.

3.5 Scenarios

The FI-GM-PHD filter is evaluated in three scenarios: (I) tracking decoupled from formation control, where robots do not use the tracking data for control,

Table 2. OSPA metrics for Scenarios I-III.

p_{md}	Scenario I					p_{md}	Scenario II-A				
	Std	FSys					Std	FSys			
	1.0	0.0	0.5	0.9	1.0		1.0	0.0	0.5	0.9	1.0
OSPA mean	0.63	0.41	0.49	0.56	0.57	OSPA mean	0.64	0.50	0.57	0.62	0.63
OSPA std	0.24	0.11	0.16	0.18	0.19	OSPA std	0.21	0.14	0.18	0.19	0.19
$e_M[m]$	Scenario II-B					p_{md}	Scenario III				
	FSys						Std	FSys			
	0.0	0.1	0.3	0.6	1.0		1.0	0.0	0.5	0.9	1.0
OSPA mean	0.50	0.48	0.49	0.52	0.55	OSPA mean	0.74	0.53	0.63	0.67	0.67
OSPA std	0.14	0.14	0.14	0.14	0.14	OSPA std	0.23	0.16	0.20	0.20	0.20

(II) tracking for formation control, where we alter the quality of communication (i.e. the message drop probability) and simulate an augmented detection error (i.e. the measurement error), and (III) realistic navigation, where robots move among obstacles scattered in the environment. Our methods are compared with the standard GM-PHD filter and with respect to the baseline formation control with fully reliable communication and no tracking.

Scenario I: Multi-robot Tracking. We collect a dataset (raw sensor data, positioning information and the formation state) in the baseline experiment with formation relying on ideal communication conditions. The three-robot formation follows an eight-shape trajectory, forming a triangle shape with the inter-robot spacing of 1.75 m. We perform multi-robot tracking with the collected data *offline*, with the standard GM-PHD filter, and with the FI-GM-PHD filter with emulated message drop probabilities of $p_{md} = 0.0$ (i.e. ideal communication), $p_{md} = 0.5$, $p_{md} = 0.9$ and $p_{md} = 1.0$ (i.e. no communication). For each experiment, we perform 11 sequential runs, each lasting 120 s.

Scenario II: Tracking for Formation Control. In contrast to Scenario I, in the following experiments, tracking is running *online*, and used for formation control directly. In other words, the performance of the tracking system affects the formation error, which in turn has an effect on tracking through the coalition step. We distinguish two sub-scenarios:

Scenario II-A: Message Drop Probability where we vary the message drop probability as in Scenario I (i.e. $p_{md} \in \{0.0, 0.5, 0.9, 1.0\}$) and compare to an ideal communication baseline case.

Scenario II-B: Measurement Error where we manipulate the precision of the robot detection by adding a random uniform error of magnitude $e_M = \{0.0, 0.3,$

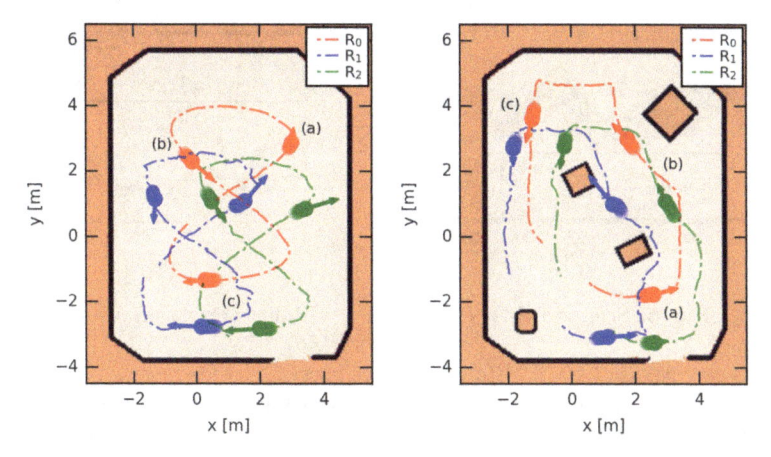

Fig. 4. Trajectories of the robots using the FI-GM-PHD filter with $p_{md} = 1.0$, i.e. with no communication. (Left) Scenario II-A trajectories of the robots plotted at (a) $t = 15$ s, (b) $t = 55$ s, (c) $t = 95$ s. (Right) Scenario III trajectories of the robots plotted at (a) $t = 15$ s, (b) $t = 45$ s, (c) $t = 75$ s.

$0.6, 1.0\}$ m to the original measurement. The probability of message drop is zero (i.e. ideal communication), in order to decouple the effects of communication and sensing quality. The experimental settings, including the number of robots, the desired formation shape and the trajectory are identical to Scenario I.

Scenario III: Realistic Environment. In the final set of experiments, the robots move in a triangular formation with the inter-robot spacing of 1.6 m in an arena scattered with obstacles. The leader robot plans the trajectory using a Fast Marching Method [23]. For each experiment we perform 11 sequential runs of approximately 100 s. We perform an ideal communication baseline experiment and runs with varying communications quality with $p_{md} \in \{0.0, 0.5, 0.9, 1.0\}$.

4 Results

We use the following acronyms for labeling the methods. *NT* stands for the baseline experiments with the formation relying on ideal communication and no tracking, *Std* stands for the standard GM-PHD filter and *FSys* stands for the full FI-GM-PHD algorithm.

4.1 Scenario I: Multi-robot Tracking

The OSPA performance is summarized in Table 2, from which we draw two conclusions. First, the tracking performance of the FI-GM-PHD filter degrades gracefully with the drop of the communication quality. Compared to when the positioning data is received 10 Hz, in the case of no communication the performance of FI-GM-PHD method is only 37% worse. Second, in the case of

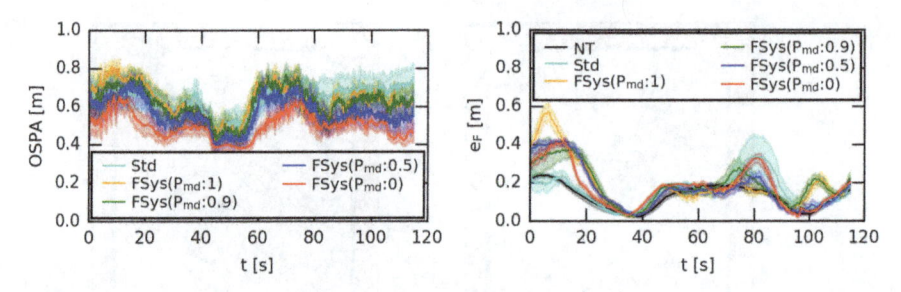

Fig. 5. Scenario II-A: (Left) OSPA. (Right) Formation error.

$p_{md} = 1.0$, i.e. no communication, the FI-GM-PHD filter outperforms the standard GM-PHD filter. This is a fair comparison, as both methods rely on the same data, but the FI-GM-PHD filter performs an additional coalition step.

4.2 Scenario II: Tracking for Formation Control

Scenario II-A: Tracking-Based Formation Control with Varying Message Drop Probability. An example of a trajectory of the FI-GM-PHD filter in the $p_{md} = 1.0$ case is shown in Fig. 4. The formation error, shown in Fig. 5, remains bounded for all the tested cases. It oscillates between as low as 0 m and up to 0.4 m, with a short-term peak at 0.6 m in the *FSys* case without communication ($p_{md} = 1.0$). Higher values of e_F are resulting from the fact that during part of the experiment the leader robot is situated behind the followers, and the "pushing" forces it exerts have a smaller effect than the "pulling" ones (they act against the follower-to-follower forces, not with them). For the majority of the run duration, the formation error of all the methods follows that of the *NT* baseline.

Shown in Fig. 5, on average the OSPA error is the lowest for the *FSys* method and it gracefully degrades with the reduction of the communication throughput. As summarized in Table 2, the rise of the OSPA error with respect to the p_{md} is moderate, with the difference between the $p_{md} = 0.0$ and $p_{md} = 1.0$ amounting to 27%. This confirms the results we obtain in Scenario I, but in this case the tracking is performed online. On average, the OSPA error of the FI-GM-PHD with no communication ($p_{md} = 1.0$) is almost identical to that of the standard GM-PHD filter. However, during our experiments, the *Std* method resulted in *three formation failures* out of the total of 11 runs. A run is labeled as failed when at least one of the robots stops keeping the formation with the other robots and falls behind. This phenomenon is typically caused by a lost estimate, an estimate mistakenly associated to a static object in the area, a mistaken role association, or a combination of the above. No failures occur in the *FSys* case, even when no communication is allowed.

Scenario II-B: Tracking-Based Formation Control with Varying Measurement Error. Based on the results summarized in Table 2, we can deduce

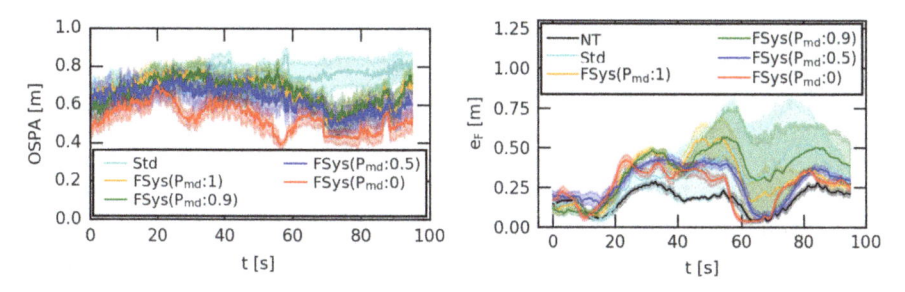

Fig. 6. Scenario III: (Left) OSPA. (Right) Formation error.

Fig. 7. Failed scene of *FSys* with $p_{md} = 1.0$. (a) R_1 and R_2 follow R_0. (b) R_1 tries to avoid an obstacle, but it takes time. (c) R_1 is left behind.

that once the communication quality is high (i.e. $p_{md} = 0$), the measurement error has little effect on the performance of our tracking method. Recall that our preliminary evaluation determined that the baseline detection error of our setup with two LRFs is around 0.33 m (see Table 1). An addition of a random uniform error of less than that value (as in the $e_M = 0.1$ and $e_M = 0.3$ cases) has no effect on the tracking performance, while the injection of an error as high as 1 m (one and a half times the robot diameter) results in a 14% increase of the OSPA error compared to the $e_M = 0.0$ case, confirming the robustness of the FI-GM-PHD filter to sensing noise.

4.3 Scenario III: Realistic Environment

The experimental setup with the obstacles scattered around the arena and the robot trajectories recorded during one run of the FI-GM-PHD filter with $p_{md} = 1.0$ is shown in Fig. 4. The OSPA metrics, plotted in Fig. 6 and summarized in Table 2, once more confirm the stability of the tracking performance of our FI-GM-PHD filter, even when the formation experiences deformations resulting from the presence of obstacles. Once more, we observe the trends recognized in Scenario I and Scenario II-A, namely that the increase of p_{md} has a bounded effect on the quality of tracking (with the OSPA in the $p_{md} = 1.0$ case being 26% worse than in the $p_{md} = 0.0$ case) and that the FI-GM-PHD filter outperforms the standard filter even in the case when communication is not used (with OSPA of *FSys*, $p_{md} = 1.0$ being 10% lower than *Std*).

Robots keep the ideal formation shape at times $t = 0-20$ s, while moving sideways, as shown in the formation error in Fig. 6. Then, the error remains close to the NT baseline, with the exception of the Std and the $FSys$ with $p_{md} = 0.9$ and $p_{md} = 1.0$ conditions, while the under the same conditions, variance rapidly increases around $t = 50$ s, at the time where both algorithms experience formation failures. Out of all the tested cases, the Std and the $FSys$ with $p_{md} = 0.9$ and with $p_{md} = 1.0$ each result in failure to maintain the formation in 1 out of 11 runs. Each of these conditions involves very little (1 message per second) to no communication. Figure 7 shows the failed situation of $FSys$ with $p_{md} = 1.0$. Once R_1 falls slightly behind during a maneuver of negotiating an obstacle, it has no means to recover since the obstacle occludes the detection of the other formation members, while the impact of including the formation geometry is reduced because of the geometry drifting from the desired set point. When robots reach the left top of the arena (around $t = 70$), the formation error, except for the failure cases, get close to ideal, but afterwards the error rises since the leader robot pushes the followers.

5 Conclusion

The primary objective of the paper was to validate our original work on the FI-GM-PHD filter with real robots, their associated sensing and actuation noises, and the stochasticity of interactions with and within a real environment. We conclude that our filtering method prove to be highly robust, and does require minimal fine-tuning when moved from simulation to reality, as we have not performed any re-parametrization except for integrating a probability of missed detection for each individual robot.

The presented results consistently lead us to two conclusions. Firstly, our filtering algorithm is robust to a deterioration of the communication quality, sensory imperfections, and the complexity of the environment, with the tracking performance degrading gracefully with increasing levels of experimental challenges. Second, the FI-GM-PHD filter outperforms marginally, or, in some cases even significantly, the standard GM-PHD filter, even in the cases when no communication is available. One should note that although it may seem that the FI-GM-PHD filter has obvious advantages over the standard filter, it achieves such superior performance thanks to an increased complexity: in fact, it combines data from multiple information sources. In particular, reaching an effective fusion is nontrivial because of the inconsistencies introduced by the self-localization (incorporated in the communicated positioning information) and the detection errors. Fusion, if done inappropriately, can result in track splitting and ambiguity of estimates, which in turn can lead to erroneous role assignment, ill-defined formation, and eventually, breaking of the formation. The GM-PHD filter facilitates fusion of data from multiple heterogeneous sources, but care must be taken so that the advantageous properties of the original algorithm are not sacrificed.

Through our experimental campaign, not only we have proven the robustness of the FI-GM-PHD filter, but we also tested it in settings more challenging than what the filter has been originally designed for – situations where communication suffers from short-term outages. The FI-GM-PHD filter is shown to be able to successfully sustain the formation even in cases without inter-robot communication, keeping the probability of formation failure marginal even in environments cluttered with obstacles.

Acknowledgments. Supported by the JSPS Grant-in-Aid for Scientific Research (A) No. 19H01130, Research Institute for Science and Engineering of Waseda University, JST PRESTO No. JPMJPR1754, and the Top Global University Japan Program of the Ministry of Education, Culture, Sports, Science and Technology. Partially supported by ISR/LARSyS Strategic Funds from FCT project FCT[UID/EEA/5009/2013] and FCT/11145/12/12/2014/S. Additional information about this project can be found here https://www.epfl.ch/labs/disal/research/institutionalroboticsformations/.

References

1. Au, T., Banerjee, B., Dasgupta, P., Stone, P.: Multi-robot systems. IEEE Intell. Syst. **32**(06), 3–5 (2017)
2. Alonso-Mora, J., Knepper, R., Siegwart, R., Rus, D.: Local motion planning for collaborative multi-robot manipulation of deformable objects. In: IEEE International Conference on Robotics and Automation, pp. 5495–5502 (2015)
3. Saska, M., Krátký, V., Spurný, V., Báča, T.: Documentation of dark areas of large historical buildings by a formation of unmanned aerial vehicles using model predictive control. In: 22nd IEEE International Conference on Emerging Technologies and Factory Automation, pp. 1–8 (2017)
4. Booth, K.E.C., Mohamed, S.C., Rajaratnam, S., Nejat, G., Beck, J.C.: Robots in retirement homes: person search and task planning for a group of residents by a team of assistive robots. IEEE Intell. Syst. **32**(6), 14–21 (2017)
5. Wasik, A., Pereira, J.N., Ventura, R., Lima, P.U., Martinoli, A.: Graph-based distributed control for adaptive multi-robot patrolling through local formation transformation. In: IEEE/RSJ International Conference on Intelligent Robots and Systems, pp. 1721–1728 (2016)
6. Khan, A., Rinner, B., Cavallaro, A.: Cooperative robots to observe moving targets: review. IEEE Trans. Cybern. **48**(1), 187–198 (2018)
7. Das, K., Wymeersch, H.: A network traffic reduction method for cooperative positioning. In: 8th Workshop on Positioning, Navigation and Communication, pp. 56–60 (2011)
8. Jandaeng, C., Suntiamontut, W., Elz, N.: Throughput improvement of collision avoidance in wireless sensor networks. In: 6th International Conference on Wireless Communications Networking and Mobile Computing, pp. 1–5 (2010)
9. La, H.M., Sheng, W.: Dynamic target tracking and observing in a mobile sensor network. Robot. Auton. Syst. **60**(7), 996–1009 (2012)
10. Gautam, A., Mohan, S.: A review of research in multi-robot systems. In: IEEE 7th International Conference on Industrial and Information Systems, pp. 1–5 (2012)
11. Wasik, A., Lima, P.U., Martinoli, A.: A Robust localization system for multi-robot formations based on an extension of a gaussian mixture probability hypothesis density filter. Auton. Robots **44**, 395–414 (2020)
12. Vo, B.N., Ma, W.K.: The Gaussian mixture probability hypothesis density filter. IEEE Trans. Signal Process. **54**(11), 4091–4104 (2006)
13. Ivaldi, S., Padois, V., Nori, F.: Tools for dynamics simulation of robots: a survey based on user feedback. arXiv preprint arXiv:1402.7050 (2014)

14. Wu, W., Zhang, F.: Robust cooperative exploration with a switching strategy. IEEE Trans. Robot. **28**(4), 828–839 (2012)
15. Balakirsky, S., Carpin, S., Dimitoglou, G., Balaguer, B.: From simulation to real robots with predictable results: methods and examples. In: Madhavan, R., Tunstel, E., Messina, E. (eds.) Performance Evaluation and Benchmarking of Intelligent Systems, pp. 113–137. Springer, Boston (2009). https://doi.org/10.1007/978-1-4419-0492-8_6
16. Drumwright, E., Shell, D.A.: An evaluation of methods for modeling contact in multibody simulation. In: IEEE International Conference on Robotics and Automation, pp. 1695–1701 (2011)
17. Ahmad, A., Ruff, E., Bulthoff, H.H.: Dynamic baseline stereo vision-based cooperative target tracking. In: 19th International Conference on Information Fusion, pp. 1728–1734 (2016)
18. Falconi, R., Gowal, S., Martinoli, A.: Graph based distributed control of nonholonomic vehicles endowed with local positioning information engaged in escorting missions. In: IEEE International Conference on Robotics and Automation, pp. 3207–3214 (2010)
19. AMCL. http://wiki.ros.org/amcl
20. Wąsik, A., Ventura, R., Pereira, J.N., Lima, P.U., Martinoli, A.: Lidar-based relative position estimation and tracking for multi-robot systems. In: Robot 2015: Second Iberian Robotics Conference. AISC, vol. 417, pp. 3–16. Springer, Cham (2016). https://doi.org/10.1007/978-3-319-27146-0_1
21. SciPy Library. https://www.scipy.org/
22. Schuhmacher, D., Vo, B.T., Vo, B.N.: A consistent metric for performance evaluation of multi-object filters. IEEE Trans. Signal Process. **56**(8), 3447–3457 (2008)
23. Ventura, R., Ahmad, A.: Towards optimal robot navigation in domestic spaces. In: Bianchi, R.A.C., Akin, H.L., Ramamoorthy, S., Sugiura, K. (eds.) RoboCup 2014. LNCS (LNAI), vol. 8992, pp. 318–331. Springer, Cham (2015). https://doi.org/10.1007/978-3-319-18615-3_26

Opportunistic Multi-robot Environmental Sampling via Decentralized Markov Decision Processes

Ayan Dutta[1(✉)], O. Patrick Kreidl[2], and Jason M. O'Kane[3]

[1] School of Computing, University of North Florida, Jacksonville, USA
`a.dutta@unf.edu`
[2] School of Engineering, University of North Florida, Jacksonville, USA
`patrick.kreidl@unf.edu`
[3] Department of Computer Science and Engineering, University of South Carolina, Columbia, USA
`jokane@cse.sc.edu`

Abstract. We study the problem of information sampling with a group of mobile robots from an unknown environment. Each robot is given a unique region in the environment for the sampling task. The objective of the robots is to visit a subset of locations in the environment such that the collected information is maximized, and consequently, the underlying information model matches as close to reality as possible. The robots have limited communication ranges, and therefore can only communicate when nearby one another. The robots operate in a stochastic environment and their control uncertainty is handled using factored Decentralized Markov Decision Processes (Dec-MDP). When two or more robots communicate, they share their past noisy observations and use a Gaussian mixture model to update their local information models. This in turn helps them to obtain a better Dec-MDP policy. Simulation results show that our proposed strategy is able to predict the information model closer to the ground truth version than compared to other algorithms. Furthermore, the reduction in the overall uncertainty is more than comparable algorithms.

Keywords: Information sampling · Markov Decision Process · Gaussian mixture

1 Introduction

Coverage path planning by a group of autonomous mobile robots has many real-world applications including area cleaning, painting, and precision agriculture and the problem has been extensively studied in the literature. The goal in this task is to cover all the locations in the environment [7]. Recently, researchers have looked into more constrained scenarios, in which the robots can only visit a subset of points in the environment due to the budget constraints, while collecting maximal information from an unknown environment [5,10,12,14,19]. We study such a multi-robot information sampling problem in this paper. This problem is known to be NP-hard problem to solve optimally [19].

In a real-world setting, the robots not only have a budget constraint, but they also have limited communication ranges, and therefore, they are not always guaranteed to maintain a global communication network unless the underlying control mechanism makes them do so continuously or periodically [1]. We use a less restrictive model in which connectivity is *opportunistic*—the robots form ad-hoc local networks with the nearby robots whenever possible. The robots *locally* share their history of noisy sensor measurements for better developing the global information model. This significantly reduces the communication overhead and execution time compared to continuous connectivity models [5,12]. On the other hand, for applications in extreme environmental conditions, e.g., ocean surface mapping [3], the control of the robots becomes stochastic. Our presented solution gracefully handles this uncertainty by modeling the planning problem as a Decentralized Markov Decision Process (Dec-MDP), where the coordinating robots share a joint reward system while their state and action spaces are independent. Simulation results show that our proposed approach is up to 71.68% faster than a comparable continuous connectivity approach while performing at par in terms of the modeling of the underlying information field.

Our primary contributions in this paper are two-fold:

- First, to the best of our knowledge, this is the first work that employs a decentralized MDP technique for multi-robot information collection under control uncertainty.
- Secondly, we address another practical challenge, i.e., limited communication ranges of the robots, by developing an opportunistic connectivity-based novel decentralized coordination mechanism.

2 Related Work

Autonomous mobile robots are used for information collection in real-world applications such as precision agriculture, search and rescue, monitoring, among others. One of the first approaches is due to Krause et al. [10], who proposed greedy strategies to find the informative locations to place a set of sensors, utilizing Gaussian Processes to model the phenomena [16]. Singh et al. [19] proposed the first informative path planning solution for mobile robots. A decentralized multi-robot online informative sampling method is proposed by Viseras et al. [20]. Similar to ours, the sensing is assumed to be noisy. However, the robots exchange a significant amount of information (e.g., past visited locations and corresponding measurements, next locations, etc.), which might be infeasible to achieve in a real-world setting. Similar to our work, Luo and Sycara partition the environment *a priori* and assign each robot to a unique Voronoi cell. Region partitioning has also been used for multi-robot information collection in [4,6,9]. A multi-robot information collection approach with dynamic goal location planning is proposed in [14]. Most of these studies introduce centralized methods, which do not take robots' communication constraints into account. In the real world, the robots have limited communication ranges, which poses a challenge for coordinated environmental sampling. A survey of various connectivity strategies is presented in [1]. Three primary connectivity methods are found: periodic [17], continuous [5], and no requirement, e.g., opportunistic connectivity [4]. The first two requirements are more stringent—the robots have to plan their future locations jointly. However, in an opportunistic setting, the robots' primary goal is to collect maximal information. When two or more robots come within each other's communication ranges, they share their findings in order to make more informative decision in the future [4]. Although real robots exhibit stochastic motion in applications such as underwater monitoring, most of the prior work on multi-robot information collection does not handle control uncertainty. The only informative sampling work for a single robot, to the best of our knowledge, that models stochastic motion using a Markov Decision Process (MDP) is due to Ma et al. [13]. In this work, we consider n robots instead of one, and thus we propose a decentralized (Dec) MDP-based coordination technique for information collection under control uncertainty. An optimal solution for Dec-MDP is proposed in [2], but it is not scalable to large multi-robot systems due to its NEXP-completeness. A heuristic solution is presented in [15]. Our Dec-MDP solution is greedy for better scalability, similar to the ones proposed in [11,18].

3 Problem Setup and Basic Algorithm

A homogeneous team of n mobile robots $r_1, r_2, \ldots, r_n$ moves through a shared planar environment. Each robot is equipped with sufficient on-board sensing (i.e. GPS) to localize itself within the environment, a sensor that measures a phenomenon of interest at the robot's current location, and a communication device that enables limited-range communication with other nearby robots. Let $\mathcal{V}$ denote a given finite set of information collection points, or *nodes*, that cover the environment in a grid pattern. Each robot r_i, starting from a unique node s_0^i in $\mathcal{V}$, is responsible for a subset of nodes $\mathcal{V}_i$ containing s_0^i such that $\mathcal{V} = \cup_{i=1}^n \mathcal{V}_i$ and, for every other robot r_j, the subsets $\mathcal{V}_i$ and $\mathcal{V}_j$ are disjoint. We use k-medoids clustering to achieve such a partitioning [8]. The centroids of the partitions are selected to be the start nodes. (Other partitioning techniques such as Voronoi partitioning or k-means can also be used without affecting our presented solution.) The robots, having common knowledge of this size-n partition, each move sequentially over time in the cardinal directions to adjacent nodes within their own regions, until a given movement budget B expires. The outcome of each action in finite set $\mathcal{U}$ is stochastic, modeled by a transition probability function $f : \mathcal{V} \times \mathcal{U} \times \mathcal{V} \to [0,1]$, under which $f(s, u, s')$ represents the probability for arriving at node s' upon executing action u at node s. For example, this transition probability function f should assign high probability to cardinal movements in the intended direction, and smaller probabilities to movements that represent imperfect movements. The robots are interested in some ambient real-valued phenomenon, which varies across the environment. We model this phenomenon as a random vector $\mathbf{X}$, so that component X_s denotes the value of the phenomenon at node s. *The multi-robot objective is to navigate the environment*

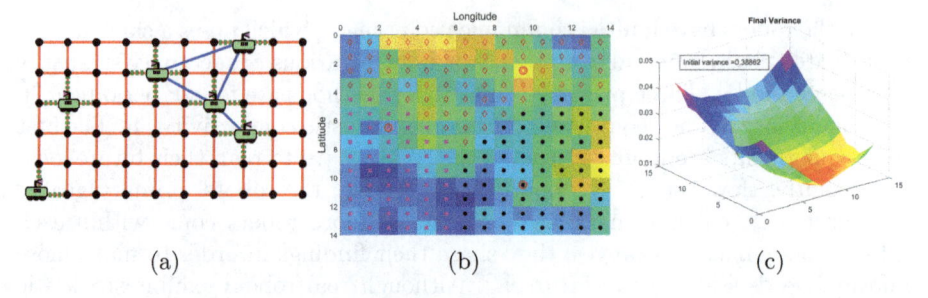

Fig. 1. (a) Illustration of the problem setup with six robots moving on a grid of nodes from which to gather information. Nearby robots may communicate either directly or via multiple hops. (b) A specific 14-by-14 node grid (with the spatially-varying "ground truth" phenomenon conveyed by the blue/yellow shading) to be explored by three robots with given start nodes (red circles) and component static partitions (color-coded node markers). (c) An instance of the initial and final per-node variances, quantifying the reduction in prediction uncertainty, using our proposed solution ($n = 3$).

in order to maximally reduce prediction uncertainty in **X** subject to the budget constraint. Figure 1 illustrates (a) the conceptual setup as well as (b) a specific partition instance of the problem and (c) the character of its solution in our simulation experiments. The following subsections detail the basic models and algorithms by which the robots pursue this objective *in the absence of inter-robot communication*, leaving the extensions to leverage communication for Sect. 4.

3.1 Prediction via Gaussian Processes (GPs)

We use the Gaussian Process (GP) to model the uncertain environment, specifically assuming that (i) the phenomenon of interest at every node takes a scalar real value and (ii) all nodes generate information according to a length-$|\mathcal{V}|$ Gaussian random vector **X** with known (prior) mean vector μ and covariance matrix Σ. It is well known that the optimal prediction, in the sense of minimum mean square error, is the mean vector μ for which the covariance matrix Σ characterizes the prediction's uncertainty [16]. Its (differential) entropy, a volumetric measure of that uncertainty, is given by $H(\mathbf{X}) = \frac{1}{2}\log|\Sigma| + \frac{|\mathcal{V}|}{2}\log(2\pi e)$, where $|\mathcal{V}|$ denotes set cardinality but $|\Sigma|$ denotes matrix determinant.

Each robot's sensing process is imperfect, specifically assuming that measurements are corrupted by zero-mean stationary additive white Gaussian noise, independently and identically distributed (across robots and nodes) with variance σ_n^2. We suppose that (i) all robots are initialized with the same prior model $GP^0 = \{\mu, \Sigma\}$ and (ii) each robot r_i takes measurement y_0^i at its start node s_0^i. It follows that, before any movement decisions are made, the model GP_0^i local to robot r_i is given by (posterior) statistics

$$\Sigma_0^i = \Sigma - \Sigma \mathbf{C}\left(s_0^i\right)' \left(\mathbf{C}\left(s_0^i\right) \Sigma \mathbf{C}\left(s_0^i\right)' + \sigma_n^2 \mathbf{I}\left(s_0^i\right)\right)^{-1} \mathbf{C}\left(s_0^i\right) \Sigma \quad (1)$$
$$\mu_0^i = \mu + \Sigma_0^i \mathbf{C}\left(s_0^i\right)' \left(y_0^i - \mathbf{C}\left(s_0^i\right)\mu\right)/\sigma_n^2$$

where $\mathbf{C}\left(s_0^i\right)$ denotes the length-$|\mathcal{V}|$ row vector of all zeros except for a one in component s_0^i, $\mathbf{C}\left(s_0^i\right)'$ is its matrix transpose (i.e., the analogous column vector) and $\mathbf{I}\left(s_0^i\right)$ denotes the 1-by-1 identity matrix (i.e., the scalar value of one).

The matrix notation within (1) prepares for processing measurements in batch. That is, suppose $\mathbf{y}$ denotes a length-p column vector representing a sequence of collected measurements and let $\mathbf{s}$ be the associated sequence of visited nodes (possibly with repetition). Then, the posterior second-order statistics $\mu_{\mathbf{X}|\mathbf{Y}}$ and $\Sigma_{\mathbf{X}|\mathbf{Y}}$ are computed via (1) upon letting $\mathbf{C}(\mathbf{s})$ denote the p-by-$|\mathcal{V}|$ matrix whose rows consist of successive unit vectors, each having direction corresponding to the successive components of $\mathbf{s}$, while $\mathbf{I}(\mathbf{s})$ denotes the p-by-p identity matrix. Under our noise assumptions, whether (1) is implemented in batch or recursively over measurements gives the same statistics: the resulting mean $\mu_{\mathbf{X}|\mathbf{Y}}$ is the minimum mean-square-error predictor of $\mathbf{X}$, given $\mathbf{Y} = \mathbf{y}$, and the resulting covariance $\Sigma_{\mathbf{X}|\mathbf{Y}}$ analogously implies posterior entropy $H(\mathbf{X}|\mathbf{Y})$.

A final remark concerns the determinant $|\Sigma|$ being an expensive computation, scaling roughly cubically with matrix dimensions. It is common (e.g., in kernel-based parametrizations of GPs) that there are diminishing correlations among nodes as the distance between them grows. This motivates approximation of its determinant by the product of the per-node variances $\sigma_1^2, \sigma_2^2, \sigma_3^2, \ldots$ along its diagonal e.g., $\log |\Sigma| \approx \sum_{s=1}^{|\mathcal{V}|} \log(\sigma_s^2)$. The approximation is, in fact, an upper bound on the true determinant (via Hadamard's inequality), achieving equality if and only if the matrix is truly diagonal. It follows for posterior entropy that

$$H(\mathbf{X}|\mathbf{Y}) \leq \sum_{s=1}^{|\mathcal{V}|} H(X_s|\mathbf{Y}) \quad \text{with} \quad H(X_s|\mathbf{Y}) = \frac{1}{2} \log\left(2\pi e \sigma_{s|\mathbf{Y}}^2\right). \tag{2}$$

We utilize these per-node entropies $\{H(X_s|\mathbf{Y}); s \in \mathcal{V}\}$ to define a reward function that directs the robots' movements, as described in the next section.

3.2 Control via Markov Decision Processes (MDPs)

Before deployment, the robots are provided with a common set of initial training data $\mathcal{D}$ to generate their local initial GP models, GP^i, and calculate the local hyper-parameters. The initial rewards are then calculated based on GP^i using (2). In a single-robot system, to handle the control uncertainty, we can model the problem as a Markov Decision Process (MDP), where the states are represented by the node set $\mathcal{V}$ and the reward can be calculated using (2). We can use standard algorithms such as value iteration or policy iteration to solve the MDP and compute an optimal policy π, which the robot can then follow to collect information. At the start of the multi-robot deployment, similar to a single-robot system, each robot r_i generates such an initial policy π^i by which control $u_0^i \in \mathcal{U}$ is selected and the first (stochastic) move from node s_0^i to node s_1^i is realized. We assume temporarily (for the simplified presentation in this section) that *no communication is available*, so multi-robot system reduces to n independent single-robot systems, each having its local π^i and GP^i. When a new node is visited by robot r_i, it senses the data at the node, incorporates the new observation into its local GP^i and estimates the posterior statistics given by (1), and finally executes the next control, following the policy π^i. This *sense-estimate-move* cycle repeats until the budget of r_i is exhausted. As this process proceeds, r_i will adapt its policy π^i, by re-planning using its updated local model GP^i, after every τ cycles to reflect its improved knowledge of the phenomenon.

4 Algorithm with Opportunistic Communication

Section 3 characterized the basic algorithms by which each robot r_i explores its own subset of nodes $\mathcal{V}_i$ in the absence of inter-robot communication. We certainly expect improved prediction and control when *full* inter-robot communication during navigation is permitted, but we consider the challenge of a connectivity model that is only *opportunistic*, meaning we neither require the robots to remain in continuous communication with each other nor to plan to establish communication after a certain time interval. Instead, the robots focus primarily upon exploring in essentially the same distributed fashion described in Sect. 3, communicating during exploration only when within each other's communication ranges. Let CR denote the maximum communication range and define $R_k^i = \{r_j \mid \|s_k^i - s_k^j\| \leq CR\}$ as the set of robots that are within r_i's communication range in stage k, where s_k^i denotes the location of r_i at stage k. Thus $\bar{R}_k^i = R_k^i \cup \{r_i\}$ forms a connected communication graph such that any two robots in this network can send/receive messages from each other either directly or via hops. Note that, because communication is assumed to be symmetric, all the robots in $\bar{R}_k^i$ form the same local communication graph.

4.1 Prediction via GP Mixtures

With respect to prediction, the main challenge raised by distributed control with only opportunistic communication is that robots, because each explores a unique subset of nodes, are likely to meet possessing GPs that imply different posterior statistics. We thus seek a method by which to combine, or "fuse," different Gaussian statistics. The approach taken in [12], albeit under continuous connectivity, casts the problem as learning parameters within a set of Gaussian mixtures from data via an Expectation-Maximization (EM) algorithm. We shall adopt the same approach, but suitably modified for only opportunistic communication.

Recall from Sect. 3 that, in the basic algorithm, our use of GPs results in two essential outputs: the mean vector that represents the minimum-mean-square-error prediction of process **X**, and the per-node variances that determine the per-node entropies $H(X_s)$. Consider assigning a length-n probability vector $(q_s^1, q_s^2, \ldots, q_s^n)$ to every node s in $\mathcal{V}$, where q_s^i represents the probability that component X_s is described by the Gaussian statistics of robot r_i. Denoting for each i the associated mean and variance by m_s^i and v_s^i, respectively, the Gaussian mixture's statistics describing X_s are given by

$$m_s^* = \sum_{i=1}^n q_s^i m_s^i \quad \text{and} \quad v_s^* = \sum_{i=1}^n q_s^i \left(v_s^i + (m_s^i)^2 - (m_s^*)^2 \right). \tag{3}$$

The statistics in (3) may be used to obtain the same essential outputs discussed in the basic algorithm: the mean vector $\mathbf{m}^* = (m_1^*, m_2^*, \ldots, m_{|\mathcal{V}|}^*)$ represents the fused minimum-mean-square-error prediction of process $\mathbf{X}$, while the per-node variances v_s^* permit approximate per-node entropies via $H(X_s) \approx \frac{1}{2} \log(2\pi e v_s^*)$.

We proceed to describe the EM algorithm that determines (i) the per-node mixture probabilities $\{q_s^i\}$ as well as (ii) the associated n per-robot Gaussian statistics i.e., per-node means $\{m_s^i\}$ and variances $\{v_s^i\}$ of each robot r_i. The data in the algorithm is a batch of p measurements, in the sense discussed for (1) in Sect. 3 i.e., length-p column vectors $\mathbf{y}$ and $\mathbf{s}$ that represent, respectively, the collected measurements (possibly by multiple robots) and the associated sequence of visited nodes (possibly with repetition). The algorithm is iterative, initialized assuming n distinct GPs are available, each associated with a mean vector μ^i and covariance matrix Σ^i representing the prior statistics local to robot r_i. For every node s in $\mathcal{V}$ and index $i = 1, 2, \ldots n$, assign

$$m_s^i := \mathbf{C}(s)\mu^i, \quad v_s^i := \mathbf{C}(s)\Sigma^i\mathbf{C}(s)' \quad \text{and} \quad q_s^i := \begin{cases} 1 - \epsilon, & \text{if } s \in \mathcal{V}_i \\ \epsilon/(n-1), & \text{otherwise} \end{cases}$$

with matrix $\mathbf{C}$ as defined in (1) and $0 < \epsilon \ll 1$ denoting a "small" probability. The algorithm then repeats the following two-step procedure until convergence:

Expectation: For every node s in $\mathcal{V}$, denote by $\mathbf{y}_s$ the subvector of batch measurements $\mathbf{y}$ collected at node s and, for every index $i = 1, 2, \ldots n$, assign q_s^i proportional to $1/v_s^i$ if $\mathbf{y}_s$ is empty and otherwise assign q_s^i proportional to $q_s^i L_s^i$, where $L_s^i(\mathbf{y}_s)$ denotes the likelihood of $\mathbf{y}_s$ assuming independent measurements under a (univariate) Gaussian PDF with mean m_s^i and variance v_s^i.

Maximization: For every index $i = 1, 2, \ldots n$, denote by $\mathbf{y}^i$ the length-p^i subvector of batch measurements collected by robot r_i and by $\mathbf{s}^i$ the associated sequence of visited nodes. If $\mathbf{y}^i$ is empty, assign $\Lambda^i := \Sigma^i$ and $\nu^i := \mu^i$; otherwise,

$$\begin{aligned}
\Lambda^i &:= \Sigma^i - \Sigma^i \mathbf{C}\left(\mathbf{s}^i\right)' \left(\mathbf{C}\left(\mathbf{s}^i\right) \Sigma^i \mathbf{C}\left(\mathbf{s}^i\right)' + \sigma_n^2 \mathbf{Q}\left(\mathbf{s}^i\right)^{-1}\right)^{-1} \mathbf{C}\left(\mathbf{s}^i\right) \Sigma^i \\
\nu^i &:= \mu^i + \Lambda^i \mathbf{C}\left(\mathbf{s}^i\right)' \mathbf{Q}\left(\mathbf{s}^i\right) \left(\mathbf{y}^i - \mathbf{C}\left(\mathbf{s}^i\right) \mu^i\right) / \sigma_n^2
\end{aligned} \quad (4)$$

with $\mathbf{Q}(\mathbf{s}^i)$ denoting the p^i-by-p^i diagonal matrix of probabilities q_s^i ordered along the diagonal in correspondence with subvector $\mathbf{s}^i$. Then, for every node s in $\mathcal{V}$ and index $i = 1, 2, \ldots n$ assign $m_s^i := \mathbf{C}(s)\nu^i$ and $v_s^i := \mathbf{C}(s)\Lambda^i\mathbf{C}(s)'$.

The EM initialization assumes all robots' prior models $\{GP_{k-1}^i\}$ are available, which is not necessarily the case under only opportunistic communication. For example, consider the stage-0 perspective of a particular robot r_i, who will know only its local model GP_0^i and the prior model GP^0 from which it evolved. It can simply assign $GP_0^j := GP^0$ for every $j \neq i$ and proceed in subsequent stages as described. Of course, until robot r_i enters a stage k in which communication with another robot r_j is possible (i.e., until R_k^i is non-empty), the non-local models will not change (i.e., the maximization step renders $GP_k^j := GP_{k-1}^j$) and thus be mismatched from the stage-k prior model local to robot r_j. Stages k with R_k^i non-empty begin with synchronizing the prior models $\{GP_{k-1}^i \mid r_i \in \bar{R}_k^i\}$ and sharing the local measurements $\{(\mathbf{y}^j, \mathbf{s}^j) \mid r_j \in \bar{R}_k^i\}$ to form batch data $(\mathbf{y}, \mathbf{s})$ upon which each connected robot proceeds to locally execute its EM algorithm. It should be noted that the algorithm concludes with synchronised posterior models $\{GP_k^j \mid r_j \in \bar{R}_k^i\}$, but if any models $\{GP_k^j \mid r_j \notin \bar{R}_k^i\}$ over disconnected robots remain mismatched then the statistics in (3) will also likely differ.

4.2 Control via Decentralized MDPs

For multi-robot systems, where one robot's reward might be affected by the observations made by the other robots, we can extend the n independent MDP model to Decentralized MDPs (Dec-MDP). We consider a factored-state (the local states are unique to the robots, e.g., the node set is partitioned into n subsets), transition-independent, and non-reward-independent Dec-MDP model [2,11,15]. This is due to the fact that the robots are placed in unique regions in the environment, but following (2), one robot's local reward is affected by other robots' sensed information. Unfortunately, this has been proved to be a NP-complete problem to solve optimally [2], and therefore, we adopt a greedy strategy to solve it [18]. The pseudocode for the approach is presented in Algorithm 1. Essentially, each robot $r_j \in R_k^i \cup r_i$ updates its GP model following the formulation in Sect. 4.1. Next, r_j augments its local MDP with the updated joint rewards calculated using (2) and generates a new local optimal policy to follow.

5 Evaluation

We have tested our proposed opportunistic online information sampling planner in simulation using MATLAB. The experiments are run on a laptop computer with a 1.80 GHz. Intel Core i7-8500U Processor, 16 GB RAM. We varied the number of robots between $\{2, 4, 6\}$. The robots were placed in a 4-connected grid environment of size 14×14 m having unit-length square cells. Each robot was given a budget of 20 m of travel. The policy update frequency, τ, is set to 5, a fraction of B. We use the Value Iteration algorithm available in the MDP-toolbox (https://bit.ly/38PPPcf) to obtain the policies. Our ground truth environment is modeled by a zero-mean Gaussian random vector $\mathbf{X} = (X_1, X_2, \ldots, X_{196})$ with covariance matrix built upon an exponential kernel function: specifically, for any pair of nodes s and t at spatial locations $\mathbf{p}_s$ and $\mathbf{p}_t$, respectively, let $\text{Cov}(X_s, X_t) = \beta^2 \exp(-||\mathbf{p}_s - \mathbf{p}_t||/\ell)$, where hyperparameters $\beta > 0$ is the local standard deviation and ℓ (in meters) is the exponential rate of diminishing covariance between increasingly distant states. Our experiments assume $\beta = 1$ and $\ell = 25$ and then sample the resulting isotropic Gaussian Markov random field to simulate ground truth; Fig. 1(b) depicts such an instance. The noisy sensing process at any node is simulated by adding to its ground truth value a sample from the zero-mean univariate Gaussian distribution with variance $\sigma_n^2 = 0.25$. The probability parameters in the EM algorithm are set to $\epsilon = 10^{-3}$ and $\delta = 10^{-4}$.

Algorithm 1: Decentralized Environmental Sampling Using Opportunistic Connectivity and Dec-MDP

1 Procedure `sampleInformation()`
   Input: $\mathcal{V}_i \leftarrow$ robot r_i's unique, static partition – calculated offline.
   $B \leftarrow$ Exploration budget of the robots.
2 $k \leftarrow$ the number of total moves, initially set to 0.
3   Each robot r_i will follow the exploration scheme – <Move, Sense, Connect, Estimate, Adapt> – within $\mathcal{V}_i$:
4   *Sense* data in the starting node and add it to the initial training set.
5   Each robot 1) begins having the same prior GP learned from the initial training set, 2) then updates to GP_0^i using the measurement at the start node and 3) predicts the initial per-node entropies.
6 $\pi_0^i \leftarrow r_i$'s initial MDP policy based on initial rewards.
7   **while** $k < B$ **do**
8 $\quad k \leftarrow k + 1$.
9 $\quad R_k^i \leftarrow \emptyset$.
10 $\quad$ *Move* to the next node s_k^i following the local policy π_{k-1}^i.
11 $\quad$ *Sense* data in the current node; add the observation to the training set.
12 $\quad$ Broadcast message and update R_k^i (Sect. 4.1).
13 $\quad$ **if** $R_k^i \neq \emptyset$ **then**
14 $\quad\quad$ **if** r_i *has previously encountered with some robots* $R'' \subseteq R_k^i$ **then**
15 $\quad\quad\quad$ Use the mixture parameters from this last encounter along with the newly observed data to update the local Gaussian model GP_k^i.
16 $\quad\quad$ **else**
17 $\quad\quad\quad$ Share all observed data with $r_j \in R_k^i$ and use the EM algorithm to update GP_k^i.
18 $\quad\quad$ Update the rewards (2) based on GP_k^i.
19 $\quad\quad$ *Adapt* by executing $solveDecMDP(R_k^i)$ and updating the local MDP policy $\pi_k^i \leftarrow \pi_i^*$.
20 $\quad$ **else**
21 $\quad\quad$ *Estimate* Use the newly locally observed data and update GP_k^i.
22 $\quad$ *Adapt* the local policy π_k^i based on revised rewards from GP_k^i after every τ cycles.

23 Procedure `solveDecMDP()`
   Input: $R' \leftarrow$ robots that are within CR distance of robot r_i
   Output: Solution of the local MDP of robot r_i
24 $MDP_i^* \leftarrow$ Augmented local MDP with the joint reward function for $\forall r_j \in R'$.
25 $\pi_i^* \leftarrow$ Solve MDP_i^* using the Value/Policy Iteration algorithm.
26 return π_i^*.

We measured the performance of our approach by testing five other variants: 1) control strategies were varied between Dec-MDP and greedy—in the Dec-MDP variant, robots' actions are decided based on the MDP policy and in

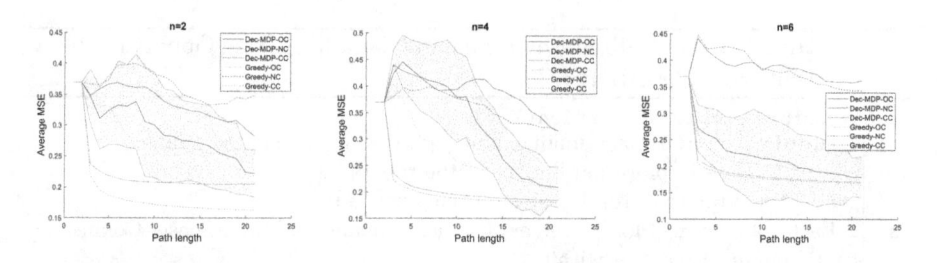

Fig. 2. Comparison of the MSE metric among the algorithms [lower is better].

the greedy variant, a greedy action, i.e., that maximizes the one-step reward is chosen; 2) connectivity strategies were varied between no communication (NC), opportunistic (OC), and continuous communication (CC) among the robots. In cases of CC and NC, the robots were assumed to have infinite and zero communication ranges respectively. Note that the greedy-CC strategy is similar to the centralized technique proposed in [12]. In case of OC, the CR is set to 0.3 times of the environment's diagonal. Ten trials were conducted for each scenario.

Results. First, we are interested in investigating the most important metric to measure the performance of the proposed multi-robot information sampling approach – Mean Square Error (MSE), which depicts how closely the robots could model the underlying information field. The average MSE between the final predicted measurements and the ground truth measurements for different robot counts are shown in Fig. 2. The standard deviation of our algorithm's yielded MSE is also shown in shaded blue. As can be observed, for $n \in \{2, 4, 6\}$, average MSE over time has reduced. Although this is true for all the implemented algorithms, for our proposed one (Dec-MDP-OC), the final MSE is one of the lowest amongst all. As expected, if there is no communication available, regardless of the control strategies, the MSE values are the highest, indicating a worse difference between the predicted and the ground truth information model. Similarly, if there is continuous communication, regardless of the control strategy, the average MSE is always better (lower) than our approach. However, the maximum final difference is only 11% with $n = 4$. Recall that this approach requires continuous communication among the robots, which is not only non-trivial to maintain [5], but also incurs high run-time cost. Our algorithm outperforms the comparable greedy strategy except for $n = 4$, since our approach looks into an infinite horizon to find the solution as opposed to the one-step look ahead used by the greedy strategy.

A similar trend can be seen for the variance metric as well. The results for this metric are presented in Fig. 3. We have calculated the per-node variance after every GP prediction and the average across n robots are shown here. As discussed earlier, the variance indicates the uncertainty in the Gaussian Process prediction. Similar to MSE, the average variance decreases as the robots visit more nodes in the environment. In case of NC, the variance does not decrease as sharply as in the OC and CC cases since there is no Gaussian mixture process involved. On

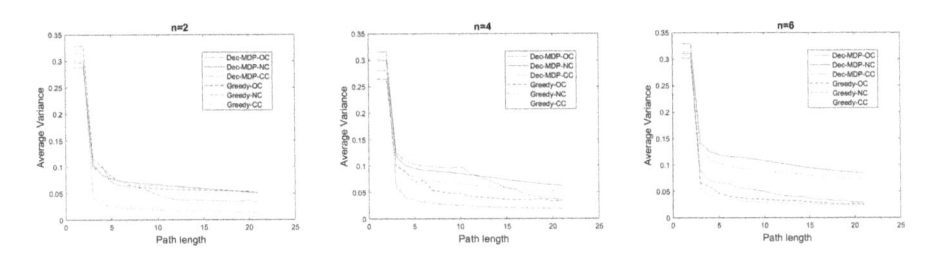

Fig. 3. Comparison of the variance metric among the algorithms [lower is better].

the other hand, the proposed Dec-MDP-OC strategy performs better or at par with the greedy-OC method while averaging very close to the CC strategies (the minimum being only 50% higher than the Dec-MDP-CC solution with $n = 6$).

Next, we investigated the run times of the proposed solution and compare it against other implemented algorithms. As can be seen in Fig. 4(a), the run times of the proposed Dec-MDP-OC strategy are low and grow in a quadratic fashion with n. For example, with $n = 6$, the run time for the proposed approach is only 27.87 s, whereas for the Dec-MDP and greedy-CC approaches they are 35.78 and 32.89 s respectively. A *connected component* is defined as a maximal set of robots for which the member robots can communicate directly or via multiple hops. These are the robots that participate in the proposed Gaussian mixture-based coordination strategy. The lower-bound on the number of connected components in a communication graph is 1, which essentially represents a CC strategy at that time. However, we have found that in the presented OC strategy, the connected component count is always greater than 1, which indicates a lower communication and coordination overhead than the centralized CC strategy such as used in [12]. This is also supported by the fact that the greedy and Dec-MDP CC strategies take more time than our OC technique. The number of messages required to be sent for the Gaussian mixture algorithm to be converged is also negligible, as shown in Fig. 4(b).

The average reward collected by the robots is reported in Fig. 4(c). This result shows that our proposed Dec-MDP-based control mechanism always collects higher reward than the greedy approach with the greatest difference occurring with $n = 4$. Finally, we visually compare the ground truth information field against the final predicted model across various algorithms. We can notice in Figs. 5 that the proposed Dec-MDP-OC approach makes more fine-grained and close-to-reality prediction than that of the greedy-OC and the NC strategies. This observation is supported by the numerical MSE data presented in Fig. 2.

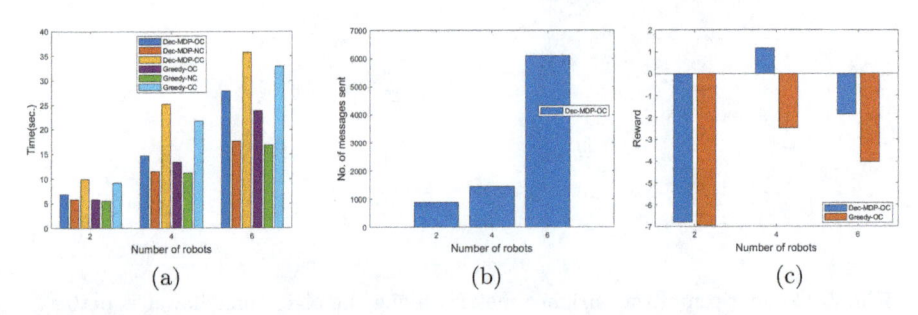

Fig. 4. Comparison of a) Run times, b) Average number of messages sent by each robot, and c) Collected average rewards.

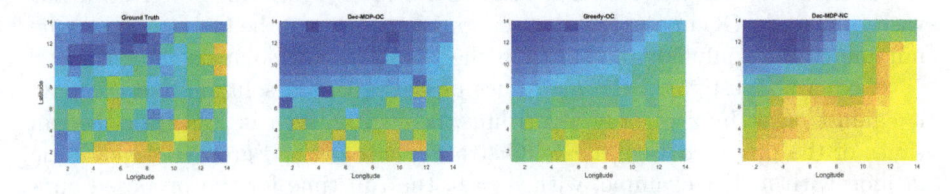

Fig. 5. Solution instance: visual comparison of predicted models by various algorithms against the ground truth measurements ($n = 2$).

6 Conclusions and Future Work

We have proposed an online multi-robot information sampling technique for an unknown environment. Our presented approach gracefully handles real-world constraints such as limited communication ranges and stochastic motion of the robots. The proposed strategy relies on robots' opportunistic communication patterns instead of forcing the robots to stay in continuous communication or connect after regular intervals. Results show that this technique can reduce the robots' run times significantly while performing at par in terms of modeling the underlying information field and reducing the uncertainty in the environment, compared to (potentially centralized) techniques that maintain continuous communication connectivity. In the future, we plan to investigate a better mixture method, in which each robot will maintain a history of past coordinated Gaussian mixtures. Finally, we plan to test the proposed solution with real robots for applications in domains such as precision agriculture and underwater robotics.

Acknowledgements. A. Dutta and O.P. Kreidl are partially supported by NSF CPS grant #1932300.

References

1. Amigoni, F., Banfi, J., Basilico, N.: Multirobot exploration of communication-restricted environments: a survey. IEEE Intell. Syst. **32**(6), 48–57 (2017)

2. Becker, R., Zilberstein, S., Lesser, V., Goldman, C.V.: Solving transition independent decentralized Markov decision processes. JAIR **22**, 423–455 (2004)
3. Delight, M., Ramakrishnan, S., Zambrano, T., MacCready, T.: Developing robotic swarms for ocean surface mapping. In: ICRA (2016)
4. Dutta, A., Bhattacharya, A., Kreidl, O.P., Ghosh, A., Dasgupta, P.: Multi-robot informative path planning in unknown environments through continuous region partitioning. Int. J. Adv. Rob. Syst. **17**(6), 1–18 (2020)
5. Dutta, A., Ghosh, A., Kreidl, O.P.: Multi-robot informative path planning with continuous connectivity constraints. In: ICRA, pp. 3245–3251 (2019)
6. Fung, N., III, J.G.R., Nieto, C., Christensen, H.I., Kemna, S., Sukhatme, G.S.: Coordinating multi-robot systems through environment partitioning for adaptive informative sampling. In: ICRA, pp. 3231–3237. IEEE (2019)
7. Galceran, E., Carreras, M.: A survey on coverage path planning for robotics. Robot. Auton. Syst. **61**(12), 1258–1276 (2013)
8. Kaufmann, L.: Clustering by means of medoids. In: Proc. Statistical Data Analysis Based on the L1 Norm Conference, Neuchatel, 1987, pp. 405–416 (1987)
9. Kemna, S., Rogers, J.G., Nieto-Granda, C., Young, S., Sukhatme, G.S.: Multi-robot coordination through dynamic voronoi partitioning for informative adaptive sampling in communication-constrained environments. In: ICRA, pp. 2124–2130. IEEE (2017)
10. Krause, A., Singh, A., Guestrin, C.: Near-optimal sensor placements in gaussian processes: Theory, efficient algorithms and empirical studies. JMLR **9**(Feb), 235–284 (2008)
11. Kumar, R.R., Varakantham, P., Kumar, A.: Decentralized planning in stochastic environments with submodular rewards. In: AAAI, pp. 3021–3028 (2017)
12. Luo, W., Sycara, K.P.: Adaptive sampling and online learning in multi-robot sensor coverage with mixture of gaussian processes. In: ICRA, pp. 6359–6364. IEEE (2018)
13. Ma, K., Liu, L., Sukhatme, G.S.: An information-driven and disturbance-aware planning method for long-term ocean monitoring. In: IROS, pp. 2102–2108. IEEE (2016)
14. Ma, K.C., Ma, Z., Liu, L., Sukhatme, G.S.: Multi-robot informative and adaptive planning for persistent environmental monitoring. In: DARS, pp. 285–298. Springer (2018)
15. Peshkin, L., Kim, K.E., Meuleau, N., Kaelbling, L.P.: Learning to cooperate via policy search. UAI, pp. 489–496. Morgan Kaufmann Publishers Inc. (2000)
16. Rasmussen, C.E.: Gaussian processes in machine learning. In: Summer School on Machine Learning, pp. 63–71. Springer (2003)
17. Ruiz, A.V., Xu, Z., Merino, L.: Distributed multi-robot cooperation for information gathering under communication constraints. In: ICRA, pp. 1267–1272. IEEE (2018)
18. Unleashing dec-mdps in security games: Shieh, E.A., Jiang, A.X., Yadav, A., Varakantham, P., Tambe, M. Enabling effective defender teamwork. In: ECAI **263**, 819–824 (2014)
19. Singh, A., Krause, A., Guestrin, C., Kaiser, W.J., Batalin, M.A.: Efficient planning of informative paths for multiple robots. In: IJCAI, vol. 7, pp. 2204–2211 (2007)
20. Viseras, A., et al.: Decentralized multi-agent exploration with online-learning of Gaussian processes. In: ICRA (2016)

A PHD Filter Based Localization System for Robotic Swarms

R. A. Thivanka Perera[✉], Chengzhi Yuan, and Paolo Stegagno

ICRobots Lab, University of Rhode Island, Kingston, RI 02881, USA
{thiva,cyuan,pstegagno}@uri.edu

Abstract. In this paper, we present a Probability Hypothesis Density (PHD) filter based relative localization system for robotic swarms. The system is designed to use only local information collected by onboard lidar and camera sensors to identify and track other swarm members within proximity. The multi-sensor setup of the system accounts for the inability of single sensors to provide enough information for the simultaneous identification of teammates and estimation of their position. However, it also requires the implementation of sensor fusion techniques that do not employ complex computer vision or recognition algorithms, due to robots' limited computational capabilities. The use of the PHD filter is fostered by its inherent multi-sensor setup. Moreover, it aligns well with the overall goal of this localization system and swarm setup that does not require the association of a unique identifier to each team member. The system was tested on a team of four robots.

Keywords: Localization · PHD filter · Swarm · Sensor fusion · Multi-robot

1 Introduction

In recent years, robotic swarms have been receiving increasing attention thanks to many potential applications [1]. Tasks as target search and tracking [2], search and rescue [3], exploration [4], information gathering, clean up of toxic spills [5], and construction [6] have been proposed throughout the years. Most works focus on control algorithms, however, many control laws and collaborative swarm behaviors require the ability to identify other robots in the environment, and compute an estimate of their position.

To retrieve this information many localization algorithms in different operative conditions have been proposed for multi-robot systems. In cooperative localization (e.g., [7]), the robots communicate each other's odometry and relative measurements to compute the location of each team member in a common frame of reference, usually through an online Bayesian filter (e.g., [8]) or estimator (e.g., [9]). However, the assumption of a common frame of reference accounts as a form of centralization and should be avoided in a robotic swarms.

In relative localization algorithms, the assumption of a common frame of reference is eased and the goal of each robot is to estimate the pose of other robots in its attached frame of reference. This has been addressed through Bayesian filters [10], geometrical arguments [11], or a combination of both [12]. Usually, both relative and cooperative localization algorithms require not only some position, bearing or distance measurements, but also that each measurement comes with the unique identifier of the measured robot. Typical approaches include visually tagging each robot and extracting the tag through cameras [13], using dedicated infrared systems [14] or RFIDs [15].

However, tagging and ID exchange in many cases could be difficult or undesirable. It could be technically unfeasible, particularly in case of large numbers of robots or with sensors, as laser scanners, that do not allow for unique identification capabilities. It also accounts as a form of centralization, meaning that all robots need to know the same set of IDs. Last but not least, it may jeopardize the task to make explicit the identity of each robot, if the swarm is for example in an escorting or disguising mission. In a number of papers, the problem of computing an estimate of other robot's location with untagged measurements has been referred to as localization with anonymous measurements [12,16], or unknown data association. In [16], using odometry and untagged relative measurements communicated from other robots, the robots were able to produce tagged (i.e., associated with a unique identifier) relative pose estimates.

However, associating ids to each robot is not a mandatory condition to perform cooperative tasks as formation control [17], encircling [18], and connectivity maintenance [19], as long as each robot is able to identify that some entities in the environment are generically teammates, and compute an estimate of their relative positions. Moreover, in a robotic swarm setup, robots could have limited or no communication capabilities, and should rely only on local self-gathered measurements to perform their tasks. In this situation, the choice of the sensor equipment endowed to the robots becomes even more crucial. On the one hand, the sensors should provide enough information to allow (non-unique) identification of other robots, and quantitative estimation of their relative position. On the other hand, robotic swarms are usually composed by relatively small and cheap robots, featuring limited computational capabilities that are not compatible with expensive informative rich sensor equipment. Single sensor approaches are limited by the sensing technology. Using distance sensors as lidars, robots can be easily mistaken for obstacles of similar size, and vice versa. On the other hand, camera sensors would be able to identify robots more reliably, but they would directly provide only bearing information, and distance estimates could be affected by consistent noise, have long convergence time, and require persisting excitation conditions [20,21]. RGB-D sensors offer the best of both worlds but usually have limited fields of view, while the robots should be aware of teammates and obstacles in entirety of their surroundings.

In this paper we propose a multi-modal approach in which we employ multiple sensors – fisheye cameras and laser scanners – to combine the recognition capability of the first with the accuracy of the second. However, this approach requires non-trivial data fusion techniques. Given the multi-target multi-sensor tracking nature of the proposed problem, a natural choice would be the employment of a Probability Hypothesis Density (PHD) filter. The PHD filter was first proposed in [22] as a recursive filter for multi-target multi-sensor tracking. The filter in its theoretical form would require infinite computational power. However, some authors have proposed Gaussian mixture [23] and particle based implementations [24] among others.

PHD filters have already been employed in multi-robot localization. In [25], the authors presented a PHD filter to incorporate absolute poses exchanged by robots and local sensory measurement to maintain robots' formation when communication fails. In [26], a team of mobile sensors was employed to cooperatively localize an unknown number of targets via PHD filter. However, in these two works a common frame of reference was assumed, which is not compatible with our setup. In [27], the authors implemented a PHD filter to fuse ground robot (UGV) odometry and aerial camera measurements to estimate the location and identity of the UGVs. However, only the aerial robot computes the position of the other robots, and not every team member. In [28], the authors used two different visual features to describe the target of interest enhancing the PHD-based tracking. However, this is an example of video tracking and the metric pose of the targets are not estimated.

None of the setups discussed in literature is compatible with the needs of a robotic swarm and our settings. Therefore, in this paper we propose a novel robo-centric implementation of the PHD filter for the fusion of lidar and camera measurements in a swarm setup. The ultimate objective of the filter, which runs independently on each robot, is to compute an estimate of other teammates positions, while simultaneously discarding possible obstacles in the environment.

The rest of the paper is organized as follows. In Sect. 2 we formally introduce the problem. In Sect. 3 we provide some background on the PHD filter. In Sect. 4 we propose two different implementations of the PHD filter for tracking of other robots. In Sect. 5, we provide an experimental validation of our methods, and in Sect. 6 we conclude the paper.

2 Problem Setting

The system we consider (Fig. 1) consists of n UGVs $\{\mathcal{R}^1, \mathcal{R}^2, ..., \mathcal{R}^n\}$ in a 2D space, with n and unknown and time-variant. The generic robot $\mathcal{R}^j$ is modeled as a rigid body moving in 2D space and is equipped with an attached reference frame $\mathcal{F}_j = \{O_j, X_j, Y_j\}$ whose origin coincides with a representative point of the robot. Let $q_h^j \in \mathbb{R}^2$ be and $\psi_h^j \in SO(2)$ respectively the position and orientation of $\mathcal{R}^h$ in $\mathcal{F}_j$, and let o_h^j be the position of $\mathcal{O}_h$ in $\mathcal{F}_j$. In the following, we indicate with $R(\phi)$ the elementary 2D rotation matrix of an angle ϕ $R(\phi) = \begin{bmatrix} cos(\phi) & -sin(\phi) \\ sin(\phi) & cos(\phi) \end{bmatrix}$. Robot $\mathcal{R}^j$ is equipped with multiple sensors. First, the odometry module of $\mathcal{R}^j$ provides, at each time k, a measurement $U_k^j = [\Delta x_k^j \ \Delta y_k^j \ \Delta \psi_k^j]^T \in \mathbb{R}^2 \times SO(1)$ of the robot linear and angular displacement between two consecutive sampling instants $k-1$ and k on the XY plane.

$\mathcal{R}^j$ is also equipped with a lidar sensor. Lidar measurements are processed with a feature extraction algorithm that identifies all objects in the scan (including robots) whose size is comparable with the size of the robots. In general, we

Fig. 1. Left: the robotic swarm used to validate our localization system. Center: one of the robots used in this work. Right: a representation of the problem setting: triangles are robots, false positives are circles, × are lidar measurements l_k^h, straight dashed lines are camera measurements c_k^h, blind spots in the lidar measurements are represented as shaded areas.

assume that there is an unknown number of objects in the environment that will be detected in the lidar as possible robots. Therefore, at each time step k the algorithm provides a set of l_k relative position measurements $L_k = \{l_k^1, ..., l_k^{l_k}\}$ in $\mathcal{F}_j$, representing the position of robots or obstacles in the field of view of the sensor. The sensor is affected by false positive (some measurements may not refer to actual objects) and false negative measurements (some object or robot may not be detected) due to obstructions and errors of the feature extraction.

Lastly, two fisheye cameras are mounted on $\mathcal{R}^j$, one oriented towards the front of the robot, and one towards the back. This setup allows to identify robots in a 360° field of view. The images from the cameras are processed using a feature extraction algorithm which has the capability of identifying a generic robot based on color. The algorithm does not uniquely identify and label each robot. At time step k, the cameras provide a set of c_k of bearing measurements $C_k = \{c_1, ...c_k^{c_k}\}$ in $\mathcal{F}_j$. Also in this case, there may be false positive (non-robots identified as robots) and false negative (missed robot detections) measurements. In the following, the camera and lidar measurements collected at time k will be denoted together as $Z_k = \{L_k, C_k\}$. Note that camera and lidar have different rates and in general are not synchronized. Therefore, without loss of generality, for some k it may be either $L_k = \emptyset$, or $C_k = \emptyset$, or both. A representation of all sensor readings is provided in Fig. 1.

The objective of $\mathcal{R}^j$ is to compute at each time step k an estimate of the number $n(k)$ and positions of all robots in the environment.

3 Multi-sensor PHD Filter

This Section provides the necessary background on the PHD filter and is mostly based on [22–24]. Assuming that there are n (with n unknown and variable over time) targets living in a space $\mathcal{X}$, the goal of the standard PHD filter is to compute an estimate of the PHD of targets in $\mathcal{X}$. The PHD $f_k(x)$ at time k is defined as the function such that its integral over any subset $S \subseteq \mathcal{X}$ is the expected number of targets $N(S)$ in that subset, i.e., $N(S) = \int_S f_k(x)dx$.

The PHD filter is a recursive estimator composed of two main steps: a time update and a measurement update. The time update is meant to produce a prediction of the PHD $f_{k|k-1}(x)$ at time step k given the estimate $f_{k-1|k-1}(x)$ at time $k-1$, through the time update equation:

$$f_{k|k-1} = b_{k|k-1}(x) + \int [P_s(x')f_{k|k-1}(x|x') + b_{k|k-1}(x|x')]f_{k-1|k-1}(x')dx' \quad (1)$$

where $b_{k|k-1}(x)$ is the probability that a new target appears in x between times $k-1$ and k, $P_s(x')$ is the probability that a target in x' at time $k-1$ will survive into step k, $f_{k|k-1}(x|x')$ is the probability density that a target in x' moves to x, and $b_{k|k-1}(x|x')$ is the probability that a new target spawns in x at time k from a target in x' at time $k-1$.

Note that both $f_{k-1|k-1}(x)$ and $f_{k|k-1}(x)$ are computed considering only the measurements up to time $k-1$. Measurements Z_k are incorporated in the estimate through the measurements update to compute the posterior PHD:

$$f_{k|k}(x) = f_{k|k-1}(x)\left[1 - P_D(x) + \sum_{z \in Z_k} \frac{P_D(x)g(z|x)}{\lambda c(z) + \int P_D(x')g(z|x')f_{k|k-1}(x')dx'}\right] \quad (2)$$

where $P_D(x)$ is the probability that an observation is collected from a target with state x, $g(z|x)$ is the sensor likelihood function, and $\lambda c(z)$ expresses the probability that a given measurement z is a false positive.

Although elegant, Eqs. (1) and (2) cannot be implemented in practice for generic functions. A popular approximation, the Gaussian Mixture PHD filter (GM-PHD) considers all PHD functions $f_{k-1|k-1}(x)$, $f_{k|k-1}(x)$, and $f_{k|k}(x)$ to be sums of weighted Gaussian functions in the form:

$$f_{*|*}(x) = \sum_i f^i_{*|*}(x) = \sum_i w^i_{*|*} \mathcal{N}(x; m^i_{*|*}, p^i_{*|*}) \quad (3)$$

where $f^i_{*|*}(x)$ is the generic i-th component, $w^i_{*|*}$, $m^i_{*|*}$, and $p^i_{*|*}$ are respectively the weight, mean and covariance matrix of the i-th component. Introducing the GM representation (3) in Eq. (1), and assuming that the probability of survival can be approximated as a constant for each component ($P_s(x') \simeq P^i_s$), the spawning probability $b_{k|k-1}(x|x')$ is zero, and the system model $f_{k|k-1}(x|x')$ and target birth probability $b_{k|k-1}(x)$ are Gaussian functions, the GM-PHD filter time update equation becomes:

$$f_{k|k-1} = b_{k|k-1}(x) + \sum_i w^i_{k-1|k-1} P^i_s \int f_{k|k-1}(x|x')f^i_{k-1|k-1}(x')dx' \quad (4)$$

Therefore, the PHD prediction will have a component for each component in the PHD posterior $f_{k-1|k-1}(x)$. Moreover, the integral term will be the same as a prediction step of the standard Kalman filter, so every component of $f_{k|k-1}(x)$ will be a Gaussian function, and it will be possible to compute the PHD prediction by simply applying component-wise the time update of a Kalman filter.

Introducing the GM representation (3) in Eq. (2), assuming that the probability of detection can be approximated as a constant $P_D(x) \simeq P_D^i$ for each component $f_{k+1|k}^i(x)$, and $c(z) = 0$, the GM-PHD filter measurement update equation becomes:

$$f_{k|k}(x) = \sum_i f_{k|k-1}^i(x)\left(1 - P_D^i\right) + \sum_i \sum_{z \in Z_k} \frac{P_D^i f_{k|k-1}^i(x) g(z|x)}{\sum_i \int P_D^i g(z|x') f_{k|k-1}^i(x') dx'}. \quad (5)$$

showing that, if Z_k contains m measurements, each component $f_{k|k-1}^i(x)$ generates $m+1$ components in $f_{k|k}(x)$. Moreover, if $g(z|x)$ is a Gaussian function, the last term is a sum of Gaussian functions, each function being the result of a single-component Kalman filter measurement update step.

An additional pruning step is needed to limit the number of components in the PHD. In fact, if all components were kept at each time step, their number would grow exponentially with the number of measurements. Therefore, all components whose weight is below a given threshold at the end of the measurement update are eliminated.

It is clear from its formulation that the PHD filter is inherently multi-sensor. When multiple sensors are present, multiple measurement updates can be applied consecutively, each one as a component-wise Kalman filter update step.

4 PHD-Filter Based Relative Localization Module

Following the scheme presented in the Sect. 3, our localization module consists of a time update step and two measurement update steps. Note that the module is asynchronous, so there is no particular order or sequence in which these steps are performed. While the time update is periodically performed, the measurement updates are performed if and when measurements become available. The estimated state of the target robots is their position in $\mathcal{F}_j$, $m_{k|k}^i = q_*^j(k) \in \mathbb{R}^2$, where the $*$ expresses the concept that the i-th component may refer to any of the tracked target robots. The covariance of $m_{k|k}^i$ is therefore $p_{k|k}^i \in \mathbb{R}^{2 \times 2}$.

4.1 Time Update

During the time update, the owner's $\mathcal{R}^j$ odometry $U_k^j = [\Delta x_k^j \; \Delta y_k^j \; \Delta \psi_k^j]^T$ is used to update the mean and covariance of all components of the PHD. The kth time update for the generic i^{th} component is given by:

$$m_{k|k-1}^i = R(\Delta \psi_k^j)(m_{k-1|k-1}^i - [\Delta x_k^j \; \Delta y_k^j]^T) \quad (6)$$

$$p_{k|k-1}^i = R(\Delta \psi_k^j) p_{k-1|k-1}^i R(\Delta \psi_k^j)^T + R(\Delta \psi_k^j) Q_{k-1} R(\Delta \psi_k^j)^T \quad (7)$$

$$w_{k|k-1}^i = P_s^i w_{k-1|k-1}^i \quad (8)$$

where P_s^i is the survival probability from time step $k-1$ to the time step k of the ith component $f_{k-1|k-1}^i$, and Q_{k-1} is the system noise.

Ideally, the survival probability P_s^i, depends on the real probability that a target disappear. In a robotic swarm context, this probability would be extremely low in the whole domain. Therefore, we have used it as a design parameter to meet the objectives of the localization module. Coherently with the task and motivation of this paper, only local information is required and available to each robot. Using P_s^i, we prefer to let too far components fade. At this aim, we use an inverse sigmoid function to compute P_s^i:

$$P_s^i = \frac{1}{(1.05 + e^{4(||m_{k-1|k-1}^i||-4)})} \tag{9}$$

This creates a circular area in around R^j in which it tracks other robots. To account for targets that enters into this area from outside, a birth target component $b_{k|k-1}(x)$ is added at each time update, such that its mean, covariance and weight are respectively: $m^b = \begin{pmatrix} 0 & 0 \end{pmatrix}$, $p^b = \begin{pmatrix} 4 & 0 \\ 0 & 4 \end{pmatrix}$, $w^b = 0.001$. The assigned weight is very low so if there is no correspondence with the measurements (i.e., at least one measurement without a good correspondence with one or more components of the PHD prior), $b_{k|k-1}(x)$ will be pruned immediately.

The choice of limiting the area in which each robot tracks its teammates is also beneficial for the scalability of the method. In fact, even if the swarm was comprised of hundreds of agents, each robot would only track the ones that are closer to it, therefore linking the computational complexity of the filter to the density of the swarm rather than to the total number of robots.

4.2 Lidar Measurement Update

After the time update, the lidar measurement update is performed only when new measurements are available. We assume that the lidar at time k collect l_k position measurements $l_k^h \in L_k, h = 1, ..., l_k$ in $\mathcal{F}_j$. Following Eq. (5), each component $f_{k|k-1}^i$ of $f_{k|k-1}$ generates $l_k + 1$ components $f_{k|k}^{i(l_k+1)}$, $f_{k|k}^{i(l_k+1)+h}$, $h = 1, ..., l_k$ in $f_{k|k}$. One component $f_{k|k}^{i(l_k+1)}$ has the same mean $m_{k|k}^{i(l_k+1)} = m_{k|k-1}^i$ and same covariance $p_{k|k}^{i(l_k+1)} = p_{k|k-1}^i$ of the original component, while the weight is updated as:

$$w_{k|k}^{i(l_k+1)} = (1 - {}^L P_d^i) w_{k|k-1}^i \tag{10}$$

where ${}^L P_d^i$ is the probability that a target corresponding to component $f_{k|k-1}^i$ is detected by the lidar. The other l_k components are created using measurement update equations of the Kalman filter:

$$m_{k|k}^{i(l_k+1)+h} = m_{k|k-1}^i + {}^L K_k^i (l_k^h - m_{k|k-1}^i) \tag{11}$$

$$p_{k|k}^{i(l_k+1)+h} = (I - {}^L K_k^i {}^L H_k) p_{k|k-1}^i \tag{12}$$

$$w_{k|k}^{i(l_k+1)+h} = {}^L P_d^i w_{k|k-1}^i \mathcal{N}\{l_k^h; m_{k|k-1}^i, p_{k|k-1}^i\} \tag{13}$$

where ${}^L H_k$ is the lidar observation matrix and K_k is the associated Kalman gain:

$$^{L}H_k = \begin{pmatrix} 1 & 0 \\ 0 & 1 \end{pmatrix}, \; ^{L}K_k^i = p_{k|k-1}^i {}^{L}H_k^T ({}^{L}H_k p_{k|k-1}^i {}^{L}H_k^T + {}^{L}R_k)^{-1} \quad (14)$$

where $^{L}R_k$ is the covariance of the noise on the lidar measurements, that is determined experimentally as $^{L}R_k = \begin{pmatrix} 0.0025 & 0 \\ 0 & 0.0025 \end{pmatrix}$.

The probability of detection $^{L}P_d^i$ is a key parameter for the success of the filter. For each component $f_{k|k-1}^i$, $^{L}P_d^i$ is calculated considering four factors that limit the lidar sensor ability to detect objects. Distance, blind spots caused by the camera holders, obstruction of a robot by another robot, and interference caused by other lidar sensors. The final $^{L}P_d^j$ is the product of all those factors:

$$^{L}P_d^j = {}^{L}P_{d|dis}^i *^{L}P_{d|cs}^i *^{L}P_{d|b}^i *^{L}P_{d|in}^i \quad (15)$$

The first factor is the distance of each component $f_{k|k-1}^i$ which is related to the lidar sensor range. If a component is located beyond the range of the lidar, then it will not be detected. In our particular case, the range of the lidar is limited to 2 m. A sigmoid function was used to calculate $^{L}P_{d|dis}^i$:

$$^{L}P_{d|dis}^i = \frac{1}{\left(1.02 + e^{8*(||m_{k+1|k}^i||-1.5)}\right)}. \quad (16)$$

The second factor is due to the two pillars that support the fish eye cameras on $\mathcal{R}^j$, that create four blind spots in the field of view (FOV) of the lidar, whose centers $\alpha_i, i = 1, \ldots, 4$ and angular width $\beta_i, i = 1, \ldots, 4$ were determined experimentally. Denoting with $\theta_{k|k-1}^i$ the bearing angle of the mean of the i-th component, a sum of Gaussian functions is implemented to calculate $^{L}P_{d|cs}^i$:

$$^{L}P_{d|cs}^j = \sum_{i=1}^{4} \left(1 - \mathcal{N}\{\theta_{k|k-1}^i; \alpha_i, \beta_i/2\}\right). \quad (17)$$

The third factor $^{L}P_{d|b}^i$ models the situation in which robots block each other from the FOV of the lidar, that is therefore unable to collect a measurement for the robot that is behind. Hence the probability of detection of each component is reduced incorporating a zero mean Gaussian function $^{L}P_{d|b}^i$ based on i) the angle difference θ_{diff} between pairs of components; and ii) their Euclidean distance $||m_{k+1|k}^i||$ from R^j. When θ_{diff} becomes close to zero for some pair, the robot which has the shortest Euclidean distance from $\mathcal{R}^j$ will block the other robot. For the generic component $f_{k+1|k}^i(x)$, using all the components, $^{L}P_{d|b}^i$ is:

$$\theta_{diff} = |\theta_{k|k-1}^j - \theta_{k|k-1}^i| \quad (18)$$

$$^{L}P_{d|b}^i = \sum_{\forall \{i,j\}: ||m_{k|k-1}^j|| < ||m_{k|k-1}^i||} (1 - w_{k|k-1}^j \mathcal{N}(\theta_{diff}; 0, 3deg)) \quad (19)$$

The last factor $^LP^i_{d|in}$ models the interference caused by the lidar sensors mounted on the other robots, that we noticed during the testing phase. Whenever two lidar sensors are pointing at each other, their readings record null (invalid) measurements in correspondence to the other robots. Given the rotational nature of the internal mechanical structure of the lidars, this interference manifested itself as a loss of a measurement associated with the appearance of null measurements with a pseudo-periodic pattern. To model this interference, we considered that, along with the measurements, the lidar provides the intensity of the returning laser beam. When interference occurs, it will zero the lidar intensity $l^\theta_{int} \in L_k$, $\vartheta = 0, 0.5, \ldots, 360$. Therefore we used the intensity readings to calculate $^LP^i_{d|in}$ for $f^i_{k+1|k}(x)$:

$$^LP^i_{d|in} = \sum_{\forall\{l^\vartheta_{int}=0, \vartheta=0,0.5,\ldots,360\}} \frac{1 - 0.6 * e^{-(\theta^i_{k|k-1} - \angle l^\vartheta_{int})^2}}{2 * c^2} \tag{20}$$

where c denotes the covariance of each $l^\vartheta_{int} = 0$.

4.3 Camera Measurement Update

Similar to the lidar measurement update, the camera update is performed only when new camera measurements $C_k = \{c^1_k, \ldots c^{c_k}_k\}$ are available. Each c^i_k is provided as a 2D normalized vector pointing in the direction of a target. Following Eq. (5), each component $f^i_{k|k-1}$ generates $c_k + 1$ components in $f_{k|k}$. As for the lidar measurements, one component $f^{i(c_k+1)}_{k|k}$ has the same mean $m^{i(c_k+1)}_{k|k} = m^i_{k|k-1}$ and covariance $p^{i(l_{k+1})}_{k|k} = p^i_{k|k-1}$ of the original component, while the weight is updated as:

$$w^{i(c_k+1)}_k = (1 - {^CP^i_d})w^i_{k|k-1} \tag{21}$$

The other c_k components are computed using the Kalman filter equations:

$$m^{i(c_k+1)+h}_{k|k} = m^i_{k|k-1} + {^CK^i_k}(c^h_k - m^i_{k|k-1}) \tag{22}$$

$$p^{i(c_k+1)+h}_{k|k} = (I - {^CK^i_k}{^CH_k})p^i_{k|k-1} \tag{23}$$

$$w^{i(c_k+1)+h}_{k|k} = {^CP^i_d}w^i_{k|k-1}\mathcal{N}\{c^h_k; m^i_{k|k-1}, p^i_{k|k-1}\} \tag{24}$$

where $^CP^i_d$ is the camera probability of detection, H_k is the observation matrix, and K_k is Kalman gain:

$$H_k = \left(\frac{-y^i}{||m^i_{k|k-1}||}, \frac{x^i}{||m^i_{k|k-1}||}\right), \quad K^i_k = p^i_{k|k-1}H^T_k(H_k p^i_{k|k-1}H^T_k + {^CR_k})^{-1} \tag{25}$$

where $m^i_{k|k-1} = [x^i \; y^i]^T$ and $^CR_k = 25deg$ is the covariance of the noise of the camera measurements.

The probability of detection $^{C}P_d^i$ is computed as the product of two factors, distance from $\mathcal{R}^j$ and obstruction of a robot by another robot:

$$^{C}P_d^i = {}^{C}P_{d|dis}^i * {}^{C}P_{d|b}^i. \tag{26}$$

The first factor is computed with an inverse sigmoid function:

$$^{C}P_{d|dis}^i = \frac{1}{(1.1 + e^{5*(||m_{k+1|k}^j||-4)})} \tag{27}$$

while for the second factor, $^{C}P_{d|b}^i = {}^{L}P_{d|b}^i$.

5 Experiments

The relative localization system have been tested on robot experiments with four Unmanned Ground Vehicles (UGV). The UGVs are constructed using a commercially available four wheeled differential drive robot platform, the DFRobot Cherokey (22.5 cm × 17.5 cm). Each UGV is equipped with wheel encoders and a Romeo V2 (an Arduino Robot Board (Arduino Leonardo) with Motor Driver). The Romeo V2 processes and executes the low level control to follow desired velocity commands. For processing higher level tasks, an Odroid-XU4 - a small single board computer - is mounted on the robot. The Odroid-XU4 hosts an Exynos5422 Cortex™-A15 2 Ghz Quad core and a Cortex™-A7 1.5 Ghz Quad core CPUs with Mali-T628 MP6 GPU. The Odroid runs a GNU-Linux OS along with Robot Operating System (ROS) to manage sensor data collection and real time processing.

Each UGV is equipped with a Lidar and two omni directional cameras. We use an YDLIDAR X4 which is a lightweight, belt driven, 360-degree two-dimensional range finder with 7 Hz frame rate. The two cameras are standard USB web cameras equipped with a 180 °C fish-eye lens. Each camera provides two megapixel 1920 × 1080 resolution images at 30 fps rate. All three sensors are connected to the Odroid using standard USB ports. A camera holder was designed and 3D printed to hold the cameras above the lidar at an height of 130 mm from UGV and a 45 downward degree angle. This specific height and angle is designed to maximize the horizontal FOV of the combined camera images. With this setup, both the lidar and the cameras are fixed in the origin O_j of $\mathcal{F}_j$ and aligned with the X axis, eliminating rotational complexities during the image and scan processing. All the UGVs have been equipped with a red strip around the camera holder in order to allow camera tracking via color extraction.

The testing area is a 3 m × 3 m square space with raised walls to avoid disturbances from the external environment and keep the robots in proximity of each other. However, it is also larger than the FOV of the lidar to allow robots to exit and re-enter the tracking radius. Ground truth of the actual position of the robots is provided at each time by an Optitrack 6D of motion tracking system.

During a typical experiment, the UGVs will run a simple pseudo-random motion with obstacle avoidance. All computations are done on the on-board Odroid and using on-board sensors. One UGV performs the estimation algorithm. The final estimates are published on a ROS topic and recorded to a ROS bag along with ground truth provided by Optitrack. In order to highlight the benefit of the proposed multi-sensor approach, we have run offline on the same dataset the method introduced in Sect. 4 with (Lidar+Camera - LC) and without (Lidar only - LO) providing the camera measurements.

5.1 Results

Figure 2 shows the distance error for each robot from the closest component whose weight is greater than 0.1. Overall, the LC method performs well except for some instants near time 150 s and 190 s in which the robot performing the estimate was consistently in a corner of the arena, hence with a limited field of view, effectively leading to robot's UGV 07 position not being measured for several tens of seconds. The plots show also that the LC method outperforms the LO method being able to keep the error bounded for most of the time when measurements are available.

A numerical comparison between the LC and LO methods is provided in Fig. 3. In order to quantify the better performance of the LC filter, we have

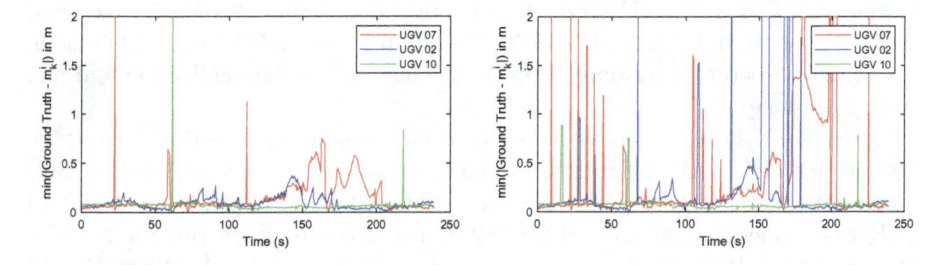

Fig. 2. Distance error of the three UGVs with LC (left) and LO (right).

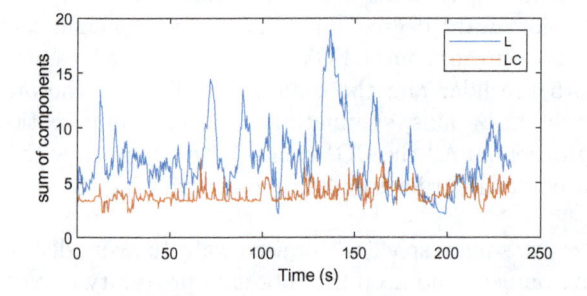

Method	LO	LC
UGV 07	20.7%	12.7%
UGV 02	10.9%	2.6%
UGV 10	1%	0.2%

Fig. 3. Comparison between the LO and LC. Left: sum of the weights of all the components with LO (blue) and LC (red). Right: percentage of time that the error on the position of each robot is greater than 30 cm with LO and LC.

computed the percentage of time for which each robot's distance error is greater than 30 cm. The values, reported in the table in Fig. 3 (left), show how the employment of camera measurements in addition to the lidar greatly reduces the error time of a factor 2 to 5. Finally, in Fig. 3 (right) we report the total sum of the weight of all components with respect to time during the whole experiment. From this plot, it is possible to establish that the LC method is more effective in correctly estimating the number of robots, and therefore in eliminating estimates that refer to objects in the environment and not robots.

6 Conclusions

In this paper we have presented a multi-sensor relative localization system for robotic swarms based on the PHD filter. Our system has been tested with real robot experiments, and evaluated against a single-sensor method based on the same principle. The results show that the multi-sensor approach performs better than the single-sensor method. In the future, on the one hand we plan on improving the relative localization and include negative information measurements to simultaneously track robots and obstacles. On the other hand, we plan to pair the localization system with a decentralized formation control methods to perform real-world tasks as exploration, SLAM, and human-swarm interaction.

Acknowledgments. This work was supported in part by the National Science Foundation under Grant RII Track-2 FEC 1923004.

References

1. Zheng, Z., Tan, Y.: Research advance in swarm robotics. Def. Technol. **9**, 18–39 (2013)
2. Senanayake, M., Senthooran, I., Barca, J.C., Chung, H., Kamruzzaman, J., Murshed, M.: Search and tracking algorithms for swarms of robots: a survey. Robot. Auton. Syst. **75**, 422–434 (2016). http://www.sciencedirect.com/science/article/pii/S0921889015001876
3. Bakhshipour, M., Ghadi, M.J., Namdari, F.: Swarm robotics search & rescue: a novel artificial intelligence-inspired optimization approach. Appl. Soft Comput. **57**, 708–726 (2017). http://www.sciencedirect.com/science/article/pii/S1568494617301072
4. McGuire, K.N., De Wagter, C., Tuyls, K., Kappen, H.J., de Croon, G.C.H.E.: Minimal navigation solution for a swarm of tiny flying robots to explore an unknown environment. Sci. Robot. **4**(35), (2019). https://robotics.sciencemag.org/content/4/35/eaaw9710
5. Zahugi, E., Shanta, M., Prasad, T.: Oil spill cleaning up using swarm of robots. Adv. Intell. Syst. Comput. **178**, 215–224 (2013)
6. Kayser, M., et al.: Design of a multi-agent, fiber composite digital fabrication system. Sci. Robot. **3**(22), (2018). https://robotics.sciencemag.org/content/3/22/eaau5630
7. Roumeliotis, S.I., Bekey, G.A.: Distributed multirobot localization. IEEE Trans. Robot. Autom. **18**(5), 781–795 (2002)

8. Huang, G., Trawny, N., Mourikis, A., Roumeliotis, S.: Observability-based consistent EKF estimators for multi-robot cooperative localization. Auton. Robots **30**, 99–122 (2011)
9. Nerurkar, E.D., Roumeliotis, S.I., Martinelli, A.: Distributed maximum a posteriori estimation for multi-robot cooperative localization. In: 2009 IEEE International Conference on Robotics and Automation, pp. 1402–1409 (2009)
10. Howard, A., Mataric, M.J., Sukhatme, G.S.: Putting the 'I' in 'team': an ego-centric approach to cooperative localization. In: 2003 IEEE International Conference on Robotics and Automation, pp. 868–874 (2003)
11. Zhou, X.S., Roumeliotis, S.I.: Determining 3-d relative transformations for any combination of range and bearing measurements. IEEE Trans. Rob. **29**(2), 458–474 (2013)
12. Franchi, A., Oriolo, G., Stegagno, P.: Mutual localization in multi-robot systems using anonymous relative measurements. Int. J. Robot. Res. **32**(11), 1302–1322 (2013). https://doi.org/10.1177/0278364913495425
13. Ye, M., Anderson, B.D.O., Yu, C.: Bearing-only measurement self-localization, velocity consensus and formation control. IEEE Trans. Aerosp. Electron. Syst. **53**(2), 575–586 (2017)
14. Falconi, R., Gowal, S., Martinoli, A.: Graph based distributed control of non-holonomic vehicles endowed with local positioning information engaged in escorting missions. In: 2010 IEEE International Conference on Robotics and Automation, pp. 3207–3214 (2010)
15. Katić, D., Rodić, A.: Intelligent multi robot systems for contemporary shopping malls. In: IEEE 8th International Symposium on Intelligent Systems and Informatics, pp. 109–113 (2010)
16. Stegagno, P., Cognetti, M., Oriolo, G., Bülthoff, H.H., Franchi, A.: Ground and aerial mutual localization using anonymous relative-bearing measurements. IEEE Trans. Rob. **32**(5), 1133–1151 (2016)
17. Rashid, A., Abdulrazaaq, B.: A survey of multi-mobile robot formation control. Int. J. Comput. Appl. **181**, 12–16 (2019)
18. Franchi, A., Stegagno, P., Oriolo, G.: Decentralized multi-robot encirclement of a 3d target with guaranteed collision avoidance. Auton. Robots **40**, 245–265 (2015)
19. Siligardi, L., et al.: Robust area coverage with connectivity maintenance. In: 2019 International Conference on Robotics and Automation (ICRA), pp. 2202–2208 (2019)
20. Hossein Mirabdollah, M., Mertsching, B.: Bearing only mobile robots' localization: observability and formulation using SIS particle filters. In: 2011 International Conference on Communications, Computing and Control Applications (CCCA), pp. 1–5 (2011)
21. Martinelli, A., Siegwart, R.: Observability analysis for mobile robot localization. In: Proceedings of the IEEE/RSJ International Conference on Intelligent Robots and Systems, January 2005
22. Mahler, R.: The multisensor PHD filter: I. general solution via multitarget calculus. In: Proceedings of SPIE - The International Society for Optical Engineering, May 2009
23. Vo, B., Ma, W.: The Gaussian mixture probability hypothesis density filter. IEEE Trans. Signal Process. **54**(11), 4091–4104 (2006)
24. Junjie, W., Lingling, Z., Xiaohong, S., Peijun, M.: Distributed computation particle PHD filter (2015)

25. Wasik, A., Lima, P., Martinoli, A.: A robust localization system for multi-robot formations based on an extension of a Gaussian mixture probability hypothesis density filter. Auton. Robots **44**, 395–414 (2020)
26. Dames, P., Kumar, V.: Autonomous localization of an unknown number of targets without data association using teams of mobile sensors. IEEE Trans. Autom. Sci. Eng. **12**(3), 850–864 (2015)
27. Stegagno, P., Cognetti, M., Rosa, L., Peliti, P., Oriolo, G.: Relative localization and identification in a heterogeneous multi-robot system. In: 2013 IEEE International Conference on Robotics and Automation, pp. 1857–1864 (2013)
28. Wu, J., Wang, Y., Hua, S.: Adaptive multifeature visual tracking in a probability-hypothesis-density filtering framework. Signal Process. **93**(2915–2926), 850–864 (2013)

An Innate Motivation to Tidy Your Room: Online Onboard Evolution of Manipulation Behaviors in a Robot Swarm

Tanja Katharina Kaiser(✉), Christine Lang, Florian Andreas Marwitz, Christian Charles, Sven Dreier, Julian Petzold, Max Ferdinand Hannawald, Marian Johannes Begemann, and Heiko Hamann

Institute of Computer Engineering, University of Lübeck, Lübeck, Germany
{kaiser,hamann}@iti.uni-luebeck.de

Abstract. As our contribution to the effort of developing methods to make robots more adaptive and robust to dynamic environments, we have proposed our method of 'minimal surprise' in a series of previous works. In a multi-robot setting, we use evolutionary computation to evolve pairs of artificial neural networks: an actor network to select motor speeds and a predictor network to predict future sensor input. By rewarding for prediction accuracy, we give robots an innate, task-independent motivation to behave in structured and thus, predictable ways. While we previously focused on feasibility studies using abstract simulations, we now present our first results using realistic robot simulations and first experiments with real robot hardware. In a centralized online and onboard evolution approach, we show that minimize surprise works effectively on Thymio II robots in an area cleaning scenario.

Keywords: Online onboard evolution · Evolutionary swarm robotics · Object manipulation · Innate motivation · Task-independent fitness

1 Introduction

Mobile robots and multi-robot systems can be programmed with state-of-the-art methods to work reasonably in a majority of situations. However, a small remaining fraction of non-anticipated situations may cause erratic behavior or even system breakdowns. As an alternative approach to classical techniques of machine learning, we propose our concept of 'minimal surprise' [1]. We use methods of evolutionary computation [2] to evolve the weights of actor-predictor pairs of artificial neural networks (ANN). That is, we provide our robots with a special form of a world model (predictor) in addition to a standard controller (actor). The robot swarm receives task-independent rewards for correct prediction of future sensor input. This biases the evolutionary dynamics towards 'boring' behaviors that simplify the prediction. As there is no task-specific reward in our standard

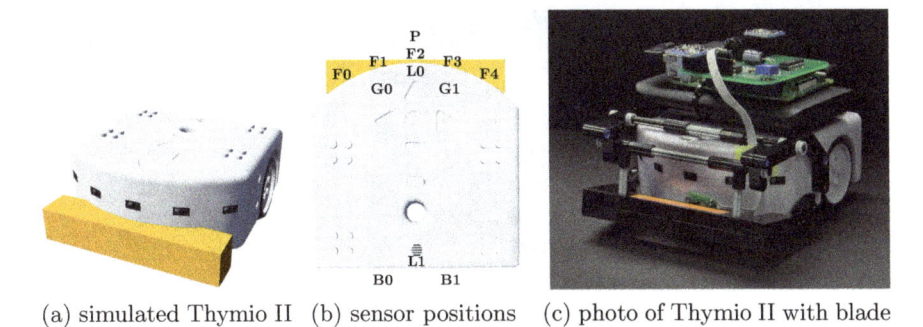

(a) simulated Thymio II (b) sensor positions (c) photo of Thymio II with blade

Fig. 1. Extended Thymio II robot in the Webots simulator (a, b) and in reality (c). The robot has 7 horizontal IR sensors ($F0, \ldots, F4, B0, B1$), 2 ground IR sensors ($G0, G1$), 2 light sensors ($L0, L1$; invisible in simulation) and one force sensor (P) [8].

minimize surprise approach, a variety of robot swarm behaviors emerges during the evolutionary process that then needs to be reviewed and selected for task-specific applications. Previously, we have shown that we can successfully generate typical swarm robot behaviors (aggregation, dispersion, flocking) [1] but also more complex behaviors, such as self-assembly [3] and collective construction [4]. However, all these results were based on rather abstract simulations of 1D and 2D discrete and continuous worlds. Here, we report our first results of using realistic robot simulations and real robot hardware. Our main contribution is the proof that minimize surprise works in the real world. A secondary contribution is the implementation of an evolutionary approach adapted to the real-world setting. Before, we used a genetic algorithm that required many simulated evaluation runs. On real robots, that is infeasible as it would be too time consuming and wearing down our robots. Instead, we now use a genetic algorithm in a centralized online and onboard architecture with (1+1)-selection that was shown to be feasible for such applications [5] and, for example, was already used to evolve foraging behaviors in a swarm of Thymio II robots [6]. We also find non-trivial prediction behaviors in the evolved predictor networks. Previously we had mainly found trivial predictions, such as outputting the current state of sensor input as prediction. We have extended four Thymio II robots [7] with bulldozer blades in a bumper style to enable them to push objects.

2 Experiment Design

2.1 Robots in Simulation and Hardware

We use extended Thymio II robots [7] in the Webots simulator [9] and in physical robot experiments. The standard version of the robot includes seven horizontal infra-red (IR) proximity sensors. Five are located at the front ($F0, \ldots, F4$) and two are located at the back ($B0$ and $B1$) of the robot as marked in Fig. 1b. Also, the Thymio II has two IR ground sensors ($G0$ and $G1$). The sensors are updated

(a) standard arena   (b) gradient arena   (c) real arena

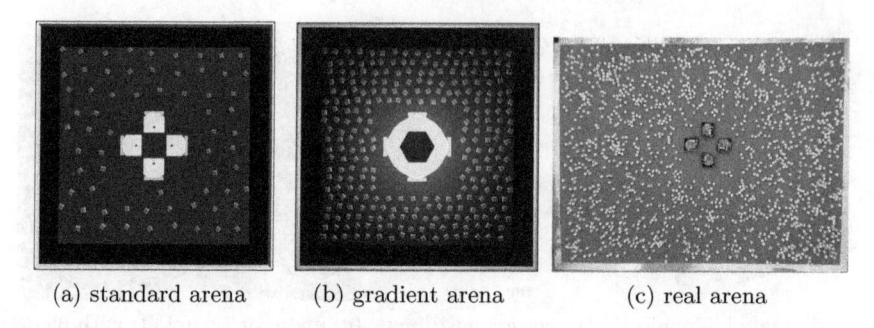

Fig. 2. Test environments in simulation (a, b) and in reality (c). The starting position of the robots is fixed for all scenarios while boxes are distributed randomly. (a) illustrates the low and (b) the high box density setting [8].

with 10 Hz on the real robot and every 10 ms (simulated time) in Webots. The robot's two differential drive motors allow for a maximum linear velocity of 20 $\frac{cm}{s}$. For more gentle usage of the motors, we restrict the maximum speed to 10.6 $\frac{cm}{s}$ on the real robots and to 12.6 $\frac{cm}{s}$ in simulation. We extend simulated (Fig. 1a) and real robots (Fig. 1c) with a bulldozer blade in the form of a bumper equipped with a force sensor (P), and two light sensors ($L0$ and $L1$) on top of each robot. The force sensors allow to measure forces when pushing objects with the robot's blade while we use the light sensors to detect light gradients (here only used in simulation). In Webots, we extend the robot design by modifying the open-source PROTO-files of the robot model. The real robot's blade is built of LEGO® parts and mounted to the attachment points for LEGO® on the Thymio II. The force sensor (HSFPAR303A) and the two light sensors (TSL45315) are connected to a Raspberry Pi 3B (RPi) that is mounted on top of the robot with a sandwiched external battery between. The RPi is then connected to the Thymio II through USB and its D-Bus interface. We program the robot in Python 3.6.9.

2.2 Environment

We use two different environments in simulation (Figs. 2a, b) and one with real robots (Fig. 2c). The simulated arenas have a size of 1.1 m × 1.1 m to allow for fast simulation and the real arena has a size of 2.2 m × 2.9 m. All have a carpeted floor and boundaries that are detectable by the ground IR sensors, that is, the edges where the carpeted floor stops in simulation and mirror film in the real arena. The simulated environments have additional walls in distance to the arena's real boundaries to ensure that robots and boxes cannot leave the arena completely. These walls are only detectable by the robots if they crossed the real arena's boundaries which may occur when robots drive backwards. In the real environment, this is prevented by a rim formed by the mirror film or, in some cases, we had to put a lost robot back into the arena. We randomly distribute wooden cubes (boxes) in the arena. Their dimensions are 2.5 × 2.5 × 2.5 cm³ and they weigh ca. 10 g. In simulation, we adjusted their weights to 2 g to compensate

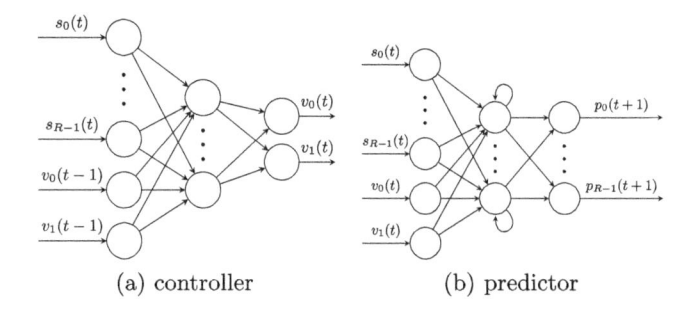

Fig. 3. ANN pair of each robot in minimize surprise. Inputs are R sensor values $s_0(t), \ldots, s_{R-1}(t)$ at time step t, the motor values $v_0(t-1)$ and $v_1(t-1)$ for the left and right wheel of the Thymio robot of time step $t-1$ or $v_0(t)$ and $v_1(t)$ of time step t, respectively. Outputs are a pair of motor speeds $v_0(t)$ and $v_1(t)$ or R sensor value predictions $p_0(t+1), \ldots, p_{R-1}(t+1)$ for time step $t+1$ [8].

Table 1. Parameters for (1+1)-evolution. Values in brackets give the adjusted parameters for the real robot scenario.

Parameter	Value
Mutation rate	0.1
Evaluation length (time steps)	1,000 (100)
Post-evaluation length (time steps)	10,000 (1,000)
Max. evaluations	1000 (350)
Re-evaluation probability	0.2
Re-evaluation weight α	0.2

for differences between simulated and real environment. We use a box density of ca. 14% (i.e., $\approx 220 \frac{\text{boxes}}{\text{sqm}}$) in simulation and in the real arena. Additionally, we test a box density of ca. 3.6% (i.e., $\approx 55 \frac{\text{boxes}}{\text{sqm}}$) in simulation only. The boxes are too small to be detected by the horizontal IR sensors. In summary, a robot can discriminate robots, boxes, and the arena boundaries using its sensors.

We vary the light settings in the two simulated environments to investigate the effect of light on emerging behaviors. The *standard arena* (Fig. 2a) is uniformly illuminated rendering light sensors irrelevant for this setting. The *gradient arena* (Fig. 2b) has a simulated light bulb above the center of the arena with light intensity decreasing gradually towards the arena boundaries.

2.3 Evolution of Robot Controllers

We apply our minimize surprise approach [1] to this area clearing scenario to evolve robot controllers for a homogeneous swarm using an innate, task-independent motivation. Each robot is equipped with an ANN pair. A feedforward network serves as the robot controller (Fig. 3a) and outputs normalized

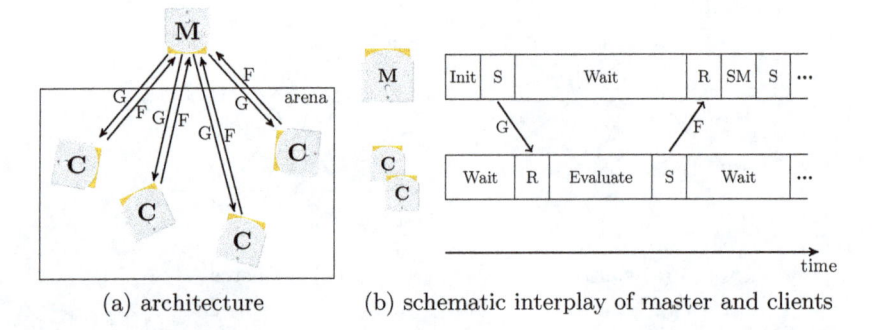

Fig. 4. Centralized online evolution architecture with one external master robot (M) running the evolutionary process and overseeing the four client robots (C) in the arena. The master initializes the first genome (G) and sends (S) it to the clients. The clients receive (R) the genome, evaluate it and reply with their individual fitness (F). The master calculates the total fitness, selects the current best genome and decides whether to re-evaluate it or to create offspring by mutation (SM). The chosen genome will be send to the clients and the process continues until termination [8].

speeds $v \in [-1.0, 1.0]$ for the two differential drive motors. When sent to the robot's motors, the normalized values v are scaled with the maximum speed $v_{\max}$. The controller ANN is paired with a recurrent neural network that serves as a predictor for the next sensor values (Fig. 3b). Inputs for both ANNs are the current sensor values that are normalized by their maximum possible value and the last set motor speeds for the controller network and the next speeds for the predictor network. New motor values and sensor predictions are determined with every sensor update of the Thymio II, that is, every 100 ms on the real robot and every 10 ms in simulation. Thus, time is discretized into steps of 100 ms and 10 ms, respectively, and allows for discrete fitness calculation. Fitness is given to an ANN pair for high prediction accuracy and normalized to a maximum of 1. We define the fitness function as

$$F = \frac{1}{NRT} \sum_{n=0}^{N-1} \sum_{r=0}^{R-1} \sum_{t=0}^{T-1} 1 - |p_r^n(t) - s_r^n(t)|, \qquad (1)$$

with swarm size N, number of sensors R, time steps T, real value $s_r^n(t)$ of and prediction $p_r^n(t)$ for sensor r of robot n at time step t.

We use an evolutionary algorithm with (1+1)-selection [2]. The weights of the network pairs are encoded in the genome of the individuals. An evaluation lasts for 1,000 time steps in simulation and for 100 time steps on the real robots, that is, 10 s. Theoretically, this online evolution approach could be run infinitely for an open-ended adaptation process, but we restrict our experiments to 1,000 evaluations in simulation and 350 in the real arena per evolutionary run. Evaluated is either the current best individual with a 20% chance or its offspring. Offspring is created by mutation of the best individual, that is, by adding a uniformly random value to each gene with a probability of 0.1. In the case of

(a) low box density

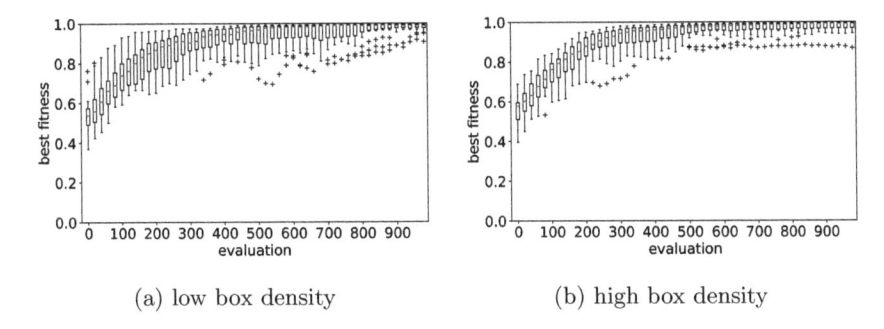

(b) high box density

Fig. 5. Current best fitness over evaluations in simulation for the standard arena and both box densities over 20 runs. We print the boxes for every 20 evaluations for a clearer illustration. Medians are indicated by red bars [8].

re-evaluations, the fitness of the best individual is calculated as an exponentially weighted mean $F_t = \alpha \hat{f}_t + (1-\alpha) F_{t-1}$ with $\alpha = 0.2$ of the fitness value $\hat{f}_t$ reached during re-evaluation and the previous best fitness F_{t-1}. After the experiment, we run the last best individual for 100 s in a post-evaluation to store sensor values, predictions, and, in simulation, box positions and robot trajectories to analyze emergent behaviors. We reset the arena for post-evaluation by placing the robots at their initial positions and the boxes at randomly distributed positions in the arena. The parameters for our (1+1)-evolution are summarized in Table 1.

We realize learning with a centralized online evolution architecture [2] as shown in Fig. 4a. One Thymio II robot is used as the central master guiding the evolutionary process, that is, it distributes genomes to a swarm of $N = 4$ Thymio II robots, collects individual fitness and calculates the total fitness, selects the best individual, and creates offspring by applying mutation. This master robot is placed outside the arena and is not evaluating genomes itself. The swarm members (clients) receive the genome from the master, evaluate it for the evaluation length, and send their individual fitness back to the master. Figure 4b illustrates the interplay of master and clients schematically. The communication between the robots is realized with the Emitter and Receiver provided by Webots in simulation. On the real robots, a TCP connection and WiFi are used. In case of transmission errors, a genome is evaluated again in the next evaluation. We do 20 independent evolutionary runs per scenario in simulation and five independent runs with the real robots in the standard arena with the high box density.

2.4 Hardware Protection

We implement a hardware protection layer to prevent robot damages by stopping potentially harmful actions. Fitness evaluation continues during hardware protection as usual. Here, we implement a simple hardware protection that leads to an escape behavior if the front or back IR sensors detect a too close obstacle. We limit the allowed amount of pushed boxes to about 10 boxes (detected by the pressure sensor) to avoid motor damage. If an above-threshold pushing force is detected, the robot turns on spot away from the boxes. As we cannot measure a robot's pushing force when driving backwards, we stop robots after 9 s of constantly going backwards. This limit is reset once positive motor values occur. Furthermore, robots are prevented from leaving the arena by forcing the robot to turn when detecting the arena's boundaries. When driving backwards, robots detect the boundaries when they are mostly out of the arena already (ground sensors are in front). In simulation, the outer wall triggers the escape behavior and thus, robots drive back into the arena. In the real experiments, the experimenter has to trigger the behavior manually by activating the back IR sensors which could be automated by adding walls to the real arena, too.

3 Results

3.1 Simulation

Figure 5 shows the increase of best fitness for the standard arena for both box densities and exemplifies the fitness curves observed in all experiments. The median best fitness (prediction accuracy) in the last evaluation is 0.99 in the standard arena and 0.97 in the gradient arena for both box densities. The high prediction accuracy shows that our approach is successful, but provides little informative value about the robot system as such. Hence, we investigate the resulting robot and swarm behaviors, and in particular their diversity. All figures and a video illustrating emergent behaviors can be found online [8].

Measures of Behavioral Diversity. We quantify resulting behaviors based on box displacement and distance covered by robots during the post-evaluation run, see Fig. 6. We define the distance covered by robots d_R as mean accumulated robot displacement over runtime T as given by

$$d_R = \frac{1}{N} \sum_{n=0}^{N-1} \sum_{t=0}^{T-1} ||l_n(t+1) - l_n(t)||_2, \qquad (2)$$

with N robots and positions $l_n(t)$ and $l_n(t+1)$ of robot n at time steps t and $t+1$, respectively. Robots can cover a theoretical maximum distance d_R of 12.6 m when constantly driving with maximum linear speed of 0.126 $\frac{m}{s}$ during 10,000 post-evaluation time steps of 10 ms each. We define box displacement

$$d_B = \frac{1}{B} \sum_{b=0}^{B-1} ||l_b(T) - l_b(0)||_2, \qquad (3)$$

as the mean Euclidean distance between the starting positions $l_b(0)$ and final positions $l_b(T)$ of the B boxes. The theoretical maximum displacement of a box is the arena's diagonal, but we obviously expect a lower effective mean distance. We find by qualitative analysis that a threshold of 0.1 in box displacement d_B distinguishes behaviors that lead to the pushing of boxes from behaviors with limited or no box manipulation.

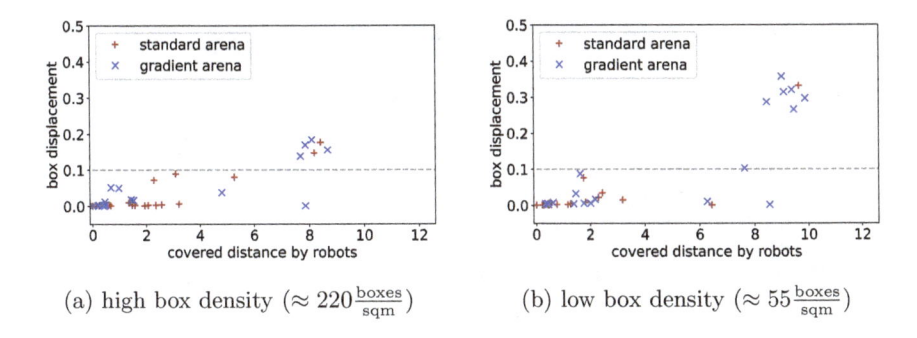

(a) high box density ($\approx 220 \frac{\text{boxes}}{\text{sqm}}$)

(b) low box density ($\approx 55 \frac{\text{boxes}}{\text{sqm}}$)

Fig. 6. Distance covered by robots d_R (Eq. 2) vs. box displacement d_B (Eq. 3) for the standard and gradient arena and both box densities. The dashed gray line marks a threshold between behaviors leading to the pushing of boxes and other behaviors [8].

Circling Behaviors. The majority of emergent behaviors has robots go in circles ranging from small circles (Fig. 7a) with short covered distances d_R to larger circles with robots following each other ('circle dance') resulting in larger covered distances. Boxes are mostly pushed when robots avoid each other. The evolutionary process seems to exploit hardware protection behaviors with robots distributing themselves by triggering the escape behavior. In most cases, these escape behaviors are only executed shortly at the beginning of the run for limited duration. In rare cases, robots evolve intrinsic obstacle avoidance.

Reverse Driving. Other behaviors with $d_B < 0.1$ lead to robots driving backwards until they are stopped by hardware protection (Fig. 7b). Boxes are pushed rarely or not at all. When all robots are stopped by hardware protection, the environment is static and thus, easily predictable.

Behaviors Leading to the Pushing of Boxes. Behaviors with box displacements d_B larger than 0.1 emerge more often in the gradient arena for both box densities. For the low box density setting, they lead to robots clearing the arena from boxes or, in one run, to robots pushing boxes around in the arena. In the high box density setting, robots form small box clusters (Fig. 7c). This difference is probably caused by hardware protection as with increasing box density, robots exceed the threshold of maximum pushed boxes faster and turn away. For high box densities, clusters form while for low densities, boxes get easily pushed out of the arena and higher box displacement is achieved. The robots implement a random walk that implements behaviors leading to the pushing of boxes by exploiting the hardware protection's boundary avoidance behavior.

Sensor Value Predictions. A good interplay between controller and predictor allows for high prediction accuracy. A controller generating predictable behaviors simplifies the predictor's job. Thus, studying predictions and real sensor values of the emergent behaviors may provide insights into the evolutionary process. The mean sensor values and predictions of a circling behavior in the gradient

(a) driving in circles (b) reverse driving (c) pushing of boxes

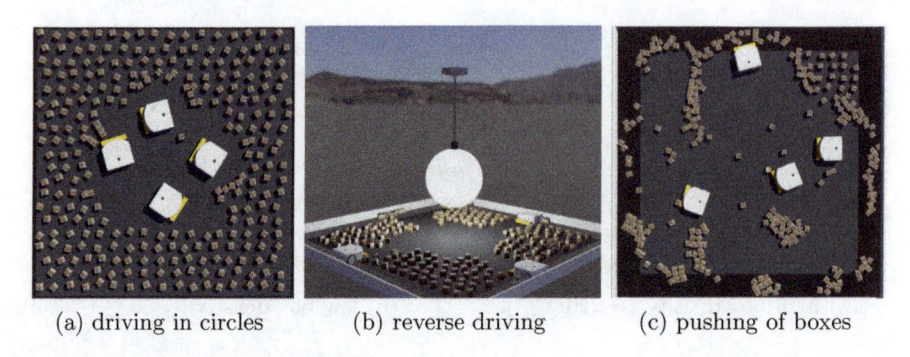

Fig. 7. Examples of emergent behaviors in simulation. The pictures show box and robot positions at the end of the post-evaluation runs [8].

arena are shown in Fig. 8a and serve as representative example for most emergent behaviors. In general, we find that robots do not predict or sense other robots nearby (i.e., $F0, \ldots, F4, B0, B1 \approx 0$). Exceptions are the 'circle dance' behavior that leads to driving collectively in a circle based on the front left IR sensor values and the detection of the outer arena walls with the back IR sensors in reverse driving. The real and predicted ground IR sensor values ($G0, G1$) match the light reflected from the arena's carpet. In all but one run, the pressure sensor (P) values and predictions are below 0.08. This is an effect of hardware protection as robots are forced to turn when pushing more than 10 boxes, that is, $P > 0.25$. Thus, all of these sensors allow for trivial predictions. However, the two additional light sensors ($L0, L1$) in the gradient arena do not generally allow for trivial predictions as light intensity fluctuates based on the distance from the arena's center. Repetitive behaviors, such as circling, can lead to steady fluctuations of light intensity as depicted in Fig. 8b. For most behaviors, we find that predictions follow these fluctuations roughly but with an offset. In a few cases (as seen in Fig. 8b for $L0$), we find sophisticated predictor outputs. Constant light intensity is reached in reverse driving behaviors as robots are stopped by hardware protection. Usually, robots are then located at the arena's boundaries where light intensity is also low. As the light sensors account only for $\frac{1}{6}$ of the total fitness when predicting all sensors, we studied whether better light sensor predictions emerge when forcing the predictor to specialize for light and pressure sensors only (i.e., no prediction of IR sensors $F1, \ldots, F4, B0, B1, G0, G1$). We run 20 independent runs in the gradient arena with high box density. As before, we find predictions following the fluctuations but with lower values than the real sensors. Furthermore, the number of behaviors leading to the pushing of boxes decreased compared to the previous runs.

Robot-Environment-Feedback Loop. We find that the fitness of the post-evaluation runs is statistically significantly lower than the best fitness at the end of the evolutionary runs for all scenarios (Mann-Whitney U Test, $p < 0.05$). This implies a feedback loop during evolution: individuals change the environment and the altered environment leads to adapted individuals. As we reset

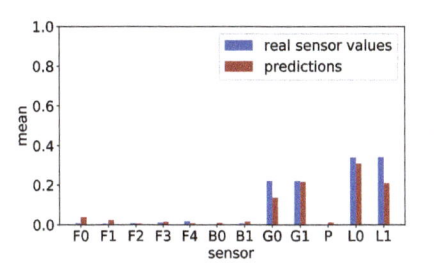

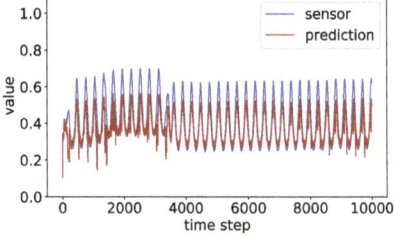

(a) mean predicted and real sensor values

(b) non-trivial predictions of L0

Fig. 8. Predicted and real sensor values of one behavior that lets robots drive in circles in the gradient arena with front IR sensors $F0,\ldots,F4$, back IR sensors $B0$ and $B1$, ground IR sensors $G0$ and $G1$, pressure sensor P and light sensors $L0$ and $L1$ [8].

the arena for post-evaluation, the individuals may not be adapted to this 'new' environment anymore. We test this hypothesis by doing 20 evolutionary runs predicting light and pressure sensors in the gradient arena with high box density without resetting the arena for post-evaluation. Thus, the last best individual is post-evaluated in the same environment as at the end of the evolutionary run. Here, we do not find statistically significantly different fitness values. Robots frequently move to the arena's edges and sense only low light intensities. Thus, adaptation to fluctuations in light intensity is not rewarding as those are only small at the arena's boundaries. Resetting the arena leads to higher light intensities and fluctuations. Thus, prediction accuracy is lower.

3.2 Experiments on Physical Robots

We have done five real-world experiment runs in the arena (Fig. 2c) with four extended Thymio II robots (Fig. 1c) evaluating the individuals and one robot serving as master for centralized online evolution. One of the runs terminated early due to connection errors. We include the data of the four complete runs in our evaluation. The increase of best fitness over the 350 evaluations is shown in Fig. 9a. The median best fitness reached in the last evaluation is 0.91.

The best individuals of the last generation of three out of four runs lead to robots driving in circles as was also mainly found in the simulated arenas. One run leads to constant reverse driving of robots. A video illustrating one complete evolutionary run leading to robots driving in circles can be found online [8]. Over time, robots clear space from boxes and early in the evolutionary runs, the circling behavior emerges as this allows for easy sensor predictions. Similarly to the simulation results, we find that the sensor values for all horizontal proximity sensors ($F0,\ldots,F4$, $B0$, $B1$) are approximately zero for most of the time and low values are also predicted (Fig. 9b). As before, predictions and real values for the ground IR sensors ($G0, G1$) match the light reflected from the arena's carpet (≈ 0.4). The pressure sensor P has low values as robots do not push boxes while driving in circles.

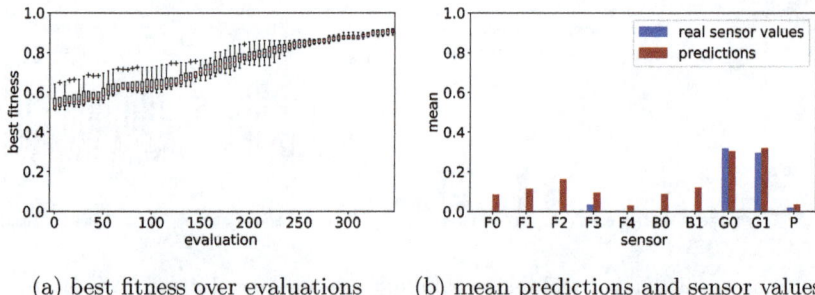

(a) best fitness over evaluations   (b) mean predictions and sensor values

Fig. 9. (a) Current best fitness over evaluations in the real arena of 4 runs. We print boxes for every 5 evaluations for a clearer illustration. Red bars indicate the median. (b) Bar chart of mean predicted and real sensor values of one evolutionary run with front IR sensors $F0,\ldots,F4$, back IR sensors $B0$ and $B1$, ground IR sensors $G0$ and $G1$ and pressure sensor P [8].

4 Conclusion

We have shown that our minimal surprise approach effectively leads to the emergence of diverse behaviors when applied to a real robot swarm in an area clearing scenario. Using a distributed implementation of centralized onboard (1+1)-evolution, we have also shown that robots can adapt to a changing environment at runtime as it is manipulated by themselves. The robots show an innate motivation to tidy areas in order to simplify their predictions. Here, our focus was on proving that minimize surprise is feasible on real robot swarms. In future work, we will transfer our previous methods [3] to the real-robot case and engineer self-organization towards desired behaviors.

In our master-client approach of centralized distributed evolution, we have placed the master robot out of the arena to simplify handling and maximize reliability during robot experiments. The computational power provided by the RPi-extension board is, however, sufficient to also allow the master robot to operate in the arena while managing the evolutionary algorithm. In future work, we plan to implement a fullscale distributed online onboard evolutionary approach.

The observed emergent behaviors in our scenario classify rather as tidying than construction. Structure formation of boxes is unlikely here because robots tend to push boxes out of the arena instead of forming box clusters. A more sophisticate approach would require a box-pull action or even a gripper [10]. In our previous work [4], agents lived on a torus which simplified the scenario and allowed for more structures to emerge.

In summary, the minimize surprise approach implements one option to deal with dynamic environments in real-world multi-robot systems. We argue that the approach increases reliability of the system in an open-ended process of adaptation because the robots autonomously adapt to any, possibly non-anticipated, situation. This happens due to the doctrine of accurate predictions and we speculate that predictable robot behaviors are to be considered generally as safe and reliable behaviors as also argued by Friston et al. for the biological case [11].

In future work, we want to do further robot experiments with more robots in different scenarios to continue showing the diversity of emerging behaviors and the real-world capability of minimize surprise. We also hope to give proof for the hypothesized connection between easy-to-predict behaviors and reliable robot behaviors for real-world mobile robot applications in dynamic environments.

References

1. Hamann, H.: Evolution of collective behaviors by minimizing surprise. In: Sayama, H., Rieffel, J., Risi, S., Doursat, R., Lipson, H. (eds.) 14th International Conference on the Synthesis and Simulation of Living Systems (ALIFE 2014), pp. 344–351. MIT Press (2014)
2. Eiben, A.E., Smith, J.E.: Introduction to Evolutionary Computing. Springer, Heidelberg (2015). https://doi.org/10.1007/978-3-662-44874-8_1
3. Kaiser, T.K., Hamann, H.: Engineered self-organization for resilient robot self-assembly with minimal surprise. Robot. Auton. Syst. **122**, 103293 (2019). https://doi.org/10.1016/j.robot.2019.103293
4. Kaiser, T.K., Hamann, H.: Self-organized construction by minimal surprise. In: 2019 IEEE 4th International Workshops on Foundations and Applications of Self* Systems (FAS*W), pp. 213–218 (2019). https://doi.org/10.1109/FAS-W.2019.00057
5. Bredeche, N., Haasdijk, E., Eiben, A.E.: On-line, on-board evolution of robot controllers. In: Collet, P., Monmarché, N., Legrand, P., Schoenauer, M., Lutton, E. (eds.) EA 2009. LNCS, vol. 5975, pp. 110–121. Springer, Heidelberg (2010). https://doi.org/10.1007/978-3-642-14156-0_10
6. Heinerman, J., Zonta, A., Haasdijk, E., Eiben, A.E.: On-line evolution of foraging behaviour in a population of real robots. In: Squillero, G., Burelli, P. (eds) EvoApplications 2016. LNCS, vol. 9598, pp. 198–212. Springer, Cham (2016). https://doi.org/10.1007/978-3-319-31153-1_14
7. Riedo, F., Chevalier, M., Magnenat, S., Mondada, F.: Thymio II, a robot that grows wiser with children. In: 2013 IEEE Workshop on Advanced Robotics and its Social Impacts, pp. 187–193 (2013). https://doi.org/10.1109/ARSO.2013.6705527
8. Kaiser, T.K., et al.: An innate motivation to tidy your room: online onboard evolution of manipulation behaviors in a robot swarm (plots and video) (2020). https://doi.org/10.5281/zenodo.4293487. CC BY 4.0
9. Michel, O.: Webots: professional mobile robot simulation. J. Adv. Robot. Syst. **1**(1), 39–42 (2004)
10. Allwright, M., Zhu, W., Dorigo, M.: An open-source multi-robot construction system. HardwareX **5**, e00050 (2019). https://doi.org/10.1016/j.ohx.2018.e00050
11. Friston, K., Thornton, C., Clark, A.: Free-energy minimization and the dark-room problem. Front. Psychol. **3**, 130 (2012). https://doi.org/10.3389/fpsyg.2012.00130

Multi-agent Reinforcement Learning and Individuality Analysis for Cooperative Transportation with Obstacle Removal

Takahiro Niwa[✉], Kazuki Shibata, and Tomohiko Jimbo

Data Analytics Research Domain, Toyota Central R&D Labs., Inc.,
41-1 Yokomichi, Nagakute, Aichi, Japan
niwa-takahiro@mosk.tytlabs.co.jp

Abstract. Cooperative transportation is one of the essential tasks for multi-robot systems to imitate the decentralized systems of social insects. However, in a situation involving an obstacle on the pathway, multiple robots need to realize transportation and obstacle removal simultaneously. To address this multitasking problem, we first introduce a learning scenario and train robots' decentralized policies via multi-agent reinforcement learning. Next, we propose two virtual experiments with blindfold teams and homogeneous teams to analyze the individual behaviors of the trained robots. The results showed that three robots with different policies performed two tasks simultaneously as a team. One robot's policy tended to perform obstacle removal, and the other robots' policies tended to perform cooperative transportation. Further, the first robot's policy had the potential to perform two tasks simultaneously depending on the situation. Finally, we demonstrated the trained policies with three ground robots to show the feasibility of the system.

Keywords: Division of labor · Multi-robot system · Multi-agent reinforcement learning · Individuality · Cooperative transportation

1 Introduction

Division of labor among individual workers is a crucial characteristic of social insects. The individual workers have initially homogeneous characteristics. To meet the demands of the whole colony, the individual workers become specialists in one task or generalists in the other tasks through interactions with other workers and the environment [1]. Although the colony does not have a centralized decision-making system, it can adapt to the environment.

To imitate these decentralized systems, researchers have actively investigated multi-robot systems [2–6]. Dorigo et al. [3] experimentally demonstrated that 12 homogeneous robots could complete a foraging task with complex forms of division of labor by designing finite-state machines. Ferrante et al. [7] demonstrated that, like social insects, robots could adaptively specialize in different tasks by artificially evolving to maximize team performance in a foraging task.

Cooperative transportation via multiple autonomous robots is an expected task to be realized in various fields, such as delivery, manufacturing, construction, rescue, and debris removal. Cooperative transportation has been studied extensively [8]. For example, a leader-follower [9–11], optimal control [12], sliding mode control [13], and occlusion-based architecture [14] are popular. Furthermore, in cooperative transporta-

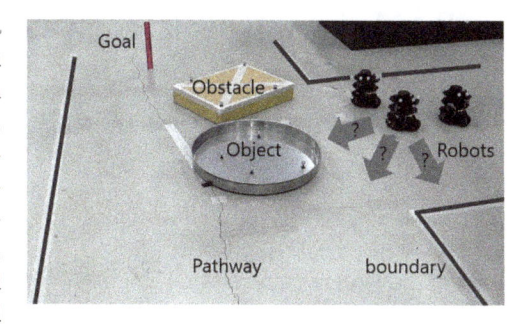

Fig. 1. Cooperative transport problem with obstacle removal

tion with multiple subtasks, such as foraging, the controller is mainly designed as a collection of predefined subtask behaviors and transition rules for the division of labor [3–5]. However, for complex tasks, the simultaneous design of desired behaviors and rules for the division of labor could be challenging because of the various situations to be considered. To address this challenge, methods are also utilized to automate a design process by optimizing neural network-based controllers [15,16]. In particular, multi-agent reinforcement learning (MARL) optimizes the controllers to maximize their long-term performance, and MARL has been adopted in the study of cooperative transportation [17–21].

However, in practical situations, a transportable pathway may not always be available because of obstacles, as shown in Fig. 1. Therefore, we address the cooperative transport problem in this study by creating a pathway in an environment where obstacles sometimes block the road. To address this problem, we exploit MARL to automate the design of decentralized controllers with cooperative behaviors and division of labor, which underpins effective transportation. First, the decentralized controllers that determine the behavior of the robots are described as policies. Second, the policies with initial homogeneity are trained under the learning scenarios via Multi-Agent Actor-Critic Deep Deterministic Policy Gradient (MADDPG) [22]. Finally, the behaviors of the trained individual robots are analyzed. The contributions of our study are summarized as follows:

- We introduce learning scenarios and reward settings;
- We propose two virtual experiments with blindfold teams and homogeneous teams to analyze the individual behaviors of the trained robots; and
- We demonstrate the trained policies in an experimental system using three ground robots.

The remainder of this paper is organized as follows. In Sect. 2, we first formulate the cooperative transport problem and specifications of the policies. In Sect. 3, we describe the policy optimization method and introduce learning scenarios. In Sect. 4, we evaluate the trained policies and analyze the individuality of the three robots. In Sect. 5, we demonstrate the policies with the three ground robots. Finally, in Sect. 6, we conclude the paper.

2 Problem Statement

The problem originated from small robots cooperatively transporting a large object in a narrow pathway with occasional obstacles, such as in a factory. We simplify and consider the cooperative transport problem with obstacle removal as shown in Fig. 2. A cylindrical target object has to be pushed toward a goal using a homogeneous team of N robots in a $\mathbb{R}^2$ environment with an obstacle. In this study, three robots ($N = 3$) were considered. A T-shaped road with a width of 1 m was set up in a 2×2 m square environment. The robots' radius was 5 cm, and the target object's and the obstacle's radii were 20 cm. In the real world, the target object and obstacle are assumed to be carts with stoppers. Therefore, they cannot be moved when they are not in contact with the robots. The movement of the target object requires cooperative transportation by two or more robots, and the obstacle can be moved by the pushing of one or more robots.

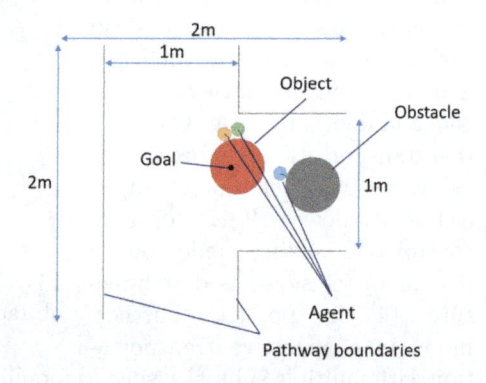

Fig. 2. Numerical experiment environment

In such an environment, an obstacle sometimes blocks the road, making it infeasible to transport the target object. Two subtasks are involved in this problem: the transportation and obstacle removal tasks. The transportation task involves getting the target object to the goal while avoiding collision with the obstacle. The obstacle removal task involves the robots pushing the obstacle to the pathway boundary to make the target object transportation feasible.

3 Policy Optimization

In this study, decentralized policies are trained. The policy of robot i generates the action by using the observation of robot i alone. The policy input of robot $i \in \mathcal{I} := \{1, \cdots, N\}$, i.e., the observation, includes the positions of the single target object $x_i^t \in \mathbb{R}^2$, single obstacle $x_i^o \in \mathbb{R}^2$, goal $x_i^g \in \mathbb{R}^2$, the other two robots $j, k (\neq i)$ $x_i^{others} := [x_i^{j\mathrm{T}}, x_i^{k\mathrm{T}}]^{\mathrm{T}} \in \mathbb{R}^4$ in the body coordinate system, and the distance to the pathway boundaries of the surrounding eight directions $d_i := [d_i^1, \cdots, d_i^8] \in \mathbb{R}^8$ in the body coordinate system. We assume that all observations are retained. The policy outputs translational velocity v (m/s) and angular velocity ω (degree/s) for the robots' control inputs, which are in the range $[-0.4, 0.4]$ and $[-150, 150]$, respectively. The control cycle of the robots lasts for 100 ms.

We exploit the optimization method and framework of the decentralized execution and centralized training proposed in MADDPG [22]. Each robot has two separate, fully connected neural networks with different parameters: a policy network that generates an action based on its observation and a critic network that predicts discounted future rewards from all the robots' observations and actions. During execution, each robot inputs its observation into its policy network and then receives instructions for its own action. During training, based on the policy gradient theorem [23], the policy network is optimized to maximize the predicted Q-value, and the critic network is optimized to minimize the temporal difference error (TD-Error).

It is difficult to learn many reinforcement learning tasks from scratch. Curriculum learning is a method of learning simple tasks, and building on the knowledge to solve more challenging tasks [20,24]. To realize the cooperative transport problem with obstacle removal, we introduce the following learning scenarios as curriculum learning.

- Scenario 1: Transportation with no obstacles
- Scenario 2: Transportation while avoiding collision with an obstacle
- Scenario 3: Simultaneous transportation and obstacle removal

Hereafter, we refer to the trained policies for the three robots of Scenario h as $\pi_h := (\pi_h^1, \pi_h^2, \pi_h^3)$. The trained π_1 and π_2 are used as the initial policy for the training of π_2 and π_3, respectively. Scenario 1 requires the simplest setup, in which three robots cooperatively transport a target object with no obstacles present on the road. The initial positions of the goal, target object, and robots are randomly generated on the road. As described in Sect. 2, we considered one obstacle in the robot's observations. In this scenario, a fixed value $[0.75, 0.75]$ is provided as a dummy observation for the obstacle position. In Scenario 2, three robots cooperatively transport the target object to the goal while avoiding an obstacle on the road. As in Scenario 1, the initial positions of the goal, target object, and robots are randomly generated on the road. The initial position of the obstacle is generated randomly along the pathway boundaries, if necessary. In Scenario 3, three robots cooperatively transport the target object to the goal while pushing an obstacle to the pathway boundary. The initial positions of the goal, target object, obstacle, and robots are randomly generated on the road. It should be noted that Scenarios 2 and 3 account for a single obstacle.

We design a reward for robot i under Scenario h, and it is expressed as

$$r_i = \begin{cases} r_i^{trans} + r_i^{penal} & \text{Scenario 1} \\ r_i^{trans} + r_i^{obs} + r_i^{penal} & \text{Scenarios 2 and 3} \end{cases} \quad (1)$$

where r_i^{trans}, r_i^{obs}, and r_i^{penal} represent the reward of the transportation task, reward of the obstacle removal task, and penalty, respectively. The rewards of the transportation task can be defined as follows:

$$r_i^{trans} = \begin{cases} \omega_1 & \text{if } x_i^t \text{ on goal} \\ -(\omega_2 \|x_i^t\| + \omega_3 |\theta_i^t| + \omega_4 \|x_i^t - x_i^g\|) & \text{otherwise} \end{cases} \quad (2)$$

where $\theta_i^t \in \mathbb{R}^1$ denote the angular position of the target object in the body coordinate system of robot i, and ω_1 to ω_4 are positive weight parameters. The reward of the obstacle removal task is calculated as follows:

$$r_i^{obs} = \begin{cases} \omega_5 & \text{if } x_i^o \text{ on pathway boundary} \\ -(\omega_6 \|x_i^o\| + \omega_7 |\theta_i^o| + \omega_8 |\min(d_i)|) & \text{otherwise} \end{cases} \quad (3)$$

where $\theta_i^o \in \mathbb{R}^1$, $\min(d_i)$, and ω_5 to ω_8 represent the angular position of the obstacle in the body coordinate system of robot i, the closest distance to the pathway boundary, and positive weight parameters, respectively. The penalty is computed as

$$r_i^{penal} = -(\omega_9 c_t^o + \omega_{10} c_t^b + \omega_{11} c_i^b + \omega_{12} c_i^{out}) \quad (4)$$

$$c_l^{l'} = \begin{cases} 1 & \text{if } l \text{ contacts } l' \\ 0 & \text{otherwise} \end{cases} \quad (5)$$

where ω_9 to ω_{12}, and $c_l^{l'}$ denote positive weight parameters, and the contact of l and $l' \in \{t, o, b, i, out\}$, which represent the target object, obstacle, pathway boundary, robot i, and outside of the environment, respectively.

4 Numerical Experiments

In this section, we describe the policy evaluation index, discuss the results, present the analyze of the individuality of multiple robots.

4.1 Validation Test

For evaluation, we implemented each policy 1,000 times for each scenario, and evaluated each policy in the validation Scenario 4, as shown in Fig. 3. In Scenario 4, the road has a different shape from those of Scenarios 1, 2, and 3. We randomly set up two obstacles and a waypoint with a goal on the pathway in Scenario 4. As described in Sect. 2, we set up the policy input to include an obstacle and a goal. In this scenario, an obstacle closest to the target object is used as the policy's input. Furthermore, the waypoint position is given as an input to the policy before the target object reaches the waypoint.

4.2 Results

Table 1 shows the conditions of the numerical experiments. The parameters were determined by trial and error. The training was performed using a desktop computer equipped with a 6-core Intel® Core™ i7 (3.20 GHz) and 16.0 GB of RAM. Under each scenario, the policies were trained until the cumulative rewards did not increase any further. A total of 315,000 episodes were required for training π_1; 78,000 for π_2, and 74,000 for π_3. The evaluation index is the success rate when the robots transport the target object to the goal. In Scenarios 3 and 4, we also measured the success rate when the robots simultaneously transport the target object to the goal and push the obstacle to the pathway boundary.

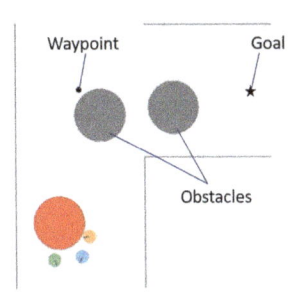

Fig. 3. Validation Scenario 4

Table 1. Numerical experimental conditions

Variable	Value
Control period [s]	0.1
Time step size of dynamics [s]	0.1
Number of steps per episode	200
Number of hidden layers (critic)	4
Number of hidden layers (policy)	4
Number of units per layer	64
Learning rate	1.0×10^{-3}
Discount factor	0.99
Batch size	5120
Replay buffer size (Number of steps)	1.0×10^{6}
ω_m ($m = 2, 3, 4, 6, 7, 8$)	1
ω_m ($m = 1, 5, 9, 10, 11$)	5
ω_{12}	10

Figure 4 shows the success rate of cooperative transportation achieved per episode for policy π_h trained under Scenario h. Robots with π_h reached goals with the highest rate for Scenario h. Furthermore, π_3 had a lower rate of reaching goals for Scenario 1 than π_1 and for Scenario 2 than π_1 and π_2. This result is attributed to the presence/absence of the obstacle and its different positions in each scenario.

In Scenario 3, π_3 showed 88.4% success rate of cooperative transportation. The simultaneous success rate of cooperative transportation and obstacle removal was 29.9% for π_3, whereas it was less than 1% for both π_1 and π_2. Note that the success rate of π_1, which was trained only for cooperative transportation, was 84.5% because the obstacle assigned randomly did not always block the road. On the other hand, in Scenario 4 with two obstacles, the success rate of π_1 was 37.6% because the obstacles often blocked the road. Even in Scenario 4, robots with π_3 transported the target object to the goal similarly to Scenario 3.

4.3 Individuality Analysis

We assessed the individuality of policy π_h^i for robot i trained under Scenario h, as discussed in Sect. 4.2. To evaluate the individuality of policy π_h^i, we first conducted a blindfold experiment, as shown in Fig. 5, and confirmed whether policy π_h^i tends to perform the cooperative transportation task or the obstacle removal task. Second, we experimented with the homogeneous policy teams, as shown in Fig. 6. Finally, we compared the weight parameter vectors of policies π_h^1, π_h^2, and π_h^3 under Scenario h to analyze the attention to the target object, obstacle, and goal.

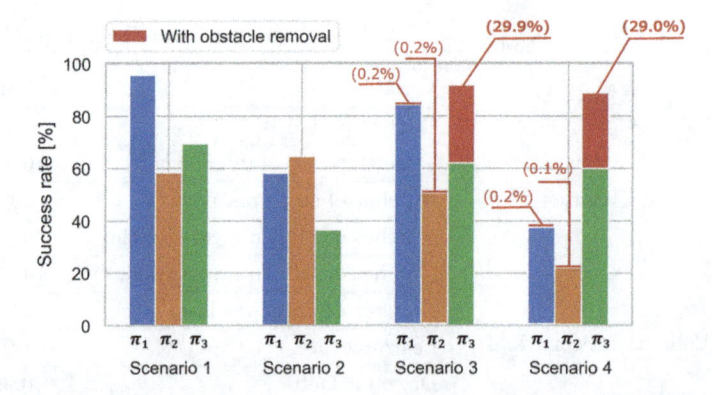

Fig. 4. Success rate of cooperative transportation for policy π_h trained under Scenario h. The numbers in parentheses in the figure are the success rates of simultaneously achieving cooperative transportation and obstacle removal.

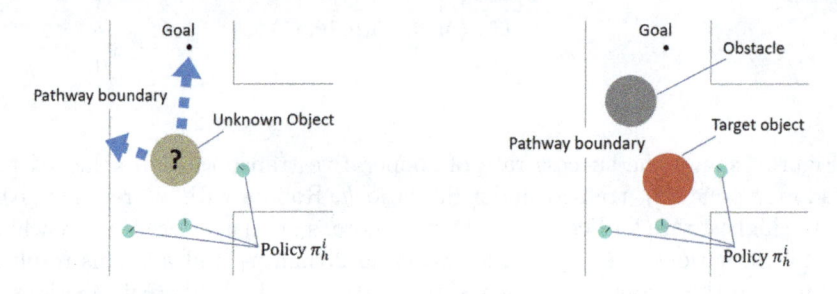

Fig. 5. Blindfold experiment

Fig. 6. Experiment with homogeneous teams

Virtual Experiment with Blindfold Teams. The blindfold experiment shown in Fig. 5 was conducted under the following configurations.

- The three robots do not know whether an object in the environment is a target object or an obstacle.
- The unknown object's position is utilized as the positions of the target object and obstacle contained in the robots' policy inputs.
- To eliminate the influence of robots j and k, we configure a team of three robots with the same policy as robot i.

With the above configurations, we ran 1,000 episodes where the three robots were randomly assigned in the T-shaped road, and we measured the number of instances where the robots successfully transported the unknown object to the goal and to the pathway boundary.

Figure 7 shows the total number of times the goal and/or pathway boundary were reached using π_3^i trained under Scenario 3. These results indicate that each robot's policy had a different task specialization. Policies π_3^2 of robot 2 and π_3^3 of robot 3 strongly tended to perform the cooperative transportation task. In contrast, policy π_3^1 of robot 1 tended to perform the obstacle removal task.

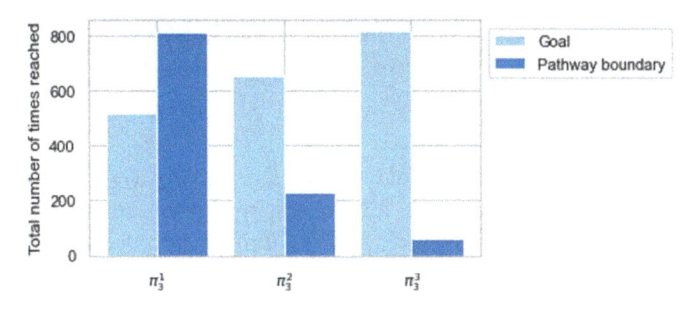

Fig. 7. Total number of times the goal and pathway boundary are reached when robot i executes π_3^i

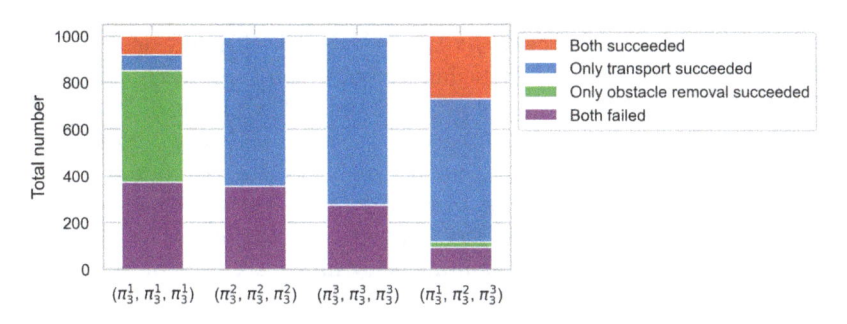

Fig. 8. Total number of successes when teams execute the homogeneous policies

Virtual Experiment with Homogeneous Teams. We conducted a second experiment for individuality analysis under the following configurations to measure the abilities of a team consisting of a single policy, as shown in Fig. 6.

– A team of three robots are configured with the homogeneous policy π_3^i of robot i.
– The experiment is conducted in the environment of Scenario 3.
– The three teams with the homogeneous policies are compared with π_3 trained under Scenario 3.

We ran 1000 episodes where three robots were randomly assigned in the T-shaped road and measured the total number of successes that the robots transported the target object to the goal and the obstacle to the pathway boundary, respectively.

Figure 8 illustrates the total number of successes with the homogeneous policy π_3^i for the three robots. These results show that the policies differed in the tasks they could perform, as with the blindfold experiment. Policies π_3^2 of robot 2 and π_3^3 of robot 3 were mainly specialized to perform the cooperative transportation task. By contrast, policy π_3^1 of robot 1 tended to perform the obstacle removal task. It should be noted that the team with π_3^1 was able to perform both the tasks simultaneously. Consequently, we confirmed that the three robots showed different individual characteristics to complete the two tasks as a team with $(\pi_3^1, \pi_3^2, \pi_3^3)$ simultaneously.

Comparison Among Policy Networks. To compare the characteristics of the task selection of policy π_h^i, we focused on the norms of weight parameter vectors for the target object, obstacle, and goal of the input layers of policy π_h^i.

The results are shown in Fig. 9. The horizontal axes correspond to the positions of the target object, obstacle, and goal; this information is included in the policy inputs of robot i. As shown in Figs. 9(a) and 9(b), under Scenarios 1 and 2, the policies of robot 1 were not significantly different from those of robots 2 and 3 in terms of the norms. In contrast, as shown in Fig. 9(c), policy π_3^1 of robot 1 under Scenario 3 shows a markedly larger norm for x_i^o and a smaller norm for x_i^t than those of policy π_3^2 of robot 2 and π_3^3 of robot 3, respectively.

These results indicate that policy π_3^1 of robot 1 trained under Scenario 3 had different characteristics from those of robots 2 and 3. π_3^1 determined the action with a relatively higher dependence on the obstacle, whereas it was less dependent on the target object, as compared to π_3^2 and π_3^3. We concluded that different characteristics emerged between robot 1 and robots 2 and 3 because two tasks had to be performed simultaneously under Scenario 3.

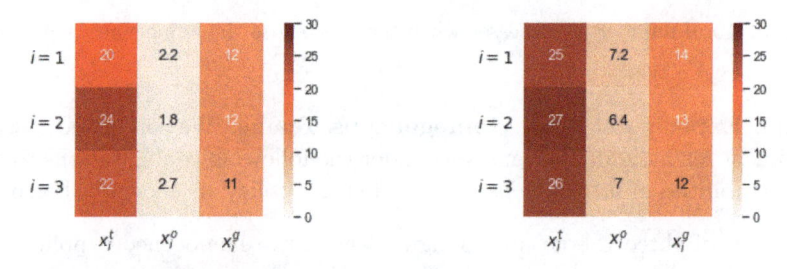

(a) π_1^i of robot i trained under Scenario 1 (b) π_2^i of robot i trained under Scenario 2

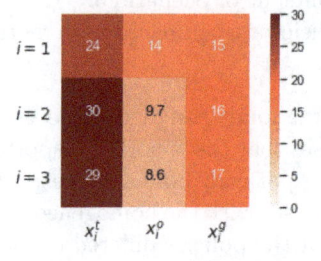

(c) π_3^i of robot i trained under Scenario 3

Fig. 9. Norms of weight parameter vectors of input layers

5 Demonstration

In this section, we demonstrate π_3 with three ground robots. The experimental configuration is shown in Fig. 1. In this case, the robots need to push the obstacle out of the pathway while transporting the target object. We set the initial positions of the robots to the right-center of the environment, as shown in the figure. The robots' observations were acquired via a motion capture system operating 120 Hz to avoid recognition errors. The actions v and ω ware calculated using a personal computer with an 8-core Intel® Core™ i7 (2.80 GHz) and 32 GB of RAM using π_3; the actions were sent to the robots at a frequency 10 Hz via Wi-Fi communication.

The results of the actual robot demonstrations are shown in Fig. 10. At the beginning of the demonstration, from their initial positions, two robots headed for the target object while one robot headed for the obstacle, as shown in Fig. 10(a). Subsequently, as shown in Fig. 10(b), the two robots heading for the target object moved to a position where they could push the target object toward the goal and made contact with the target object. On the other hand, the robot heading for the obstacle made contact with the obstacle. As illustrated in Fig. 10(c), the three robots began transporting the target object and obstacle almost simultaneously. The tasks were completed in 24 s, as shown in Fig. 10(d).

(a) 0–6 s: Approaching

(b) 6–12 s: Beginning contact

(c) 12–18 s: Transporting

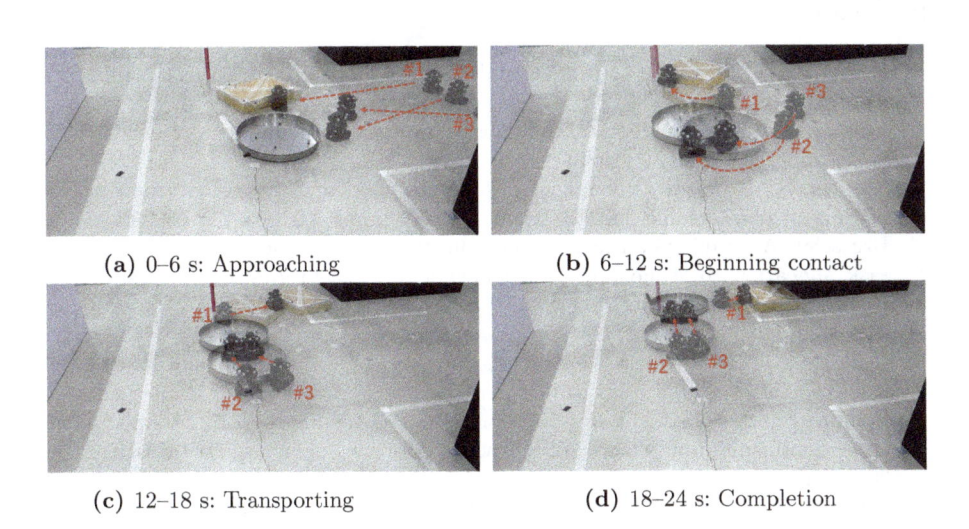

(d) 18–24 s: Completion

Fig. 10. Real robot demonstration

6 Conclusion and Future Work

In this study, we addressed the cooperative transport problem involving obstacle removal. We first introduced learning scenarios using MARL to develop decentralized policies for three robots. From individuality analyses, we confirmed that three robots' policies specialized in different tasks to simultaneously complete the two tasks as a team. The results showed that only one robot had the potential to perform two tasks simultaneously, depending on the situation. Finally, we demonstrated the trained policies with three ground robots.

The trained policies were validated in an additional scenario to ensure that the performance was maintained. To make the policies work well in various environments (e.g., where the shape of the object, obstacle, and pathway differ significantly), we plan to introduce methods that are known to improve policy generalization (e.g., dropout, L2 regularization, data augmentation, batch normalization, and environment's stochasticity) [25]. The current policy inputs include other robots' positions to learn cooperative behavior better, making it inflexible to changes in the number of robots after training. In the future, it is essential to extend a method for training policies to work cooperatively independent of the number of robots.

References

1. Beshers, S.N., Fewell, J.H.: Models of division of labor in social insects. Ann. Rev. Entomol. **46**(1), 413–440 (2001)
2. Robinson, G. E., Page, R. E.: Genetic basis for division of labor in an insect society. Genet. Soc. Evol. 61–80 (1989)
3. Nouyan, S., Gross, R., Bonani, M., Mondada, F., Dorigo, M.: Teamwork in self-organized robot colonies. IEEE Trans. Evol. Comput. **13**, 695–711 (2009)
4. Brutschy, A., Pini, G., Pinciroli, C., Birattari, M., Dorigo, M.: Self-organized task allocation to sequentially interdependent tasks in swarm robotics. Auton. Agent. Multi-Agent Syst. **28**(1), 101–125 (2012). https://doi.org/10.1007/s10458-012-9212-y
5. Garattoni, L., Birattari, M.: Autonomous task sequencing in a robot swarm. Sci. Robot. **3** (2018)
6. Bonabeau, E., Sobkowski, A., Theraulaz, G., Deneubourg, J.L.: Adaptive task allocation inspired by a model of division of labor in social insects. In: BCEC, pp. 36–45 (1997)
7. Ferrante, E., Turgut, A.E., Duéñez-Guzmán, E., Dorigo, M., Wenseleers, T.: Evolution of self-organized task specialization in robot swarms. PLoS Comput. Biol. **11**(8), e1004273 (2015)
8. Tuci, E., Alkilabi, M.H., Akanyeti, O.: Cooperative object transport in multi-robot systems: a review of the state-of-the-art. Front. Robot. AI **5**, 59 (2018)
9. Kosuge, K., Oosumi, T.: Decentralized control of multiple robots handling an object. In: Proceedings of IEEE/RSJ International Conference on Intelligent Robots and Systems, IROS 1996, vol. 1, pp. 318–323. IEEE (1996)
10. Wang, Z., Schwager, M.: Kinematic multi-robot manipulation with no communication using force feedback. In: 2016 IEEE International Conference on Robotics and Automation (ICRA), pp. 427-432. IEEE, May 2016

11. Wang, Z., Yang, G., Su, X., Schwager, M.: OuijaBots: omnidirectional robots for cooperative object transport with rotation control using no communication. In: Groß, R., et al. (eds.) Distributed Autonomous Robotic Systems. SPAR, vol. 6, pp. 117–131. Springer, Cham (2018). https://doi.org/10.1007/978-3-319-73008-0_9
12. Culbertson, P., Schwager, M.: Decentralized adaptive control for collaborative manipulation. In: 2018 IEEE International Conference on Robotics and Automation, pp. 278–285. IEEE (2018)
13. Farivarnejad, H., Wilson, S., Berman, S.: Decentralized sliding mode control for autonomous collective transport by multi-robot systems. In: 2016 IEEE 55th Conference on Decision and Control, pp. 1826–1833. IEEE (2016)
14. Chen, J., Gauci, M., Li, W., Kolling, A., Gross, R.: Occlusion-based cooperative transport with a swarm of miniature mobile robots. IEEE Trans. Rob. **31**(2), 307–321 (2015)
15. Gross, R., Dorigo, M.: Towards group transport by swarms of robots. Int. J. Bio-Inspir. Comput. **1**(1–2), 1–13 (2009)
16. Alkilabi, M.H.M., Narayan, A., Tuci, E.: Cooperative object transport with a swarm of e-puck robots: robustness and scalability of evolved collective strategies. Swarm Intell. **11**(3–4), 185–209 (2017)
17. Rahimi, M., Gibb, S., Shen, Y., La, H.M.: A comparison of various approaches to reinforcement learning algorithms for multi-robot box pushing. In: Fujita, H., Nguyen, D.C., Vu, N.P., Banh, T.L., Puta, H.H. (eds.) ICERA 2018. LNNS, vol. 63, pp. 16–30. Springer, Cham (2019). https://doi.org/10.1007/978-3-030-04792-4_6
18. Zhang, L., Sun, Y., Barth, A., Ma, O.: Decentralized control of multi-robot system in cooperative object transportation using deep reinforcement learning. IEEE Access **8**, 184109–184119 (2020)
19. Wang, Y., de Silva, C.W.: Cooperative transportation by multiple robots with machine learning. In: 2006 IEEE International Conference on Evolutionary Computation, pp. 3050–3056. IEEE (2006)
20. Gupta, J.K., Egorov, M., Kochenderfer, M.: Cooperative multi-agent control using deep reinforcement learning. In: Sukthankar, G., Rodriguez-Aguilar, J.A. (eds.) AAMAS 2017. LNCS (LNAI), vol. 10642, pp. 66–83. Springer, Cham (2017). https://doi.org/10.1007/978-3-319-71682-4_5
21. Hernandez-Leal, P., Kartal, B., Taylor, M.E.: Agent modeling as auxiliary task for deep reinforcement learning. In: Proceedings of the AAAI Conference on Artificial Intelligence and Interactive Digital Entertainment, vol. 15, no. 1, pp. 31–37 (2019)
22. Lowe, R., Wu, Y., Tamar, A., Harb, J., Abbeel, P., Mordatch, I.: Multi-agent actor-critic for mixed cooperative-competitive environments. Adv. Neural. Inf. Process. Syst. **30**, 6379–6390 (2017)
23. Sutton, R. S., McAllester, D. A., Singh, S. P., Mansour, Y.: Policy gradient methods for reinforcement learning with function approximation. Adv. Neural Inf. Process. Syst. 1057–1063 (2000)
24. Bengio, Y., Louradour, J., Collobert, R., Weston, J.: Curriculum learning. In: Proceedings of the 26th Annual International Conference on Machine Learning, pp. 41–48 (2009)
25. Cobbe, K., Klimov, O., Hesse, C., Kim, T., Schulman, J.: Quantifying generalization in reinforcement learning. In: International Conference on Machine Learning, pp. 1282–1289. PMLR (2019)

Battery Variability Management for Swarms

Grace Diehl[✉] and Julie A. Adams

Oregon State University, Corvallis, OR 97331, USA
diehlg@oregonstate.edu

Abstract. The Defense Advanced Research Projects Agency's (DARPA's) OFFensive Swarm-Enabled Tactics (OFFSET) program aims to develop a system architecture and algorithms for conducting urban missions with heterogeneous spatial swarms of up to 250 unmanned air and ground vehicles. Swarms' scale logistically prohibits modeling and tracking individual batteries, while highly variable battery lives make it difficult to determine, without such modeling, whether a vehicle has sufficient power to complete its tasks. The Swap algorithm manages battery variability by autonomously exchanging swarm vehicles with depleted batteries for ones with fresh batteries. Simulation-based evaluation demonstrates that Swap substantially increases mean task completion and reduces variance, thus increasing mission success and outcome consistency.

Keywords: Swarm robotics · Distributed system · Power management

1 Introduction

DARPA's OFFensive Swarm-Enabled Tactics (OFFSET) program aims to develop a system architecture and algorithms that enable a heterogeneous spatial swarm of up to 250 unmanned air (UAVs) and ground vehicles (UGVs) to perform urban missions [3]. Missions will span up to six hours and ten city blocks by the program's conclusion. Program progress is evaluated at live semiannual field exercises in relevant environments. The December 2019 field exercise incorporated one to two hour missions in a 45,000 square meters (m) area of operation with twenty-one single to five story buildings and other urban features (e.g., street signs and power lines). OFFSET's Command and Control of Aggregate Swarm Tactics (CCAST) team conducted six missions with swarms of up to sixty vehicles. These missions generated valuable data, including the useful battery lives of, and tasks performed by, each vehicle throughout each mission. This data enables post-exercise analysis of the swarm's mission performance and the opportunity to evaluate potential system improvements by replicating mission conditions in simulation. This manuscript considers the impact of individual battery lives on the swarm's mission performance. UAVs are emphasized, as the CCAST UGVs' batteries have historically lasted the entire mission.

The CCAST swarm utilized 3DR Solo quadcopters [1] with different payload configurations. Commercial quadcopters' ease of deployment facilitates scaling to swarms; however, their battery lives last only 10–30 min; thus, managing their battery use is critical. Ideally, an UAV accepts as many tasks as it can complete; however, an UAV's exact available energy is unknown during task allocation, even with smart batteries, as a battery's performance may degrade with age or environmental conditions. Modeling individual batteries a priori is infeasible, due to swarm's scale. The CCAST swarm uses three types of batteries and has 3–5 times more batteries than vehicles, making modeling time-consuming and tracking individual batteries during deployment unrealistic. Additionally, the power consumption required to complete a task is highly variable and can be increased by extended hovering or environmental factors, such as high winds or low temperatures. Thus, an UAV cannot accurately assess during task allocation whether it has sufficient energy to complete a task. Instead, a swarm requires dynamic task reassignment when an UAV's power level is too low, where reassignment must minimally disrupt performance and adhere to safety protocols.

The CCAST UAVs are used primarily for reconnaissance tasks, since they are able to access the tops and high exterior walls of buildings and traverse the area of operation quickly. The UAVs' tasks are not persistent, as they can be interrupted and resumed; however, significant time may be lost to the logistical challenges of refueling swarm UAVs and task resumption, particularly when the UAVs must traverse long distances (e.g., 700 m) to refuel. A less time-consuming power management method is desirable. Numerous algorithms address power management by incorporating task reallocation into vehicles' control models to accommodate refueling [6,10,11], enabling refueling during task performance [13,19], and determining when vehicles will abandon tasks to refuel [4,15,17]. However, these algorithms were designed for homogeneous systems performing a single task type, while the OFFSET missions incorporate heterogeneous tasks and vehicles. Additionally, the existing algorithms were evaluated on either small, real vehicle systems or in simulations that do not model battery variability.

This manuscript introduces the Swap algorithm, which enables swarms to autonomously reallocate tasks during mission execution from vehicles with depleted batteries to ones with fresh batteries, leveraging the fact that many swarm vehicles have the same or similar capabilities [5]. Swap applies to any vehicle type, but this analysis is focused on UAVs, due to their short battery lives. Data from the December (Dec.) 2019 field exercise was analyzed and used to replicate the mission conditions in a 3D simulator. Simulation-based evaluation demonstrates that Swap substantially increases the percentage of tasks completed and reduces variance in mission outcomes, making Swap a valuable tool to address battery variability for swarms performing real-world tasks.

2 Related Work

UAVs' and UGVs' individually variable, finite power supplies must be considered when performing real-world missions. This consideration is especially important for UAVs, as many real-world scenarios require task durations exceeding commercial UAVs' battery lives. An ideal power management system permits refueling that minimally disrupts tasks of uncertain and arbitrary lengths. The least disruptive power management approaches use mobile recharging stations for mid-task recharging; however, many commercial UAVs, including the 3DR Solos, require landing the UAV and manually exchanging the battery.

Manual battery replacement is compatible with approaches that incorporate power management into their vehicles' control models [6,10,11]. Such approaches require vehicles with low power levels to transfer tasks to neighboring vehicles prior to departing the task location for the refueling location. These approaches were designed primarily for homogeneous spatial swarms performing a single task, in which each vehicle's role is relatively localized. The heterogeneous CCAST swarm performs multiple simultaneous, heterogeneous tasks; thus, vehicles may be unable to exchange tasks with their neighbors. Further, the OFFSET missions' tasks are not localized, meaning that a vehicle's nearest compatible neighbors may be $\geq$100 m away. The potential to highly disrupt mission performance makes this approach infeasible for OFFSET missions.

Other approaches determine when vehicles will leave their tasks to refuel [12,15,17], where tasks may be abandoned [7] or transferred to nearby vehicles [16]. Erdelj *et al.*'s approach, which designates a pool of UAVs to replace those with depleted batteries, is the most relevant [4]. An UAV with low battery requests replacement with an untasked UAV and leaves to refuel upon its arrival. This approach permits UAVs to move freely throughout the environment while performing tasks, which is crucial for OFFSET missions. However, this algorithm, like others (e.g., [12,15,17]), was designed for homogeneous vehicles, each performing a single task or task type. Additionally, the simulation-based algorithm validation relied on all batteries having the same, known battery life. The CCAST swarm does not meet these assumptions. This manuscript considers data from a live field exercise, as well as variable and unknown battery lives.

3 System Architecture

The CCAST system architecture contains four primary components: a mission planner, the swarm, a central coordinator, known as the swarm dispatcher, and a human swarm commander, as shown in Fig. 1.

The mission planner is utilized prior to deployment to compose a mission plan of tasks to be performed. Individual tasks can be assigned priorities, ensuring that vehicles are assigned to mission-critical tasks. Other task requirements can be specified, such as inter-task dependencies or requiring vehicles with certain capabilities (e.g., flight). The saved mission plan is accessed by the human swarm commander who specifies during the mission when plan elements are executed.

The 2019 field exercise swarm incorporated Aion R1 and R6 UGVs [2] and 3DR Solo UAVs [1]. The vehicles are initially positioned at least 5 feet apart for deployment, due to to GPS drift, at one or more designated launch sites. Vehicles autonomously perform tasks and determine which task to perform next based on the mission plan's priorities. The vehicles perform different task types, depending on their capabilities and sensor payload. The UAVs are each equipped

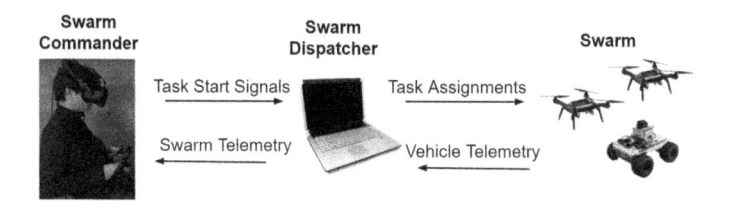

Fig. 1. OFFSET system architecture

with a Raspberry Pi Camera that is either *forward-facing* or *downward-facing*. The Solos lack sensors to detect other vehicles or obstacles; thus, deconfliction and collision avoidance are coordinated by the swarm dispatcher, based on the swarm's telemetry and a digital terrain map.

The swarm dispatcher coordinates communication among the swarm's vehicles and with the swarm commander (Fig. 1). Communication occurs via a LTE network, using a publish/subscribe system. When the swarm commander triggers the start of mission plan elements or specifies new tasks, the swarm dispatcher allocates tasks to appropriate vehicles. The swarm dispatcher also collects vehicle telemetry, including the position and battery level of each vehicle on the network, and compiles the swarm telemetry for the swarm commander.

The swarm commander initiates mission plan elements, monitors the swarm's status, and reviews gathered intelligence. The swarm commander can also generate new tasks to address unexpected events. This functionality was utilized during the 2019 field exercise, but the simulation-based experimentation assumes the swarm commander only issues components of the predefined mission plan.

4 The Swap Algorithm

The Swap algorithm addresses individual battery life variability by leveraging swarms' redundant capabilities [5]. A vehicle with a depleted battery, performing interruptible tasks relinquishes its tasks and performs its low battery safety behavior: returning to its launch site and landing (a *RTL*). The swarm dispatcher selects a replacement to assume the original vehicle's tasks. This exchange is a *swap*. The dispatcher handles parallel swaps, but may become slow if too many simultaneous swaps are requested. Swap has three phases: 1) normal task execution, 2) replacement selection on the dispatcher, and 3) task reallocation.

Vehicles in the *normal execution phase* execute tasks per the mission plan's task priorities, while monitoring their power levels. A battery level below the *swap threshold* (40%, plus 0.5%/m travel required to RTL) causes a vehicle to send the swarm dispatcher a Swap request. The vehicle continues to execute tasks until its request is answered or the *low battery threshold* is reached. The low battery threshold (30%, plus 0.5%/m to RTL) represents the point at which a vehicle must cancel its tasks and safely RTL before its battery is depleted. These thresholds are intentional overestimates, as safe RTL is a priority.

The swarm dispatcher attempts *replacement selection* upon receiving a swap request. A replacement vehicle must have a fresh battery and the capabilities required by the original vehicle's tasks. If one or more such vehicles exist, the dispatcher picks the vehicle physically closest to the original, and notifies the original vehicle. If no replacement is identified, the original vehicle cancels its tasks and RTLs, skipping the task reallocation phase.

The *task reallocation phase* commences when an active vehicle receives its replacement's identity from the swarm dispatcher. The original vehicle sends its tasks to its replacement, ceases task performance, and RTLs. The replacement vehicle begins travel to its first task site immediately. This phase relies on CCAST's tactics being interruptible. Uninterruptible tasks require the replacement vehicle to begin task performance before the original vehicle RTLs.

The Swap algorithm applies to both UAVs and UGVs without modification. This relatively simple behavior allows vehicles to contribute to missions for as long as possible and avoids requiring the swarm commander to reissue tasks manually when vehicles' battery lives are insufficient. The 2019 field exercise's swarm size was sufficiently large that it was infeasible for the swarm commander to carry out such individual interaction specifications. Further, human commanders cannot focus on individual swarm vehicles [8,9]. The autonomous Swap behavior significantly reduces the work demands on the swarm commander when very large swarms are deployed over very large areas for extended time periods, making Swap a necessary algorithm for deploying real world swarms.

5 Field Experiments

The 2019 field exercise began with small-scale testing of system capabilities, and concluded with full swarm mission scenario attempts. Table 1 summarizes the mission plans and swarms, listing (in order) the number of reconnaissance tasks in and the minimum number of UAVs required by each mission plan, the number of UAVs available (i.e., connected to the LTE network), the total swarm size during each exercise shift, whether Swap was enabled, and whether any tasks were cancelled by the swarm commander before completion. The replacement pool size is theoretically equivalent to the number of available non-required UAVs (i.e., UAVs available − UAVs required), but was generally smaller in practice, one to three UAVs, due to mission objectives.

The mission plans composite reconnaissance tasks, each focused on a region comprising two to six buildings. The swarm commander initiated tasks individu-

Table 1. Dec. 2019 field exercise mission plans and swarms.

Date	Exercise shift time	Tasks	UAVs required	UAVs available	Swarm size: UAVs & UGVs	Swap enabled	Cancelled by commander
Dec. 15	1559–1631	1	5	10	33	No	Yes
Dec. 16	0842–0951	1	5	17	39	No	No
Dec. 16	1315–1348	2	10	17	38	No	No
Dec. 17	1040–1156	3	15	33	55	Yes	No
Dec. 17	1340–1430	2	10	36	57	Yes	No
Dec. 18	0853–1000	6	35	37	60	Yes	Yes

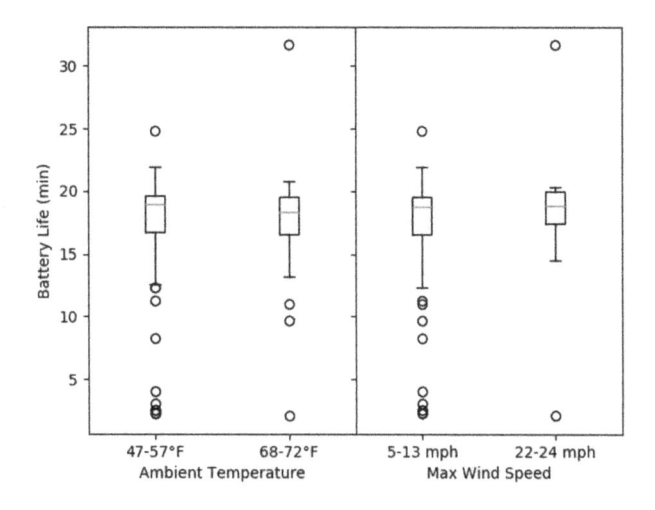

Fig. 2. Field exercise environmental conditions' impact on battery life.

ally at 0.5 to 7 min intervals based on the mission plan, triggering the dispatcher to allocate tasks to teams of untasked UAVs. Each task required four forward-facing UAVs to surveil exterior walls and one downward-facing UAV to surveil roofs. Five of the six Dec. 18 mission regions required five forward-facing and one downward-facing UAV. The dispatcher generated UAV-independent subtasks, in which a single UAV identified relevant information in its assigned area on the exterior of a building, and divided them among the team members. Each UAV autonomously performed its subtasks and RTLed due to subtask completion, a Swap request, low battery without swap, or the swarm commander's cancellation. Mission attempts concluded when all UAVs completed RTL.

5.1 Battery Life Analysis

The CCAST UAVs' useful battery lives were hypothesized to be too variable to predict if an UAV has sufficient power to perform its tasks without modeling and physically tracking the approximately 250 individual batteries. This *hypothesis* was evaluated by modeling the individual batteries from the field exercise's 104 UAV flights, using least-squares linear regression on the corresponding flight's battery telemetry. This modeling found useful battery lives between 2.06 and 34.01 minutes (min) (mean 17.21 and standard deviation (SD) 4.90).

The impact of temperature and wind speed on battery life was considered (Fig. 2) [18]. The 69 battery life models from flights with ambient temperatures of 47–57 °F had a median of 18.91 min (mean 17.01, SD 5.10), while the 34 battery life models from flights with an ambient temperature of 68–72 °F had a median of 18.27 min (mean 17.80, SD 4.50). Additionally, the 82 battery life models from flights with wind speed of 5–13 mph had a median of 18.72 min (mean 16.94, SD 4.87), while the 21 battery life models from flights with wind speeds of 22–24 mph had a median of 18.84 (mean 18.23, SD 4.99). These distributions are highly variable and left-skewed in all conditions, thus upholding the hypothesis.

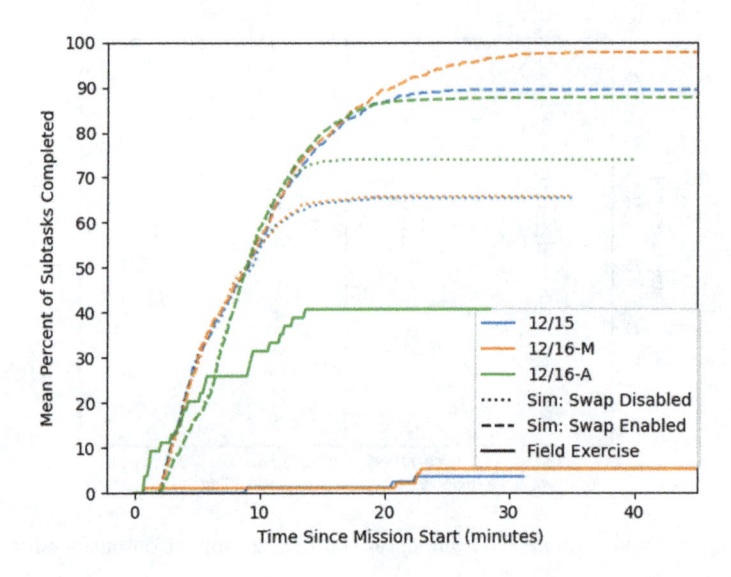

Fig. 3. The Dec. 15–16 live and simulated field exercise mission performance, where, per Table 1, 12/15 = Dec. 15, 12/16-M = Dec. 16 0842–0951, 12/16-A = Dec 16 1315–1348.

5.2 Mission Performance

Mission performance comprises the percentage of subtasks completed and subtask completion times, with the goal of completing a high percentage of subtasks quickly. Completion times were determined by each UAV's telemetry, which reported each subtask that was in progress, completed, or failed. Figures 3 and 4 provide the *mission progress*, or percent of subtasks completed over time, for each mission with and without Swap. All mission attempts plateaued by forty-five min.

Mission progress for the live December 15–16 field exercise missions, during which Swap was disabled, is shown in Fig. 3. The mission attempts finished with 3.62% (12/15), 5.32% (12/16-M), and 40.74% (12/16-A) subtask completion. Failure due to low battery accounted for only 1.20%, 15.96%, and 12.96% of subtasks, respectively. The swarm commander cancelled most 12/15 subtasks, and the swarm dispatcher never received telemetry for 52.13% of the 12/16-M and 27.78% of the 12/16-A subtasks. The unreported subtasks either completed or failed while an UAV was disconnected from the network, usually due to subtask locations outside the network coverage area. The remaining subtasks failed for other reasons, such as inaccessible subtask locations.

The Dec. 17–18 field exercise missions' progress, with Swap enabled, is provided in Fig. 4. The mission attempts finished with more subtasks completions, 47.69% (12/17-M), 10.20% (12/17-A), and 21.05% (12/18), respectively, and with fewer low battery failures, 1.54%, 4.08%, and 8.77%, respectively. No telemetry was received for 41.54%, 51.03%, and 33.92% of the subtasks, respectively, and all remaining subtasks were either cancelled by the swarm commander

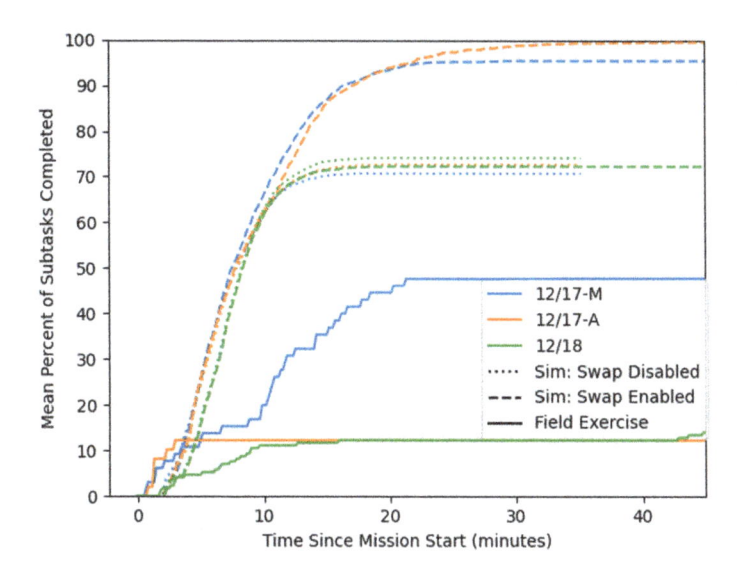

Fig. 4. Mission performance for Dec. 17–18 live and simulated field exercise missions.

or incomplete. Overall three swaps occurred; however, this analysis is insufficient to evaluate mission performance improvements.

6 Simulation Experiments

Simulation-based experiments evaluated Swap's impact on mission performance. The experiments recreated the field exercise missions in CCAST's WorldWind-based simulator [14] that includes a 3D obstacle model of the buildings, power lines, and other notable features from the field exercise's area of operation. The UAVs travelled at a constant speed of 0.5 m/s with battery lives selected from a normal distribution with a mean of 17.34 min and a SD of 5.25, based on battery life analysis that excluded UAVs flown multiple times during an exercise shift. Three experiments examining different independent variables were conducted.

6.1 Field Exercise Mission Plans with and Without Swap

The first experiment evaluated the *hypothesis* that Swap increases subtask completion without slowing the rate of mission progress. The experiment used the mission plans and number of available UAVs from the respective field exercise shifts. Twenty-five trials were conducted per mission with and without Swap.

The Dec. 15–16 missions (Fig. 3) initially progressed at the same rate. Progress without Swap ceased within 15 min, plateauing at 65.42% (SD 13.58%), 65.73% (SD 13.99%), and 73.98% (SD 11.72%) subtask completion for the 12/15, 12/16-M, and 12/16-A missions, respectively. The swarms with Swap completed

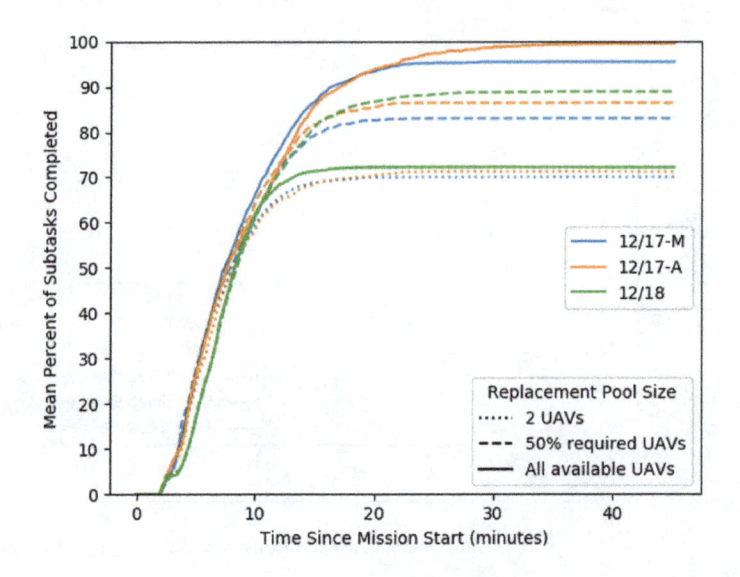

Fig. 5. Mission performance for Dec. 17–18 missions varying replacement pool size.

additional subtasks before plateauing at 89.47% (SD 7.74%), 97.73% (SD 3.29%), and 87.77% (SD 8.54%), respectively, within 40 min.

The Dec. 17–18 missions (Fig. 4) also progressed at similar rates initially. Progress without Swap ceased within 20 min, with 70.68% (SD 8.26%), 72.6% (SD 13.23%), and 74.08% (SD 7.0%) of subtasks completed for the 12/17-M, 12/17-A, and 12/18 missions, respectively. The 12/17 swarms with Swap completed almost all subtasks, 95.40% (SD 2.79%) and 99.47% (SD 1.16%), respectively, within 25 min. The 12/18 Swap trials plateaued around the same time and with fewer completed subtasks, 72.2% (SD 8.62%), than its non-Swap counterpart. The 12/18 swarm had a very small replacement pool; thus, most low battery UAVs failed to swap and RTLed at the swap threshold (40%, as compared to 30% without Swap), outweighing any benefit of a replacement pool.

6.2 Replacement Pool Size

The impact of changing the replacement pool size was examined for the Dec. 17–18 mission plans using replacement pool sizes of: 2 UAVs, 50% of the mission plan's required UAVs (7 for 12/17-M, 5 for 12/17-A, and 17 for 12/18), and all available non-required UAVs. The *hypothesis* was that increasing the replacement pool size increased the percentage of completed subtasks.

The mission progress by replacement pool size was initially similar (Fig. 5). The 12/17-M trials plateaued within 25 min at 69.97% (SD 10.08%), 82.99% (SD 8.74%), and 95.40% (SD 2.79%) for the 2 UAVs, 50% required, and all available, respectively. The 12/17-A missions plateaued within 35 min at 71.17% (SD 12.51%), 86.42% (SD 8.37%), and 99.47% (SD 1.16%), respectively, and the

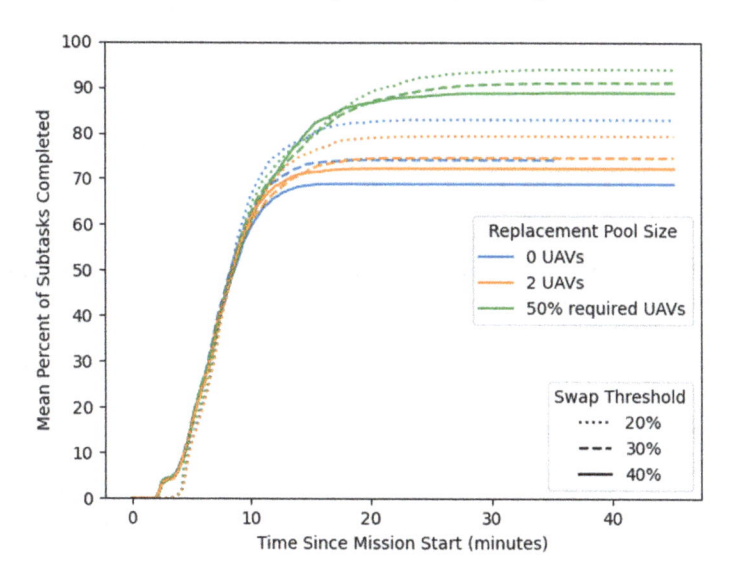

[Figure: Mission performance chart showing Mean Percent of Subtasks Completed vs Time Since Mission Start (minutes), with curves for Replacement Pool Size (0 UAVs, 2 UAVs, 50% required UAVs) and Swap Threshold (20%, 30%, 40%).]

Fig. 6. Mission performance for Dec. 18 simulated missions varying the swap threshold. 0 UAVs and a 30% threshold is equivalent to the Swap disabled results in Fig. 4.

12/18 missions plateaued within 25 min at 72.2% (SD 8.62%) for 2 UAVs (i.e., all available UAVs) and 88.92% (SD 4.68%) for 50% required.

6.3 Swap Threshold

This experiment varied the swap threshold, which determines at what power level an UAV requests a swap. Thresholds of 40% (CCAST's default swap), 30% (CCAST's default low battery RTL), and 20% were analyzed for the 12/18 mission plan with replacement pool sizes of 0, 2 (equivalent to all available UAVs), and 50% more UAVs than the mission plan required. The *hypothesis* was that decreasing the swap threshold increases the subtask completion rate.

The initial mission progressions were similar (Fig. 6) irrespective of threshold value, but diverged around 10 min. Swarms using the 20% threshold with 0 replacement UAVs plateaued at 80.43% (SD 5.60%), compared to 74.08% (SD 7.00%) for the 30% threshold and 68.84% (SD 6.08%) for the 40% threshold. Swarms with 2 replacement UAVs and a 20% threshold plateaued at 79.36% (SD 6.71%), while the same swarms with a 30% threshold plateaued at 73.64% (SD 7.35%) and a 40% threshold plateaued at 72.2% (SD 8.62%). Swarms with 50% as many UAVs in the replacement pool as required by the mission plan plateaued at 94.28% (SD 2.70%) with a 20% threshold, 91.12% (SD 4.33%) with a 30% threshold, and 88.92% (SD 4.68%) with a 40% threshold. At the 20% threshold, some UAVs had insufficient power remaining to RTL safely.

7 Discussion

The Swap algorithm was designed to address the high variability of UAVs' battery lives. Swap was hypothesized to increase the percent of subtasks completed, without excessively slowing mission progress. This hypothesis was largely upheld by experiments replicating the field exercise mission plans and swarms. All trials progressed at similar rates with and without Swap, affirming that Swap does not slow mission progress. Swap enabled additional subtask completion after the progress for trials without Swap ceased for five missions, increasing subtask completion by 13–32 percentage points and reducing performance variance substantially. The 12/18 trials' small replacement pool size resulted in no improvement in subtask completion percentage or variance with Swap.

Experiments using the Dec. 17–18 mission plans demonstrated that increasing replacement pool size substantially improves the percentage of subtask completions, as hypothesized, while also reducing variance. The increase is sublinear in the replacement pool size, because Swap requires a replacement UAV to duplicate the original UAV's work traveling to a task site. Another trade-off is that the logistics of safe refueling are complicated at the scale of swarms. A replacement pool using all available UAVs risks replacement UAVs taking off near humans who are replacing batteries, while a smaller replacement pool using a dedicated Swap launch zone can improve battery replacement safety and efficiency.

Experiments using the 12/18 mission plan examined the hypothesis that decreasing the swap threshold increases subtask completion. Decreasing the threshold did increase task success marginally; however, the difference between 30% and 40% thresholds was not meaningful and decreased as the size of the replacement pool increased, while a 20% threshold resulted in unsafe RTLs. Analysis of the 12/17 missions produced similar results. These results suggest that there is little trade-off to having a swap threshold that is higher than the low battery threshold, which is required for Swap if communication delays are expected.

This analysis focused on swarms with very limited vehicle and subtask heterogeneity (i.e., camera directions and roof vs. wall reconnaissance), but Swap can accommodate more heterogeneity without modification. During the Dec. 2019 field exercise, intelligence gathered by the UAVs directly influenced UGV tasks and the swarm's ability to complete mission objectives. Future field exercises will team UAVs and UGVs for shared tasks, and the Swap capability will be critical for maintaining heterogeneous team performance.

8 Conclusion

Modeling and tracking individual commercial UAVs' short, variable battery lives is infeasible at swarm scale, which hinders determining if each UAV can complete its tasks. Swap enables vehicles to transfer their tasks dynamically when their power levels are too low, in a manner that minimally disrupts performance. This capability, unlike prior results, can be used in highly complex missions, with heterogeneous swarms, heterogeneous tasks, and large, challenging environments.

Acknowledgements. This research was developed with funding from the Defense Advanced Research Projects Agency (DARPA). The views, opinions, and findings expressed are those of the author and are not to be interpreted as representing the official views or policies of the Department of Defense or the U.S. Government.
DISTRIBUTION STATEMENT A: Approved for public release: distribution unlimited.

References

1. 3DR: Solo (2020). www.3dr.com/company/about-3dr/solo/
2. Aion Robotics: R1 UGV (2019). www.aionrobotics.com/r1
3. Defense Advanced Research Projects Agency: OFFensive Swarm-Enabled Tactics (2019). www.darpa.mil/work-with-us/offensive-swarm-enabled-tactics
4. Erdelj, M., Saif, O., Natalizio, E., Fantoni, I.: UAVs that fly forever: uninterrupted structural inspection through automatic UAV replacement. Ad Hoc Netw. **94**, 101612 (2019)
5. Hamann, H.: Swarm Robotics: A Formal Approach. Springer, Cham (2018). https://doi.org/10.1007/978-3-319-74528-2
6. Jensen, E., Franklin, M., Lahr, S., Gini, M.: Sustainable multi-robot patrol of an open polyline. In: IEEE International Conference on Robotics and Automation, pp. 4792–4797 (2011)
7. Kernbach, S., Nepomnyashchikh, V.A., Kancheva, T., Kernbach, O.: Specialization and generalization of robot behaviour in swarm energy foraging. Math. Comput. Model. Dyn. Syst. **18**(1), 131–152 (2012)
8. Kolling, A., Walker, P., Chakraborty, N., Sycara, K., Lewis, M.: Human interaction with robot swarms: a survey. IEEE Trans. Hum.-Mach. Syst. **46**(1), 9–26 (2016)
9. Lewis, M.: Human interaction with multiple remote robots. Rev. Hum. Factors Ergon. **9**(1), 131–174 (2013)
10. Li, G., Svogor, I., Beltrame, G.: Self-adaptive pattern formation with battery-powered robot swarms. In: NASA/ESA Conference on Adaptive Hardware Systems, pp. 253–260 (2017)
11. Li, G., Svogor, I., Beltrame, G.: Long-term pattern formation and maintenance for battery-powered robots. Swarm Intell. **13**(1), 21–57 (2019)
12. Malyuta, D., Brommer, C., Hentzen, D., Stastny, T., Siegwart, R., Brockers, R.: Long-duration fully autonomous operation of rotorcraft unmanned aerial systems for remote-sensing data acquisition. J. Field Robot. **37**(1), 137–157 (2020)
13. Mathew, N., Smith, S.L., Waslander, S.L.: A graph-based approach to multi-robot rendezvous for recharging in persistent tasks. In: Proceedings of the IEEE International Conference on Robotics and Automation, pp. 3497–3502 (2013)
14. Pirotti, F., Brovelli, M.A., Prestifilippo, G., Zamboni, G., Kilsedar, C.E., Piragnolo, M., Hogan, P.: An open source virtual globe rendering engine for 3D applications: NASA World Wind. Open Geospat. Data Softw. Stand. **1** (2017). https://doi.org/10.1186/s40965-017-0016-5
15. Shin, M., Kim, J., Levorato, M.: Auction-based charging scheduling with deep learning framework for multi-drone networks. IEEE Trans. Veh. Technol. **68**(5), 4235–4248 (2019)

16. Trotta, A., Felice, M.D.F., Chowdhury, K.R., Bononi, L.: Fly and recharge: achieving persistent coverage using small unmanned aerial vehicles (SUAVs). In: IEEE ICC Mobile and Wireless Networking, pp. 1–7 (2017)
17. Trotta, A., Muncuk, U., Felice, M.D., Chowdhury, K.R.: Persistent crowd tracking using unmanned aerial vehicle swarms. IEEE Technol. Mag. **15**, 96–103 (2020)
18. Weather Underground: Moselle, MS Weather History (2020)
19. Yu, K., Budhiraja, A.K., Tokekar, P.: Algorithms for routing of unmanned aerial vehicles with mobile recharging stations. In: IEEE International Conference on Robotics and Automation, pp. 5720–5725 (2018)

Spectral-Based Distributed Ergodic Coverage for Heterogeneous Multi-agent Search

Guillaume Sartoretti[1]($\boxtimes$), Ananya Rao[2], and Howie Choset[2]

[1] Mechanical Engineering Department, National University of Singapore, Singapore 117575, Singapore
mpegas@nus.edu.sg,
http://www.marmotlab.org

[2] Robotics Institute, Carnegie Mellon University, Pittsburgh, PA 15203, USA

Abstract. This paper develops a multi-agent heterogeneous search approach that leverages the sensing and motion capabilities of different agents to improve search performance (i.e., decrease search time and increase coverage efficiency). To do so, we build upon recent results in ergodic coverage methods for homogeneous teams, where the search paths of the agents are optimized so they spend time in regions proportionate to the expected likelihood of finding targets, while still covering the whole domain, thus balancing exploration and exploitation. This paper introduces a new method to extend ergodic coverage to teams of heterogeneous agents with varied sensing and motion capabilities. Specifically, we investigate methods of leveraging the spectral decomposition of a target information distribution to efficiently assign available agents to different regions of the domain and best match the agents' capabilities to the scale at which information needs to be searched for in these regions. Our numerical results show that distributing and assigning coverage responsibilities to agents based on their dynamic sensing capabilities leads to approximately 40% improvement with regard to a standard coverage metric (ergodicity) and a 15% improvement in time to search over a baseline approach that jointly plans search paths for all agents, averaged over 500 randomized experiments.

Keywords: Multi-agent system · Distributed search · Heterogeneous teams

1 Introduction

With the rapid development of affordable robots with embedded sensing and computation capabilities, we are quickly approaching a point at which real-life applications will involve the deployment of hundreds, if not thousands, of robots [1,2]. Among these applications, significant research effort has been

---

G. Sartoretti and A. Rao—These authors contributed equally to this work.

© The Author(s), under exclusive license to Springer Nature Switzerland AG 2022
F. Matsuno et al. (Eds.): DARS 2021, SPAR 22, pp. 227–241, 2022.
https://doi.org/10.1007/978-3-030-92790-5_18

devoted to multi-agent search [3–7], where deploying numerous agents can greatly improve the time-efficiency and robustness of search. In fact, deploying robots with various motion or sensing modalities can further improve the search performance, by leveraging the natural synergies between these capabilities (see Fig. 1). Motivated by such problems, the main contribution of this work is to investigate the distribution of agents to search a domain at various spatial scales, based on their motion and sensing capabilities. Specifically, we build upon recent results in ergodic search processes [3,8–10] to propose a mapping of available agents to a spectral-based decomposition of the search problem, to best match agents' capabilities to specific classes of areas of the search domain.

The approach in this paper is based on ergodic search processes, which, similar to other information-theoretic coverage methods [3,4,8,11,12], rely on an *a priori* information distribution, representing the likelihood of finding a target at any point over the search domain, to guide the search. In practice, this information distribution can be obtained from scouting missions or from expert knowledge, and is updated during search if inaccurate. If nothing is known about the targets' whereabouts, this distribution is usually considered uniform over the whole domain, i.e., targets could be anywhere with equal probability. Using this *a priori* information distribution, ergodic search processes optimize search paths over long time horizons for all agents.

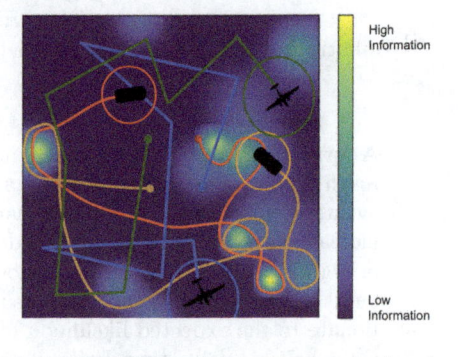

Fig. 1. Multi-agent search scenario involving two types of agents: differential-drive agents with short-range, high fidelity sensors (represented by the red and orange circles), and omnidirectional agents with long-range, low fidelity sensors (represented by the blue and green circles). The different colored lines represent the paths followed by the different agents. The underlying distribution shows the likelihood of finding targets throughout the domain.

In this work, we determine search paths via an optimization process using the ergodic metric [3]. The optimization of search paths, according to the ergodic metric, aims to drive agents to spend time in areas of the domain in proportion to the *a priori* likelihood of finding targets in these areas. This optimization is performed in the spectral domain, by minimizing the difference between the coefficients associated to the team's *time-average statistics* (i.e., fraction of the time spent in each area) and those of the information distribution. The contribution of this work is to exploit the spectral nature of the ergodic metric in search scenarios involving heterogeneous agents. To this end, we plan search paths for each agent type based on a smaller subset of the spectral coefficients associated with the information map, thus driving agents to search the domain at a spatial scale that best matches their motion and sensing capabilities. We use a distributed approach, in which agent assignment into teams, each attributed to a subset of spectral coefficient, is centralized, after which each team plans paths independently using a locally centralized path planner.

This paper is organized as follows: Sect. 2 discusses recent advances in multi-agent search and in coordination of heterogeneous multi-agent systems. We then provide a brief background of ergodic search processes in Sect. 3. Section 4 details our spectral-based decomposition of a search problem and of the available agents. We then present and discuss the results of our systematic set of experiments in Sect. 5. There, we observe that our best agent distribution approach leads to a 40% increase in coverage performance (thus generally leading to more time-efficient search), averaged over 500 randomized experiments, over a baseline that jointly plans search paths for the whole team regardless of their individual abilities. Section 6 offers concluding remarks.

2 Prior Work

2.1 Multi-agent Search

Current active search methods generally fall into one of three main categories: geometric, gradient-based, and trajectory optimization-based approaches. Geometric methods, e.g., lawnmower patterns, can be good search strategies in order to uniformly cover a domain in which there is near-uniform probability of finding a target [13,14]. Since these approaches exhaustively cover the search domain, they are also the logical choice in cases where there is no *a priori* information about the targets' locations.

An information map, or information distribution, is defined to be a probability distribution representing the likelihood of a target being found at each location in the domain. When such *a priori* information is available (and, usually, non-uniform), more advanced search processes can be created that leverage this information map in order to improve search according to some metric, such as time to find all targets.

For example, in gradient-based, or "information surfing", methods [4,11,12], agents guide their movement in the direction of the derivative of the information map around their positions to greedily maximize the short-term information gain. That is, agents are always driven in the direction of the greatest information gain, which naturally leads them to areas where the likelihood of finding a target is maximized. Information surfing can be implemented in a fully decentralized manner, since it does not require tight coordination between agents, and potential fields can be introduced to help distribute agents to different areas of the domain. However, gradient-based approaches generally do not rely on the uncertainty associated with the information distribution, which can lead to areas left unexplored, as this uncertainty can help differentiate areas of low-information that have not been explored from areas with no information to be gained. Gradient-based approaches are also very sensitive to noise in the information map, as the gradient cannot be estimated accurately in these situations, and suffer from greedily over-exploiting local information maxima.

Optimization-based approaches look at search as an information gathering maximization problem, which is then solved by planning (usually joint) paths for the agents. Several recent works in coverage methods [3,8–10] rely on sampling-based path planning, where a large number of paths are sampled and the best path is chosen based on a cost metric. Optimization-based approaches can combine both the predicted information distribution as well as its associated uncertainty into the cost function that drives the optimization. However, these approaches generally do not scale well for large multi-agent systems since they remain centralized. Even for sampling-based approaches, the number of paths that need to be sampled to find near-optimal search paths grows exponentially with the number of agents, although growing the number of samples linearly with the team size seems to experimentally provide good-quality search paths [3,9].

2.2 Heterogeneity in MAS

Most search methods developed for homogeneous groups of agents (i.e., agents with similar capabilities) do not support groups of agents with heterogeneous capabilities (for example, a set of agents with different sensing or motion capabilities), or struggle with the increased computational complexity [15–19]. Many previous works involving heterogeneous agents concentrate on offering initial, usually centralized and non-scalable solutions to the problems that they mainly focus on defining [20–23]. Some works have considered using auction-based mechanisms for task assignment in heterogeneous groups [24–26], while others have proposed agent redistribution based on given sets of their capabilities [27,28]. Other works have considered using robots with the best communication or coordination capabilities as "leader agents" to plan for and coordinate the other agents with lesser capabilities [6,29].

3 Background on Ergodic Search Processes

Ergodic search processes [8] produce trajectories for multi-agent systems, such that agents spend time in each area of the domain proportional to the expected amount of information present in this area. To this end, the spatial time-average statistics of an agent's trajectory $\gamma_i : (0,t] \to \mathcal{X}$, quantifies the fraction of time spent at a position $\boldsymbol{x} \in \mathcal{X}$, where $\mathcal{X} \subset \mathbb{R}^d$ is the d-dimensional search domain. For N agents, the joint spatial time-average statistics of the set of agents trajectories $\{\gamma_i\}_{i=1}^N$ is defined as [8]

$$C^t(\boldsymbol{x},\gamma(t)) = \frac{1}{Nt} \sum_{i=1}^{N} \int_0^t \delta(\boldsymbol{x} - \gamma_i(\tau))\, d\tau, \qquad (1)$$

where δ is the Dirac delta function.

Formally, the agents' time-averaged trajectory statistics is optimized against the expected information distribution over the whole domain, by matching their spectral decompositions. This is obtained by minimizing the ergodic metric $\Phi(\cdot)$, expressed as the weighted sum of the difference between the spectral coefficients of these two distributions [8]:

$$\Phi(\gamma(t)) = \sum_{k=0}^{m} \lambda_k \left| c_k(\gamma(t)) - \xi_k \right|^2, \qquad (2)$$

where c_k and ξ_k are the Fourier coefficients of the time-average statistics of the set of agents' trajectories $\gamma(t)$ and the desired spatial distribution of agents respectively, and λ_k are the weights of each coefficient difference. In practice, $\lambda_k = \sqrt{(1 + \|k\|^2)^{-(d+1)}}$ is usually defined to place higher weights on the lower frequency components, which correspond to larger spatial-scale variations in the information distribution.

The goal of ergodic coverage is to generate optimal controls $\boldsymbol{u}^*(t)$ for each agent, whose dynamics is described by a function $f\colon \mathcal{Q} \times \mathcal{U} \to T\mathcal{Q}$, such that

$$\begin{aligned} \boldsymbol{u}^*(t) &= \arg\min_{\boldsymbol{u}} \Phi(\gamma(t)), \\ \text{subject to } \dot{\boldsymbol{q}} &= f(\boldsymbol{q}(t), \boldsymbol{u}(t)), \\ \|\boldsymbol{u}(t)\| &\leq u_{max} \end{aligned} \qquad (3)$$

where $\boldsymbol{q} \in \mathcal{Q}$ is the state and $\boldsymbol{u} \in \mathcal{U}$ denotes the set of controls. Equation (3) can either be solved by discretizing the exploration time and solving for the optimal control input at each time-step [8], by trajectory optimization to plan feed-forward trajectories over a specified time horizon [30], or by using sampling-based motion planners [31], where it is straightforward to pose additional constraints such as obstacle avoidance. We observe that (3) computes controls centrally for all agents, the complexity of which increases exponentially with number of agents. Applying Eq. (3) individually to subteams of agents yields drastically lower-complexity optimization problems, which can be solved faster and with lower computation power.

4 Distributed Heteregeneous Ergodic Search

This work investigates the coordination of a team of heterogeneous agents during search from two key fronts. First, we look at how the spectral decomposition of the information distribution can be interpreted in order to search regions at different spatial scales. In other words, lower frequency components typically describe the distributions in broad strokes, while higher frequency ones are responsible for filling in the details. Second, we study how agents should be assigned to different spectral bands of the information decomposition, and formulate an assignment that reasons about the agents' varying capabilities to cover a domain more efficiently. Our distributed approach uses a centralized overhead to assign different spectral bands of the information decomposition to different agent teams, which then each plan paths independently using locally centralized planners.

4.1 Spectral Bands of the Information Distribution

In this work, we rely on the spectral decomposition of the information map to guide the search task assignment for agents with heterogeneous motion and sensing capabilities. We recall that, in the spectral decomposition of the information distribution Eq. (2), lower-frequency coefficients correspond to larger-scale variations in the spatial distribution of information, while higher-frequency coefficients correspond to smaller-scale variations.

Building upon this observation, we propose to define M spectral bands (i.e., sets of frequency coefficients within particular ranges), with M the number of agent types in the heterogeneous team. Each band can be seen as a separate (although not completely independent) **search subtask** that can be distributed to a specific type of agent based on its motion/sensing capabilities to search at a given spatial scale. In this work, we break down the overall set of spectral coefficients into M successive bands of equal length, but other decompositions of the set of coefficients into bands could be considered, and will be investigated in future works. To help visualize the different search subtasks, resulting from such a choice of bands, we can reconstruct partial representations of the information distribution, each based on a single band of coefficients, as shown in Fig. 2.

To formally define and use these search subtasks, we modify the ergodic metric Eq. (2) to rely on a specific band of coefficients only:

$$\Phi(\gamma(t)) = \sum_{k=c_1}^{c_2} \lambda_k \left| c_k(\gamma(t)) - \xi_k \right|^2, \qquad (4)$$

where c_1 and c_2 define the first and last coefficients of the spectral band. Note that the same result could be achieved by setting $\lambda_k = 0 \; \forall k < c_1, k > c_2$.

4.2 Assignment of Agents to Spectral Bands

In order to find an optimized set of paths according to the new ergodic coverage metric Eq. (4), we must first find an optimized assignment of agent types to spectral bands which maximizes the search performance. To this end, we note that these bands can first be distributed to heterogeneous agents based on their sensing capabilities. For example, agents with low-fidelity, high-range sensing capabilities should generally be assigned low-frequency spectral coefficients, as expressed in Eq. (5) (left) in order to perform large-scale, broad-stroke exploration. Conversely, agents with high-fidelity, low-range sensing capabilities should likely be assigned high-frequency spectral coefficients, as expressed in Eq. (5) (right) in order to perform detailed, small-scale exploration:

$$\Phi(\gamma(t)) = \sum_{k=c_1}^{\frac{c_1+c_2}{2}} \lambda_k \left| c_k(\gamma(t)) - \xi_k \right|^2 \qquad \Phi(\gamma(t)) = \sum_{k=\frac{c_1+c_2}{2}}^{c_2} \lambda_k \left| c_k(\gamma(t)) - \xi_k \right|^2, \qquad (5)$$

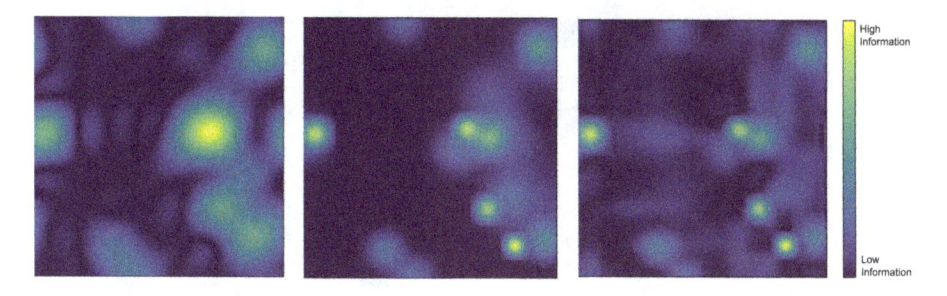

Fig. 2. Example spectral reconstruction of a given map (center), based on only its lower-order coefficients only (left), or higher-order ones only (right). Yellow regions correspond to regions of high information, while darker blue regions correspond to regions of low information (here, high/low likelihood of finding targets).

Similarly, different motion models can also be used as a basis for task distribution between the agents. For instance, we believe that faster agents, relying on lower-frequency coefficients of the decomposition, could perform a coarse exploration of the domain. On the other hand, by relying on higher-frequency coefficients, slower agents would be naturally driven to perform a smaller-scale, detailed search. A similar intuitive assignment can be made when dealing with agents with varying motion constraints. For example, given a team composed of omnidirectional and maximal-curvature-constrained agents, our hypothesis is that the former should rely on lower-frequency coefficients, while the latter can more easily chain "spots" of information obtained from high-frequency bands.

5 Results and Discussion

We present our systematic investigation of four different ways agent types can be assigned to search subtasks, by relying on a large set of simulation experiments composed of fixed, randomly generated search problems. We compare these assignment methods according to various standard search metrics, such as the time to find all targets and the effectiveness of coverage (using the ergodic metric), showing that our optimal assignment can yield up to 40% increase in these metrics. Our results rely on sampling-based trajectory optimization, but we emphasize that our investigation should extend to other optimization methods.

5.1 Agent's Sensing and Motion Models

The sensor footprint of each agent is modeled as a Gaussian distribution centered at the agent's position, whose variance prescribes a circular observation range $\rho > 0$. At each point within this observation range, we use the Gaussian probability density function to represent the likelihood of detecting a target at each time step. We consider a mix of agents with low-range, high-fidelity sensors (i.e., a Gaussian of low variance and thus higher maximal detection probability

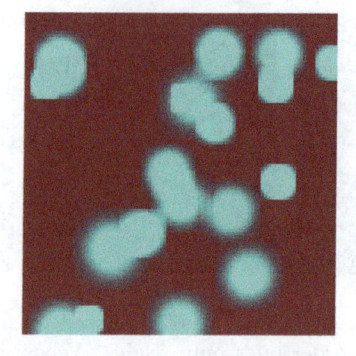

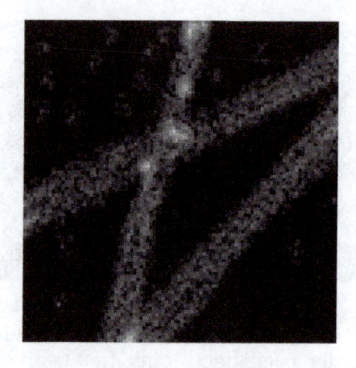

Fig. 3. Examples of the two classes (Gaussian Mixture Models (left), and road networks (right)) of randomly generated information maps for evaluating the proposed approach.

at its center), and agents with high-range, low-fidelity sensors (i.e., a Gaussian of larger variance and thus lower maximal peak detection likelihood).

In addition to the different sensing models, we also consider two types of agents' motion models. The first model is a simple first order integrator that represents omnidirectional agents, such as quad-rotor UAVs or legged ground robots. We further consider agents with differential drive constraints (i.e., resulting in curved paths with a maximum curvature), such as fixed-wing airplanes or wheeled ground vehicles. We sample paths for the agents by sequencing path primitives - straight lines of various directions and lengths for the omnidirectional agents, and curves from a finite collection with various curvatures and lengths for the differential agents. Agents plan long trajectories, execute these paths for 10 timesteps, update the map using their observations, and then replan. We further rely on a cross-entropy planner [31] to optimize the paths of all agents via 3 levels of sample refinement with a total of $15 \cdot N$ samples (where N is the total number of agents, $N = 10$ in practice).

5.2 Experiment Details

Scenarios Randomization. We compare the performance of various assignment methods, with that of a baseline that plans paths for all agents by relying on the overall distribution maps (i.e., no decomposition into bands or assignments), through 500 randomized search scenarios. These scenarios vary the locations of targets and the initial information maps (as randomly generated Gaussian mixture models, or road-network inspired information maps). Additionally, for each experiment, a randomly generated team of 10 agents is formed by selecting both the sensing and motion model with equal probability for each agent. Team compositions, starting positions, initial information maps, and target locations are kept identical among experiments with different controllers, to ensure our results are comparable.

Agent Assignments. In our experiments, we group together all agents with the same motion and sensing constraints, thus yielding four independent agent groups. We let A_0 be the set of agents with low-fidelity, high range sensing capabilities and fast, omni-directional motion models and A_1 be the set of agents with low-fidelity, high range sensing capabilities and slower, curve-constrained motion models. Similarly, we let A_2 and A_3 be the sets of agents with high-fidelity, low-range sensing capabilities and omni-directional motion models and low-fidelity, high-range sensing capabilities and curve-constrained motion models respectively. Finally, we decompose the information map into a set C of M spectral coefficient bands - $C_0, ..., C_{M-1}$. In our experiments, we fixed $M = 4$. Assignments, i.e., mappings from agent types to spectral bands can be expressed as:

$$h : A_i \longrightarrow X, \qquad X \in \{C_0, C_1, C_2, C_3\} \tag{6}$$

Our optimal assignment, based on the intuition built in Sect. 4.2, reads:

$$h_{\text{optimal}}(A_i) = C_i \tag{7}$$

In order to investigate the effectiveness of the optimal assignment Eq. (7), we compared its performance to that of more naive assignments Eq. (8), Eq. (9), as well as to that of an adversarial assignment Eq. (10). Finally, we compare these results with a baseline that assumes all agents are identical (homogeneous team), i.e., all the agents rely on all coefficients of the global information map.

$$h_{\text{naive1}}(A_i) = \begin{cases} C_0 & i = 1 \\ C_1 & i = 0 \\ C_2 & i = 3 \\ C_3 & i = 2 \end{cases} \tag{8}$$

$$h_{\text{naive2}}(A_i) = \begin{cases} C_0 & i = 2 \\ C_1 & i = 3 \\ C_2 & i = 0 \\ C_3 & i = 1 \end{cases} \tag{9}$$

$$h_{\text{adversarial}}(A_i) = C_{3-i} \tag{10}$$

Performance Metrics and Sensitivity to Hyper-parameters. We ran another set of experiments in order to investigate the sensitivity of this approach to hyper parameters. There, we focused on the ratio of agents of different capabilities, the total number of agents exploring the domain and the number of samples used in each time step of the sample-based path planner.

All of these experiments were run on Gaussian information distributions and road-network inspired information distributions (Fig. 3) in an effort to simulate potential use cases. The results of these two types of information distributions are reported separately as their metrics distributions are significantly different, although the overall performance improvement is similar.

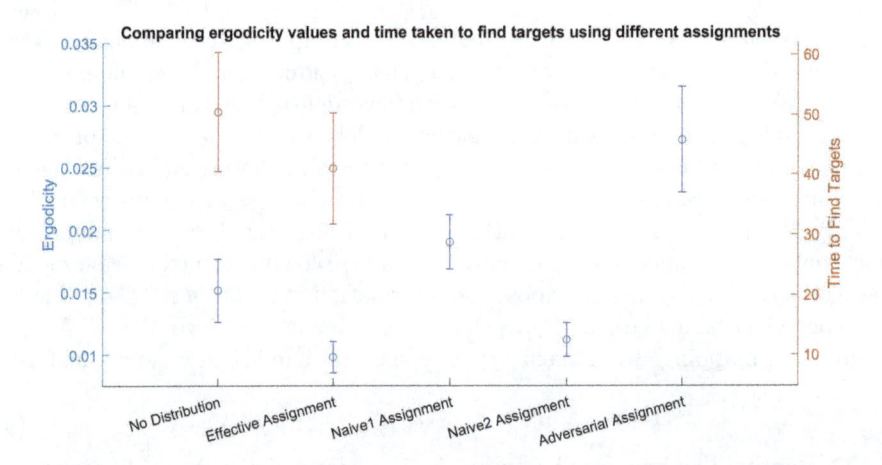

Fig. 4. Search performance comparison between the different agent assignments and the baseline, in terms of coverage performance (using the ergodic metric, lower is better) and time to find all targets (lower is better).

5.3 Experimental Results

When looking at the results of the different assignments in term of overall coverage performance, measured via the ergodic metric, the first observation we can make is that our optimal assignment, expressed in Sect. 4.2 results in approximately 40% improvement over the baseline, while naive heuristics show ±5% improvement and the adversarial heuristic yields approximately 25% deterioration in performance over the baseline approach (Fig. 4). These results verify the intuition built in Sect. 4.2. Our results also confirm an important point: more effective coverage of the domain leads to finding targets faster.

As expressed in Sect. 4.2, we believe that lower order spectral bands preserve broad domains of information. Therefore, agents with high-range sensing capabilities and less constrained, motion models would be better suited to coarse exploration as they are capable of covering larger areas quickly, with higher uncertainty. On the other hand, higher order spectral bands preserve edges and details. Lacking more general information about the map means that there will be more "false positive" areas, that is, more domains that show higher information in this scale of spatial variation but not in the original. So, agents with high-fidelity, low-range sensing capabilities seem better-suited to rely on the higher order spectral bands, because, high-fidelity might only lead to false positives that have less impact on the search. Smaller, more concentrated areas of information in this spatial scale are thus better explored by agents with curve-constrained motion models.

In the naive heuristics, agents rely on spectral bands that are well-suited to either their sensing or motion capabilities but not to both. For example, in the first naive heuristic Eq. (8), agents with low-fidelity, high range sensing capabilities and slow, curve-constrained motion models rely on the lowest frequency

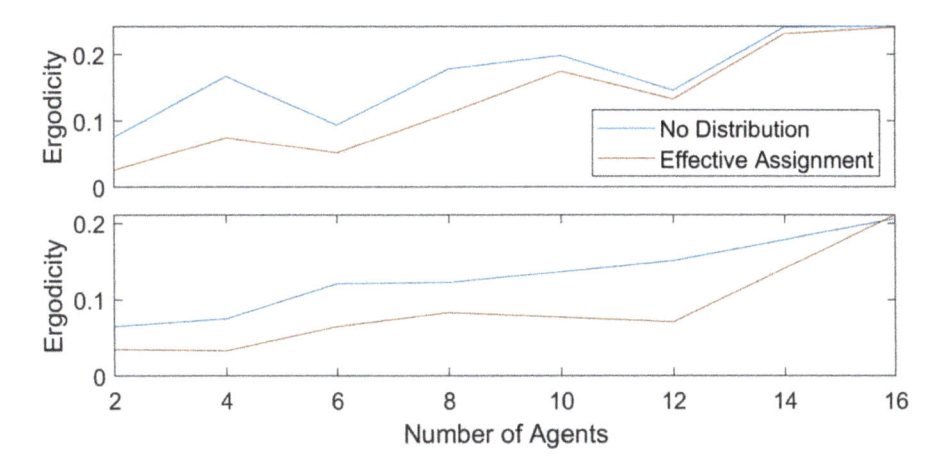

Fig. 5. Sensitivity to the team size, comparing the coverage performance between our optimal assignment and the baseline. Gaussian maps (top), and road network (bottom). As expected, note the improved performance for smaller teams.

spectral bands, which are well suited to their sensing capabilities but not to their motion model. This kind of partially suitable mapping results in a performance almost equivalent to the baseline, as this assignment drives agents to, on average, explore the information distribution in a similar manner as in the baseline (i.e., there are no advantages over the baseline, but no clear downsides either).

Finally, in the adversarial heuristic, agents rely on spectral bands worst-suited to their sensing and motion capabilities. As expected, this mismatch leads to a decrease in performance as agents struggle to search at the spatial scale assigned to them, since it doesn't match their capabilities. Some side-by-side comparison videos of these methods in example scenarios can be found at http://bit.ly/DARS21-HetMASearch.

As expected, we further note that improvement in performance over the centralized approach decreases as the number of agents covering the domain increases (Fig. 5). When there are fewer agents available to cover a domain, the coverage efficacy of each agent's path strongly influences the overall coverage of the domain by the team, since each agent is effectively responsible for a larger portion of the domain. However, when a large number of agents are covering a domain, each agent is effectively responsible for a smaller portion of the domain, so the effectiveness of the path of each agent has a negligible impact on the team's coverage of the domain.

Our results also indicate that improvement in performance over the centralized approach decreases as the number of samples taken in each step of the sample-based path planner increases (Fig. 6). We know that the path primitives sampled by each agent at each step of the sample-based path planner depend on the spatial scale at which the agent is searching. When there are fewer samples being considered at each step, the spatial distribution that the agent is relying

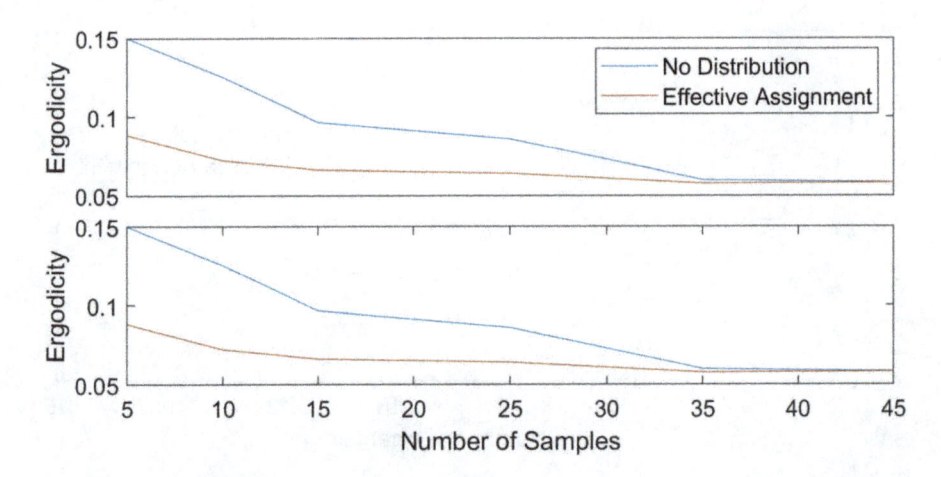

Fig. 6. Sensitivity to the number of sampled paths, comparing the coverage performance between our optimal assignment and the baseline. Gaussian maps (top), and road network (bottom). There again, and as expected, note that our distributed search approach specifically improves performance with small numbers of samples.

on has greater influence on the effectiveness of the sampled paths than when a large number of samples are drawn. That is, in the limit where sampling is performed on a near-infinite number of paths, decomposing the information map should not lead to an improved solution (since the best sampled paths will be globally optimal, and can also be found by the baseline approach). Thus, our results show that distributing the correct spectral bands (i.e., search subtasks) to the agents has more impact on the achieved coverage of a domain when planning paths using small numbers of samples. We envision this to be a significant advantage, especially for robot deployments that necessitate real-time planning and re-planning capabilities, where planning time is mainly controlled by the number of samples to be drawn.

6 Conclusion

In this paper, we investigated the idea of leveraging the spectral nature of a state-of-the-art coverage metric, the ergodic metric, to improve the heterogeneous multi-agent search of a domain by matching the agents' motion and sensing constraints to specific search subtasks. These subtasks were defined as performing search at different spatial scales, by relying on a limited subset of the spectral coefficients that represent the overall information map. After building intuition on the link between sensing and motion models and the different search scales based on subsets of coefficients, we proposed an agent assignment method to map agent types to specific search subtasks (i.e., subsets of spectral coefficients). In our systematic numerical tests, we compared our optimal assignment to naive and adversarial assignments, as well as to a baseline that plans paths for all agents regardless of their individual capabilities, and showed our distributed ergodic search approach lead to significantly improved performance (up to 40%), both in terms of coverage efficiency and time to find all targets. Additionally, our distributed approach allows sampled-based optimization methods to require a smaller number of samples to find high-quality search paths, and improves the performance of smaller agent teams, which might maximize its impact to real-world multi-robot deployments.

This work paves the way for new heterogeneous multi-agent search methods, where synergies among agents could be automatically identified and leveraged to improve the efficacy of the process. In particular, future works will approach the general problem of assigning any type of agent the right set of spectral coefficients. To this end, and for general cases where human intuition/experience cannot suffice, we believe that machine learning based methods could offer us the tool to learn such a data-driven mapping. Furthermore, the work presented in this paper considered centralized subtask assignment and path planning, but our future work will seek decentralized task assignment (and potentially ergodic path planning) solutions, to really allow such distributed heterogeneous multi-agent search methods to scale to large teams, and ultimately allow large-scale real-life deployments. Additionally, we assume in this paper that our *a priori* information map is accurate, however future work will investigate the potentially increased errors that may be caused by the use of inaccurate priors, particularly in terms of increased search times, for distributed search.

References

1. Rubenstein, M., Cornejo, A., Nagpal, R.: Programmable self-assembly in a thousand-robot swarm. Science **345**(6198), 795–799 (2014). https://doi.org/10.1126/science.1254295
2. Howard, A., Parker, L.E., Sukhatme, G.S.: Experiments with a large heterogeneous mobile robot team: exploration, mapping, deployment and detection. Int. J. Robot. Res. **25**(5–6), 431–447 (2006)
3. Ayvali, E., Salman, H., Choset, H.: Ergodic coverage in constrained environments using stochastic trajectory optimization. In: International Conference on Intelligent Robots and Systems, pp. 5204–5210. IEEE (2017)
4. Lanillos, P., Gan, S.K., Besada-Portas, E., Pajares, G., Sukkarieh, S.: Multi-UAV target search using decentralized gradient-based negotiation with expected observation. Inf. Sci. **282**, 92–110 (2014)
5. Murphy, R.R.: Disaster Robotics. MIT Press, Cambridge (2014)
6. Chand, P., Carnegie, D.A.: Mapping and exploration in a hierarchical heterogeneous multi-robot system using limited capability robots. Robot. Auton. Syst. **61**(6), 565–579 (2013)
7. Chung, T.H., Hollinger, G.A., Isler, V.: Search and pursuit-evasion in mobile robotics. Auton. Robot. **31**(4), 299–316 (2011)
8. Mathew, G., Mezić, I.: Metrics for ergodicity and design of ergodic dynamics for multi-agent systems. Phys. D **240**(4), 432–442 (2011)
9. Ayvali, E., Ansari, A., Wang, L., Simaan, N., Choset, H.: Utility-guided palpation for locating tissue abnormalities. Robot. Autom. Lett. **2**, 864–871 (2017)

10. Miller, L.M., Silverman, Y., MacIver, M.A., Murphey, T.D.: Ergodic exploration of distributed information. IEEE Trans. Rob. **32**(1), 36–52 (2016)
11. Baxter, J.L., Burke, E., Garibaldi, J.M., Norman, M.: Multi-robot search and rescue: a potential field based approach. In: Mukhopadhyay, S.C., Gupta, G.S. (eds.) Autonomous Robots and Agents. Studies in Computational Intelligence, vol. 76, pp. 9–16. Springer, Heidelberg (2007). https://doi.org/10.1007/978-3-540-73424-6_2
12. Wong, E.M., Bourgault, F., Furukawa, T.: Multi-vehicle Bayesian search for multiple lost targets. In: International Conference on Robotics and Automation, pp. 3169–3174. IEEE (2005)
13. Choset, H.: Coverage for robotics-a survey of recent results. Ann. Math. Artif. Intell. **31**(1), 113–126 (2001)
14. Ablavsky, V., Snorrason, M.: Optimal search for a moving target - a geometric approach. In: AIAA Guidance, Navigation, and Control Conference and Exhibit. AIAA (2000)
15. Yan, Z., Jouandeau, N., Cherif, A.A.: A survey and analysis of multi-robot coordination. Int. J. Adv. Rob. Syst. **10**(12), 399 (2013)
16. Dias, M.B., Zlot, R., Kalra, N., Stentz, A.: Market-based multirobot coordination: a survey and analysis. Proc. IEEE **94**(7), 1257–1270 (2006)
17. Kim, J.H., Vadakkepat, P.: Multi-agent systems: a survey from the robot-soccer perspective. Intell. Autom. Soft Comput. **6**(1), 3–17 (2000)
18. Tang, F., Parker, L.E.: ASyMTRe: automated synthesis of multi-robot task solutions through software reconfiguration. In: Proceedings - IEEE International Conference on Robotics and Automation, pp. 1501–1508 (2005)
19. Parker, L.E.: ALLIANCE: an architecture for fault tolerant, cooperative control of heterogeneous mobile robots (1994)
20. Prorok, A., Hsieh, M.A., Kumar, V.: Formalizing the impact of diversity on performance in a heterogeneous swarm of robots. In: International Conference on Robotics and Automation, pp. 5364–5371. IEEE (2016)
21. Dahl, T.S., Matarić, M., Sukhatme, G.S.: Multi-robot task allocation through vacancy chain scheduling. Robot. Auton. Syst. **57**(6–7), 674–687 (2009)
22. Jones, E.G., Browning, B., Dias, M.B., Argall, B., Veloso, M., Stentz, A.: Dynamically formed heterogeneous robot teams performing tightly-coordinated tasks. In: International Conference on Robotics and Automation, pp. 570–575. IEEE (2006)
23. Koes, M., Nourbakhsh, I., Sycara, K.: Heterogeneous multirobot coordination with spatial and temporal constraints. In: AAAI, vol. 5, pp. 1292–1297 (2005)
24. García, P., Caamaño, P., Duro, R.J., Bellas, F.: Scalable task assignment for heterogeneous multi-robot teams. Int. J. Adv. Robot. Syst. **10** (2013). https://doi.org/10.5772/55489
25. Gerkey, B.P., Matarić, M.J.: Sold!: auction methods for multirobot coordination. IEEE Trans. Robot. Autom. **18**(5), 758–768 (2002)
26. Zlot, R., Stentz, A.T., Dias, M.B., Thayer, S.: Multi-robot exploration controlled by a market economy. In: International Conference on Robotics and Automation, vol. 3, pp. 3016–3023 (2002)
27. Prorok, A., Hsieh, M.A., Kumar, V.: Fast redistribution of a swarm of heterogeneous robots. In: EAI International Conference on Bio-inspired Information and Communications Technologies, pp. 249–255. ICST (2016)
28. Halász, A., Hsieh, M.A., Berman, S., Kumar, V.: Dynamic redistribution of a swarm of robots among multiple sites. In: International Conference on Intelligent Robots and Systems, pp. 2320–2325. IEEE (2007)

29. Grabowski, R., Navarro-Serment, L.E., Paredis, C.J., Khosla, P.K.: Heterogeneous teams of modular robots for mapping and exploration. Auton. Robot. **8**(3), 293–308 (2000)
30. Miller, L.M., Murphey, T.D.: Trajectory optimization for continuous ergodic exploration. In: American Control Conference (ACC), 2013, pp. 4196–4201. IEEE (2013)
31. Kobilarov, M.: Cross-entropy motion planning. Int. J. Robot. Res. **31**(7), 855–871 (2012)

Multi-agent Deception in Attack-Defense Stochastic Game

Xueting Li[✉], Sha Yi, and Katia Sycara

Carnegie Mellon University, Pittsburgh, PA 15213, USA
xuetingl@andrew.cmu.edu

Abstract. This paper studies a sequential adversarial incomplete information game, the attack-defense game, with multiple defenders against one attacker. The attacker has limited information on game configurations and makes guesses of the correct configuration based on observations of defenders' actions. Challenges for multi-agent incomplete information games include scalability in terms of agents' joint state and action space, and high dimensionality due to sequential actions. We tackle this problem by introducing deceptive actions for the defenders to mislead the attacker's belief of correct game configuration. We propose a k-step deception strategy for the defender team that forward simulates the attacker and defenders' actions within k steps and computes the locally optimal action. We present results based on comparisons of different parameters in our deceptive strategy. Experiments show that our approach outperforms Bayesian Nash Equilibrium strategy, a strategy commonly used for adversarial incomplete information games, with higher expected rewards and less computation time.

Keywords: Attack-defense game · Multiple agent systems · Defense strategy · Game theory · Zero-sum game · Games of incomplete information · Deception

1 Introduction

In applications with adversarial opponents, collaboration between agents can improve the performance of the system, for example, pursuit-evasion games [17,19], and attack-defense games [4,15]. In such adversarial scenarios, a game-theoretic formulation provides a framework to model and reason about the benefits and trade-off between players.

Limitations and challenges arise when the game-theoretic framework is applied to infinite horizon games with incomplete information. First, state space grows exponentially with increasing number of agents in the system. Second, with a game-theoretic model, Nash Equilibrium only provides solutions to adversarial games with complete information. For incomplete information games, additional belief needs to be incorporated for the unknown game configurations. In the incomplete information setting, Bayesian Nash equilibrium is introduced to solve such game by incorporating players' beliefs. However, the state-belief space then becomes extremely high-dimensional [7], which is computationally expensive. Third, due to the sequential nature of infinite horizon games, it is intractable to compute the optimal sequence of strategies for the players. In such incomplete information games where there is a information gap between the players, one can use deception to manipulate the belief of opponents [11]. Deception strategies help the players to gain higher rewards in long time horizon, while sacrificing reward gains in short time horizon.

In this paper, we propose a deceptive game-theoretic framework for the infinite horizon, incomplete information games with multiple players. We consider an attack-defense game where there are multiple defenders, a single attacker, and multiple target assets. The true configuration of the target assets, i.e., real target values are known to the defender, but only partially known to the attacker. The attacker aims to maximize the total value of targets attacked by guessing the game configuration while avoiding getting caught by any defender. We present techniques that reduce the computational cost of solving the game compared with a standard Bayesian Nash Equilibrium method for incomplete information games. In particular, defenders and the attacker pre-compute a library of Nash strategies, based on the commonly known information, that is exploited during execution. Additionally, we employ the Monte Carlo Tree Search that samples defenders and attacker actions to further reduce the computation cost. We compare the performance and computation cost of our deceptive defense strategy with and without sampling versus a standard one-step BNE strategy.

The paper makes the following contributions: First, we enrich current attack-defense game formulations by considering (a) how defender moves can deceive the attacker and (b) by considering an infinite-time-horizon game. Second, we provide a novel framework to enable online planning for a multi-defender team in a game with incomplete information.

The paper is organized as follows. Section 2 reviews some previous research on the use of deception in artificial intelligence system and past studies on attack-defense games. In Sect. 3, we first present the formal definition of the attack-defense game we aim to study. We then review definitions of Nash strategy and Nash Equilibrium for computation of the strategy library before execution. We will also introduce Bayesian Nash Equilibrium in Sect. 3.2 as our baseline for comparison. Monte Carlo Tree Search will be introduced in Sect. 3.5 for improving scalability and computation efficiency. Section 4 shows experimental results of the defender team performance and computation cost when the defender team follows BNE or our deceptive planning against an attacker who follows BNE or the strategy described in Sect. 3.3.

2 Related Work

In multi-player incomplete information games, planning can be extremely computationally heavy due to game history and large state space [1,7,18,20]. In adversarial scenarios, where a fraction of players hold complete information while others hold incomplete information (asymmetric game), deception is often introduced to aid the planning process and improve the strategy quality [6,11,21]. Specifically, in attack-defense games, deception is majorly studied in either signal games [2,21], or repeated games [6,13]. Signal games are one-time games where the defender deceives the attacker by sending signals for one time [21]. In repeated games, players do not enter a new state after each move, which means previous actions do not influence future decisions [6]. In multi-player incomplete information sequential games, computation complexity grows exponentially with number of agents and history. Due to the high-dimensionality, deception for sequential games mainly focuses on planning for a finite-time horizon [5,10,14], which can only apply to games ending in known steps but cannot handle games that do not have an immediate reward. Planning for infinite-time horizon games [9] often has small action space.

To the best of our knowledge, the current literature does not provide an efficient algorithm to address the computational intractability of the belief state space in asymmetric two-player zero-sum stochastic games with infinite horizon.

3 Technical Approach

We first formally define our problem in Sect. 3.1. In Sect. 3.2, we review details of Nash Equilibrium of complete information games for later computation of the strategy library, and the standard Bayesian Nash Equilibrium methods of incomplete information games for comparison in Sect. 4. We will introduce our assumptions and methods of attacker belief update in Sect. 3.3, and our k-step deception planning method in Sect. 3.4. Monte Carlo game tree sampling will be discussed in Sect. 3.5 for improving computational efficiency.

3.1 Problem Formulation

We model the game as a two-player zero-sum game with incomplete-information and infinite time horizon. We discretize time and space and represent the environment as a grid world. Both the attacker and defenders have information of the target locations and can always observe the adversary's movement. The game terminates when (i) the attacker is caught by any defender, or (ii) all valuable targets have been attacked. The attacker is considered caught if it is on the same cell as any defender and a target is attacked if the attacker is on the same cell as the target. Throughout the game, the attacker will constantly try to guess the true target configuration. Meanwhile, the defender team will move in a way to deceive the attacker about the true target configuration and mislead the attacker to go to targets of lower value.

We model the game as Markov Decision Processes (MDP) $< S, A, T, \Phi, \mathbf{R}, \mathbf{b}, \gamma >$. S is the joint state space of defenders, attacker, and targets $S = S_d \times S_a \times S_w$. $S_d = S_0 \times \cdots \times S_n$ where S_i is the state of ith defender. S_i and S_a are the positions of corresponding agent in the grid world. S_w is a binary vector where $S_{w_j} = 1$ if j^{th} target has not been attacked, 0 otherwise. A is the joint action space of the defender team and attacker. $A = A_d \times A_a$ where $A_d = A_1 \times A_2 \times \cdots \times A_n$. A_i is the action space of ith defender and A_a denotes the action space of the attacker. We assume four-connected grid so the action space of each agent is $A_i, A_a = \{N, E, W, S, Z\}$, which corresponds to move to North, East, West, South, and remain in the current cell. T is the deterministic transition matrix that $T(s'|s, a) = 1$ if the following state of taking action a at state s is s', 0 otherwise.

In our game, there is a discrete set of target rewards given different target configurations Φ. Each $\phi_l \in \Phi$ represents a reward set of the lth target configuration. Each ϕ_l consists of the positions of targets and rewards for attacking different targets. The positions of targets are accessible for both players. However, only the defender team knows the rewards. The real target configuration is the one with the true rewards of targets. The reward set of the real target configuration ϕ_{l^*} is included in this set, i.e. $\phi_{l^*} \in \Phi$. We assume both teams have common knowledge on all possible reward setup. Given the number of targets, there is a finite set of possible target configuration that is known by both players.

The goal of the defenders is to catch the attacker and minimize the total value of targets being attacked. The goal of the attacker is to maximize the value of targets attacked and avoid getting caught by any defender.

R is the reward function set. For each ϕ_l, there is a distinct reward function $R_l : S \times A \rightarrow \mathbb{R}$ corresponding to that target configuration. $R_l(s, a)$ is the reward of defenders' team when joint action a has been taken in the joint state s:

$$R_l(s'|s, a) = \begin{cases} -g_{l,w}^j & \text{if } j^{th} \text{ target is attacked at state } s' \\ g_{l,c} & \text{if attacker is caught by any defender at state } s' \\ 0 & \text{otherwise} \end{cases} \quad (1)$$

g$_{l,w}$ is a vector consists of the values of different targets where $g_{l,w}^j$ is the value of jth target in l^{th} target configuration. $g_{l,c}$ is the value of catching the attacker. Since it is a zero-sum game, the reward for the attacker is thus $-R_l$. The reward function of the real target configuration is denoted as $R_{l^*} \in \mathbf{R}$. The expected future reward is discounted over an infinite time horizon where γ is the discount factor. Both players aim to maximize their expected future reward. The defender team's objective is to maximize the expected discounted reward:

$$\max_{a_d^0 \ldots a_d^\infty} \sum_{t=0}^{\infty} \gamma^t R_{l^*}(s_t, a_{d,t} \times a_{a,t}) \quad (2)$$

Not knowing the true target configuration, the attacker maintains a belief vector $\mathbf{b} = [b_0, \cdots, b_l, \cdots]$ where $b_l = prob(\phi_l = \phi_{l^*})$ denotes the probability the attacker thinks ϕ_l is the true target configuration. The attacker updates **b** based on the observed moves of the defender team, which will be introduced in Sect. 3.3. The defender team can reconstruct the attacker's belief because all information on the attacker side is available to the defender team. This is commonly used to keep track of adversary belief when using deception [9].

3.2 Nash Equilibrium and Bayesian Nash Equilibrium

Before execution, we pre-compute a strategy library based on Nash Equilibrium of all possible configurations for both the defender team and the attacker. During execution, we utilize the Nash values for deceptive planning in Sect. 3.4. Nash equilibrium is used to model the interaction between multiple agents when all players have complete information of the game and aim to maximize their own expected payoffs. The extension of Nash equilibrium for incomplete information game is Bayesian Nash equilibrium [8] where players incorporate the belief of unknown information. We will show by comparison in Sect. 4 that our k-step deception method outperforms the standard Bayesian Nash equilibrium.

By definition, a Nash equilibrium occurs when no player can do better by unilaterally changing its strategy [12]. Given the full target configuration, i.e. both locations and values, the optimal strategy for each player is to follow the Nash strategy of the complete information game. A Nash strategy can be either a pure strategy (deterministic) or a mixed strategy (stochastic). We consider mixed strategy here because in a single-stage zero-sum game with finite state-action space, a mixed strategy always exits whereas a pure strategy might not [12]. At each time step, each player selects an action based on the probability distribution of the computed mixed strategy. A mixed strategy for the defender team is denoted as $x : S \times A_d \to [0,1]$. Similarly, a mixed strategy for the attacker is $y : S \times A_a \to [0,1]$. It has been proven in [16] that in a two-player zero-sum finite-state space infinite-horizon stochastic game, there exists a unique Nash value for the complete information game and a unique mixed Nash strategy can be induced from this Nash value. This uniqueness ensures that in a given state of a known target configuration, the attacker knows the exact defender strategies. Thus, based on a specific target configuration, both the attacker and defender team can compute the strategy of one another.

Consider $V(x, y, s)$ as the defender's discounted expected reward obtained by the defenders' and attacker's strategy pair (x, y) starting at state s. The discounted expected reward for the attacker is thus $-V(x, y, s) : S \to \mathbb{R}$. We denote the Nash strategy of the defender team and attacker to be x^* and y^* correspondingly. To differentiate Nash strategies among different target configurations, we use x_l^*, y_l^* to denote the Nash strategies of l_{th} target configuration. The Nash value $V(x_l^*, y_l^*, s)$ is then the defenders' value obtained from players that follow Nash strategy x_l^* and y_l^* at state s in target configuration ϕ_l. For each $s \in S$, the Nash value satisfies:

$$V(x_l^*, y_l^*, s) \geq V(x, y_l^*, s), \forall x \in X \tag{3}$$

$$V(x_l^*, y_l^*, s) \leq V(x_l^*, y, s), \forall y \in Y \tag{4}$$

In a zero-sum game, the solution to find the Nash value at single state can be formulated as a min-max problem:

$$V(x_l^*, y_l^*, s) = \min_y \max_x \sum_{a \in A} \left[x(a_d|s) y(a_a|s) \sum_{s' \in S} T(s'|s, a) \left(R_l(s, a) + \gamma V(x^*, y^*, s') \right) \right]$$

where $a = a_d \times a_a$, $x(a_d|s)$ and $y(a_a|s)$ are the probability of the defender team taking action a_d, and attacker taking action a_a at state s correspondingly. With the above update equation, we can obtain the Nash values at each game state with different target configurations.

To enable online planning for both players in Sect. 3.3 and 3.5, we further define a Nash policy Set Π^* which contains Nash policies corresponding to different target configurations ϕ_l. More specifically, for each $\phi_l \in \Phi$, there is a $\pi_l^* = (x_l^*, y_l^*) \in \Pi^*$ where π_l^* is the Nash strategy pair of the defender team and attacker given the target configuration ϕ_{l^*}. Let $\pi_{l^*}^* = (x_{l^*}^*, y_{l^*}^*), \pi_{l^*}^* \in \Pi^*$ denote the policy corresponding to the true target configuration ϕ^*. For simplicity, let $Nash_{l^*}(s) = V(x_{l^*}^*, y_{l^*}^*, s), s = s_d \times s_a \times s_w$ denote the Nash value of a state s when all players know the true target configuration and play best strategies.

An extension of Nash equilibrium to incomplete information game is a Bayesian Nash equilibrium (BNE). BNE is not part of our approach, but will serve as a baseline of comparison with our k-step deception method in Sect. 4. Extending from Eq. (3) and (4) to the BNE formulation, the uninformed player, in our case the attacker, computes BNE based on its belief of the target configuration:

$$BNE(x_{l^*}, y_{l^*}, s, \mathbf{b}) =$$

$$\min_y \max_x \sum_{a \in A} \left[x(a_d|s)y(a_a|s) \sum_{s' \in S} T(s'|s, a) \sum_{R_l \in \mathbf{R}} \mathbf{b}_l \left(R_l(s, a) + \gamma BNE(x^*, y^*, s') \right) \right]$$

Importantly, the action spaces, the reward functions, possible target configurations, and the belief of the uninformed player are assumed to be common knowledge. That is to say, the informed player(defender team) can compute what the attacker strategy is based on attacker's belief. With that in mind, the optimal strategy of the defender team can then be simply calculated as an optimization function based on the true reward function and attacker's strategy.

3.3 Attacker Strategy and Belief Update

Since both players (defender team and the attacker) assume their adversary would play optimally and the attacker does not know whether the defender team might take deceptive action to manipulate its belief, the attacker updates its belief about each target configuration based on its observation of defenders' last actions. We assume the attacker's initial belief of the target configuration is uniformly distributed as $b_l = \frac{1}{|\Phi|}, \forall l$. After actions (a_d, a_a) have been executed at state s, the attacker updates its belief based on defenders' action a_d:

$$b_l' = prob(\phi_l|a_d, s) = prob(x_l^*|a_d, s) = \alpha \cdot b_l \cdot x_l^*(a_d|s) \quad (5)$$

$prob(\phi_l|a_d, s) = prob(x_l|a_d, s)$ because for a given target configuration ϕ_l, the defenders will play optimally, namely the defender team's Nash strategy x_l^*. α is a normalizing factor such that $||\mathbf{b}|| = 1$. We assume the attacker plays a strategy

Before execution: Precompute Nash Equilibriums to enable online planning

[Nash Policy π_0] .... [Nash Policy π_{l^*}] .... [Nash Policy $\pi_{|\Phi|-1}$]

During execution:
- Attacker chooses a Nash policy based on its belief of target configuration
- Defender team plans k steps ahead for a deterministic deceptive strategy

→ Attacker observes defenders' actions
→ Attacker updates its belief of target configuration

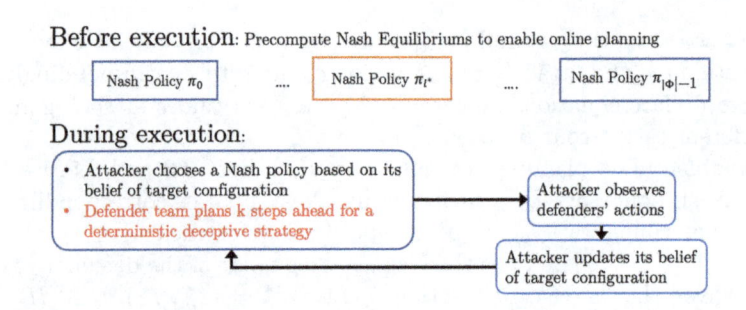

Fig. 1. Flow diagram of our method. Before execution, we pre-compute the Nash equilibrium policies. During execution, the attacker tries to play Nash strategy based on its belief. The defender team tries to deceive the attacker of the target configuration.

based on its updated belief **b** and corresponding Nash strategy of each ϕ_l (which we later referred to as the belief-based Nash Strategy):

$$y(a_a|\mathbf{b}, s) = \sum_{\pi_l \in \Pi} b_l \cdot y_l^*(a_a|s) \quad (6)$$

where $s = s_d \times s_a \times s_w$. $y_l^*(a_a|s)$ is the probability of executing a_a at state s with strategy y_l^*. $y(a_a|\mathbf{b}, s)$ is the attacker strategy during execution, which is not only based on the joint state s but also based on the attacker's belief of the target configuration. Since the defender team knows 1) the target configuration set, 2) all actions are fully observable, 3) the attacker will update its belief based on defenders' action, the defender team can perfectly reconstruct the belief of the attacker. We will utilize this feature during our k-step deception method for the defender team in the following section.

3.4 Deceptive Planning

An overview of our method is shown in Fig. 1. Before execution, we pre-compute the Nash strategy library based on all possible target configurations. During execution, at each step, the attacker takes an action based on its belief of the true target configuration. Since the defender team can perfectly reconstruct the attacker's belief, the defenders will forward simulate all actions k steps ahead and plays the deceptive action that gives the highest estimated value. After both teams take actions simultaneously, the attacker updates its belief of the true target configuration based on its observation of the defenders' actions.

Algorithm 1 shows the detailed computation of the k-step deceptive strategy. k is a user-defined constant that determines how many steps the user would like to simulate forward. This algorithm first builds a tree of depth k to explore all deterministic moves the defender can take in k time steps and all the attacker beliefs **b** induced by those actions. The attacker's actions are simulated in a stochastic way based on their belief of the target configurations and the corresponding Nash strategies. We estimate the value of the leaf states by their Nash

Algorithm 1: kStepDeception finds the best action and maximum expected value

---
Input: State s, Attacker Belief $\mathbf{b}$, Step k
Output: Best action and corresponding value
1 $maxValue \Leftarrow Nash^*_{l^*}(s)$;
2 $bestAction \Leftarrow x^*_{l^*}(s)$;
3 **if** $k > 0$ **then**
4   **for** $a_d \in A_d$ **do**
5          $\mathbf{b}' \Leftarrow updateBelief(\mathbf{b}, a_d)$;
6          $value \Leftarrow \sum_{a_a \in A_a, s' \in S} [p(a_a|\mathbf{b},s) \cdot T(s'|s, a_d, a_a) \cdot kStepDeception(s', \mathbf{b}', k-1).maxValue]$;
7   **if** $value > maxV$ **then**
8              $maxValue \Leftarrow value$;
9              $bestAction \Leftarrow a_d$
10   **end**
11   **end**
12 **end**
13 **return** $bestAction, maxValue$;

---

value in the true target configuration as if both players were to play a Nash equilibrium in complete information games from that state forward. Then the algorithm evaluates each k-depth action sequence by back-propagating the Nash values of the leaf states. The first action in the action sequence that has the highest expected value is then executed by the defender team.

To guarantee the deceptive action is not worse than the original Nash strategy, we define $Decep(s, \mathbf{b}, n)$ to be the expected discounted reward for the defender team, at state s and attacker belief $\mathbf{b}$, to take a deceptive action sequence with length n, i.e. actions within n time steps. Denote $Decep^*(s, \mathbf{b}, n)$ to be the maximum of all $Decep(s, \mathbf{b}, n)$ among all legal action sequences of length n. $Decep^*(s, \mathbf{b}, n)$ is calculated as follows:

$$Decep^*(s, \mathbf{b}, 0) = Nash_{l^*}(s) \tag{7}$$
$$Decep^*(s, \mathbf{b}, k+1) = \max_{a_d} T(s, a, s') y(a_a|\mathbf{b}, s)(R^*(s,a) + \gamma Decep^*(s', \mathbf{b}', k)) \tag{8}$$

In Theorem 1, we show that $Decep^*(s, \mathbf{b}, n)$ will always be greater or equal to the Nash value of state s.

Theorem 1. *for all* $n \in Z^+$,

$$Decep^*(s, \mathbf{b}, n) \geq Nash_{l^*}(s), \forall \mathbf{b} \tag{9}$$

The detailed proof can be accessed at https://github.com/DeniseLi123/Multi_Agent_Deception. Based on the theorem above, we optimize defender's local payoff by finding a local optimal k-step pure strategy $(a^1, a^2..., a^k)^*$ with value $Decep^*(s, \mathbf{b}, k)$. From the proof we also know there exits at least one such pure strategy $(a^1, a^2..., a^k)$.

Although the deceptive action returned from the k-step deception algorithm may result in a lower reward in the current state compared with a Nash strategy, by simulating k step forward into the future, our deceptive action gives higher reward in the longer time horizon. With this method, it is guaranteed that the action sequence returned from our k-step deception algorithm will always give higher, or equal, expected discounted reward compared with Nash strategy. However, explore the full game tree of depth k may still be computationally heavy as the state space increases. We incorporate Monte Carlo sampling methods into our approach in the coming section.

3.5 Game Tree Sampling

In Algorithm 1, we built a search tree with all valid defender and attacker moves. The tree size (number of states) grows exponentially with the action space of the agents: $|s \in Tree_k| = (|A_d| \cdot |A_a|)^k$, where $Tree_k$ is the search tree with step size k. With multiple defenders, the number of defender actions $|A_d|$ grows exponentially with the number of defenders. In order to improve the scalability of our approach, we adopt the idea of Monte Carlo Tree Search (MCTS) [3]. Inspired by MCTS, we sample m actions at each step instead of exploring all $a_d \in A_d$ (at line 4 in Algorithm 1). The size of the search tree is thus $|s \in Tree_k| = (m \cdot |A_a|)^k$. We compare two different sample criteria: (1): Sample actions based on their probability in the Nash strategy π^* : $p(a_d|s) = x^*_{l^*}(a_d|s)$. (2): Sample actions based on uniform distribution: $p(a_d|s) = \frac{1}{|A_d|}$.

Apart from the two above mentioned sample criteria for the defender team, to further improve the computation efficiency, we also sample attacker actions during our forward simulation (at line 6 in Algorithm 1). We compare two different sampling criteria: (1): Sample actions with maximum probability in Nash strategy: $a_a = \arg\max y^*_{l^*}(a_a|s)$. This decreases the tree size from $(m \cdot |A_a|)^k$ to m^k. (2): Sample the top H percent actions based on the probability of actions from largest to the lowest: $\sum y^*_{l^*}(a_a|s) = H$.

4 Results

We extensively ran simulations with two defenders based on the configurations in Fig. 2. In our experiments, we assume there is only one real target and all players are aware of this information. We set $g^j_{l,w} = 5$ when j^{th} target is the real target, 0 otherwise. The catch reward $g_{l,c} = 5$. Unless explicitly stated, $k = 2$ for all deceptive planning. We randomized agents' initial positions and compared the defender payoff and computation cost among different methods in Sect. 3.5.

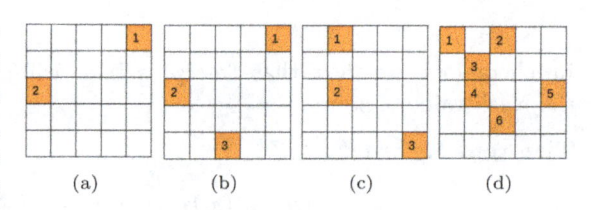

Fig. 2. Experiment configurations.

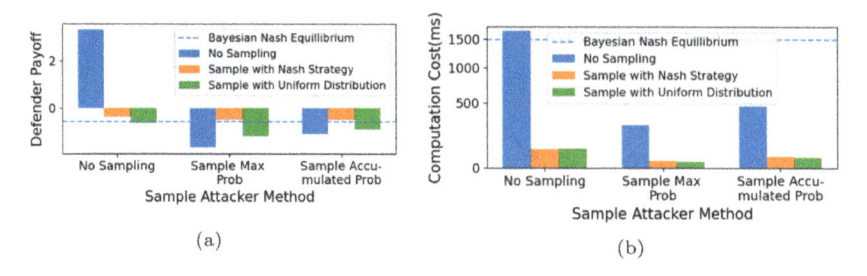

Fig. 3. Results of experiment set 1: Different action sample methods for configuration Fig. 2a: (a) defender payoff (b) computation cost

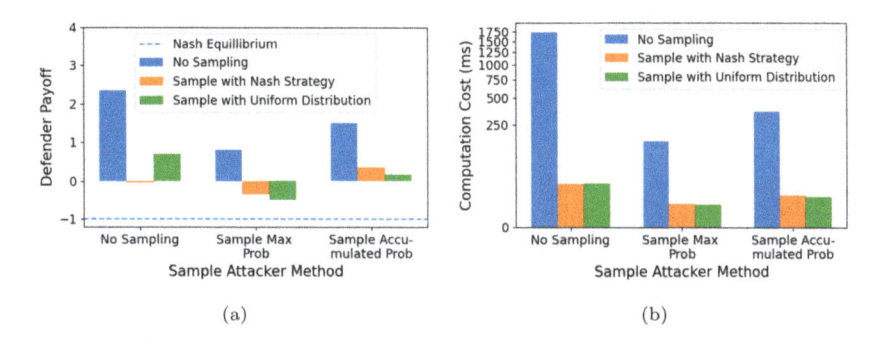

Fig. 4. Experimental results with different action sample methods for configuration Fig. 2b and c: (a) defender payoff (b) computation cost

We ran the experiments on an Intel i7 Quad-Core machine with 16 GB memory. The code is written in python and we use scipy.optimize.linprog as the linear programming solver for solving the (Bayesian) Nash equilibrium. We ran six sets of experiments, detailed settings of each set of experiments are as follows: (1) Defender team: three different deceptive strategies or one-step BNE strategy; Attacker: belief-based Nash Strategy; Configuration: Fig. 2a; (2) Defender team: three different deceptive strategy; Attacker: one-step BNE strategy or belief-based Nash Strategy; Configuration: Fig. 2a (3) Defender team: deceptive strategy with three different sampling method; Attacker: belief-based Nash Strategy; Configuration: Fig. 2b and c (4) Defender team: deceptive strategy with two different k values; Attacker: belief-based Nash Strategy; Configuration: Fig. 2b and c (5) Defender team: deceptive strategy with three different sample methods and sample size; Attacker: belief-based Nash Strategy; Configuration: Fig. 2b and c (6) Different size of possible target configuration set; Defender team: deceptive strategy without sampling; Attacker: belief-based Nash Strategy; Configuration: Fig. 2d.

First, we compare the performance and computation cost between a traditional one-step BNE approach and our $k = 2$ deceptive strategy. For the sake of fairness, we also use the pre-computed Nash values as a state estimation function for the calculation of BNE. Since the computation cost of BNE grows exponen-

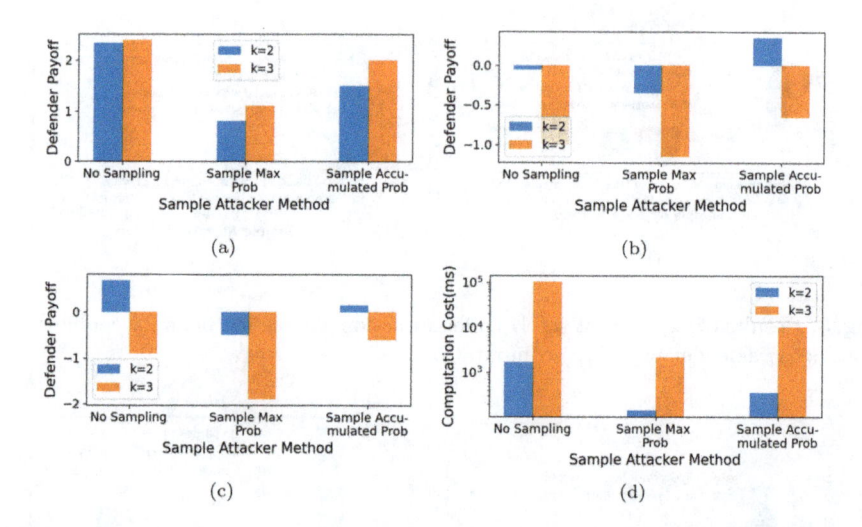

Fig. 5. Results of experiment set 4: (a)–(c) Defender payoff with different attacker action sample methods when defender sample method is (a) no sampling (b) sample based on Nash strategy (c) sample based on uniform distribution; (d) computation cost with different attacker action sample method when all defender actions are considered (no sampling)

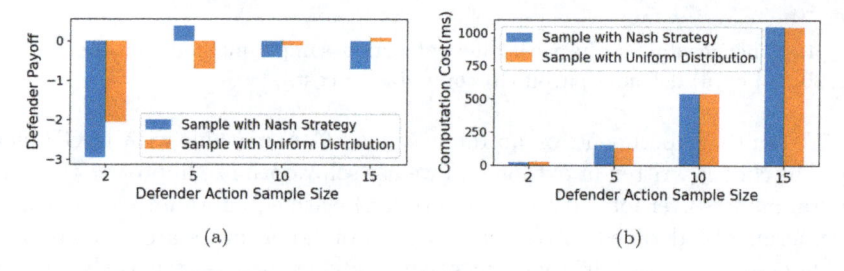

Fig. 6. Results of experiment set 5: different defender action sample method and different sample size for configuration Fig. 2c: (a) defender payoff (b) computation cost

tially with the increase of actions and target configurations, it is not solvable when there are more than two target configurations. As a result, experiment set 1 is run on configuration Fig. 2a, resulting in two possible target configurations: (1) $g_{l,w}^1 = 5, g_{l,w}^2 = 0$, (2) $g_{l,w}^1 = 0, g_{l,w}^2 = 5$. As shown in Fig. 3b, the computation cost of our 2-step full-planning with deception is similar to that of a single-step BNE. However, in Fig. 3a, the deceptive strategy has a significant performance increase with defender payoff of 3.3 while BNE only has a payoff of -0.6. The sampling methods have similar defender payoff compare to BNE.

The second experiment set was run based on configuration Fig. 2a and the same two possible target configuration as the first experiment set. When we plan without sampling, the average defender payoff decreases slightly from 3.45 to 3.3

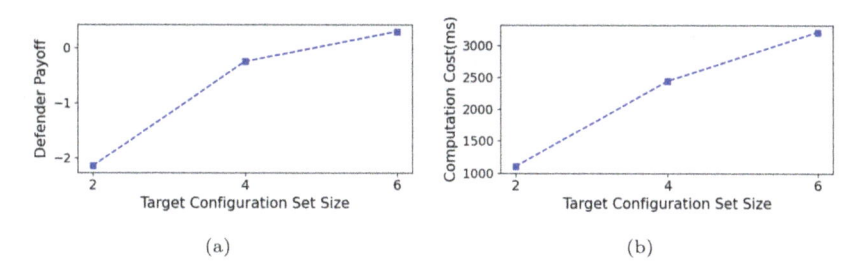

Fig. 7. Results of experiment set 6: experimental results with different target configuration set size (a) defender payoff (b) computation cost

when the attacker change strategy from belief-based Nash Strategy to BNE. This implies that our no-sampling approach is robust against a smarter adversary (an attacker who follows BNE) even we do not simulate attacker to follow BNE during planning (high computation cost). However, if we plan with sampling, the average defender payoff decreases dramatically from 1.212 to −0.868. The mismatch between the attacker strategy we assume and real attacker strategy makes the sampling methods weaker.

In the third experiment set, we compare the performance of different sampling methods. We can see that all deceptive methods have better performance than the Nash Strategy. Full exploration of defender actions show the best performance but also the highest computation cost (in Fig. 4a). Figure 4a shows that simulating single action with maximum probability has the worst performance. Simulating the actions that take 75% of the total probability has a similar performance as simulating all attacker actions. Regarding computation cost, Fig. 4b shows that the computation cost of sample the top $H = 75\%$ percent actions is approximately 20% to 40% of the cost of full simulation.

In fourth experiment set, we investigate the influence of the value of k from Fig. 5. Figure 5a shows slight performance increase with the increase of k when we explores all legal defender moves. Figure 5b and c shows that sampling methods has significant performance decrease with k value. It is because when the search tree grows with k, the sampling methods lead to less accurate evaluation of the states. Although we have some performance increase in Fig. 5a, d shows that the computation cost of $k = 3$ is approximately 100 times more than that of $k = 2$ but the performance increase is only around 2%.

In fifth experiment, we investigate the influence of different defender action sample methods and different sample size. Figure 6a shows that for Nash strategy based sampling, sample size 5 has the best performance. For uniform sampling, the performance increase with sample size. However, the computation cost also increases rapidly. Overall, Nash strategy based sampling with sample size 5 has the best performance with a low computation cost.

Last, we tested the scalability of our approach regarding the size of target configuration space. The real target in each simulation is always target 2. The possible real target indexes in each simulation are: when $|\Phi| = 2$: [2,5]; when $|\Phi| = 4$: [2,4,5,6]; when $|\Phi| = 6$: [1–6]. Figure 7b shows the computation cost increase slowly with the size of possible target configurations. Figure 7a shows defender team can gain more payoff when there are more possible configurations. The code is available: https://github.com/DeniseLi123/game_theoretical_deception.

5 Conclusion and Future Research

In this paper, we studied the attack-defense game in which the attacker has incomplete information of the game configuration. We have proposed a k-step deception algorithm in which the defender team can generate a deceptive strategy based on pre-computed Nash strategy library and forward simulation of game tree. Comparing our algorithm to the traditional Bayesian Nash Equilibrium, our algorithm can handle larger action space and larger target configuration space. Results from experiments show that our method is more computationally efficient and receives larger payoff in the game.

From our experiments, we observe that as k increases, the computation cost increases exponentially, but may not necessarily leads to higher reward gain. Choosing the value of k remains an open problem and it is worth investigating if there is a optimal k value given a specific game configuration and user preference of computation limits. Additionally, choosing the proper methods of sampling in the game tree influences the policy performances. Further research can be done with exploring more methods of sampling.

Acknowledgments. This work has been supported in part by AFOSR Award FA9550-18-1-0097 and AFOSR/AFRL award FA9550-18-1-0251.

References

1. Brown, N., Sandholm, T.: Superhuman AI for multiplayer poker. Science **365**(6456), 885–890 (2019)
2. Carroll, T.E., Grosu, D.: A game theoretic investigation of deception in network security. Secur. Commun. Netw. **4**(10), 1162–1172 (2011)
3. Chaslot, G., Bakkes, S., Szita, I., Spronck, P.: Monte-Carlo tree search: a new framework for game AI. In: AIIDE (2008)
4. Deng, Z., Kong, Z.: Multi-agent cooperative pursuit-defense strategy against one single attacker. IEEE Robot. Autom. Lett. **5**(4), 5772–5778 (2020)
5. Durkota, K., Lisý, V., Bošanský, B., Kiekintveld, C.: Optimal network security hardening using attack graph games. In: Twenty-Fourth International Joint Conference on Artificial Intelligence (2015)
6. Gilpin, A., Sandholm, T.: Solving two-person zero-sum repeated games of incomplete information. In: Proceedings of the 7th International Joint Conference on Autonomous Agents and Multiagent Systems, vol. 2, pp. 903–910 (2008)
7. Gmytrasiewicz, P.J., Doshi, P.: A framework for sequential planning in multi-agent settings. J. Artif. Intell. Res. **24**, 49–79 (2005)
8. Harsanyi, J.C.: Games with incomplete information played by "Bayesian" players, i–iii part i. the basic model. Manag. Sci. **14**(3), 159–182 (1967)

9. Horák, K., Zhu, Q., Bošanskỳ, B.: Manipulating adversary's belief: a dynamic game approach to deception by design for proactive network security. In: Rass, S., An, B., Kiekintveld, C., Fang, F., Schauer, S. (eds.) GameSec 2017. LNCS, vol. 10575, pp. 273–294. Springer, Cham (2017). https://doi.org/10.1007/978-3-319-68711-7_15
10. Li, L., Shamma, J.: LP formulation of asymmetric zero-sum stochastic games. In: 53rd IEEE Conference on Decision and Control, pp. 1930–1935. IEEE (2014)
11. Lisỳ, V., Zivan, R., Sycara, K., Pěchouček, M.: Deception in networks of mobile sensing agents. In: Proceedings of the 9th International Conference on Autonomous Agents and Multiagent Systems, vol. 1, pp. 1031–1038 (2010)
12. Nash, J.: Non-cooperative games. Ann. Math. **54**, 286–295 (1951)
13. Nguyen, T.H., Wang, Y., Sinha, A., Wellman, M.P.: Deception in finitely repeated security games. In: Proceedings of the AAAI Conference on Artificial Intelligence, vol. 33, pp. 2133–2140 (2019)
14. Píbil, R., Lisỳ, V., Kiekintveld, C., Bošanský, B., Pěchouček, M.: Game theoretic model of strategic honeypot selection in computer networks. In: Grossklags, J., Walrand, J. (eds.) GameSec 2012. LNCS, vol. 7638, pp. 201–220. Springer, Heidelberg (2012). https://doi.org/10.1007/978-3-642-34266-0_12
15. Sengupta, S., Kambhampati, S.: Multi-agent reinforcement learning in Bayesian Stackelberg Markov games for adaptive moving target defense. arXiv preprint arXiv:2007.10457 (2020)
16. Shapley, L.S.: Stochastic games. Proc. Natl. Acad. Sci. **39**(10), 1095–1100 (1953)
17. Sincák, D.: Multi-robot control system for pursuit-evasion problem. J. Electr. Eng. **60**(3), 143–148 (2009)
18. Torreño, A., Onaindia, E., Sapena, Ó.: A flexible coupling approach to multi-agent planning under incomplete information. Knowl. Inf. Syst. **38**(1), 141–178 (2014)
19. Vidal, R., Rashid, S., Sharp, C., Shakernia, O., Kim, J., Sastry, S.: Pursuit-evasion games with unmanned ground and aerial vehicles. In: Proceedings 2001 ICRA. IEEE International Conference on Robotics and Automation (Cat. No. 01CH37164), vol. 3, pp. 2948–2955. IEEE (2001)
20. Yi, S., Nam, C., Sycara, K.: Indoor pursuit-evasion with hybrid hierarchical partially observable Markov decision processes for multi-robot systems. In: Correll, N., Schwager, M., Otte, M. (eds.) Distributed Autonomous Robotic Systems. SPAR, vol. 9, pp. 251–264. Springer, Cham (2019). https://doi.org/10.1007/978-3-030-05816-6_18
21. Zhuang, J., Bier, V.M., Alagoz, O.: Modeling secrecy and deception in a multiple-period attacker-defender signaling game. Eur. J. Oper. Res. **203**(2), 409–418 (2010)

Tractable Planning for Coordinated Story Capture: Sequential Stochastic Decoupling

Diptanil Chaudhuri[1]([✉]), Hazhar Rahmani[2], Dylan A. Shell[1], and Jason M. O'Kane[2]

[1] Department of Computer Science and Engineering, Texas A&M University, College Station, TX, USA
diptanil@tamu.edu
[2] Department of Computer Science and Engineering, University of South Carolina, Columbia, SC, USA

Abstract. We consider the problem of deploying robots to observe the evolution of a stochastic process in order to output a sequence of observations that fit some given specification. This problem often arises in contexts such as event reporting, situation depiction, and automated narrative generation. The paper extends our prior work by formulating and examining the multi-robot case: a team of robots move about, each recording what they observe, and, if they manage to capture some event, communicating that fact with the group. In the end, all events from all the robots are collated to provide a cumulative output. A plan is used to decide what each robot will attempt to capture next, based on the state of the world and the events that have been captured (collectively) so far. This paper focuses on the question of how to compute effective multi-robot plans. A monolithic treatment, involving the optimal selection of joint choices, i.e., choosing the next elements to attempt to capture by all robots, is formulated where costs are minimized in an expected sense. Since such plans are prohibitive to compute, variants based on an approximation scheme based on solving a sequence of individual planning problems are then introduced. This scheme sacrifices some solution quality but requires far less computational expense; we show this permits one to scale to greater numbers of robots.

Keywords: Robot videography · Formal methods · Heuristics for cooperative planning

1 Motivation

Imagine a nature documentary. Muffled, but in his signature rasping hush, David Attenborough intones: *"We see now the baby gazelle, utterly unaware of danger*

This material is based upon work supported by the National Science Foundation under Grants 1849249 & 1849291.

© The Author(s), under exclusive license to Springer Nature Switzerland AG 2022
F. Matsuno et al. (Eds.): DARS 2021, SPAR 22, pp. 256–268, 2022.
https://doi.org/10.1007/978-3-030-92790-5_20

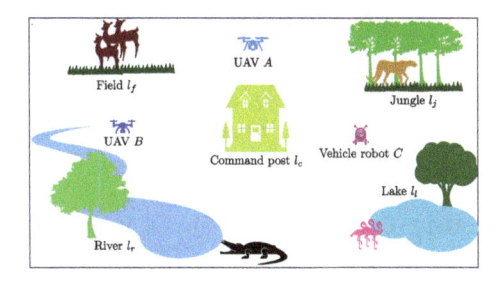

Fig. 1. A overall view of the wildlife reserve, with 5 main regions, a grass field l_f, a jungle l_j, a command post l_c, a riverside l_r, and a lakeside l_l. Notable fauna includes a cheetah, a crocodile, a herd of gazelle, and a flamboyance of flamingos; the latter two, being gregarious, remain as a group.

lurking close, as she edges toward the water's edge. Nearby, Mother gazelle is distracted, only for a moment, but..." and the wild drama ensues—tooth, claw, and all. Later, as the credits roll by, it turns out that the rare footage making up this documentary was captured not by expert human camera operators, but by a team of autonomous videographer robots. These robots, aided by traditional tags for tracking animals, have only a coarse sense of the locations of certain animals and are only able to make imperfect predictions for what activities the creatures will engage in. But they are also given a description of the sorts of events that are worth capturing, events to help describe Nature's unfolding story. Multiple robots ought to be engaged to ensure good coverage of the district, especially as there may be events of interest occurring simultaneously at different locations, such as when the flamingos take wing all at once, while a lone cheetah breaks cover from a thicket of trees elsewhere (See Fig. 1). The robots plan their movements and capture events strategically so that, ultimately, the captured events form a depiction that is an engaging record of activity within the wildlife reserve. For example, one robot captures a majestic predator silhouetted against the moon, other robots capture (many) scenes of frolicking young. Comparatively little footage represents animals lolling about during the heat of the day.

This paper formalizes settings like the preceding scenario. Section 3 formulates the problem of coordinated story capture via: (*i*) a stochastic process to model worlds that generate random events, (*ii*) a means for expressing constraints on the robots' capabilities in terms of what they may capture in succession, and (*iii*) an automaton expressing sequences of observations (or recordings) to be captured. Next, in Sect. 4, we turn to finding plans for such problems: presenting both monolithic and decoupled (or sequentialized) approaches. Section 5, thereafter, returns to the nature documentary scenario to provide performance comparisons via case studies.

2 Related Work

Most closely related is our own prior work [1], which formalizes the problem of using autonomous robots to capture videos in unpredictable environments, an idea advanced in [2]. There, we studied the problem of computing, for a single robot, a policy that minimizes the expected number of steps to record an event sequence satisfying a given specification. This paper is a multi-agent extension of that the same problem, but with the objective being replaced by minimizing cost (rather than steps) of recording a desired event sequence. Here we also introduce a new structure, termed the *valid-action automaton*, which we use not only for specifying action-related costs but also for imposing spatial, temporal, or other constraints on the robots' choices of actions.

Also, some related problems bear similarity to robot video capture. Among them is the work of Yu and LaValle [3], who studied the *story validation problem*, the aim of which is to validate whether an event sequence captured by a set of sensors in the environment is consistent with a given story or not—this is, roughly, the inverse of the problem we consider here. The other two are the *video summarization problem* [4–9] and the *vacation snapshot problem* [10] the purpose of which are, respectively, to make a summary of a given video and to make a diverse selection of samples observed by a mobile robot. The essence in the video summarization problem is to post-processes a collection of images, while the idea in our problem is to decide which images the robot should attempt to capture without knowing which images will actually be realized by the world. For research about summarization in other contexts, see [11,12] for generating commentary, and see [13,14] for producing narrative text.

3 Definitions and Problem Statement

We start defining basic model elements, then give our formal problem statement.

3.1 Worlds and Narratives: Event Model and Story Automaton

The atomic items that the robots capture are *events*, elements from a set E, which occur at specific times and places. The set of all event sequences (finite words) over E is denoted E^*. For integer m, the set of all event sequences over E with length at most m is denoted $E^{\leq m}$. We will mostly write sequences $e_1 e_2 \ldots e_m$ of events, but occasionally it helps to treat them as tuples too, like $(e_1, e_2, \ldots, e_m)$. For each $1 \leq i \leq m$, we will write $e_i \in (e_1, e_2, \ldots, e_m)$, abusing notation to treat the tuple as a set as well. Using the tuple form, $E^* = \bigcup_{j=0}^{\infty} E^j$.

Events are assumed to be generated by a stateful stochastic process, unaffected by the actions of the robot.

Definition 1 (Event Model [1]). *An event model $\mathcal{M} = (W, P, w_0, E, g)$ has (1) W, which is a nonempty finite set, is the state space of the model; (2) $P : W \times W \to [0,1]$ is the transition probability function of the model, such that for each state $w \in W$, $\sum_{w' \in W} P(w, w') = 1$; (3) $w_0 \in W$ is the initial state; (4) E is the set of all possible events; (5) $g : W \times E \to [0,1]$ is a labeling function such that for each state w and event e, $g(w, e)$ is the probability that event e happens at state w. We assume that $g(w_0, e) = 0$ for any event e, meaning that no events happens at state w.*

The model assumes that the events in each state of the event model are both mutually and temporally independent. That is, the probability of occurrence of an event e in state w at time t does not depend on the probability of occurrence of any event e' at time t, nor at any time before that.

An execution of the system starts from w_0, and then at each time step t, the system transitions from state w_t to a state w_{t+1}, which is chosen randomly based on $P(w_t, \cdot)$. Accordingly, the system's execution goes through a path $w_0 w_1 \cdots$. When the system enters state w_t, each event e occurs with probability $g(w_t, e)$. At a time step, it is possible that several events occur simultaneously.

In this paper, we consider a problem in which multiple robots are working cooperatively to record a sequence of events. As the system evolves along $w_0 w_1 w_2 \cdots$, the robots attempt to record some of the events that occur in the world to form a story $\xi \in E^*$. To specify the story, we use deterministic finite automaton (DFA) [15], called the **story automaton** $\mathcal{D} = (Q, E, \delta, q_0, F)$, in which Q is the state space, the set of events E is the alphabet, $\delta : Q \times E \to Q$ is the transition function, q_0 is the initial state, and $F \subseteq Q$ is the set of all accepting (final) states.

The language of the story automaton is denoted by $\mathcal{L}(\mathcal{D})$. Given a state $q \in Q$, we say that state q' is *reachable* from q by event sequence $e_1 e_2 \cdots e_m$ if there is a sequence of states $q_1, q_2, \ldots, q_{m+1}$ such that $q_1 = q$, $q_{m+1} = q'$, and $q_{i+1} = \delta(q_i, e_i)$ for each $1 \leq i \leq m$. We assume that each state is reachable from itself (by ϵ, the empty string).

3.2 Robot Model

The current state of the event model is assumed to be observable to all the robots, i.e., at each time t, the robots know the current state of the event model (or the world) w_t; however, they do not know what the next state, w_{t+1}, will be. (In [1], our earlier work, we considered the case where the current state of the event model was not observable: for simplicity, especially in exploring the complexities arising in the multi-robot case, the present paper utilizes MDPs throughout, rather than the less tractable POMDP model.)

Generally, robots use actions to alter their relationship with the world—here, the robots also attempt to record events. To cooperatively capture a sequence of events each of the n robots chooses an action to execute from A, the set of all possible actions. Each action is associated with the event they aim to capture via the *recording function*, $r : A \to E \cup \{\epsilon\}$. Since some actions may not involve any recording, the symbol ϵ is included, indicating that no event will ever be captured by the associated action. At every time step, each robot executes an action from A, if that action aims to capture an event and that event occurs during the execution of the action, the robot will succeed in capturing that event. We assume every event can be recorded by some action, i.e., for every $e \in E$, there is some action $a_e \in A$ such that $r(a_e) = e$. Further, the set of actions includes a *no-action* choice, $\bot \in A$, that does nothing and records no event, $r(\bot) = \epsilon$. Occasionally we will apply $r(\cdot)$ to a tuple in a point-wise fashion.

Owing to constraints, present either in the world or in the way the robots interact with the world, not all actions can be executed at all times. Hence, an action $a \in A$ with $r(a) = \epsilon$ may still be useful because, though it won't capture an event itself, it may alter what can be captured subsequently. Think, for instance, of a robot using the time step's duration to shift location, or to deploy a stalking horse. The following structure expresses such constraints and also associates costs to each action.

Definition 2 (Valid-action Automaton (VA)). *For robot $i \in \{1, \ldots, n\}$, we define its* valid-action automaton *as a 5-tuple, $\mathbb{V}^{(i)} = (V^{(i)}, v_0^{(i)}, A, \tau^{(i)}, c^{(i)})$, (1) $V^{(i)}$ is the set of vertices; (2) $v_0^{(i)} \in V^{(i)}$ is the initial vertex; (3) A, its alphabet, a set of all possible robot actions; (4) $\tau^{(i)} : V^{(i)} \times A \hookrightarrow V^{(i)}$, which could be partial, is the transition function; (5) $c^{(i)} : V^{(i)} \times A \to \mathbb{R}_{>0} \cup \{+\infty\}$ is the cost function, such that for each $(v, a) \in V^{(i)} \times A$, $c^{(i)}(v, a)$ is the cost of taking action a at vertex v. We assume that $c^{(i)}(v, a) = +\infty$ for any (v, a) such that $\tau^{(i)}(v, a)$ is not defined.*

Each robot $i \in \{1, 2, \ldots, n\}$ keeps track of the current state of its own valid-action automaton, denoted $v_t^{(i)}$. Actions are performed as follows. Robot i makes a choice, from among those actions a for which $\tau(v_t^{(i)}, a)$ is defined, to enact at time $t + 1$. We denote the action $a_t^{(i)}$, because it is chosen at time t. The world evolves from w_t to w_{t+1}, and robot i pays cost $c^{(i)}(v_t^{(i)}, a_t^{(i)})$ executing $a_t^{(i)}$ to change its circumstances, with aspects relevant for subsequent actions being represented in $v_{t+1}^{(i)}$. Finally, if $r(a_t^{(i)}) \neq \epsilon$, the robot attempts to record event $r(a_t^{(i)}) \in E$, which succeeds with probability $g(w_{t+1}, r(a_t^{(i)}))$.

For each time step $t \geq 1$, we define $X_t \subseteq E^{\leq n}$ to be the set of all event sequences, in any order, formed from all the events that were captured by the robots at time step t. For example, if at time step t_0, e_1 was captured by robot 1, e_2 was captured by robot 2, and ϵ (nothing) was captured by robot 3, then $X_{t_0} = \{e_1 e_2, e_2 e_1\}$. We also let $\mathbf{X}_t = \prod_{i=1}^{t} X_i$ be the set of all event sequences obtained by concatenating the event sequences made for the time steps $1, \ldots, t$. As an example, if $X_1 = \{e_3, e_4 e_2\}$ and $X_2 = \{e_1 e_2, e_2 e_1\}$, then $\mathbf{X}_2 = \{e_3 e_1 e_2, e_3 e_2 e_1, e_4 e_2 e_1 e_2, e_4 e_2 e_2 e_1\}$. The robots check at each time t, if there exists an event sequence $x \in \mathbf{X}_t$ such that $x \in \mathcal{L}(\mathcal{D})$ or not. If yes, then it means that the robots have successfully collected events to make a desired story, namely x, and they terminate. Note that $\mathbf{X}_t$ is the set of all event sequences the robots can make by concatenating all the events they have captured until time step t with the constraint that for times t_1 and t_2 for which $t_1 < t_2$, no event captured at t_2 precedes an event captured at t_1.

3.3 Policies and Problem Statement

The robots' choice of actions is governed by a policy $\pi(\cdot, \cdot, \cdot)$, that, at time t, based on the current state of the event model, states in the story automaton, and current states in the robots' valid-action automata, produces a n-tuple of actions, termed a joint action, telling each robot what action to execute.

Given valid-action automata $\mathbb{V}^{(i)} = (V^{(i)}, v_0^{(i)}, A, \tau^{(i)}, c^{(i)})$, $i \in \{1, \ldots, n\}$, we will write $\mathbf{V} = V^{(1)} \times \cdots \times V^{(n)}$. Similarly, for joint actions, we have $\mathbf{A} = A \times \cdots \times A = A^n$. (To lighten the notation, we assume that A is identical for every robot; no generality is lost because, should robot i be unable to execute some $a \in A$, then a simply does not appear in $\mathbb{V}^{(i)}$.)

Given $\mathbb{V}^{(i)}$ for $i \in \{1, \ldots, n\}$, we define $\mathbf{c} : \mathbf{V} \times \mathbf{A} \to \mathbb{R}_{>0}$, the *aggregate cost function*, by $\mathbf{c}((v^{(1)}, \ldots, v^{(n)}), (a_1, \ldots, a_n)) = \sum_{i=1}^{n} c^{(i)}(v^{(i)}, a_i)$. Then the total cost incurred up until time t is $J_t = \sum_{i=0}^{t-1} \mathbf{c}((v_t^{(1)}, \ldots, v_t^{(n)}), (a_t^{(1)}, \ldots, a_t^{(n)}))$. Let T be the first time such that $\mathbf{X}_T \cap \mathcal{L}(\mathcal{D}) \neq \varnothing$, then we say the story has been captured at time T and we write the cost of capturing the story as $J = J_T$.

We now define the problem we study in this paper.

Problem: Multi-Robot Recording Cost Minimization (MRRCM)

Input: An event set E; an event model $\mathcal{M} = (W, P, w_0, E, g)$; the story automaton $\mathcal{D} = (Q, E, \delta, q_0, F)$; the number of robots, n; a set of n valid-action automata $\mathbb{V}^{(i)} = (V^{(i)}, v_0^{(i)}, A, \tau^{(i)}, c^{(i)}), \forall i \in \{1, \cdots n\}$.

Output: A policy, $\pi^* : S \times 2^Q \times \mathbf{V} \to \mathbf{A}$, that minimize the expected cost, J, to capture a story in $\mathcal{L}(\mathcal{D})$.

---

4 Solving MRRCM

In this section we provide two algorithms to solve MRRCM.

4.1 Preliminary Definitions

In the initial step, the algorithm makes use of the story automaton and n, the number of robots, to construct a *footage automaton* as follows:

Definition 3 (Footage Automaton). *Let $\mathcal{D} = (Q, E, \delta, q_0, F)$ be the story automaton and n be the number of robots. We construct the footage automaton as a nondeterministic finite automaton (NFA) $\mathbf{N} = (Q, \mathbf{E}, \delta_\mathbf{N}, q_0, F)$, where (1) Q is the state space; (2) $\mathbf{E} = \{(e^{(1)}, \ldots, e^{(n)}) \mid e^{(i)} \in E \cup \{\epsilon\}, \forall i \in \{1, \ldots, n\}\}$ is its alphabet; (3) $\delta_\mathbf{N} : Q \times \mathbf{E} \hookrightarrow 2^Q$, is the transition function, such that for $q \in Q$ and $(e^{(1)}, \ldots, e^{(n)}) \in \mathbf{E}$, $\delta_\mathbf{N}(q, (e^{(1)}, \ldots, e^{(n)})) = \{q_i \mid q_i \in Q, \text{ where, in } \mathcal{D}, q_i \text{ is reachable from } q \text{ using some permutation of the tuple } (e^{(1)}, \ldots, e^{(n)})\}$; (4) q_0 is the initial state; (5) F is the set of final states.*

Each transition starting from a state in the footage automaton corresponds to at most n consecutive transitions starting from that state in the story automaton. The idea is that the footage automaton tracks the story automaton states which can be reached using the events captured by all of the robots. The next step converts the footage automaton $\mathbf{N}$ into a *deterministic footage automaton* $\mathbf{D} = (\mathbf{Q}, \mathbf{E}, \delta_\mathbf{D}, q_0, \mathbf{F})$, which is, in fact, a deterministic finite automaton, using the well-known technique of NFA to DFA conversion [15]. The number of edges in the constructed $\mathbf{N}$, of the output $\mathbf{D}$, and the work needed in this conversion step, can be reduced by fixing a canonical representative, equivalent up to permutation, for the n-tuples comprising $\mathbf{E}$. Sorting the tuples works.

We define a function $h : \mathbf{A} \times 2^E \to (E \cup \{\epsilon\})^n$ such that for each joint action $\mathbf{a} \in \mathbf{A}$, and a set of events $B \subseteq E$, $h(\mathbf{a}, B) = (d^{(1)}, \ldots, d^{(n)})$ in which for each $j \in \{1, \ldots, n\}$, $d^{(j)} = r(a^{(j)})$ if $r(a^{(j)}) \in B$, otherwise $d^{(j)} = \epsilon$.[3] Intuitively, given that only the events in B happen, the function h outputs an n-tuple of events which are captured by the action $\mathbf{a}$. We then define $o : \mathbf{A} \to 2^{(E \cup \{\epsilon\})^n}$ such that for each $\mathbf{a} \in \mathbf{A}$, $o(\mathbf{a}) = \bigcup_{B \subseteq E} \{h(\mathbf{a}, B)\}$. This function produces any tuple of events that could be captured by a joint action.

Two additional functions will be needed. Let $\varrho : \mathbf{A} \times W \times (E \cup \{\epsilon\})^n \hookrightarrow \mathbb{R}_{\geq 0}$ be a function such that for each $\mathbf{a} \in \mathbf{A}$, $w \in W$, and $\mathbf{b} \in o(\mathbf{a})$,

$$\varrho(\mathbf{a}, w, \mathbf{b}) = \left(\prod_{e_0 \in \mathbf{b}} g(w, e_0) \right) \cdot \left(\prod_{\substack{e_1 \in r(\mathbf{a}) \\ e_1 \notin \mathbf{b}}} (1 - g(w, e_1)) \right).$$

The interpretation is: assuming that at time t the robots execute joint action $\mathbf{a}$ and, at $t + 1$, the event model transitions to w, then $\varrho(w, \mathbf{a}, \mathbf{b})$ gives the probability that $\mathbf{b}$ is realized by w. Or, in other words, among those events attempted to be captured by $\mathbf{a}$, only those within $\mathbf{b}$ happened in state w.

Next, using $\mathbb{1}_A(\cdot)$ for set A's indicator function, let $\lambda : \mathbf{Q} \times \mathbf{A} \times W \times \mathbf{Q}$ be

$$\lambda(\mathbf{q}, \mathbf{a}, w, \mathbf{q}') = \sum_{\mathbf{b} \in o(\mathbf{a})} \mathbb{1}_{\{\mathbf{q}'\}}(\delta_\mathbf{D}(\mathbf{q}, \mathbf{b})) \cdot \varrho(\mathbf{a}, w, \mathbf{b}).$$

At time t, if the footage automaton is in state $\mathbf{q}$ and the robots execute $\mathbf{a}$, and thereupon the event model transitions next to state w, then $\lambda(\mathbf{q}, \mathbf{a}, w, \mathbf{q}')$ is the probability that the footage automaton transitions to $\mathbf{q}'$ at $t + 1$.

With these definitions, we now present our algorithms for solving MRRCM.

4.2 Full Joint Plan

The first step of the algorithm makes from valid-action automata of the robots, an automaton defined as follows:

Definition 4 (Joint Action Automaton (JA)). *Given valid-action automata $\mathbb{V}^{(i)} = (V^{(i)}, v_0^{(i)}, A, \tau^{(i)}, c^{(i)})$ for $i \in \{1, 2, \ldots n\}$, and the aggregate cost function $\mathbf{c} : \mathbf{V} \times \mathbf{A} \to \mathbb{R}_{>0}$, their joint action automaton is $\mathcal{V} = (\mathbf{V}, \mathbf{v_0}, \mathbf{A}, T, \mathbf{c})$, where: (1) $\mathbf{V} = V^{(1)} \times V^{(2)} \times \cdots \times V^{(n)}$ is the set of all the vertices; (2) $\mathbf{v_0} = (v_0^{(1)}, v_0^{(2)}, \ldots, v_0^{(n)})$ is the initial vertex; (3) $\mathbf{A} = A^n$ is the set of all actions; (4) $T : \mathbf{V} \times \mathbf{A} \hookrightarrow \mathbf{V}$ is the valid transitions function, such that for each $(v^{(1)}, v^{(2)}, \ldots, v^{(n)}), (w^{(1)}, w^{(2)}, \ldots, w^{(n)}) \in \mathbf{V}$ and $(a^{(1)}, a^{(2)}, \ldots, a^{(n)}) \in \mathbf{A}$,*

---

[3] Being consistent with that above, we use $a^{(j)}$ to denote the jth element of $\mathbf{a}$.

$T((v^{(1)}, v^{(2)}, \ldots, v^{(n)}), (a^{(1)}, a^{(2)}, \ldots, a^{(n)})) = (w^{(1)}, w^{(2)}, \ldots, w^{(n)})$ *if for each* $i \in \{1, \ldots, n\}$, $\tau^{(i)}(v^{(i)}, a^{(i)}) = w^{(i)}$; *(5)* $\mathbf{c} : \mathbf{V} \times \mathbf{A} \to \mathbb{R}_{>0}$ *is the aggregate cost function.*

Now, to solve the MRRCM problem jointly for all the robots, we search over all a joint action space. To do so we construct an MDP, called the *joint MDP*.

Definition 5 (Joint MDP). *For event set E and model $\mathcal{M} = (W, P, w_0, E, g)$, joint action automaton $\mathcal{V} = (\mathbf{V}, \mathbf{v_0}, \mathbf{A}, T, \mathbf{c})$, and the deterministic footage automaton $\mathbf{D} = (\mathbf{Q}, \mathbf{E}, \delta_{\mathbf{D}}, q_0, \mathbf{F})$, construct $\mathbb{M}_{\mathcal{M}, \mathbf{D}, \mathcal{V}} = (\mathbf{S}, \mathbf{s_0}, \mathbb{A}, \mathbf{P}, \mathbf{J}, \mathbf{G})$, where (1) $\mathbf{S} \subseteq W \times \mathbf{Q} \times \mathbf{V}$, is the set of states; (2) $\mathbf{s_0} = (w_0, q_0, \mathbf{v_0}) \in \mathbf{S}$ is the initial state; (3) $\mathbb{A} : \mathbf{S} \to 2^{\mathbf{A}}$ is the action function, such that for each $\mathbf{s} = (w, \mathbf{q}, \mathbf{v}) \in \mathbf{S}$, $\mathbb{A}(\mathbf{s}) = \{\mathbf{a} \in \mathbf{A} \mid \delta_{\mathbf{D}}(\mathbf{q}, r(\mathbf{a}))$ and $T(\mathbf{v}, \mathbf{a})$ are defined$\}$; (4) $\mathbf{P} : \mathbf{S} \times \mathbf{A} \times \mathbf{S} \to [0, 1]$, is the probability function, $\mathbf{P}((w, \mathbf{q}, \mathbf{v}), \mathbf{a}, (w', \mathbf{q}', \mathbf{v}')) = \mathbb{1}_{\{\mathbf{v}'\}}(T(\mathbf{v}, \mathbf{a})) \cdot P(w, w') \cdot \lambda(\mathbf{q}, \mathbf{a}, w', \mathbf{q}')$; (5) $\mathbf{J} : \mathbf{S} \times \mathbf{A} \to \mathbb{R}$, is the cost function, $\mathbf{J}((w, \mathbf{q}, \mathbf{v}), \mathbf{a}) = (1 - \mathbb{1}_{\mathbf{F}}(\mathbf{q})) \cdot \mathbf{c}(\mathbf{v}, \mathbf{a})$; (6) $\mathbf{G} = W \times \mathbf{F} \times \mathbf{V}$ is the set of goal states.*

Note that this construction is a goal MDP, meaning it is an MDP supplemented with a set of goal states. The MRRCM problem then is reduced to finding for this MDP, a policy that minimizes the expected cost of reaching goal states. Such a policy, denoted $\pi_{\mathbb{M}}^*$, is a function $\pi_{\mathbb{M}}^* : \mathbf{S} \to \mathbf{A}$ which gives the optimal action for each $\mathbf{s} \in \mathbf{S} \setminus \mathbf{G}$ using the Bellman equation, which may be computed by a variety of methods [16].

Note that for all $\mathbf{s} \in \mathbf{G}$, $V_{\mathbb{M}}^*(\mathbf{s}) = 0$. As each state $s \in \mathbf{S}$ is a triple $(w, \mathbf{q}, \mathbf{v})$, and each state $\mathbf{q}$ of the deterministic footage automaton corresponds to a set of states of the story automaton, the computed policy $\pi_{\mathbb{M}}^*$ is a solution to MRRCM.

This MDP has a state space of size $\Theta(|W||\mathbf{Q}||V^{(1)}||V^{(2)}| \cdots |V^n|)$ and an action space of size $|A|^n$. As the number of robots increases, both the state space and the action space of the MDP grows exponentially. Because of this, the next section pursues a solution to the MRRCM problem with less expense.

4.3 Sequentialized Planning

The overall idea of our second algorithm, following the classical approach in multi-agent planning [17], is to solve a sequence of MDPs, each being considerably smaller than the full joint MDP. Each MDP is constructed for a single robot in the team, the structure of each conditioned on the optimal policies of those preceding it in the sequence. Each is a goal MDP and, hence, an optimal policy can be computed via Bellman recurrences.

Suppose that the n robots are ordered: $j_1 j_2 \ldots j_n$, where $j_k \in \{1, \ldots, n\}$. If $j_1, \ldots, j_{k-1}$ have determined how they will act, robot j_k can solve an MDP with stochastic transitions incorporating the events that the other $k-1$ might record, along with the associated probabilities of the events actually occurring, as gratis contributions. Once robot j_k solves this to obtain a policy, we have policies for the first k robots, and could proceed onward to robot j_{k+1}. And so on, until j_n.

The difficulty is that, even if the $k-1$ robots do have policies, those policies involve states within $\mathbb{V}^{(j_1)}, \mathbb{V}^{(j_2)}, \ldots, \mathbb{V}^{(j_{k-1})}$, which is information that robot j_k is not privy to, so policies will not give a determination of the actions of the first $k-1$ robots. Even if the robot had that information—obtained, say, by copious broadcast communication—this would still yield a policy for j_k as a function over $W \times \mathbf{Q} \times V^{(j_1)} \times \cdots V^{(j_k)}$, which grows exponentially in n in the worse case.

We pursue the following alternative, with more attractive scaling properties. For robot j_k, we compute a policy over state space $W \times \mathbf{Q} \times V^{(j_k)}$. The construction of the j_k's MDP is as follows:

Definition 6. *For robot j_k, event model $\mathcal{M} = (W, P, w_0, E, g)$, deterministic footage automaton $\mathbf{D} = (\mathbf{Q}, \mathbf{E}, \delta_{\mathbf{D}}, q_0, \mathbf{F})$, valid-action automaton for the robot $\mathbb{V}^{(j_k)} = (V^{(j_k)}, v_0^{(j_k)}, A, \tau^{(j_k)}, c^{(j_k)})$, policies π_{j_m} for $m \in \{1, \ldots, k-1\}$, and a distribution over the valid-action automata states for the $k-1$ robots, $\Delta^{(j_m)} : V^{(j_m)} \to [0,1]$ for $m \in \{1, \ldots, k-1\}$, we construct the sequential MDP $\mathbb{M}_{j_k} = (S^{(j_k)}, s_0^{(j_k)}, A, \mathbf{P}^{(j_k)}, J^{(j_k)})$, where*

- $S^{(j_k)} \subseteq W \times \mathbf{Q} \times V^{(j_k)}$ *is the state space;*
- $s_0^{(j_k)} = (w_0, q_0, v_0^{(j_k)}) \in S^{(j_k)}$ *is the initial state;*
- A *is the action space;*
- $\mathbf{P}^{(j_k)} : S^{(j_k)} \times A \times S^{(j_k)} \to [0,1]$ *is the probability function such that*

$$\mathbf{P}^{(j_k)}(s, a, s') = \mathbb{1}_{\{v'\}}(\tau^{(j_k)}(v, a)) \sum_{\alpha \in \mathbf{A}_a^k} P(w, w') \mu_{w, \mathbf{q}}(\alpha) \lambda(\mathbf{q}, \alpha, w', \mathbf{q}')$$

with $s = (w, \mathbf{q}, v)$, $s' = (w', \mathbf{q}', v')$, and where $\mu_{w,\mathbf{q}} : \mathbf{A}_a^k \to [0,1]$ is

$$\mu_{w,\mathbf{q}}(\alpha) = \sum_{\substack{v^{(1)} \in V^{(j_1)} \\ \vdots \\ v^{(k-1)} \in V^{(j_{k-1})}}} \prod_{m=1}^{k-1} \left(\mathbb{1}_{\{a^{(m)}\}}(\pi_{j_m}(w, \mathbf{q}, v^{(m)})) \Delta^{(j_m)}(v^{(m)}) \right)$$

and $\mathbf{A}_a^k$ consists of joint actions $(a^{(1)}, a^{(2)}, \ldots, a^{(k-1)}, a, \overbrace{\bot, \bot, \cdots, \bot}^{n-k})$;
- $J^{(j_k)} : S^{(j_k)} \times A \to \mathbb{R}$ *is the cost function, such that for $s = (w, \mathbf{q}, v) \in S^{(j_k)}$ and $a \in A$, we have $J^{(j_k)}(s, a) = (1 - \mathbb{1}_{\mathbf{F}}(\mathbf{q})) c^{(j_k)}(v, a)$.*

The intuition here is that, in lieu of actual information on the state of each $\mathbb{V}^{(j_k)}$, estimates (in the form of distribution $\Delta^{(j_k)}$) are used as an approximation. In what follows, we make a maximum entropy assumption over the states of the valid-action automata, i.e., $\Delta^{(i)}(v) = \frac{1}{|V^{(i)}|}$, though cleverer choices exist.

Based on this treatment, one expects that the ordering of the robots would affect the overall solution quality. Though a random order will work, it may fail to give a good policy so we employ the following greedy heuristic to choose a favourable ordering. First, we calculate the policies for all n robots tentatively assuming each would be operating alone. Then we select the robot whose individual policy gives the least expected cost to capture the story, and use it as the robot for the first spot. Having determined j_1, we compute policies for the remaining $n-1$ robots, given j_1 and its policy π_{j_1}. The robot with the least expected cost becomes j_2, and the process is repeated but now with $\{j_1, j_2\}$ determined. This is repeated until all n have been ordered.

5 Case Study

In this section, we present results of our Python implementation of the algorithms, which we executed on an Ubuntu 16.04 computer with a 3.6GHz CPU.

We revisit the shooting of documentary in a wildlife reserve as used as motivation initially, and outlined in Fig. 1. A system-level event model is constructed from event models for the animals. Figure 2 shows, for each type of animal, the transition probability function P of its event model, in which each entry $P(l_1, l_2)$ is the probability that the animal(s) go from the current location l_1, to l_2 in the next hour. Based on these event models, we see the flamingos are only interested in the lake, while the crocodile is interested in both the river and the lake. The others roam more widely. The event model for the whole system is obtained via a Cartesian product of these individual models.

The events of interest are gazelle grazing, g_e; cheetah eating gazelle, c_g; crocodile eating gazelle, k_g; flamingos mating, f_m; and crocodile eating flamingo, k_f. Event g_e happens with probability 1 for both cases, whether the gazelle are in the field or the jungle. The probabilities that the cheetah can successfully capture and kill a gazelle in the field and the jungle are, respectively, 0.2 and 0.25. The probabilities that the crocodile can successfully capture and kill a gazelle by the river and by the lake are, respectively, 0.2 and 0.25. The probability that the crocodile can capture and kill a flamingo is 0.15. The probability that a flamingos mating event happens in an hour is 0.4.

The robots cannot track or follow animal(s) continually, and video clips are only recorded at locations l_f, l_j, l_r, and l_l. At each time step, each robot observes the current state of the event model by receiving a message from the command post, which reports the rough locations of the animals based on GPS trackers connected to tags on the animals. The robot does not know which location will be the creatures' destination in the next hour. Also, at every step t, each robot relays to the other robots whether it was successful in recording the event it attempted. Accordingly, all the robots can compute $\mathbf{X}_t$ and $\mathbf{Q}_t$. At each time step, a robot must guess the destination of an animal of interest so that it can go to that destination and prepare to capture a desired event.

Three kinds of videographer robots A, B, and C are available. The robots are camouflaged, so their presence does not change the animals' behavior, i.e., we make a non-causality assumption. Robots of type A are UAVs capable of traveling between any of those locations in a time step. Its actions are unconstrained, so it has a trivial (single state) valid-action automaton. Robots of type B are short-endurance UAVs with a simple constraint on their actions: if at time t, such a robot chooses event k_g to capture at time $t+1$, then it is forbidden from repeating it (regardless of whether it successfully captured k_g at $t+1$ or

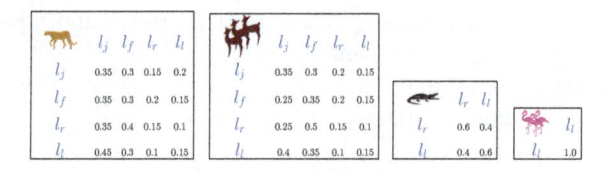

Fig. 2. Transition probabilities of the animals.

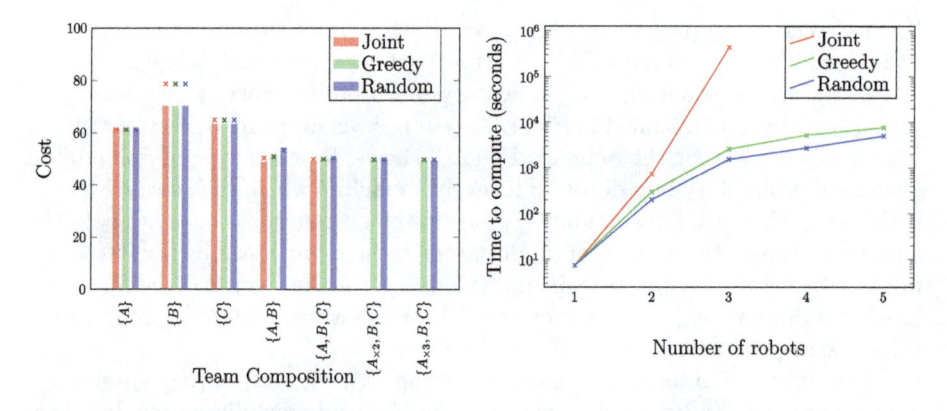

Fig. 3. Left: Cost to capture story. The theoretical prediction (expected cost for policy) is shown via × marks. Average cost for 1000 simulation is shown via bars. Right: Time to compute the policies. (The y-axis is in the logarithmic scale).

not). Robots of type C are ground vehicles. When starting out from the river or lake, these vehicles must take a detour, visiting the command post (for one time step), to be outfitted with equipment required to travel through the jungle or field. Likewise, when in the jungle/field, travel to riverside/lakeside, requires a sojourn at the command post first. Accordingly, if a robot of type C moves to l_c, then in the next time step no event will be captured by that robot.

We set the cost function in a way that the objective in the MRRCM problem becomes minimizing the expected number of hours to capture a desired story. For this purpose, for each robot i and action a, for each vertex v of the valid-action automaton of robot i, we set $c^{(i)}(v, a) = 1$. Because the joint cost of a group of robots is the sum of costs for individual robots, in reporting the results we have divided by the number of robots, each of the expected and the average costs obtained for a joint plan so that these figures represent, respectively, the excepted number of hours and the average number of hours to record a story. For the sequential plan, no division is needed and we use the expected and average costs obtained for the last MDP in the sequential plan directly.

In this case study we are interested in capturing a sequence that is a supersequence for both the sequences $g_e f_m c_g k_f$ and $g_e f_m k_g k_f$, each of which chronicles both a gazelle's life and a flamingo life. Once a desired sequence was captured, we post-process it to make two videos $g_e f_m c_g k_f$ and $g_e f_m k_g k_f$ from it, each for a TV channel. Note that the language of the specification DFA in this case is infinite. We considered several scenarios in which different numbers of each type of robot are tasked to capture a desired story. For our implementations, we use the value-iteration method to solve the underlying MDPs. For each scenario, we solved the MRRCM problem using the joint approach and the sequential approach with the random and greedy strategies. Also for each scenario, we generated 1000 simulations of executing the event model and for each simulation the robot(s) use the computed policy to capture a story. For each case, we computed the average cost of capturing a desired story over the 1000 simulations. Figure 3 shows the expected number of hours and the average number of hours for those simulations. As it was expected, a robot of type A outperformed the other two robot types B and C in yielding a smaller expected cost. Also for each experiment, the expected cost was very close to the average cost for 1000 simulations. We were able to generate a joint plan only for up to three robots; it took approximately 14 h to generate a joint plan for three robots, while generating a sequential plan for three robots with each of the random and the greedy strategies took approximately 43 min and 85 min respectively.

6 Conclusion

This paper considered the problem of computing a policy, for a team of robots, minimizing the expected cost of recording a sequence of events that happen unpredictably. The problem is reduced to that of computing an optimal policy for a joint MDP of the robots. To overcome the computational complexities of solving the joint MDP, we proposed to solve the problem via a sequence of MDPs, using a greedy heuristic to order that sequence, and finally we presented our implementation results via a wild life case study.

References

1. Rahmani, H., Shell, D.A., O'Kane, J.M.: Planning to chronicle. In: Workshop on the Algorithmic Foundations of Robotics (WAFR XIV) (2020)
2. Shell, D.A., Huang, L., Becker, A.T., O'Kane, J.M.: Planning coordinated event observation for structured narratives. In: IEEE ICRA (2019)
3. Yu, J., LaValle, S.M.: Story validation and approximate path inference with a sparse network of heterogeneous sensors. In: IEEE ICRA (2011)
4. Lee, Y.J., Ghosh, J., Grauman, K.: Discovering important people and objects for egocentric video summarization. In: IEEE CVPR (2012)
5. Hong, R., Tang, J., Tan, H.-K., Ngo, C.-W., Yan, S., Chua, T.-S.: Beyond search: event-driven summarization for web videos. ACM Trans. Multimed. Comput. Commun. Appl. **7**(4), 35 (2011)
6. Potapov, D., Douze, M., Harchaoui, Z., Schmid, C.: Category-specific video summarization. In: ECCV (2014)
7. Feng, L., Li, Z., Kuang, Z., Zhang, W.: Extractive video summarizer with memory augmented neural networks. In: ACM Multimedia (2018)

8. Chang, P., Han, M., Gong, Y.: Extract highlights from baseball game video with hidden Markov models. In: IEEE ICIP (2002)
9. Kolekar, M.H., Sengupta, S.: Event-importance based customized and automatic cricket highlight generation. In: IEEE ICME (2006)
10. Girdhar, Y., Giguere, P., Dudek, G.: Autonomous adaptive exploration using real-time online spatiotemporal topic modeling. Int. J. Robot. Res. **33**(4), 645–657 (2014)
11. H. Hajishirzi, J. Hockenmaier, E. T. Mueller, and E. Amir, "Reasoning in Robocup Soccer Narratives," in *UAI*, 2011
12. Rosenthal, S., Selvaraj, S.P., Veloso, M.: Verbalization: narration of autonomous robot experience. In: IJCAI (2016)
13. Riedl, M.O., Young, R.M.: Narrative planning: balancing plot and character. J. Artif. Intell. Res. **39**, 217–268 (2010)
14. Barot, C., et al.: Bardic: generating multimedia narrative reports for game logs. In: Working Notes of the AIIDE Workshop on Intelligent Narrative Technologies (2017)
15. Hopcroft, J.E., Ullman, J.D.: Introduction to Automata Theory. Languages and Computation. Adison-Wesley. Reading, Mass (1979)
16. Puterman, M.: Markov Decision Processes: Discrete Stochastic Dynamic Programming. Wiley, Hoboken (1994)
17. Ephrati, E., Rosenschein, J.S.: Divide and conquer in multi-agent planning. In: AAAI-94 (1994)

Errors in Collective Robotic Construction

Jiahe Chen[1(✉)], Yifang Liu[1], Adam Pacheck[2], Hadas Kress-Gazit[2], Nils Napp[1], and Kirstin Petersen[1]

[1] School of Electrical and Computer Engineering, Cornell University, Ithaca, NY 14853, USA
{jc3472,yl892,nnapp,kirstin}@cornell.edu
[2] Sibley School of Mechanical and Aerospace Engineering, Cornell University, Ithaca, NY 14853, USA
{akp84,hadaskg}@cornell.edu

Abstract. We investigate the effect of errors in collective robotic construction (CRC) on both construction time and the probability of correctly completing a specified structure. We ground our investigation in the TERMES distributed construction system, which uses local sensing and stigmergic rules that enable robots to navigate and build 3D structures. We perform an in depth analysis and categorization of action failures in CRC systems. We present an approach to mitigating action failures and preventing errors that prohibit completion of a structure by adding predictive local checks. We show that the predictive local checks can increase the probability of success by orders of magnitude in large structures. This work demonstrates the need to consider both construction time and the effect of errors in collective robotic construction.

1 Introduction

Collective robotic construction (CRC) involves multiple robots collaborating to build structures much larger than themselves [1]. This approach can enable construction by teams of dispensable robots in places dangerous or inaccessible to humans, such as disaster sites or extraterrestrial environments. CRC is a growing research field spanning complex industrial robots cooperating through detailed global planners to simple, locally-aware robots cooperating through a combination of global plans and stigmergic rule sets [2–7]. The latter are of special interest in task settings that provide little pre-existing infrastructure yet demand scalability in deployment, high redundancy, and speed through parallelism. Applications range from building pre-determined structures using traditional materials (e.g., clay bricks and concrete) or custom bricks to ease robot manipulation [8], to functional structures such as access paths built out of amorphous materials [9,10]. Most of these systems focus purely on additive manufacturing, which makes occasional errors especially problematic since they cannot be undone.

Due to the relatively small scale of robot-built structures shown in the past, errors and error propagation in CRC have been largely overlooked. The majority of prior work has focused on decreasing construction time through improving system efficiency and maximizing parallelism [3–5,7] while relying on systems where errors are mitigated through engineering effort or simulations that do not incorporate errors. However, in

---

The first three authors contributed equally to this work.

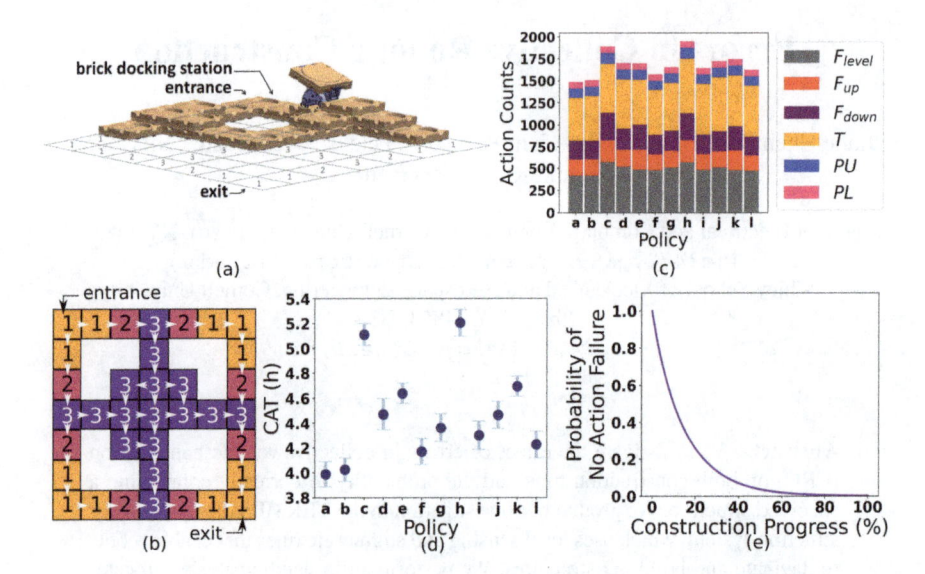

Fig. 1. (a) Sketch of a TERMES robot assembling a structure. b) A blueprint with an example policy overlaid. (c) The composition of actions required to complete (b) given different policies. (d) The cumulative action time (robot hours) required to build (b) given different policies. (e) The probability of completing (b) without having an action failure, using policy 'a'.

bigger structures and larger collectives, even unlikely errors are bound to occur and it is critical to understand how these affect the overall system. Similar tendencies were shown in the related field of swarm robotics when they grew in numbers from tens to a thousand [11]. Comparatively, errors in CRC often have bigger repercussions because robots with limited sensing and motion capabilities are physically and permanently altering their environment. Therefore, when evaluating CRC systems, the probability of successful structure completion must be considered in addition to construction time.

In this work, we focus explicitly on CRC of pre-determined structures, with local stigmergic rules that have a non-zero probability of action failure (Fig. 1). When actions fail, the effect on the collective construction effort varies. While some action failures make no difference beyond the wasted effort, others can lead to a partial structure that is impossible to complete, a situation which we term a *fatal error*. We show that by analyzing which action failures are most likely to cause fatal errors, we can make informed modifications to the system and improve the probability of success. First, we discuss an intuitive way to increase the probability of success by using policies that minimize the number of required sequential actions. We show that this is closely related to finding the policy that results in the lowest construction time, and investigate the difference between a parallel policy that minimizes the number of sequential actions, but may lead to wasted trips without assembly actions, versus a sequential policy where robots are always able to perform assembly actions. The approach of minimizing action sequences, however, still has a fundamental limit based on structure size and robot reliability. We further show that we can improve the probability of success by adding predictive local checks to detect when an action fails, thereby preventing errors from cascading. We show that while adding predictive local checks can substantially improve

Table 1. Action success rate and execution time in our simulated TERMES system. We chose values proportional to those reported for the real TERMES system [8], but decreased the failure probability by 20% to enable more interesting outcomes.

Action	Success rate ($Pr(a)$)	Execution time
Move forward (F_{level})	0.998	7.6 s
Climb up (F_{up})	0.999	18.96 s
Climb down (F_{down})	0.9999	10.88 s
Turn 90 °C (TU)	0.986	6.4 s
Pick up a brick (PU)	1	5.6 s
Place a brick (PL)	0.998	21.6 s

reliability, it too has a fundamental limit to its practical application due to an increase in construction time. While preventing fatal errors, structures may never be completed because robots abandon trips when action failures occur.

We study these implications in the context of the TERMES system [6] (Fig. 1(a)), an autonomous multi-robot system with distributed control in which robots use only local sensing and follow a stochastic plan to place bricks to build a structure. Although we use the TERMES system as a platform to ground our findings, many of the insights gained translate to other CRC systems, especially the critical importance of considering the system error characteristics when robots with limited sensing and motion constraints manipulate a shared environment.

Contributions: **I** An in depth analysis and categorization of action failures and fatal errors in CRC systems; **II** A general approach to mitigating action failures and preventing fatal errors by using predictive local checks; **III** An evaluation of construction times and success rates for the TERMES system based on published error rates as well as examples of predictive local checks that can increase the probability of success by orders of magnitude, especially in large structures.

2 Terminology

This section briefly summarizes background information and terminology. We focus on CRC systems in which robots with local information move over and assemble predetermined structures, as in [6] and shown in Fig. 1. For clarity, we define the following terms, and include examples of how these apply to the TERMES system:

Blueprint: The desired structure is an N × M grid, where each grid cell, or *location*, is associated with a number representing the number of bricks at that location, and if they serve as a structure entrance or exit (Fig. 1(b)). For simplicity, we assume that the entrance and exit are always located in the upper left and lower right corner, respectively. Note that the robot always picks up a brick at the entrance. In the following, an N × N × 1 blueprint is a square structure of height 1 without any holes.

Path: The sequence of locations visited by the robot as it navigates through a structure.

Policy: The blueprint is passed to a compiler which generates one or more *solutions* specifying possible transitions between locations (Fig. 2(a)). The policy combines solutions with *transition probabilities*, such that transition probabilities out of a location always sum to one [7]. Robots travel on top of the structure from the entrance to the exit by following these stochastic policies.

Child and Parent Locations: The parents of a location are those for which the robot has a non-zero probability of transitioning from; the children are those which the robot has a non-zero probability of transitioning to. Parents and children are specific to policies.

Assembly Rules: The assembly rules define valid *assembly locations*, where it is valid to place a brick, based on locally available information. A TERMES robot can place a brick in a location if and only if [6]: 1) The location height is less than the blueprint-specified height; 2) All parents are of height greater than the location, or have reached their blueprint-specified height; 3) All children are of equal height to the location, or their blueprint-specified height differ from the location blueprint-specified height by more than 1. If these are true, the robot executes a sequence of placement actions.

Actions: Robots have a set of actions. In TERMES, these actions correspond to "move forward", either on level bricks or by climbing one brick up or down (F), "turn 90 °C left or right" (TU), "pick up a brick" (PU), or "place a brick" (PL). The sequence of placement actions is always either $TU-F-TU-TU-PL$ or $F-TU-TU-PL$.

Action Failure: Physical robots have a non-zero probability of failing to correctly perform an action. The TERMES failure characteristics are reported in [6] and listed in Table 1, where $Pr(a)$ is the probability of action a executing successfully. For simplicity, we assume that failure of locomotion actions means that the robot believes it has performed the action correctly, but physically remains in its current pose. For errors in the placement action (PL), the robot is assumed to have made an error placing that brick and the structure is no longer traversable.

Cumulative Action Time (CAT): The total robot time taken to complete a structure, given by the average action execution time ([6], Table 1). Note that the actual construction time will depend on the number of robots deployed.

Probability of Success (PoS): Probability that a structure will be completed, given the particular failure characteristics of the robots and policy.

Productive and Wasted Trips: In a productive trip, a robot finds a legal assembly location and correctly places a brick; in a wasted trip the robot does not place a brick.

3 Influence of Policy Choice on System Performance

The number of policies can grow exponentially with the size of the structure, e.g. a $3 \times 3 \times 1$ structure has 5 policies while a $5 \times 5 \times 1$ structure has 1010 policies. Each policy may lead to a different CAT and PoS due to different lengths of action sequences and number of wasted trips. We modified the TERMES simulator from [7] to investigate which policies perform the best given the failure characteristics in Table 1. We pay special attention to two policies: *parallel* and *sequential* (Fig. 2(a)). Parallel policies

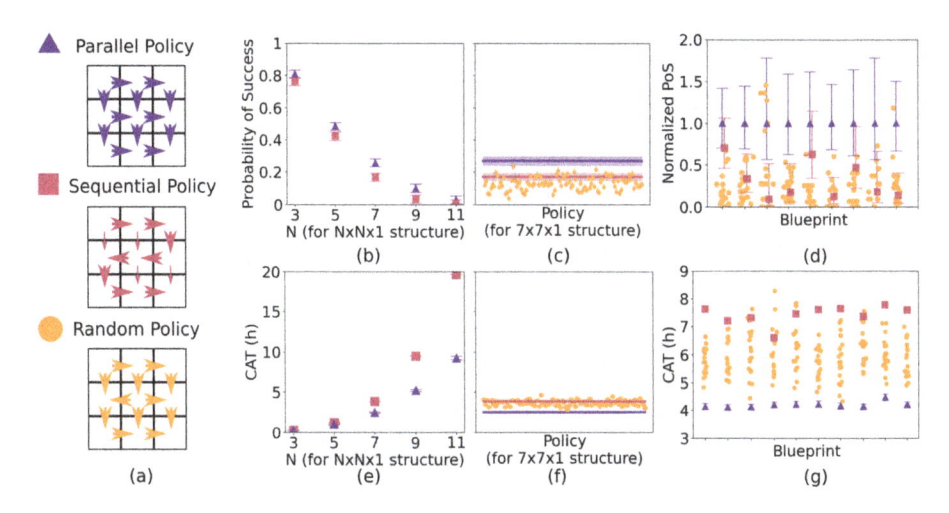

Fig. 2. The PoS and CAT for different policies, based on 1000 and 100 simulations per policy respectively. (a) Example transitions between locations for a parallel, sequential, and random policy. Smaller arrows denote small (∼0) transition probabilities. (b, e) Parallel versus sequential policy for N × N × 1 blueprints. (c, f) Parallel, sequential, and 100 random policies for a 7 × 7 × 1 blueprint. Purple and pink lines denote the parallel and sequential policies; shaded regions the 95% confidence interval (in (f) the error bars are too small to be seen). (d, g) Parallel, sequential, and 19 randomly selected policies for ten 7 × 7 blueprints with random heights between 0 and 3.

minimize the longest path robots can take through the structure, and further facilitate the maximum number of simultaneously legal assembly locations. Sequential policies have a single path and thus force robots to visit every assembled location on every trip but eliminate wasted trips. Note that these policies are not necessarily unique, and that not every blueprint has a sequential policy. For example, a square structures with sides of even length and opposing entrance and exit locations lacks a sequential policy.

We found that for larger structures, the benefit of minimizing the path lengths outweighs the risk of wasting trips, making the parallel policies superior to the sequential policies in terms of both PoS and CAT, as shown in Fig. 2(b, e). Figure 2(c, f) show that the parallel policy also consistently outperforms other randomly selected policies for a 7 × 7 × 1 blueprint in terms of both PoS and CAT. Similarly, in Fig. 2(d) we show that the parallel policy has a higher PoS than 19 randomly selected policies and the sequential policy for eight out of ten 7 × 7 blueprints with random heights. In Fig. 2(g) we show that the CAT of the parallel policy is lower than other policies on all tested blueprints. As parallel policies are (usually) the better option in terms of both CAT and PoS, this will be our focus for the remainder of the paper.

For the structures and policies we tested, the parallel policy has the lowest CAT and usually the highest PoS, but still only allows practical construction of relatively small structures; a 9 × 9 × 1 structure, for example, has 10.1% PoS. For comparison, an average American family house consists of 8,000 bricks [7]. To enable such large-scale structures, we need to investigate how errors are caused and methods to mitigate them.

4 Error Analysis

In this section, we investigate how action failures reduce the PoS. It is important to note that the probability of completing a structure without any action failures happening is much smaller than the PoS. The PoS for a $9 \times 9 \times 1$ structure is 10.1%, as determined by 1000 simulations. However, on average it takes 2885 actions to complete the same structure and the probability of each of these actions completing successfully is approximately $0.999^{2885} \approx 0$. The reason for this mismatch is that not all action failures prevent a structure from being completed; only fatal errors prevent structure completion. A *fatal error* occurs when a brick is mistakenly placed such that robots are no longer able to physically move through the structure to place additional bricks, or because the combination of the construction state and assembly rules prevent future placements. In the TERMES system, fatal errors are due to the fact that robots cannot climb more than one brick at a time or fill gaps that are restricted on both sides by other bricks.

To further analyze the impact of action failures and fatal errors on the PoS, we define three categories for the consequences of action failures:

Category I: No brick is placed, either because the robot never finds a valid assembly location or because the robot is capable of detecting the failure and leaves the structure. Both lead to a wasted trip.

Category II: A brick is placed without violating the assembly rules, either by luck (Fig. 3(a)) or the robot detecting a failure and finding a new valid assembly location.

Category III: A brick is placed causing a fatal error, either because the placement itself failed or because one or more prior action failures led to a brick placement that violates the assembly rules. This category can be further divided into 3 cases:

1. Motion constraint violation, because the newly placed brick forms a cliff;
2. Manipulation constraint violation, because the newly placed brick forms an unfillable gap in the structure (Fig. 3(b)), or because the robot attempts to place the brick at a different height from where it is standing (Fig. 3(d));
3. Child locations are assembled before their parents (Figs. 3(c, e, f)); although a robot is physically capable of resolving the error, doing so would violate the assembly rules.

Among these categories, our primary interest is in category III, as the consequences of failures that lead to this category directly impact the PoS. It is important to note that

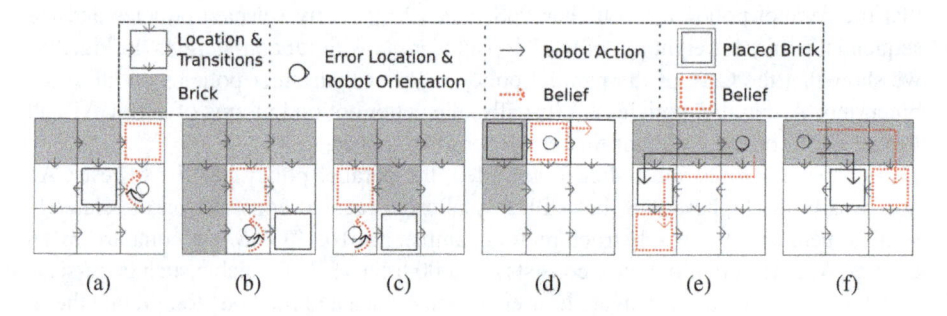

Fig. 3. Examples of action failures. (a–c) show failure to turn (TU_e); (d–f) show failure to move forward (F_e). (a) does not lead to a fatal error; (b–f) lead to fatal errors.

an action failure may occur several steps before the fatal error occurs. For example, in Fig. 3(f) the robot attempts to move forward, but fails. It then takes another step forward, turns, and places a brick at what it thinks is the beginning of a row, but is actually the middle of a row. This results in a fatal error, because a robot can never legally place a brick at the far right location of the middle row.

Using our simulator, we recorded which action failures most commonly lead to fatal errors given different policies (Table 2 rows 2–4). We distinguish between action failures that occur during navigation and the brick placement action sequence. Failures to turn (TU_e) (Table 2 columns 4 and 5) result in more than 70% of the fatal errors across different policies. The number of placement failures (PL_e) is stable since it only depends on the number of attempted brick placements. The rest of the fatal errors are caused by failures to move forward (F_e), either during navigation or placement. Different policies also have different error distributions. For the parallel policy, the percentage of fatal errors caused by F_e failures during navigation is 0.78%, as opposed to 17.3% for the sequential policy. This is because the sequential policy enforces much longer paths, increasing the probability of F_e failures.

Based on analyzing the action failures that lead to fatal errors through the examples shown in Fig. 3 and Table 2, we can determine if and how a robot can detect and mitigate the effects of action failures. If a robot can predict what the surroundings should look like after an action, then it can recognize action failures that cause a mismatch between the expected surroundings and what is physically sensed, and react accordingly. In the next section, we discuss how to mitigate errors by adding such local predictive checks based on the failures shown in Fig. 3.

5 Mitigating Errors Through Predictive Local Checks

The ability to sense and reason about legal assembly locations given knowledge of only the local environment is critical in stigmergy coordinated CRC. Here, we suggest several additional local checks a robot can perform to detect if an action failure has occurred, either immediately or after a sequence of actions. The goal of these checks is to convert action failures from category III to category I before they lead to a fatal error. Theoretically, turning category III failures into category II failures would be a better choice, but that requires significant additional effort in relocalizing, estimating the true assembly state, and probabilistic reasoning, so we leave this for future work. In this section, we first discuss predictive local checks to improve the PoS, then discuss how additional wasted trips caused by detected action failures impact the CAT.

5.1 Predictive Local Checks

We propose predictive local checks that compare what the robot expects to sense following a successful action to what it actually senses. This allows a robot to detect some failures to move forward and turn. Here, we assume that the robot can navigate off the structure once a failure is detected without changing the current construction state even if its localization is wrong. The difficulties in designing predictive local checks are that sometimes the local view of the construction state is the same before and after an action, and that the exact construction state in non-sequential policies is not known as actions

Table 2. We had a robot repeatedly construct a 7 × 9 × 1 blueprint until it successfully placed 30,000 bricks. This table shows the number of action failures that directly lead to fatal errors, without (rows 2–4) and with (rows 5–7) predictive local checks. Subscript e indicates an action failure, and subscript p and n means that the failure occurred during brick placement or navigation, respectively.

		Solution	$TU_{e,n}$	$TU_{e,p}$	$F_{e,n}$	$F_{e,p}$	PL_e	Unsuccessful builds	Successful builds
2	TERMES baseline	Parallel	0	768	7	62	57	894	166
3	TERMES baseline	Sequential	2	837	198	53	56	1146	104
4	TERMES baseline	Random	106	770	130	65	54	1125	105
5	w/predictive checks	Parallel	0	392	3	2	53	450	304
6	w/predictive checks	Sequential	0	224	165	10	55	454	326
7	w/predictive checks	Random	0	284	81	13	50	428	322

can be completed in many different orders. As shown in Table 2, the majority of fatal errors stemming from failures to turn occur during brick placement, which means these must be detected immediately. Conversely, we do not need to immediately detect fatal errors stemming from failures to move forward as the majority of such failures happen during navigation, leaving more steps before a brick is placed. We do not attempt to avoid errors stemming from placement failures (PL_e), as these are only associated with hardware reliability and cannot be mitigated by better execution plans. Considering the effort and cost of re-designing the hardware, we focus on local checks that can be accomplished with minimal hardware changes or entirely in software. Specifically, we propose the following two predictive local checks to avoid fatal errors:

Front Checking: This check is implemented entirely in software. The TERMES robots have two sensors that permit them to reason about the height of the current and neighboring location: a tilt-sensor that indicates when the robot is climbing up or down and a downward-facing sensor mounted on the claw that detects the relative height of a neighboring location. Using these sensors, we can ensure that the robot never attempts to place a brick at a height different from its own location. This means that the types of errors shown in Fig. 3(d) are eliminated. Similarly, the robot can detect the action failures shown in Fig. 3(e, f). In Fig. 3(f), when the robot believes it is facing right at the top right corner, the height of the front location is outside of the structure and should be lower than the robot's current height. If not, the robot can determine that one or more actions failures occurred. Similarly, in Fig. 3(e), when the robot is at the top right corner facing downwards, it can detect that going forward should result in a climb down, when it doesn't, the robot knows an error occurred.

Side Checking: By adding three distance sensors to the left, right, and rear of the robot, similar to the front distance sensor that already exists, the robot can detect discrepancies associated with failed turns, such as the ones shown in Fig. 3(b).

By incorporating front and side checking, we are able to reduce the number of fatal errors and increase the PoS. As shown in Table 2 rows 5–7, the number of F_e and TU_e decreases. Most significantly, the number of turn failures during placement decreased by over 45%. Additionally, no fatal errors resulted from turn failures during navigation in all tested policies. The addition of predictive local checks increases the PoS for all policies (Fig. 4). The PoS for parallel and sequential policies on N × N × 1 struc-

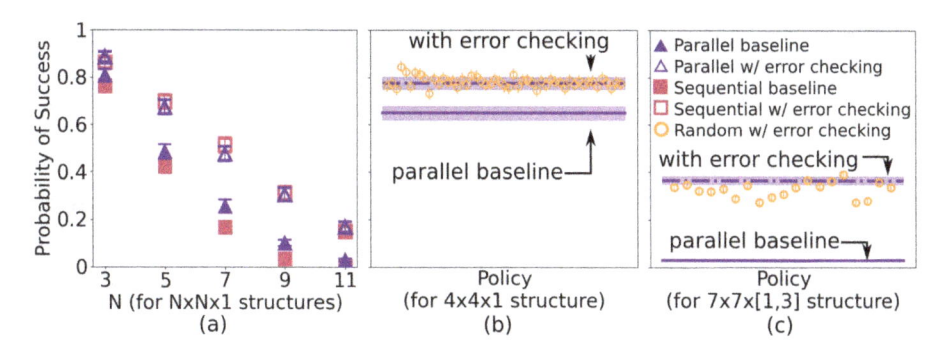

Fig. 4. Adding predictive local checks improves the PoS. Each policy was simulated 1000 times (except for the $11 \times 11 \times 1$ blueprint, which was simulated 2000 times) (a) The PoS for both parallel and sequential policies increases with the additional checks. (b) The PoS for all policies is higher than the original PoS of the parallel policy (solid purple line, shaded region is error), and similar to the parallel policy with additional checks (dashed purple line) for a $4 \times 4 \times 1$ blueprint. (c) Additional checks also improve the PoS of a blueprint with randomly varying heights ($7 \times 7 \times [1, 3]$).

tures improved by an average of $2.6(\pm 1.7)$ times and $9.7(\pm 12.2)$ times, respectively (Fig. 4(a)). The PoS of random policies becomes much closer to that of parallel polices, with some random policies even outperforming the parallel policy. The PoS of policies for a $7 \times 7 \times [1, 3]$ random structure as shown in Fig. 4(c) improved by $100.5(\pm 95.3)$ times. This occurs because most action failures that lead to fatal errors in the parallel policy arise from turn failures like the one in Fig. 3(c), which the robot cannot detect right away, as all the surrounding locations have the same relative height, while all other turn failures that occur more often in random and sequential policies can be detected immediately. Notice that not all action failures can be detected by the robot, even with the virtually updated hardware. For instance, the robot cannot immediately detect a failure to move forward unless the front location has a different height. Even in this case, the robot may still detect that a failure has occurred later on. The robot also cannot immediately detect a failure to turn if all the surroundings have the same relative height.

5.2 Impact of Predictive Local Checks on Construction Time

While predictive local checks significantly improve the PoS, they also impair the CAT by introducing additional wasted trips. To study this impact, we developed an analytical expression for the parallel policy CAT. This method not only allows us to quickly reason about the number of robots to deploy for a given structure; it also allows us to study how the CAT scales with the structure size given predictive local checks, enabling us to consider structures that are too large to effectively simulate.

To introduce the model, we define a snapshot of the structure in time, as it is being built, as a *construction state*. Given a construction state, robots navigate through the structure until they reach a location where a brick still needs to be placed; there can be multiple such locations per construction state. If a robot reaches a location that complies

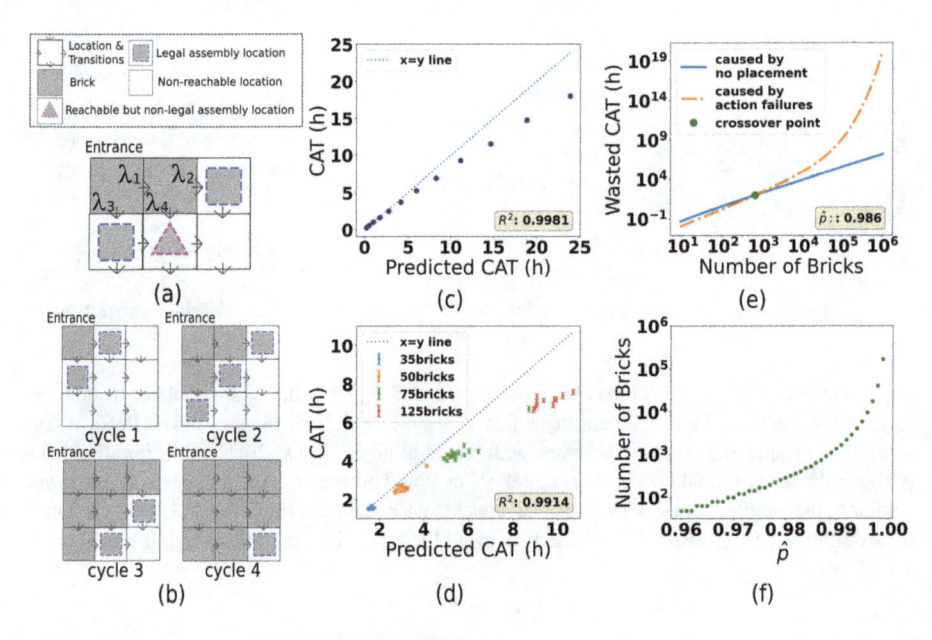

Fig. 5. (a) A possible construction state where λ is the transition probability. (b) Construction cycles for a $3 \times 3 \times 1$ blueprint, as assumed in the general model proposed in Eq. 1. (c) Simulated versus estimated CAT obtained from the simplified model in Eq. 2 of $N \times N \times 1$ ($3 \leq N \leq 14$) blueprints simulated 100 times. (d) Simulated vs estimated CAT obtained from the general model in Eq. 1 of $N \times N$ ($5 \leq N \leq 7$) blueprints with random heights between 0 and 3 simulated 100 times. (e) Wasted CAT caused by action failures and no placement at different structure size. $\hat{p}$ is set to the action success rate of turning $90°$ (Table 1). (f) Crossover point at different $\hat{p}$.

with the assembly rules (a *reachable legal assembly location*), it places a brick and leaves the structure. If a robot reaches a location that violates any assembly rules (a *reachable but non-legal assembly location*), it leaves the structure. The probability of following a particular path in the structure is the product of the transition probabilities between the corresponding locations. We define the duration of a path, excluding the time to place a brick, as *trip time*. The time to place a brick after the arrival is counted separately, as the number of placements is fixed. We do not include the time it takes for the robot to return to the entrance after exiting the structure into account. For a location that can be reached through multiple valid paths, the weighted average trip time is used.

To approximate the CAT, we split the construction process into *construction cycles*. In each cycle, the robot has to visit all reachable legal assembly locations at least once, and a brick will be placed at each location (Fig. 5(b)). Modeling the construction process as a sequence of construction cycles leads to an over-approximation, since in reality once some of the locations have been visited and assembled, new locations can become reachable and "legal", but visiting them still counts as wasted trips in our calculations.

We can now produce a closed form expression for a CAT over-approximation. For each construction cycle d, we denote the weighted average trip time of all reachable legal assembly locations and all reachable but non-legal assembly locations as $t_{I,d}$ and $t_{J,d}$ respectively; we denote the probability of visiting any reachable legal assembly location as $p_{I,d}$. We consider two random variables: X_d as the number of trips needed to visit all reachable legal assembly locations at least once given that only reachable legal assembly locations will be visited, and Y_d as the number of trips needed until a reachable legal assembly location is visited. Since visiting a location does not affect the probability of visiting any other location, X_d and Y_d are independent. The expected number of trips to visit all reachable legal assembly locations at least once is then $E[X_d]E[Y_d]$. The expectation of Y_d is $1/p_{I,d}$. Since only one out of the $E[Y_d]$ number of trips goes to a reachable legal assembly location, the trip time is $(E[Y_d] - 1)t_{J,d} + t_{I,d}$. Computation of $E[X_d]$ can be cast as a general *coupon collector's problem*. Each location can be considered as a coupon with certain probability to collect it and $E[X_d]$ is the expected number of trials needed to collect all of them. A closed form solution has been proposed in [12] (Eq. 14b) and is used in our model. Then for a construction process that has D construction cycles, the estimated expected CAT $E[T]$ is given by the general model:

$$E[T] = T_{PL} + \sum_{d=1}^{D} E[X_d]\left(\left(\frac{1}{p_{I,d}} - 1\right)t_{J,d} + t_{I,d}\right), \quad (1)$$

where $T_{PL} = 48.4 N_B$ is the total brick placement time in seconds and N_B is the total number of bricks that need to be placed in the structure. To place a brick the robot will execute 1 move, 1 placement, and at most 3 turns which amounts to a maximum of 48.4 s (Table 1). Figure 5(d) compares the expected CAT obtained from the general model (Eq. 1) and simulation results and shows that although the model is an over-approximation of the expected CAT, it has a linear relationship with the simulation result and can be used to predict CAT growth as a function of the structure size.

For flat, square structures with the parallel policy, each construction cycle contains only reachable legal assembly locations, and each location has equal probability of being visited (Fig. 5(b)), enabling a direct mapping to a closed-form solution from [12]: for n locations, the expected number of trips until all locations have been visited at least once is nH_n, where $H_n = \sum_{k=1}^{n} \frac{1}{k}$ is the nth harmonic number. For an $N \times N \times 1$ structure, there are $2(N-1)$ construction cycles and the number of reachable legal assembly locations in each cycle is $\{2, 3, ..., N-1, N, N-1, ..., 2, 1\}$. For each construction cycle d, a trip contains 1 pickup, d moves and on average $2 + d/4$ turns. Thus, for an $N \times N \times 1$ structure with the parallel policy, the CAT over-approximation $E[T_{N \times N \times 1}]$ is:

$$E[T_{N \times N \times 1}] = T_{PL} + \sum_{d=1}^{2(N-1)} E[X_d]\, t_{I,d},$$

with $\quad E[X_d] = \begin{cases} (d+1)H_{d+1} & \text{for } 1 \leq d \leq N-1 \\ (2N-1-d)H_{2N-1-d} & \text{for } N \leq d \leq 2(N-1) \end{cases}, \quad (2)$

$$t_{I,d} = \tau_F\, d + \tau_{TU}(2 + d/4) + \tau_{PU},$$

where τ_F, τ_{TU} and τ_{PU} are the execution time of move, turn, and pickup, respectively (Table 1). Figure 5(c) shows that the simplified model still maintains a linear relationship with the simulation results. We will use this simplified model to study the scaling behavior of the CAT due to its computation efficiency. We can also express the expected time of trips during which a robot actually places a brick (*productive CAT*) for an N × N × 1 structure with a parallel policy in Eq. 3.

$$E[T_{N\times N\times 1}^{\text{productive}}] = T_{PL} + \sum_{d=1}^{2(N-1)} n_{I,d}\, t_{I,d}, \tag{3}$$

where $n_{I,d}$ is the number of reachable legal assembly locations in construction cycle d; $t_{I,d}$ is the same as in Eq. 2. The time over the productive CAT is *wasted CAT*.

To examine the effect of predictive local checks, we assume an ideal scenario where a robot can detect all action failures and then leave the structure. For simplicity, we assume that every action has the same average success rate $\hat{p}$. For each construction cycle d, we define a random variable Z_d as the number of trials needed until a "successful" trip, i.e. a trip without failures, occurs. Z_d and the two random variables X_d and Y_d defined in the general model (Eq. 1) are independent. Therefore, the expected number of trips needed to visit all reachable legal assembly locations at least once successfully is $E[X_d]E[Y_d]E[Z_d]$. We denote the weighted average number of actions in the trips to all reachable legal assembly locations as $a_{I,d} = 3 + 5d/4$. Then $E[Z_d] = 1/\hat{p}^{a_{I,d}}$. To further simplify the model, we assume that action failures do not occur during brick placement since that time only amounts to a small portion of the total CAT. Under these assumptions, the CAT over-approximation with predictive local checks $E[T_{N\times N\times 1}^{\text{checks}}]$:

$$E[T_{N\times N\times 1}^{\text{checks}}] = T_{PL} + \sum_{d=1}^{2(N-1)} E[X_d]\, \frac{1}{\hat{p}^{a_{I,d}}}\, t_{I,d}, \tag{4}$$

Equation 4 tells us that the CAT can be wasted due to both action failures, and the situation where there is "no placement" because the robot does not find a valid assembly location. The wasted CAT caused by action failures is $E[T_{N\times N\times 1}^{\text{checks}}] - E[T_{N\times N\times 1}]$ and the wasted CAT caused by no placement is $E[T_{N\times N\times 1}] - E[T_{N\times N\times 1}^{\text{productive}}]$. Figure 5(e) compares the wasted CAT associated with each, and shows that as the structure size increases, the effect of action failures grows and ultimately surpasses the effect of trips without placement. We define the structure size at which the action failures and the trips without placements lead to the same wasted CAT as the *crossover point*. Beyond this point, action failures become the dominant cause of wasted CAT and the scaling behavior of the CAT becomes undesirable. Therefore the crossover point can also be interpreted as the structure size limit under which predictive local checks are a good strategy. Figure 5(f) shows the crossover point at different $\hat{p}$ and supports the intuitive notion that improving the action success rate can significantly shift the crossover point higher, making the system more capable of building larger structures within reasonable time span.

6 Conclusion

In this work, we investigated the effect of errors in CRC. Grounding our work in the TERMES system, we have shown that the parallel policy, compared to a sequential or random policy, is in general the policy that will yield the highest probability of success and lowest cumulative action time. Additionally, we proposed a categorization of mistakes into action failures and fatal errors, showing how certain action failures are more likely to yield fatal errors. Using insights gained from the investigation of errors, we proposed predictive local checks requiring software modifications and minimal hardware changes. We showed how these predictive local checks substantially increase the probability of success, in some cases by an order of magnitude. Adding these checks also reduces the differences in success probability between the various policies, so that the worst ones are the most improved. In addition, we showed that these local predictive checks can result in an increase in cumulative action time.

This work has many potential future directions. Here, we considered action failures that do not change the robot state; next, it would be interesting to investigate more complex classes of failures or sensing errors. We showed that when constructing large structures, local predictive checks increase the probability of success, while increasing the cumulative action time. It would also be interesting to develop localization methods that allow the robots to recover from failures instead of simply detecting them.

Acknowledgments. This project was funded by the Packard Fellowship for Science and Engineering, GETTYLABS, and the National Science Foundation (NSF) Grant #1846340 and #2042411.

References

1. Petersen, K.H., Napp, N., Stuart-Smith, R., Rus, D., Kovac, M.: A review of collective robotic construction. Sci. Robot. **4**(28) (2019)
2. Solly, J., et al.: ICD/ITKE research pavilion 2016/2017: integrative design of a composite lattice cantilever. In: Proceedings of IASS Annual Symposia, pp. 1–8. International Association for Shell and Spatial Structures (IASS) (2018)
3. Augugliaro, F., et al.: The flight assembled architecture installation: cooperative construction with flying machines. IEEE Control Syst. Mag. **34**(4), 46–64 (2014)
4. O'hara, I., et al.: Self-assembly of a swarm of autonomous boats into floating structures. In: 2014 IEEE International Conference on Robotics and Automation (ICRA), pp. 1234–1240. IEEE (2014)
5. Lindsey, Q., Mellinger, D., Kumar, V.: Construction of cubic structures with quadrotor teams. In: Proceedings of the Robotics: Science & Systems VII (2011)
6. Werfel, J., Petersen, K., Nagpal, R.: Designing collective behavior in a termite-inspired robot construction team. Science **343**(6172), 754–758 (2014)
7. Deng, Y., Hua, Y., Napp, N., Petersen, K.: A compiler for scalable construction by the TERMES robot collective. Robot. Auton. Syst. **121** (2019)
8. Petersen, K., Nagpal, R., Werfel, J.: TERMES: an autonomous robotic system for three-dimensional collective construction. Robot. Sci. Syst. **7**, 257–264 (2012)
9. Napp, N., Nagpal, R.: Distributed amorphous ramp construction in unstructured environments. Robotica **32**(2), 279–290 (2014)
10. Soleymani, T., Trianni, V., Bonani, M., Mondada, F., Dorigo, M.: Bio-inspired construction with mobile robots and compliant pockets. Robot. Auton. Syst. **74**, 340–350 (2015)
11. Rubenstein, M., Cornejo, A., Nagpal, R.: Programmable self-assembly in a thousand-robot swarm. Science **345**(6198), 795–799 (2014)
12. Flajolet, P., Gardy, D., Thimonier, L.: Birthday paradox, coupon collectors, caching algorithms and self-organizing search. Discret. Appl. Math. **39**(3), 207–229 (1992)

Optimal Multi-robot Perimeter Defense Using Flow Networks

Austin K. Chen[1(✉)], Douglas G. Macharet[2], Daigo Shishika[3], George J. Pappas[1], and Vijay Kumar[1]

[1] GRASP Lab at the University of Pennsylvania, Philadelphia, USA
{akchen,pappasg,kumar}@seas.upenn.edu
[2] Department of Computer Science, Universidade Federal de Minas Gerais, Belo Horizonte, Brazil
doug@dcc.ufmg.br
[3] Mechanical Engineering Department, George Mason University, Fairfax, USA
dshishik@gmu.edu

Abstract. In perimeter defense, a team of defenders seeks to intercept a team of intruders before they reach the perimeter. Though the single defender case is relatively well studied, with multiple defenders significant complexity is introduced because coordination must also be considered. In this work, we present a formulation of the perimeter defense problem as an instance of the min-cost-max-flow problem for flow networks, and leverage existing efficient algorithms for network flows to solve both the task assignment and routing problems for perimeter defense concurrently. When considering homogeneous defender robots, the computed solution is optimal for any individual timestep. Additionally, we detail a deconflict-based strategy for dealing with heterogeneous defenders, and show in simulation that the proposed solutions match or outperform a naive greedy baseline.

Keywords: Multi-robot systems · Perimeter defense · Flow networks

1 Introduction

In the perimeter defense task, a team of defenders seeks to intercept intruders before they reach a predefined boundary. Many variants of the problem have been studied with a differential-game formulation [8,10,11], which considers both the strategy of the intruders and the strategy of the defenders.

In this work, we consider the case where intruders move with constant velocity. Then the scenario is no longer a game, as the intruder strategy is fixed. The defenders must plan and execute multiple captures of intruders, which appear at random locations and times. With a single defender, the optimal assignment and route of intruders to capture is then equivalent to solving the longest path problem [12]. This work has also been extended for radial boundaries [3], showing that strategies developed on one kind of perimeter can often be adapted

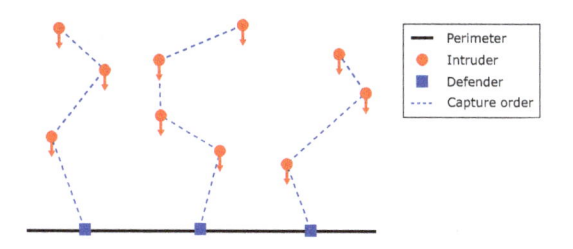

Fig. 1. Visualization of a homogeneous perimeter defense scenario. Intruders (red circles) move towards the perimeter (black line), while defenders (blue squares) move horizontally to intercept the intruders. Defender assignments and capture routes are denoted with the dashed blue lines.

for other kinds of perimeters. A visualization of the scenario considered in this paper is presented in Fig. 1.

It is not yet clear, however, how to find the optimal assignment and routing for the perimeter defense task with multiple defender robots in a computationally tractable manner. One way to solve this problem is to partition the environment based on the spatial proximity of the robots to tasks, i.e. assigning each intruder to the closest defender [5]. The partition boundaries between robots can also change over time [7]. However, these solutions essentially decompose the overall task into a set of sub-tasks with minimal coordination, and so may not scale well with the number of robots.

We handle task assignment and robot routing concurrently in this work through the use of flow networks, which are directed graphs where each edge has a corresponding capacity. Flow networks have been studied extensively for many years [2]. By converting the perimeter defense problem to an instance of the min-cost-max-flow flow problem, we can make use of existing algorithms to solve the perimeter defense problem efficiently. As we will later show, an optimal min-cost-max-flow solution can give an optimal perimeter defense solution for homogeneous defenders. Additionally, our formulation can handle intruders that approach at different velocities and intruders with different capture rewards. We also present an extension of our method to heterogeneous defenders, and show that our proposed method outperforms a naive greedy baseline strategy.

2 Related Work

Perimeter defense and area monitoring [1,9] are commonly studied tasks in robotics. Previous work has formulated these tasks as a pursuit-evasion game, where both the defender and intruder strategies are considered [8,10,11]. A maximum matching problem between defenders and intruders is solved a single time for the assignment, since it is assumed that each defender can only execute one capture.

In [12] the authors consider a variation of the perimeter defense task where a single defender can execute multiple captures while moving upon a linear perimeter. Intruders also move radially towards the perimeter with constant speed so there is no game aspect. The authors introduce the concept of a *reachability graph*, which models all possible transitions given the agents' velocities. The longest path on this graph represents the optimal policy. Based on the same idea, [3] tackled the case where the vehicle is able to move freely in the environment and must defend a circular perimeter against radially incoming intruders.

One way to extend these single-robot solutions to multi-robot systems is to partition the environment and allocate robots to specific regions [5]. In [7], an adaptive partitioning strategy based on intruder arrival estimation is proposed. Within each partition, defenders follow an independent routing policy based on [12]. In this context, other coordination strategies have also used simple reactive threshold-based methods [16] or metrics utilizing Bayes risk [4].

For multi-agent path planning, an interesting alternative to TSP-based solutions is a conversion to a network flow problem [14,15]. A combined target-assignment and path-finding problem is tackled in [6], which is closely related to our problem. The authors use a min-cost-max-flow algorithm formulation, allowing them to individually assign agents to targets and plan their paths.

In this work, we design a unified reachability graph that takes all agents into account. This graph is then converted into a min-cost-max-flow problem such that the solution gives optimal capture paths for all defenders.

3 Problem Formulation

Given an obstacle-free rectangular environment $\mathcal{E}$ of fixed size, consider a line $\mathcal{P}$ which lies on the bottom edge of $\mathcal{E}$. Intruders appear at the top edge and move straight downwards towards the bottom edge. A team of defender robots must intercept the intruders before they reach $\mathcal{P}$. The distance between $\mathcal{P}$ and the arrival location of the intruders may be interpreted as the sensing horizon of the defenders.

Definition 1 (Perimeter). *Without loss of generality, we assume $\mathcal{P}$ is a line segment located on the x-axis (using Cartesian coordinates) with length w. Thus, the perimeter is defined as the line segment between the points $(0,0)$ and $(w,0)$.*

Definition 2 (Defenders). *A team of defenders $\mathcal{D} = \{\mathcal{D}_1, ..., \mathcal{D}_N\}$ is distributed over $\mathcal{P}$. The defender's position is represented in Cartesian coordinates: $\mathbf{p}_i^d = [x_i^d, y_i^d]^T$. Defenders are constrained to move on perimeter, i.e., $y_i^d = 0$, and with limited velocity ν^d, i.e., $|\dot{x}_i^d| \leq \nu_i^d$. We assume the defender's dynamics to be first order and so the velocity of the defender can be controlled directly.*

Definition 3 (Intruders). *An intruder $\mathcal{A}_j$ is an agent approaching $\mathcal{P}$, and its position is represented by Cartesian coordinates $\mathbf{p}_j^a = [x_j^a, y_j^a]^T$. Intruders move perpendicular to $\mathcal{P}$ and approach the perimeter with constant speed, i.e., $\nu_j^a = -\dot{y}_j^a$.*

The insertion of intruders in the environment occurs sequentially over time accordingly to a Poisson process with rate λ. The horizontal location (i.e. the position x^a) of new intruders is randomly chosen accordingly to some unknown distribution ϕ^* within the width of the perimeter, but at a fixed distance η from the perimeter (i.e. $y^a = \eta$ for a newly generated intruder). We assume that as soon as a new intruder is inserted it starts moving towards $\mathcal{P}$ in a straight line with constant linear velocity ν^a.

An intruder j is *captured* (intercepted) by a defender i when $||\mathbf{p}_j^a - \mathbf{p}_i^d|| = 0$, otherwise, it has *escaped* if it was able to reach the perimeter, i.e., $||y_j^a|| \leq 0$, without being captured.

Problem 1 (Cooperative Perimeter Defense). Given a perimeter $\mathcal{P}$ and a group of N defender robots placed on the perimeter, design a defense strategy that minimizes the number of escaped intruders.

The overall performance of the proposed methodology is assessed by the fraction of intruders that are captured [3,12], i.e.,

$$C = \lim_{t\to\infty} \mathbb{E}\left[\frac{n_{\text{cap}}(t)}{\lambda t}\right], \tag{1}$$

considering the arrival of new intruders is governed by a Poisson process with rate λ, and with $n_{cap}(t)$ denoting total captures at time t.

4 Flow Network Formulation

The perimeter defense problem as stated in Sect. 3 may be considered as an assignment (of defenders to intruders) and routing (how to move defenders to assigned intruders) problem. In this section, we detail the formulation of the assignment and routing problem for perimeter defense as an instance of the min-cost-max-flow problem.

4.1 Graph Generation

To convert the perimeter defense problem into a network flow problem, we first start with the generation of a flow network. We aim to create a network such that solving the min-cost-max-flow problem for this graph will give us an optimal solution to the perimeter defense problem. In contrast to previous work that studied the single-defender case [12], this graph approach allows us to solve the perimeter defense problem for multiple homogeneous defenders.

More formally, a flow network is a graph $G = (V, E)$ that has a set of vertices V, edges E, and corresponding capacity $c(e)$ for every edge $e \in E$. Since we solve the min-cost-max-flow problem, the edges must also have a cost attribute $a(e)$ for $e \in E$. We can write the min-cost-max-flow problem as

$$\begin{aligned}
\underset{f}{\text{minimize}} \quad & \sum_{e \in E} a(e) \cdot f(e) \\
\text{subject to} \quad & f(e) \leq c(e) \quad \forall e \in E \\
& \sum_{e:(s,v)\in E} f(e) \text{ is maximized},
\end{aligned}$$

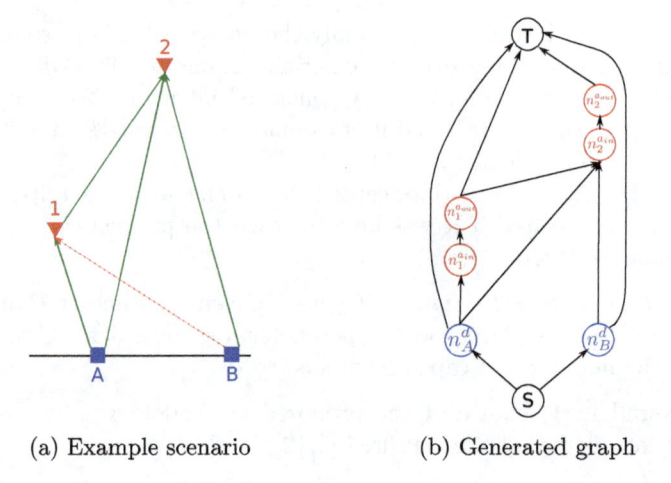

Fig. 2. An instance of the perimeter defense scenario (a) and its corresponding flow network (b). In this example, defender A can reach intruders 1 and 2 (green lines), but defender B cannot reach intruder 1 (red dotted line). In addition, intruder 2 is reachable after capturing intruder 1. The generated flow network reflects this reachability information, with edges existing when capture is feasible.

where s is the source node and $v \in V$. In other words, we select the minimum cost solution among all maximum flows.

To generate the graph, start with two nodes S and T. These will act as the start and target nodes for our flow network, respectively. Then, add one node for each of the defenders, which we will call node n_i^d for defender i. Finally, add two nodes for each intruder, which we will call nodes $n_j^{a_{in}}$ and $n_j^{a_{out}}$.

Next, we detail the edges in the graph, which all have unit capacity. First, we connect each of the defender nodes to the start node S with cost 0. Next, connect each $n_j^{a_{in}}$ to each corresponding $n_j^{a_{out}}$ with negative cost proportional to the reward for capturing that intruder. That is, the higher the importance of capture, the higher the magnitude of the negative cost should be. For intruders that are all equally important (or to maximize the total number of intruders captured), the cost can be set to -1. Then, create edges from all defender nodes n_i^d and all intruder $n_j^{a_{out}}$ nodes to the target node T with cost 0.

Now, we explain the process for creating edges between defender nodes and intruder nodes, and between different intruder nodes. Edges from defender nodes to intruder nodes are based upon the dynamic feasibility of capture. For defender i to capture intruder j, it must be able to traverse the difference in x position in less time than it takes for the intruder to reach the perimeter, i.e.,

$$\frac{|x_j^a(t) - x_i^d(t)|}{\nu^d} \leq \frac{y_j^a(t)}{\nu_j^a}. \qquad (2)$$

In cases where this condition is true, add an edge to the graph between n_i^d and $n_j^{a_{in}}$ with capacity 1 and cost 0. Otherwise, do not add an edge. To allow for consecutive captures, we must also add edges between intruder nodes corresponding to different intruders. The feasibility condition is very similar to that of (2), and is written as

$$\frac{|x_j^a(t) - x_k^a(t)|}{\nu^d} \leq \frac{y_j^a(t) - y_k^a(t)}{\nu_j^a}. \tag{3}$$

If (3) holds, an edge with cost 0 is added between $n_k^{a_{out}}$ and $n_j^{a_{in}}$. An example scenario and its corresponding flow network are presented in Fig. 2.

4.2 Intuition Behind Constructed Graph

The edges between S and the defender nodes constrain the flow through each defender node to be 1. In fact, for any min-cost-max-flow solution of the graph G as constructed in Sect. 4.1, the max flow from S to T is equal to $|\mathcal{D}|$.

Lemma 1. *For any min-cost-max-flow solution to a graph G as constructed according to Sect. 4.1, the maximum flow from S to T is equal to $|\mathcal{D}|$.*

Proof. Since the flow from S may reach T only through the defender nodes, the maximum flow is upper bounded by the number of defender nodes and therefore the number of defenders. Suppose, towards a contradiction, that the maximum flow is lower than the number of defenders. Then, at least one of the edges e^* from S to n^{d^*} must have $f(e^*) = 0$. However, there always exists a path from n^{d^*} to T with cost 0 and capacity 1 by construction, so sending flow from S to T through n^{d^*} will increase flow without changing cost. Hence, the candidate solution is not a min-cost-max-flow solution. Therefore, we have a contradiction, and the maximum flow must be equal to the number of defenders.

Since the only edges with nonzero cost are those between the intruder $n^{a_{in}}$ and $n^{a_{out}}$ nodes, the min-cost-max-flow solution will send flow through as many of these edges as possible (assuming equal costs for intruders). By tracing these flows back to the defender nodes, we will be able to determine capture assignments and routes for each defender. Note that since the capacity is constrained to be 1 on all edges, each intruder will be assigned to at most one defender.

Because the creation of edges between defender nodes and intruder nodes is contingent upon the dynamic feasibility of the defender performing a capture, the resulting network flow solution is also guaranteed to provide flows that correspond to valid assignments. The same logic holds for consecutive captures, as the edges between intruders are constructed from the same principles.

4.3 Recovery of Intruder Assignments

Given a valid flow $f^*(e)$, we recover the ordered intruder assignment for defender i by following the flow through the corresponding node n_i^d. If an outgoing edge from n_i^d has nonzero flow to an intruder node $n_j^{a_{in}}$, we assign intruder j to defender i. We then repeat this process from $n_j^{a_{out}}$ until we reach the terminal node T, at which point we have the assignments $(j, k, \dots)$ for defender i.

As mentioned previously, these assignments are guaranteed to be dynamically feasible because of the construction of graph G. For defender control, it suffices to set the velocity of defender i in the direction of its next capture of intruder j (i.e. $\dot{x}_i^d = \nu_i^d \text{sgn}(x_j^a - x_i^d)$ while $x_i^d \neq x_j^a$).

4.4 Optimality of Intruder Assignments

Since the only nonzero costs exist between the intruder nodes $n^{a_{in}}$ and $n^{a_{out}}$ and because the cost between these nodes is always negative, the minimum cost maximum flow is achieved when maximum flow is sent through the intruder nodes. This corresponds exactly to guaranteed (i.e. dynamically feasible) intruder captures, so an optimal min-cost-max-flow solution gives an optimal ordered intruder assignment for any specific time.

4.5 Complexity

To construct the graph, we add $|\mathcal{D}|$ defender nodes and $2|\mathcal{A}|$ intruder nodes, along with 2 nodes for the start and target. To check for reachability of the intruders, it is necessary to check (2) and (3) for all relevant defender-intruder pair and intruder-intruder pairs, or $\mathcal{O}(|\mathcal{D}||\mathcal{A}| + |\mathcal{A}|^2)$ times. Thus, construction of the graph takes $\mathcal{O}(|\mathcal{D}||\mathcal{A}| + |\mathcal{A}|^2)$ time.

To solve the min-cost-max-flow problem, we may use the network simplex algorithm. This algorithm has time complexity $\mathcal{O}(|V||E|\log|V|\log(|V|C))$ where C is the maximum magnitude of any edge cost [13]. If we consider the case of homogeneous intruders (i.e. $C = 1$), the overall complexity is $\mathcal{O}(|V||E|\log^2|V|)$ where $|V| = (|\mathcal{D}| + 2|\mathcal{A}| + 2)$ and $|E|$ is bounded from above by $(|\mathcal{D}||\mathcal{A}| + |\mathcal{A}|^2)$.

Finally, tracing through the min-cost-max-flow solution to recover the intruder assignment for one defender takes $\mathcal{O}(|\mathcal{A}|)$ time. Consequently, recovering intruder assignments for all defenders takes $\mathcal{O}(|\mathcal{D}||\mathcal{A}|)$ time.

5 Extension to Heterogeneous Defenders

Note that in (2) and (3), the maximum defender speed ν^d must be identical across all defenders. If the defenders have different maximum velocities, then the construction of G according to Sect. 4.1 is not possible since the constructed graph might vary from defender to defender.

One possible solution is to set ν^d equal to the lowest maximum velocity of the defender team. However, this conservative approach can lead to very poor performance, as we demonstrate in Sect. 6. For this reason, we propose a deconflict-style approach, where we first solve the perimeter defense problem individually for each defender and then iteratively refine the team solution.

5.1 Deconflict Assignment

First, a graph G_i is generated for each defender i according to Sect. 4.1. During the generation, only one defender is considered and the speed of the individual defender is taken into account. Then, the min-cost-max-flow problem is solved for each graph to give every defender an initial assignment.

To combine these individual solutions, we repeatedly modify the individual assignments. First, the initial assignments are stored as candidates. Next, within graph G_i of defender i, all intruder nodes that are assigned to other defenders are removed from G_i and the min-cost-max-flow problem is solved again for each graph. The defender with the lowest cost flow solution in its modified graph then has its candidate solution changed to that of its modified graph, all removed nodes are restored, and the process is repeated. The algorithm terminates when the total number of captures remains the same between iterations.

An example of the deconflict assignment process is shown in Fig. 3. In this example, we consider two defenders A and B. Though defender A has more intruders in its reachable set, it is initially assigned the same intruders as defender B because it maximizes captures over its own graph. Through the deconflicting process, the assignment for defender A changes and the team achieves a higher overall score.

(a) Initial Assignment   (b) Trial Assignment   (c) Updated Assignment

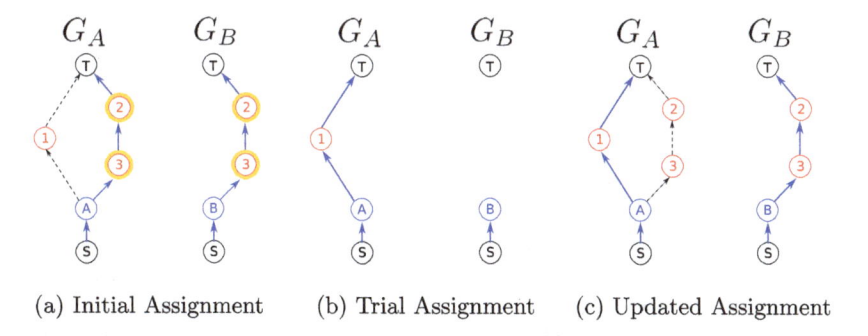

Fig. 3. Example of deconflict assignment. For visual clarity, we omit edges from the defender nodes to the target node and combine the two nodes for each intruder. In (a), the initial min-cost-max-flow solutions are shown by the solid blue lines. Intruders 2 and 3 (orange highlight) are overprovisioned. (b) shows recomputed assignments with the overprovisioned intruder nodes removed, and (c) shows the updated assignments after the augmentation round is complete.

5.2 Intuition Behind Deconflict Assignment

Due to the lack of coordination between the individual solutions, defenders may overprovision for the capture of a single intruder. Deconflict assignment seeks to improve a collection of individual assignments by greedily seeking the best improvement at every iteration. Intruder nodes which are assigned to more than one defender are removed from the graphs when searching for improvements; since these nodes are overprovisioned, not visiting that intruder node will not degrade the team's performance if only one defender's assignment is modified. Because of this, at every iteration the overall team performance will never decrease and defenders are assigned to intruders that have not yet been captured.

5.3 Complexity

The complexity analysis for deconflict assignment is similar to that presented in Sect. 4.5, except each defender now has its own graph and the min-cost-max-flow problem must be solved for each graph. Thus, the complexity for each min-cost-max-flow solve is $\mathcal{O}(|\mathcal{A}|^3 \log |\mathcal{A}| \log(|\mathcal{A}|C))$. To get overprovisioned intruders we compute a set intersection, which has worst case runtime $\mathcal{O}(|\mathcal{A}|^2)$. So computing intruder overlap for each defender takes $\mathcal{O}(|\mathcal{D}||\mathcal{A}|^2)$ time. Finally, since the total number of captures (or achieved capture score for heterogeneous intruders) must increase for every round of augmentations, the number of iterations is limited by the maximum captures (or score). Thus, the complexity of deconflict assignment is $\mathcal{O}\Big((|\mathcal{A}|C)(|\mathcal{A}|^3 \log |\mathcal{A}| \log(|\mathcal{A}|C) + |\mathcal{D}||\mathcal{A}|^2)\Big)$.

6 Results

6.1 Experimental Setup

In our experiments we use $w = 45$ m and $\eta = 25$ m, with parameters defined according to Sect. 3. During each run, intruder arrival events occur for 30 s at an average rate $\lambda = 2$, and the simulation ends when all intruders are either captured or escaped. We consider a team of four defenders (either homogeneous or heterogeneous) that are initially distributed evenly across the perimeter. We consider homogeneous and heterogeneous intruder velocities and capture priorities, which are detailed in the following sections. We report the average and standard deviation for all results over 100 runs. Every run is random but the same intruder arrivals are given to each algorithm across an individual run.

As part of our results, we include a baseline greedy strategy (denoted *Greedy*). In the greedy algorithm, the defenders individually maximize the number of intruder assignments or score over the set of currently unassigned intruders (according to some arbitrary order). We compare this greedy strategy with our min-cost-max-flow formulation from Sect. 4 (denoted *Maxflow*) and the heterogeneous deconflicting solution from Sect. 5 (denoted *Deconflict*).

Defenders are separated into two cases. For homogeneous defenders, denoted *Homogeneous*, the defenders each have maximum velocity $\nu^d = 5$ m/s. For heterogeneous defenders, denoted *heterogeneous*, the maximum velocities for the defenders are linearly spaced between 1 and 6. For example, for four defenders, the velocity for each defender would be one of $[1, 2.67, 4.33, 6]$. Snapshots of an experimental run are shown in Fig. 4.

(a) $t = 5s$

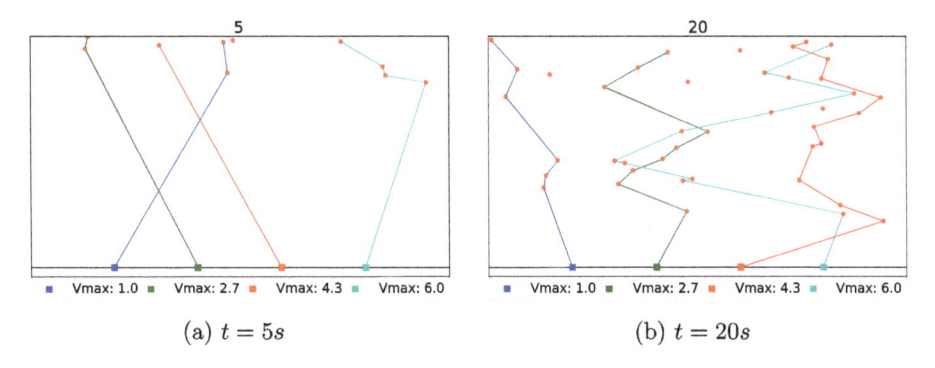

(b) $t = 20s$

Fig. 4. Trial run with heterogeneous defenders. Defender capture assignments are shown with the colored lines. As time progresses and new intruders appear, the intruder assignments are recalculated (note the differences between $t = 5$ and $t = 20$).

6.2 Discussion of Results

First, we consider homogeneous intruders where the velocity of each intruder is identical and equal to 1 (i.e. $\nu^a = 1$). The "Homogeneous Intruder $\mathcal{C}$" column in Table 1 shows the resulting capture rates. *Maxflow* yields the best results when the defenders are homogeneous, as solving the min-cost-max-flow problem for the converted graph directly corresponds with solving the perimeter defense problem at any timestep. Performance drops off severely when the defenders are heterogeneous, as the defenders are no longer able to leverage faster-moving robots. In this case, *Deconflict* degrades in performance far less when compared to the homogeneous case. In all tests, both proposed approaches are similar (within two percent) or superior to *Greedy*.

Table 1. Capture rates ($\mathcal{C}$) and weighted scores for tested algorithms, reported as mean ± standard deviation over 100 runs.

Defender Type	Algorithm	Homogeneous Intruder $\mathcal{C}$	Heterogeneous Intruder $\mathcal{C}$	Heterogeneous Intruder Score
Homogeneous Defenders	*Greedy*	74.3 ± 6.1	68.8 ± 7.0	85.1 ± 4.62
	Maxflow	86.5 ± 4.0	88.5 ± 4.3	89.1 ± 4.3
	Deconflict	83.5 ± 4.9	64.9 ± 5.9	84.36 ± 5.2
Heterogeneous Defenders	*Greedy*	55.9 ± 10.6	52.6 ± 9.38	79.8 ± 5.0
	Maxflow	54.0 ± 4.2	58.0 ± 5.6	59.7 ± 5.4
	Deconflict	77.8 ± 4.9	60.7 ± 5.8	78.6 ± 5.9

To model heterogeneous intruders, we generate each intruder's velocity from a uniform distribution with a minimum of .5 and maximum of 3. Results are shown in the "Heterogeneous Intruder $\mathcal{C}$" column of Table 1. As before, *Maxflow* achieves a superior capture rate when the defenders are homogeneous. However, *Deconflict* achieves a lower average capture than the baseline *Greedy*. This is perhaps due to the lack of a penalty for distance travelled, so robots may move out of position for captures of unseen intruders. Because of the higher average intruder velocity, the difference between *Maxflow* and *Deconflict* for heterogeneous defenders is small, but *Deconflict* still achieves the best capture rate. In general, the variance for *Greedy* is higher than that of the other two algorithms, so *Greedy* is also more likely to provide poor solutions on individual runs.

Finally, we perform a series of experiments where the rewards for capturing intruders are nonuniform. We use heterogeneous velocities for the intruders as before, but also set the reward for capturing each intruder equal to its speed. Results can be seen in the "Heterogeneous Intruder Score" column of Table 1. Note that the numbers reported here are not capture rates, but the percentage of the maximum possible score achieved. For homogeneous defenders, all three tested algorithms perform reasonably well, but *Maxflow* again achieves the best performance. For the heterogeneous case, *Maxflow* degrades significantly in performance, while both *Greedy* and *Deconflict* perform well.

7 Conclusion

In this paper, we consider the perimeter defense problem with multiple defenders. The main challenges of assignment and routing arise when multiple defenders are used. To address these challenges, we propose a graph formulation that converts the perimeter defense problem to an instance of the min-cost-max-flow problem. This formulation allows us to leverage existing algorithms to efficiently solve the perimeter defense problem. While in this paper we only consider intruders that move perpendicular to the perimeter, our method can be applied as long as the arrival time and location of intruders at the perimeter are known.

Through numerical experiments, we show that the proposed method outperforms a benchmark method (based on greedy search) when the defenders are homogeneous. We also propose and test an extension of our method to accommodate defenders that have heterogeneous speeds. Though this method is not optimal, we show that it roughly matches or exceeds the performance of the greedy baseline when defenders are heterogeneous.

For future work, we would like to consider other methods of dealing with heterogeneous defenders, since the proposed method is greedy and not optimal. In addition, we would like to further extend this formulation to other regimes, e.g. radial boundaries or defenders that can move anywhere in the environment. Implementation on real robots also poses a unique challenge, as many of the assumptions (e.g. global knowledge of intruder locations) will need to be reconsidered. This may be aided by considering a more decentralized approach, as we currently assume global communications and synchronized sensing. In particular, we may extend *Deconflict* or use a similar market-based algorithm to synchronize robots that are in communication range.

References

1. Agmon, N., Kraus, S., Kaminka, G.A.: Multi-robot perimeter patrol in adversarial settings. In: IEEE International Conference on Robotics and Automation (ICRA), pp. 2339–2345 (2008)
2. Ahuja, R.K., Magnanti, T.L., Orlin, J.B.: Network Flows (1993)
3. Bajaj, S., Bopardikar, S.D.: Dynamic boundary guarding against radially incoming targets. In: IEEE International Conference on Decision and Control (CDC), pp. 642–649 (2019)
4. Bays, M.J., Shende, A., Stilwell, D.J.: An approach to multi-agent area protection using Bayes risk. In: IEEE International Conference on Robotics and Automation (ICRA), pp. 642–649 (2012)
5. Fu, J.G.M., Bandyopadhyay, T., Ang, M.H.: Local Voronoi decomposition for multi-agent task allocation. In: IEEE International Conference on Robotics and Automation, pp. 1935–1940 (2009)
6. Ma, H., Koenig, S.: Optimal target assignment and path finding for teams of agents. In: Proceedings of the 2016 International Conference on Autonomous Agents & Multiagent Systems (AAMAS), pp. 1144–1152. International Foundation for Autonomous Agents and Multiagent Systems, Richland (2016)
7. Macharet, D.G., Chen, A.K., Shishika, D., Pappas, G.J., Kumar, V.: Adaptive partitioning for coordinated multi-agent perimeter defense. In: IEEE/RSJ International Conference on Intelligent Robots and Systems (IROS) (2020)
8. Shishika, D., Kumar, V.: Local-game decomposition for multiplayer perimeter-defense problem. In: IEEE Conference on Decision and Control (CDC), pp. 2093–2100 (2018)
9. Shishika, D., Macharet, D.G., Sadler, B.M., Kumar, V.: Game theoretic formation design for probabilistic barrier coverage. In: IEEE/RSJ International Conference on Intelligent Robots and Systems (IROS), pp. 11,703–11,709 (2020)
10. Shishika, D., Paulos, J., Hsieh, M.A., Kumar, V.: Team composition for perimeter defense with patrollers and defenders. In: IEEE Conference on Decision and Control (CDC), pp. 7325–7332 (2019)
11. Shishika, D., Paulos, J., Kumar, V.: Cooperative team strategies for multi-player perimeter-defense games. IEEE Robot. Autom. Lett. **5**(2), 2738–2745 (2020)
12. Smith, S.L., Bopardikar, S.D., Bullo, F.: A dynamic boundary guarding problem with translating targets. In: Proceedings of the 48h IEEE Conference on Decision and Control (CDC) held jointly with 2009 28th Chinese Control Conference, pp. 8543–8548 (2009)
13. Tarjan, R.E.: Dynamic trees as search trees via Euler tours, applied to the network simplex algorithm. Math. Program. **78**(2), 169–177 (1997)
14. Yu, J., LaValle, S.M.: Multi-agent path planning and network flow. In: Frazzoli, E., Lozano-Perez, T., Roy, N., Rus, D. (eds.) Algorithmic Foundations of Robotics X. STAR, vol. 86, pp. 157–173. Springer, Heidelberg (2013). https://doi.org/10.1007/978-3-642-36279-8_10
15. Yu, J., LaValle, S.M.: Optimal multirobot path planning on graphs: complete algorithms and effective heuristics. IEEE Trans. Rob. **32**(5), 1163–1177 (2016)
16. Zhang, Y., Meng, Y.: A decentralized multi-robot system for intruder detection in security defense. In: IEEE/RSJ International Conference on Intelligent Robots and Systems (IROS), pp. 5563–5568 (2010)

Classification-Aware Path Planning of Network of Robots

Guangyi Liu[1(✉)], Arash Amini[1], Martin Takáč[1,2], and Nader Motee[1]

[1] Lehigh University, Bethlehem, PA 18015, USA
{gliu,ara416,motee}@lehigh.edu
[2] MBZUAI, Masdar City, Abu Dhabi, United Arab Emirates

Abstract. We propose a classification-aware path planning architecture for a team of robots in order to traverse along the most informative paths with the objective of completing map classification tasks using localized (partial) observations from the environment. In this method, the neural network layers with parallel structure utilize each agent's memorized history and solve the path planning problem to achieve classification. The objective is to avoid visiting less informative regions and significantly reduce the total energy cost (e.g., battery life) when solving the classification problem. Moreover, the parallel design of the path planning structure reduces the training complexity drastically. The efficacy of our approach has been validated by a map classification problem in the simulation environment of satellite campus maps using quadcopters with onboard cameras.

Keywords: Network of robots · Distributed classification · Path planning · Machine learning

1 Introduction

The distributed multi-robot system has shown tremendous performance improvement and potential in the problems involving the perception and the classification of an environment with machine learning. In most problems, robots can only sense their local environment [1], highlighting the importance of covering the most informative regions. In this work, we consider the main *challenge* as developing a path planning architecture that guarantees high performance, efficient sampling, and low energy cost from the distributed system's perspective. Then, we focus on the *problem* as classifying the map of the target environment and navigating robots to cover the most informative regions with downward-facing cameras. In this problem, robots can only access the local visual observations as a sequence of images from the environment.

The challenge of learning an action policy for classification purposes is evident as the reward will only be available at the end of the task, and it is highly dependent on the classifier's performance. This will make learning the path planning policy intractable, especially for the tasks with long operation time. Even if the classifier has optimal performance, the distributed system must adopt an effective path planning method to cover the most informative regions for effective classification. Otherwise, a terrible selection of local observations will drive the classifier into poor prediction results.

To address these challenges, we propose a classification-aware (CL-aware) path planner, which consists of a parallel structure of linear classifiers. Using the proposed approach will tackle the problem of path planning by solving a finite-label classification problem. The output feature history from LSTM cells will be classified as a sequence of actions by the CL-aware path planner, and the robot will adopt those actions while traversing the environment. Implementing the CL-aware policy helps the system to learn both long-term and short-term dependencies with the classification reward since it plans for a fixed path for each time interval and specifies the action for each time step. We refer the details to Sect. 3.3.

Various applications of distributed classification can be extended from the proposed approach. For example, identifying search and rescue regions after natural disasters (e.g., earthquake or tsunami), which changes the appearance of the region, enhancing existing maps of a region in a city for traffic and crowd control, and constructing soil map in geoscience during or after flooding to prevent and control potential subsequent disasters in a short time [2].

2 Related Work

The area of path planning in an unknown environment has been well studied in the past decades. In the context of path planning involving learning, research work [12] reviewed path-planning strategies for a map-based problem, in which robots will learn the surrounding maps. A probabilistic-based approach is presented by [6] for motion planning of robots in static work-space by learning a probabilistic road-map computed with a simple and fast local planner. Our work highlights the significant difference from the regular path planning works: for the classification purpose, the information gathered at each time step along the path is equivalently essential, contrary to solely focusing on reaching the final destination.

The idea of converting the path planning problem into a classification problem is inspired by [8], in which the author proposed a novel deterministic approach for path planning and obstacle avoidance by using Q-Learning in the grid-world. In the area of path planning with visual information, the author of [17] and [18] proposed a real-time path planning approach for the unknown environment, in which full CNN-based models and non-linear inverse reinforcement learning are utilized. A similar problem involving classification and navigation is studied by [11]. The author proposed using a color stereo camera and a single-axis ladar camera to classify the environmental terrain for autonomous navigation. In the time-varying environment, the author of [5] uses deep object detection models and a diver-following algorithm adaptive to the real-time and severe underwater conditions to solve the real-time path planning problem. Finally, for the path planning with partial observations, [3] proposed a method named LeTS-Drive that solves the problem of driving in a busy traffic intersection by using Partially Observable Markov Decision Process (POMDP) and online belief-tree search.

In the previous work [10, 13, 14], the authors solved the path planning problem for distributed multi-robot classification with single-action and goal-action policies that is learned with the classification reward. These policies highlight either long-term or short-term dependencies with the delayed reward, which may drive the robot away from

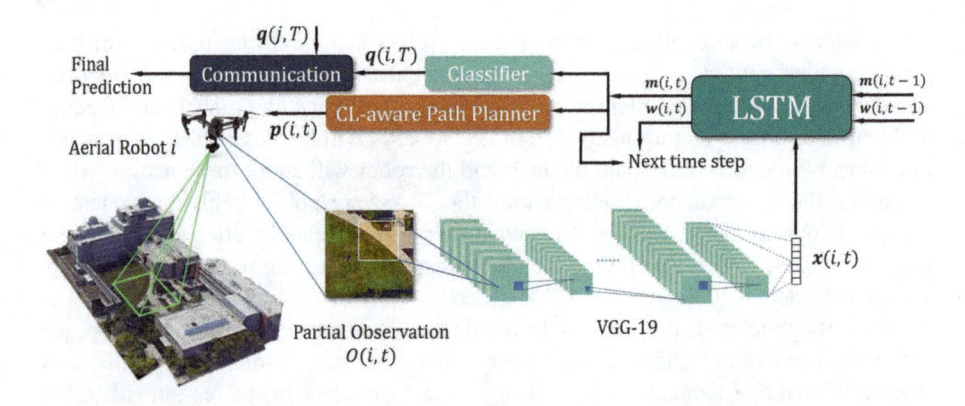

Fig. 1. Distributed classification architecture with the CL-aware path planner.

the informative sampling region. A classification-aware parallel NN structure will be used to learn both long and short-term dependencies in this work.

3 Distributed Classification Architecture

In this section, we present a distributed multi-robot map classification architecture which is a variant of the architecture shown in [10]. We refer the diagram to Fig. 1. The proposed architecture consists of four fundamental components: 1) A feature extraction section that processes the localized visual input and extracts classification features with a VGG-19 model. 2) An LSTM unit that stores the past perception features recursively. 3) A classification-aware path planning module that classifies the feature history into a pre-defined finite path set. 4) A communication network that collects the prediction among all robots and a map classifier that predicts the target region's label. Each of these components will be introduced in detail in the following subsections.

3.1 Preliminaries and Feature Extraction

We assume there are N robots operating in the target environment with the time index set $\mathbb{T} = \{1, 2, ..., T\}$. Each robot is capable of capturing a localized visual input from the environment, $O(i,t)$, in which $i \in \{1, \ldots, N\}$ denotes the label of the robot. The spatial position of the i'th robot is denoted as $l(i,t)$, and it is assumed in this work that $l(i,t) \in \mathbb{R}^2$.

For the i'th robot, the localized visual input $O(i,t)$ is fed into a pre-trained VGG-19 [16] model V_{θ_1}. The output of the VGG-19 model is presented by a feature vector $x(i,t) = V_{\theta_1}(O(i,t))$, in which θ_1 is the trainable parameter vector of the VGG-19 model. The output feature vector $x(i,t)$ contains the extracted classification features from i'th robot at time t.

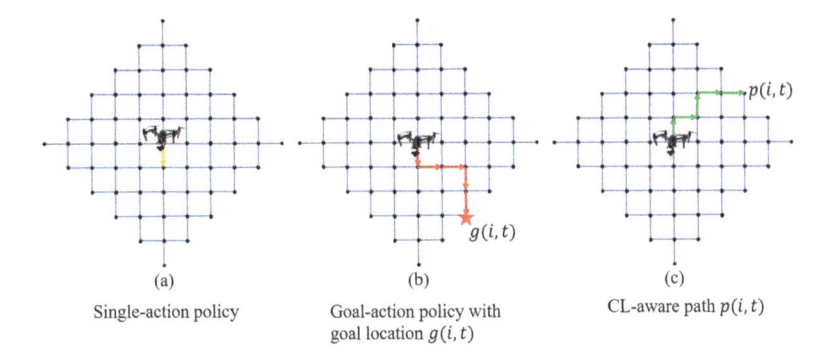

Fig. 2. Comparison of single-action policy, goal-action policy, and CL-aware path with $t_p = 5$. Actions in (b) are governed by a stochastic process.

3.2 LSTM Feature Encoding

Due to the hardware and environmental limitations (e.g., inside a building), robots can only capture localized observations at each time step. To learn the interconnection among the localized observations, robots need to keep track of the history of localized features efficiently. To achieve this, we adopt the same structure as in our previous work [13], which utilizes an LSTM [4] cell to save the history of features. Using the LSTM cell, robots are trained to learn the interrelations among history features to conclude a correct classification.

The feature vector $x(i,t)$ is treated as the input of the LSTM cell, while the hidden states $m(i,t)$ are utilized as the output for both classification and path planning purposes. The dynamics of the LSTM is governed by

$$\left[m(i,t)^T, w(i,t)^T\right]^T = F_{\theta_2}\bigl(m(i,t-1), w(i,t-1), x(i,t)\bigr), \qquad (1)$$

in which $w(i,t)$ denotes the cell state of the LSTM cell, and θ_2 is the trainable parameter vector. At each time horizon t, robots inherit the cell and hidden states from $t-1$ and fuses them with the new feature input $x(i,t)$ into history of features $m(i,t)$.

3.3 Classification-Aware Path Planning

Learning map classification with a distributed multi-robot system does not resemble the conventional classification problems in which the input is the entire environment, and the reward is independent of the actions of robots. Instead, in this problem, robots must learn a path planning policy that navigates the robot to collect environmental features efficiently with a delayed classification reward.

Single-Action and Goal-Action Policies. An intuitive approach is proposed in our previous work [13], in which the robot learns an action policy that generates one-step greedy actions at each time horizon (denoted as single-action policy), as shown in Fig. 2 (a). The performance of single-action policy is good with MNIST [9] dataset, but not

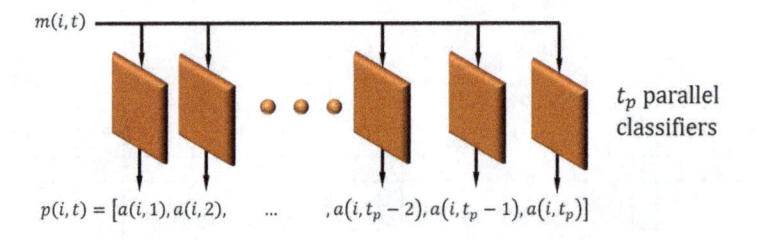

Fig. 3. The parallel structure of the CL-aware path planner.

with comparable performance in the large environment (e.g., maps). An optimal single-action policy could not be quickly learned with a delayed reward in those environments since it is only effective for short-term dependencies. Moreover, the greedy actions that ignore the long-term reward (e.g., the informative region is far away) will possibly drive the robots away from the global optimality.

To address the effect of the long-term dependencies between a path and the classification reward, the authors of [14] proposed a goal-based navigation architecture (denoted as goal-action policy) that decides an intermediate goal location $g(i,t) \in \mathbb{R}^2$ on the map for every $t_g \in \mathbb{R}$ time step. This method navigates the robot toward the goal location. We refer to Fig. 2 (b). The goal-based model generates a goal location instead of an action, and it shows a significant performance improvement in the large-dimension environment compared to the single-action method. However, the navigation between the goal location $g(i,t)$ and the robot's spatial position $l(i,t)$ is governed by a stochastic process for every t_g time step. In other words, the action policy between these two locations is independent of the classification reward. The uncertainty from the action policy would pose a potential threat to the performance when there exist obstacles in the environment or the landmark features are complicated.

CL-Aware Path Planning. Inspired by path planning with Q-Learning for obstacle avoidance [8], we consider our target environment in a 2-D grid world. We introduce a path as $p(i,t)$, which consists of a sequence of actions:

$$p(i,t) = [a(i,1), a(i,2), ..., a(i,t_p)], \qquad (2)$$

in which $t_p \in \mathbb{R}$ denotes the length of a path in the grid world, and $a(i,j), j \in \{1,...,t_p\}$ denotes the action vector for i'th robot at time $t + j$. The action vectors are defined in the discrete grid world such that $a(i,j) \in \{\leftarrow, \rightarrow, \uparrow, \downarrow\}$. If the i'th robot adopts a path $p(i,t)$, then it will traverse the environment from its current location $l(i,t)$ by taking the sequence of actions $\{a(i,1), ..., a(i,t_p)\}$, as shown in Fig. 2(c).

We denote the set of all paths with length t_p as $\mathcal{P}$, with the cardinality as $|\mathcal{P}| = 4^{t_p}$. For a finite t_p, the total possible paths in $\mathcal{P}$ are finite as well. In this manner, the path planning can be treated as a classification problem with 4^{t_p} finite outputs. Unlike the map classifier that will classify $m(i,t)$ into map labels, the CL-aware path planner will classify the feature history $m(i,t)$ into 4^{t_p} pre-defined path classes. Each class label corresponds to a unique path $p \in \mathcal{P}$, and the robot will adopt the corresponding path once the CL-aware path planner has concluded the "classification". Given the feature history $m(i,t)$, the robot will generate a deterministic path with length t_p by using the CL-aware path planner.

Since the cardinality of the set $\mathcal{P}$ grows exponentially with t_p, it will be extremely expensive to train a network with 4^{t_p} outputs. Moreover, one will need to retrain the entire network if t_p changes. To address this problem, we consider using t_p parallel linear NN layers with bias served as classifiers, as shown in Fig. 3. All the layers take $m(i,t)$ as an input and output a classification vector in $\mathbb{R}^4$. In this manner, the problem of training an extensive neural network can be boiled down into training t_p independent neural networks with relatively small output sizes.

Remark 1. For the j'th parallel layer, its classification output represents the decision of the j'th action in (2).

We denote the collection of all parallel classifiers as the CL-aware path planner such that

$$p(i,t) = P_{\theta_3}(m(i,t)),$$

in which θ_3 denotes the trainable parameters.

To learn the long-term dependencies with the delayed classification reward, robots will plan the next path only after the current path is terminated. During the time interval $[t, t+t_p]$, the action of robot will only be determined by the current path $p(i,t)$. For each task, a robot will plan a total of $\lceil T/t_p \rceil$ paths accordingly, in which $\lceil \cdot \rceil$ denotes the round-up operation. Even though each classifier in the path planner is independent of another, the learned policy for j'th action will depend on the previous actions since they share the same classification reward.

3.4 Map Classifier and Global Classification Reward

Robots also classify the target environment with the feature history vector $m(i,T)$ at the end of each task. The map classifier consists of a Softmax function and linear NN layers with bias, and it generates the belief vector as $q(i,T) = C_{\theta_4}(m(i,T))$, in which $q(i,t) \in \mathbb{R}^M$, M is the total number of classes, and θ_4 denotes the trainable parameters of the map classifier. We refer to the pseudo-code of the map classification with the CL-aware path planning to Algorithm 1.

Remark 2. Each robot will only generate a new path from the classification-aware path planner after every t_p time step. Otherwise, the robot will keep the current path and take actions along the path, as shown in Lines 6–10.

In the training stage, we assume that the all to all communication is available. All robots are capable of exchanging their prediction vector, such that the global prediction for classification is defined as

$$q = \frac{1}{N}\sum_{i=1}^{N} q(i,T).$$

Algorithm 1: Multi-Robot Map Classification with CL-aware Path Planning
---
Result: Predicted *Label*
1 Initialize $l(i,0)$, $m(i,0)$ and $w(i,0)$;
2 **while** $t \in \mathbb{T}$, **do**
3   **while** $i \in [1, N]$ **do**
4        $x(i,t) \leftarrow V_{\theta_1}(O(i,t))$;
5        $[m(i,t)^T, w(i,t)^T]^T \leftarrow F_{\theta_2}(m(i,t-1), w(i,t-1), x(i,t))$;
6   **if** $t \mod t_p = 0$ **then**
7           $p(i,t) \leftarrow P_{\theta_3}(m(i,t))$;
8   **else**
9   Keep using the current path;
10   **end**
11   Update the location with $a(i, t \mod t_p + 1)$;
12       $i \leftarrow i + 1$;
13   **end**
14    $t \leftarrow t + 1$;
15 **end**
16 $q(i,T) \leftarrow C_{\theta_4}(m(i,T))$;
17 $q \leftarrow \frac{1}{N}\sum_{i=1}^{N} q(i,T)$;
18 $Label \leftarrow \arg\max(q)$;

The predicted label is generated with an arg max function such that $Label = \arg\max(q)$. The global classification reward is evaluated by an log-sum-exp (LSE) loss as following,

$$r = -LSE(Q_k, q), \qquad (3)$$

in which Q_k is the k'th Euclidean basis in $\mathbb{R}^M$ and it denotes the ground truth vector for k'th label.

4 Training Preliminaries

We denote the collection of all trainable parameters in the proposed learning architecture as $\Theta = \{\theta_1, \theta_2, \theta_3, \theta_4\}$. We address the learning procedure as a constrained nonlinear optimization problem with cost function $J(\Theta) = r$, and the implementation with RL implies maximizing $J(\Theta)$ subject to finite time horizon T. The trainable parameters in Θ will be initialized arbitrarily before the training starts.

In this work, the training with RL is divided into two stages to ensure robots actually learn the CL-aware path planning. In the first training stage, robots are only allowed to take random actions, and θ_3 is detached from the gradient. In this stage, robots will learn how to classify a map by only updating the parameters of the VGG model, LSTM, and the map classifier. The first training stage will terminate once further training does not imply any improvement.

In the second stage, robots are only allowed to optimize the parameters in the CL-aware path planner. In other words, θ_1, θ_2, and θ_4 are detached from the gradient while robot could only update θ_3. In order to achieve a better performance, robots need to learn the CL-aware path planning policy and adopt the most informative paths.

(a) Variation over years.   (b) Variation over seasons.   (c) Variation over hours (clouds).

Fig. 4. Examples from the Campus Map Dataset showing the variations within the same label (Lehigh University).

Table 1. Optimal performance (%) with $T = 15$.

Action policy	1 robot	5 robots	10 robots	20 robots	VGG-19 w/ full map	Average optimality gap (%)
Single-action	67.59	91.68	93.34	95.77	99.43	12.33
Goal-action	72.42	97.30	97.56	98.14		8.08
CL-aware	81.95	98.38	98.68	99.21		4.88

5 Case Studies

5.1 Simulation with the Campus Map Dataset

We validate the proposed method with the Campus Map Dataset [10], the dataset contains maps of 10 university campuses over 40 years, we refer the examples to Fig. 4. The training and testing is performed in PyTorch [15] with ADAM [7] and a learning rate $l_r = 0.0001$. The model is trained with $T = 15$ and $t_p = 5$ among 5 robots throughout both training stages using ten random seeds. The two-stage training curve is shown in Fig. 5 (a). In the first stage [0,150], robots learn to classify the map with random actions. It is also shown that the performance can not be further optimized with random actions (blue). After 150 epochs, the training enters the second stage (orange), and robots learn the CL-aware path planning policy to achieve better performance by planning to cover the most informative regions on the map.

We evaluate the CL-aware path planner's optimal performance with a various number of robots, as shown in Table 1. Single-action and goal-action policy models are also trained to optimality and serve as a benchmark for the distributed map classification. Finally, we evaluate the performance of the VGG-19 model with a full map as the input and denote its performance as optimal classification performance. It is shown that the CL-aware path planner significantly decreases the optimality gap compared to the existing distributed map classification policies. A snapshot from real-world experiments of map classification with UAV and CL-aware path planner is presented in Fig. 6.

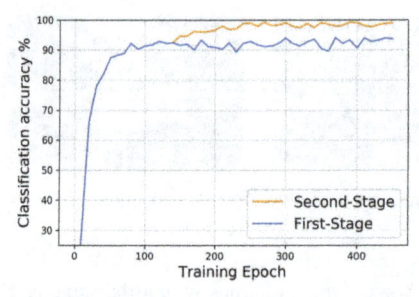

(a) The learning curve for the first and the second training stage, sampled at every 10 epochs.

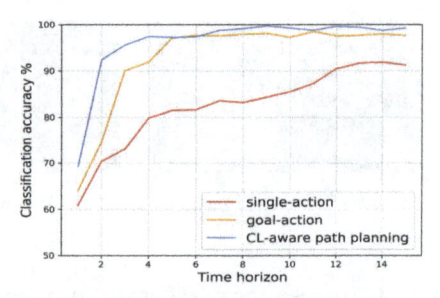

(b) The classification performance (w.r.t. time) of various path planning policies.

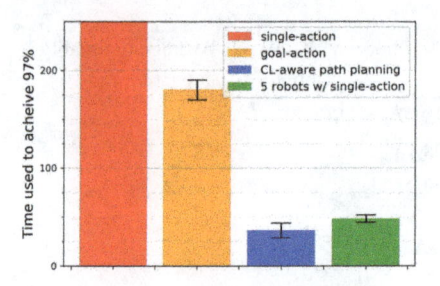

(c) The average time used to achieve 97% accuracy.

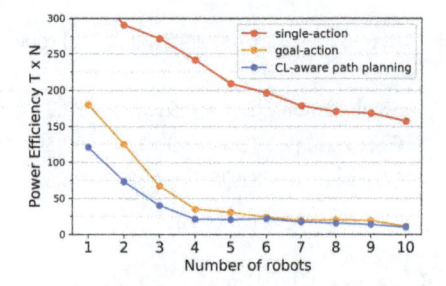

(d) The global power efficiency of various path planning methods.

Fig. 5. Simulation results on the Campus Map Dataset.

5.2 Efficient Distributed Classification

In Fig. 5 (b), we evaluate the performance (w.r.t time) among all three path planning methods. The proposed CL-aware path planner (blue) preserves a better performance than the existing methods for most task times. This implies that by using the CL-aware path planning policy, the multi-robot system will collect more classification-relevant features than the existing distributed classification path planning policies.

An efficient path planning policy usually achieves a specific classification accuracy in the shortest time. To validate this, we test the performances of a single robot adopting all policies over 100 maps. All the testing trail starts with the same initial conditions. We collect the average time used first to achieve 97% accuracy, as shown in Fig. 5 (c). The robot taking single-action policy has never reached 97% within $T \in [0, 400]$. It is shown that the CL-aware method takes the shortest amount of time to reach 97% and is almost as efficient as five robots using the single-action policy. This result is intuitive as the proposed CL-aware policy takes less time to collect adequate informative features.

5.3 The Coverage of Informative Regions

A well-learned and effective policy usually covers the most informative regions on the map (e.g., the region that is not covered by the cloud) to maximize the collection of clas-

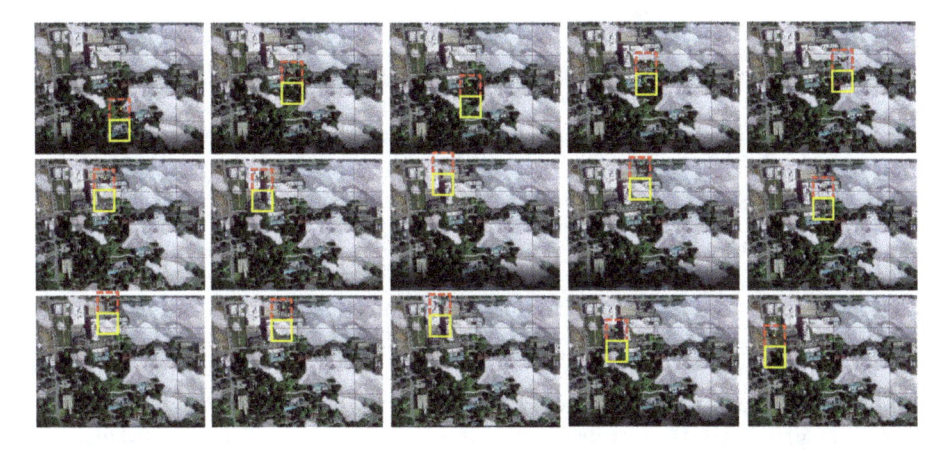

Fig. 6. A snapshot from map classification using Qualcomm Flight Pro with the CL-aware path planner. The red box denotes the location of the robot and the yellow box denotes the observation from the embedded on-board camera. *(Reading direction: first row →, second row ←, last row →)*.

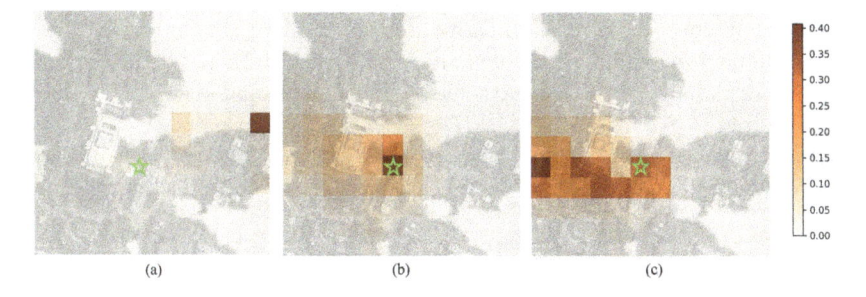

Fig. 7. Heat map of path distribution with: (a) Single-action; (b) Goal-action; (c) CL-aware path planning. The initial location is denoted as the green star.

sification information. To evaluate this, we test all methods with a single robot and the same initial location (green star) on the same map. Each method is tested for 100 trails among 20 maps, and an example heat map showing the path distribution of all testing trails is presented Fig. 7. Since the area covered with clouds will not provide any helpful classification information, an effective policy should reveal the region without clouds as much as possible. As shown in the heat map, the single-action method performs poorly in large environments, while the goal-action method could navigate the robot to the unclouded region. However, the stochastic navigation in the goal-action policy will not lead to a very effective exploration. On the other hand, the CL-aware method generates paths that avoid the cloud region and navigate the robot to the unclouded area with the highest probability (frequency).

The efficient coverage of informative regions usually indicates a shorter time needed to accomplish a specific classification task. We assume the aerial robots are identical, and $T \times N$ will measure the total energy cost (e.g., battery life) of the multi-robot system. We collect the average energy cost of all policies to achieve 97% (93% for single-action) with a various number of robots as shown in Fig. 5 (d). The result implies that by covering mostly informative regions(CL-aware method), the distributed multi-robot classification system will consume less energy system-wise to accomplish the same task.

6 Conclusions

We address the challenges in the path planning problem for a distributed multi-robot classification system to successfully learn a policy that will cover the most informative regions in the map and learn both long and short-term dependencies with the delayed classification reward. To address these challenges, a CL-aware path planning method is proposed together with the distributed classification architecture. The proposed methodology solves the path planning problem in a classification manner. Parallel linear classifiers are used to generate the action policy of each step in a planned path. We evaluate the usefulness of CL-aware path planning with a map classification problem. Using the existing action policies as the benchmark, the proposed method shows significant improvement in the classification performance, systemic energy efficiency. It also navigates robots to cover the most informative area on the map. The future work includes but is not limited to enabling the communication of path information and optimizing the CL-aware path planner to remove some redundant paths in the look-up library $\mathcal{P}$.

Acknowledgement. This work was supported in parts by the AFOSR FA9550-19-1-0004, ONR N00014-19-1-2478.

References

1. Ben-Afia, A., et al.: Review and classification of vision-based localisation techniques in unknown environments. IET Radar Sonar Navig. **8**(9), 1059–1072 (2014)
2. Bock, M., et al.: XV. Methods for creating functional soil databases and applying digital soil mapping with SAGA GIS. In: JRC Scientific and technical Reports, Office for Official Publications of the European Communities, Luxemburg (2007)
3. Cai, P., et al.: LeTS-drive: driving in a crowd by learning from tree search (2019). arXiv:1905.12197 [cs.RO]
4. Hochreiter, S., Schmidhuber, J.: Long short-term memory. Neural Comput. **9**, 1735–80 (1997). https://doi.org/10.1162/neco.1997.9.8.1735
5. Islam, Md.J., Fulton, M., Sattar, J.: Toward a generic diverfollowing algorithm: balancing robustness and efficiency in deep visual detection. IEEE Robot. Autom. Lett. **4**(1), 113–120 (2018)
6. Kavraki, L.E., et al.: Probabilistic roadmaps for path planning in high-dimensional configuration spaces. IEEE Trans. Robot. Autom. **12**(4), 566–580 (1996)
7. Kingma, D.P., Ba, J.: Adam: a method for stochastic optimization (2014). arXiv:1412.6980 [cs.LG]
8. Konar, A., et al.: A deterministic improved Q-learning for path planning of a mobile robot. IEEE Trans. Syst. Man Cybern. Syst. **43**(5), 1141–1153 (2013)
9. LeCun, Y., et al.: Gradient-based learning applied to document recognition. Proc. IEEE **86**(11), 2278–2324 (1998)

10. Liu, G., et al.: Distributed map classification using local observations. arXiv preprint arXiv:2012.10480 (2021)
11. Manduchi, R., et al.: Obstacle detection and terrain classification for autonomous off-road navigation. Auton. Robots **18**(1), 81–102 (2005). https://doi.org/10.1023/:AURO.0000047286.62481.1d. ISSN 1573-7527
12. Meyer, J.-A., Filliat, D.: Map-based navigation in mobile robots: II. A review of map-learning and path-planning strategies. Cogn. Syst. Res. **4**(4), 283–317 (2003)
13. Mousavi, H.K., et al.: Multi-agent image classification via reinforcement learning. In: 2019 IEEE/RSJ International Conference on Intelligent Robots and Systems (IROS), pp. 5020–5027 (2019). https://doi.org/10.1109/IROS40897.2019.8968129
14. Mousavi, H.K., et al.: A layered architecture for active perception: image classification using deep reinforcement learning (2019). arXiv:1909.09705 [cs.LG]
15. Paszke, A., et al.: Automatic differentiation in pytorch (2017)
16. Simonyan, K., Zisserman, A.: Very deep convolutional networks for large-scale image recognition. arXiv preprint arXiv:1409.1556 (2014)
17. Wulfmeier, M., Wang, D.Z., Posner, I.: Watch this: scalable cost-function learning for path planning in urban environments. In: 2016 IEEE/RSJ International Conference on Intelligent Robots and Systems (IROS), pp. 2089–2095. IEEE (2016)
18. Wulfmeier, M., et al.: Large-scale cost function learning for path planning using deep inverse reinforcement learning. Int. J. Robot. Res. **36**(10), 1073–1087 (2017)

Monitoring and Mapping of Crop Fields with UAV Swarms Based on Information Gain

Carlos Carbone[1(✉)], Dario Albani[1,2,7], Federico Magistri[3], Dimitri Ognibene[5,6], Cyrill Stachniss[3], Gert Kootstra[4], Daniele Nardi[1], and Vito Trianni[2]

[1] DIAG, Sapienza University of Rome, Rome, Italy
carbone@uniroma1.it
[2] ISTC, National Research Council, Rome, Italy
[3] University of Bonn, Bonn, Germany
[4] Wagening University and Research, Wageningen, The Netherlands
[5] University of Milano-Bicocca, Milan, Italy
[6] University of Essex, Colchester, UK
[7] Technology Innovation Institute, Abu Dhabi, UAE

Abstract. Monitoring crop fields to map features like weeds can be efficiently performed with unmanned aerial vehicles (UAVs) that can cover large areas in a short time due to their privileged perspective and motion speed. However, the need for high-resolution images for precise classification of features (e.g., detecting even the smallest weeds in the field) contrasts with the limited payload and flight time of current UAVs. Thus, it requires several flights to cover a large field uniformly. However, the assumption that the whole field must be observed with the same precision is unnecessary when features are heterogeneously distributed, like weeds appearing in patches over the field. In this case, an adaptive approach that focuses only on relevant areas can perform better, especially when multiple UAVs are employed simultaneously. Leveraging on a swarm-robotics approach, we propose a monitoring and mapping strategy that adaptively chooses the target areas based on the expected information gain, which measures the potential for uncertainty reduction due to further observations. The proposed strategy scales well with group size and leads to smaller mapping errors than optimal pre-planned monitoring approaches.

Keywords: Swarm robotics · Precision farming · Information gain · UAV

1 Introduction

Precision farming requires high-quality data from the field in order to support operational and strategic decisions [11]. In this respect, unmanned aerial vehicles (UAVs) present a very flexible tool for remote sensing, as they can be deployed on-demand and can quickly monitor large areas [26]. However, the platform limitations in terms of payload and flight time are often constraining, limiting the resolution of the acquired images and requiring multiple flights to cover extensive fields. At the same time, the heterogeneity of agricultural fields often requires high-resolution data only in certain areas where relevant features are present. In contrast, less interesting areas could be monitored on a coarser resolution. This is often the case, for instance, for weed-management practices, in which the density of weeds is not uniform across the field, as weeds appear in patches. To support efficient management of the field, a greater effort should be dedicated to those areas where relevant features are actually present. In this way, the limited energy budget of each UAV is properly allocated, avoiding dissipating it by flying over uninteresting areas. Given that the feature distribution over the field is a priori unknown, the monitoring and mapping strategies should be adaptive, responding to the observed field features in order to determine the best flight pattern, therefore ruling out predefined mission plans that do not take into account the actual distribution of relevant features over the field [5,8].

UAV swarms have been proposed to address the adaptive monitoring and mapping of extensive fields. A swarm-robotics approach can indeed improve efficiency thanks to parallel monitoring of the field by different UAVs, reducing the operation to a fraction of the time required by a single-UAV approach [1,3,10]. Additionally, accuracy can be improved thanks to collaboration among UAVs in their feature-detection task [17]. Finally, self-organised deployment strategies can be envisaged to leverage the ability of UAVs to estimate which region within the field is of greater interest (e.g., estimating weed density from high-altitude flights [15]) and perform accurate monitoring and mapping only where relevant (e.g., by collecting high-resolution images while flying at low altitude [2]). With such an approach, UAVs can autonomously decide to enter or leave a given region based on estimates of the monitoring activity's completion level and energy constraints. Consequently, the monitoring and mapping strategy of a single region needs to be flexible and scalable to adapt to changes in the actual number of UAVs that are concurrently operating. Approaches that a-priori divide the field to be monitored among the available UAVs are viable but do not address adaptive deployment requirements into areas of interest.

Improving over previous work [1,3], we propose a fully decentralised strategy to monitor a region or field based on reinforced random walks (RRW [25]), which maximises the monitoring effort only on areas that likely provide relevant information. To this end, we exploit the Information Gain (IG), an information-theoretic measure of the expected reduction in uncertainty from additional observations of a specific area. The usage of information theory for exploration and mapping has been demonstrated across application domains [6,16,20–22], and precision agriculture in particular [14,24]. Here, we exploit IG to support exploration and coordination among robots. To this end, we build a model of the weed-density uncertainty in a given area that accounts for detection errors of a convolutional neural network (CNN) trained on a real dataset. To enable real-time onboard execution, we reduce the neural classifier's complexity and compensate for increased error by allowing multiple observations of the same area of the field [17]. On such basis, we compute the expected reduction in uncertainty as to the IG from repeated independent observations made by UAVs within the swarm on the same area. Then, we exploit IG to prioritize areas of interest to be observed next and to determine if these areas are likely to be targeted by other UAVs within the swarm. These two aspects are combined to guide the random selection of the next location to visit. Our results with multi-UAV simulations show that the swarm manages to quickly monitor those areas of the field that require more attention, minimising the observation error faster than approaches based on potential fields [1,3], and better than a baseline approach based on a predefined flight plan that uniformly covers the entire field. Thus, the main contribution of this research is a parameter-free IG-based mapping strategy that achieves an improved reduction of observation error during the early stages of inspection, leading to the generation of reliable maps when the time/energy budget may be limited.

2 Problem Description

We consider a field-monitoring and weed-mapping problem in which areas of high weed infestation need to be identified, creating a prescription map that can be exploited for weed control (e.g., variable-rate application of herbicides). Specifically, we focus on identifying volunteer potatoes that infest sugar-beet fields—a common benchmark for precision agriculture [15,19]—and use automatic object classification to inform the monitoring and mapping strategy. Our goal is to deploy a swarm of UAVs that can rapidly minimize the error in detecting weeds within the field. To this end, we exploit a simulated scenario to evaluate the proposed strategy's effectiveness and scalability.

2.1 World Model and UAV Swarm Simulation

We consider the field as divided into small areas forming a grid of cells, and each cell is fully contained within the camera field-of-view of a UAV hovering over its centre. Without loss of generality, we consider here a square field divided in a grid of $C \times C$ square cells, each with side l_c. Each UAV travels at a cruise speed of $v = 0.1 l_c$ m/s at an altitude h sufficient to observe the whole cell given the camera footprint (e.g., $h \geq l_c/2$ if the camera aperture is $\frac{\pi}{2}$). Whenever moving over a cell, a UAV takes an RGB image used for crop/weed classification, leading to an estimation of the number of weeds present in the cell (see Sect. 2.2). After each observation, the UAV updates its local world representation, that is, a $C \times C$ map of the field (see Sect. 2.3). Additionally, UAVs can communicate with each other by broadcasting short messages exploiting a radio link, with range Rl_c. When the communication range is sufficiently large, any UAV can receive the information shared by any other UAV. Otherwise, UAVs apply a simple re-broadcasting protocol to maximise the reach of information shared within the swarm. Upon reception of a message, a UAV re-broadcasts the message once and then puts it in a blacklist hence avoiding overloading the communication channel.

UAVs exploit communication to share information about the observations made on visited cells, and to also share their absolute position—available from some GNSS positioning system—hence allowing collision avoidance (here implemented using ORCA [4]) as well as collaboration for monitoring and mapping.

While crops are uniformly distributed over the field, weeds mostly appear in patches. We consider here C_p patches, each extending on a square of $n_p \times n_p$ cells, more densely distributed in the center than in the periphery following a Gaussian distribution. Some isolated weeds are also present within C_i additional cells. Overall, the number of cells with some weed is $C_p n_p^2 + C_i$. Each cell can contain at most N_W weeds, i.e., the maximum value observed over the field.

2.2 Model of Weed-Classification Uncertainty

To build a model of the weed-classification uncertainty, we consider a CNN to detect individual plants and label them as crops or weeds. In other words, given an input image, the CNN returns a list of bounding boxes that enclose the detected plants, together with the identified class of the plant. We use the state-of-the-art framework *Faster R-CNN* [23] but limit its computational requirements by employing a shallow backbone (instead of the deep backbones normally used, such as ResNet-101) to match the constraints imposed by the limited power available on UAVs (see Fig. 1a). The input image is passed through a convolutional block followed by 4 residual blocks [9]. The convolutional block is composed of a 7×7 convolutional layer followed by a 3×3 convolutional layer (represented in light blue in Fig. 1a), each layer having 64 filters with stride 2 and ReLU activation functions. Each residual block is composed of a residual connection and two 3×3 convolutional layers with ReLU activation functions (represented in orange in Fig. 1a). The first residual block has 64 filters. Afterward, each block doubles the previous number of filters. The head of the network is the same as in the original work, predicting the location, size, and class of object detections. We exploit a dataset collected at Wageningen University, composed of 500 images—400 used as the training set and 100 as the test set—labelled to represent sugar beets in the crop class and volunteer potatoes in the weed class (see Fig. 1b). We train our network for 10,000 epochs, using the ADAM optimizer [12] with a learning rate of 0.01. We evaluate our CNN using the average precision (AP) metric as defined in the MS-COCO challenge [13] obtaining an AP score of 89.6 for the crop and 56.1 for the weed. While the AP for the crop class is in line with the literature [7,18], the lower accuracy for the weed class is mainly due to miss-classifications of plants as background (Fig. 1b).

We model the uncertainty in the weed classification to obtain a realistic observation—in terms of detected weeds—every time a UAV processes the information of a cell, as well as to update the internal knowledge about the world, as discussed in the following. Starting from the trained CNN classification output, our model associates a probability distribution to all possible observations given the actual number of weeds present in a cell. Considering that each cell can contain a discrete and small number of weeds (from the available data, we estimated $N_W = 12$ weeds for $l_c = 4$ m), we build a table that associates the

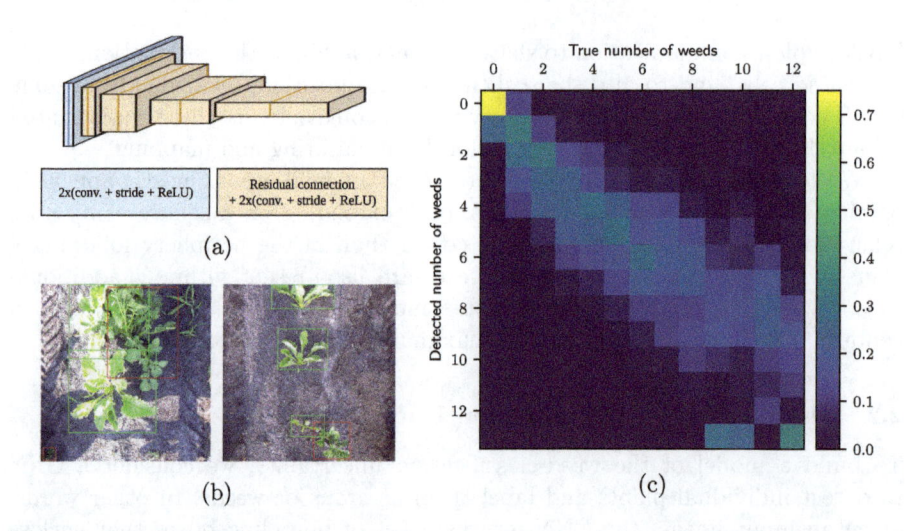

Fig. 1. (a) The Faster R-CNN framework with a shallow backbone was used while the head of the CNN remained the same as in the original work. (b) a few samples of CNN classification of volunteer potatoes (red) and sugar beets (green). (c) Sensor model derived from the CNN detection performance.

actual number of weeds w with each possible observed number o, and gives for each combination a probability of occurrence $P(o|w)$. This table is estimated from the available data exploiting the trained CNN. Specifically, we compare the number of detected weeds against the true number of weeds. The element $T(o,w) = P(o|w)$ of the table is the relative frequency of detecting o weeds when the true number is w. Having several false positive cases, we also consider into the table the case of detecting more weeds than the given maximum N_W, that is, $o \geq N_W + 1 \triangleq N_W^+$ represents the case in which the number of weeds detected is larger than N_W. The trained CNN model as obtained from the available data is presented in Fig. 1c. Both false positives and false negatives are possible, with errors becoming more frequent when the number of weeds is larger, as one would expect.

2.3 Uncertainty Reduction from Multiple Observations

For each cell c, each UAV maintains a knowledge vector $p_c = [p_c(0), ..., p_c(N_W)]$ that represents the expected probability distribution of all possible number of weeds considered. Taking a conservative approach, when no observation for a cell has been performed and no prior knowledge is available, the probability distribution is assumed to be uniform, with every element $p_c(w) = \frac{1}{N_W+1}$. Whenever a UAV observes cell c at time t, it detects a number of weeds o_c^t that depends on the actual value of weeds w_c present in c. Following the observation o_c^t, the probability $P_c(w|o_c^t)$ represents the updated knowledge vector for each possible value w. This can be easily computed exploiting Bayes' theorem:

$$\forall w, p_c(w) \leftarrow P_c(w|o_c^t) = \frac{p_c(w)P(o_c^t|w)}{P(o_c^t)} = \frac{p_c(w)T(o_c^t, w)}{\sum_{j=0}^{N_W} p_c(j)T(o_c^t, j)} \quad (1)$$

The residual uncertainty about the weeds present in the cell corresponds to the information entropy of the knowledge vector:

$$H_c(W) = \sum_{w=0}^{N_W} -p_c(w)\log(p_c(w)) \quad (2)$$

The residual uncertainty is updated at every additional observation. We consider that sufficient observations have been performed for a given cell c when the entropy decreases past a low threshold that we heuristically set at $\hat{H} = -\frac{N_W-1}{N_W}\log\frac{N_W-1}{N_W} - \frac{1}{N_W}\log\frac{1}{N_W}$, which corresponds to the case in which there are only two remaining alternatives, one of which is much more likely than the other. When the threshold is reached, cells are marked as mapped and are not considered anymore for further observations. Note that other UAVs can also share observations, allowing to update the residual uncertainty about a cell also when others have visited it. In this way, UAVs try to maintain aligned their local representation of the field.

The residual uncertainty of a cell is also the first component for the calculation of the IG of a cell, which represents the expected reduction in entropy from any possible additional observation o. A UAV can compute the IG of a cell c as follows:

$$\text{IG}_c(W) = H_c(W) - H_c(W|O) \quad (3)$$

where $H_c(W|O)$ is the conditional entropy of the same cell given that additional observations will be performed next:

$$H_c(W|O) = -\sum_{o=0}^{N_W^+} P_c(o) \sum_{w=0}^{N_W} [P_c(w|o)\log(P_c(w|o))] \quad (4)$$

$$= -\sum_{o=0}^{N_W^+} \left[\sum_{j=0}^{N_W} T(o,j)p_c(j)\right] \sum_{w=0}^{N_W} [P_c(w|o)\log(P_c(w|o))] \quad (5)$$

Hence, based on the available knowledge, each UAV can compute the IG for each cell of the field and quantify the information gathered from a new observation, that is, the utility of visiting it.

3 Reinforced Random Walks for Monitoring and Mapping

A RRW is an exploration strategy that exploits available information to bias the random selection of the targets [25]. The available information can derive from world knowledge (e.g., avoid visiting cells that are marked as mapped) or from other UAVs (e.g., others' location to avoid interference). A wise combination of both aspects can produce efficient monitoring and mapping strategies that focus on relevant areas, minimising the detection error.

3.1 Neighbourhood Selection Strategy

Whenever a UAV i has to determine the next cell to visit, it considers only those in its neighbourhood $\mathcal{N}_i$ as targets, which proved to be the best strategy to avoid long relocations that may be costly [1,3]. Additionally, a UAV tries to select cells that are *valid*, i.e., not yet marked as mapped and not currently targeted by other UAVs to avoid interference. We define the neighbourhood $\mathcal{N}_i^d$ of UAV i as the cells at Chebyshev distance d. A UAV first considers the neighbourhood $\mathcal{N}_i^1$, and only if there are no valid cells available, it considers the neighbourhood $\mathcal{N}_i^2$. If this does not contain valid cells, the UAV selects a cell in $\mathcal{N}_i^2$ to move away from the current position, limiting the choice among those cells that are already marked as mapped. With this strategy, a set of neighbouring cells $\mathcal{V}_i$ is selected for a decision to be taken. This strategy, being purely local, does not guarantee that all cells are eventually visited, even if, in the long run, this can always happen in practice. Nevertheless, in this work, we focus on reducing the mapping error with a limited time/energy budget. Hence complete coverage is not a requirement.

IG-Based RRW. The cell selection to target is based on a utility measure computed starting from the IG. Given a cell c_k, each UAV i determines the $\text{IG}_{c_k}^i$ based on the currently available knowledge—different among UAVs in case of constrained communication—and, independently from any other UAV, assigns a probability of selecting the cell proportional to the relative IG:

$$P_{i,c_k}^{\text{IG}} = \frac{\text{IG}_{c_k}^i(W)}{\sum_{z \in \mathcal{V}_i} \text{IG}_{c_z}^i(W)} \qquad (6)$$

To take into account the presence of other UAVs that are concurrently monitoring the field, the utility of selecting a cell is computed from P_{i,c_k}^{IG} considering the likelihood that the cell is *not* chosen by other UAVs:

$$u_{i,c_k} = P_{i,c_k}^{\text{IG}} \prod_{j \neq i} \left[1 - P_{j,c_k}^{\text{IG}}\right] \qquad (7)$$

Here, P_{j,c_k}^{IG} is computed considering the neighbourhood $\mathcal{V}_j$ of UAV j, although such computation is performed exploiting i's private knowledge. To limit the computational complexity of this calculation, the product is extended only to those UAVs j that could potentially target the cell c_k at the same time, that is, the UAVs within the neighbourhood $\mathcal{N}_{c_k}^d, d \leq 2$ from cell c_k. By reducing the likelihood of choosing cells in reach from other UAVs, the proposed RRW strategy implements an implicit coordination mechanism that allows UAVs to share the monitoring burden and efficiently map the relevant areas. Should UAVs decide to target the same cell, the ORCA method would prevent collisions, and a timeout mechanism allows to resolve possible—albeit unlikely—deadlocks. Given Eq. (7), we propose different heuristics to choose the next cell to visit:

G: a cell c_k is chosen greedily selecting the one with highest utility u_{i,c_k}. In case of cells with identical utility, one is chosen at random.
R: a cell c_k is randomly chosen proportionally to the utility u_{i,c_k}.
S$_\gamma$: a cell c_k is chosen according to a softmax function of u_{i,c_k} with base e^γ.

3.2 Baseline Strategy Based on Optimal Pre-planned Trajectories

As a baseline to confront the proposed strategy, we assume an optimal reference point **B** based on the uniform coverage of the field, i.e., N UAVs visit each cell of the field a fixed number of times. This optimal benchmark could be approximated by pre-planning all UAVs' trajectories, although, in reality, several factors reduce the ideal performance proposed here. Considering the speed of the UAV and the distance between cells, the time for a single UAV to fully cover the field is given by $T_1 = C^2 l_c / v = 10C^2$. We assume that N UAVs can optimally partition the field, hence $T_N = T_1/N$. To allow for independent observations of the field by different UAVs, we consider repeated passages that entail longer times, that is, MT_N for M independent passages. After each observation, the residual uncertainty of a cell c is computed following Eq. (2).

3.3 Baseline Strategy Based on Potential Fields

In addition, we consider a second baseline $\mathbf{B_{PF}}$ that exploits a RRW based on potential fields (PF). Here, the target cell is selected by UAV i according to a directional bias given by an attraction vector $\vec{a}_i$ toward areas where weed was detected, and a repulsion vector $\vec{r}_i$ from other agents to avoid overcrowding. This strategy is adapted from previous versions [1,3] to reduce its computational demands, limiting the information exploited for the computation of potential fields. Attraction and repulsion vectors are computed as follows:

$$\vec{r}_i = \sum_{j \neq i} S(\vec{x}_i - \vec{x}_j, \sigma_r), \quad \vec{a}_i = \sum_c \frac{\hat{w}_c}{N_W} S(\vec{x}_c - \vec{x}_i, \sigma_a), \quad S(\vec{v}, \sigma) = 2e^{i\angle \vec{v}} e^{-\frac{|\vec{v}|}{2\sigma^2}}, \quad (8)$$

where $\vec{x}$ represents the position of an agent/cell, and $S(\vec{v}, \sigma)$ returns a vector in the direction of $\vec{v}$ with a Gaussian length with spread σ. With respect to [1], we reduce the number of agents considered to those belonging to the neighbourhood $\mathcal{N}_i^4$, which are ultimately considered also for the IG-based RRW. Additionally, we consider attraction to the cells $c \in \mathcal{N}_i^d, d \leq 2$ but discounting the force by the number $\hat{w}_c$ of weeds detected in the last observation. Cells are not considered for attraction when they are marked as mapped. The selection of the next cell to visit by UAV i is performed randomly using the vector $\vec{v}_i = \vec{r}_i + \vec{a}_i$ as a bias. Specifically, the cell $c_k \in \mathcal{V}_i$ is selected randomly proportionally to the utility u_{i,c_k}, which is computed according to the angular difference $\theta_{i,c_k} = \angle(\vec{x}_{c_k} - \vec{v}_i)$:

$$u_{i,c_k} = C\left(\theta_{i,c_k}, 0.9\left(1 - e^{\beta|\vec{v}_i|}\right)\right), \quad C(\theta, p) = \frac{1}{2\pi} \frac{1 - p^2}{1 + p^2 - 2p\cos\theta} \quad (9)$$

where $C(\cdot)$ is the wrapped Cauchy density function with persistence p. In this way, the length of the vector—smoothed through an exponential ceiling— determines the relevance according to the bias: the smaller the module, the lower the directional bias in the cell choice. This strategy has several parameters— namely the Gaussian parameters σ_a and σ_r, and the exponential constant β— which depend on the UAV swarm size N and need to be carefully tuned for appropriate performance [1,3].

4 Experimental Results

We performed several simulations with a square field of $C \times C = 2500$ cells, having $C_q = 4$ square weed patches of $n_p \times n_p = 49$ cells randomly positioned within the field, plus additional $C_i = 40$ isolated cells randomly scattered. In total, less than 10% of the field presents cells containing weeds. As already mentioned, we consider $N_W = 12$. At initialisation, UAVs start at random positions within the field, with no prior knowledge of the weed distribution.[1] We focus on the ability to monitor and map the field minimising the mapping error. We perform 50 runs for each experimental condition obtained varying swarm size $N \in \{10, 20, 30, 40, 50\}$, the communication range $R \in \{10, \infty\}$, the heuristics exploited for the information gain (**G**, **P** and $\mathbf{S}_\gamma$ with $\gamma \in \{1, 5\}$) and the parameters $\sigma_r, \sigma_a \in \{0, 2, 4, 8, 16, 32\}, l_c$ for the PF-based RRW with $\beta = 1$. With ideal communication ($R = \infty$), each broadcasted message reaches every other UAV in the swarm.

To evaluate the system performance, we consider an aggregated world map resulting from the individual UAV maps by considering for each cell c and UAV i the knowledge vector $p_c = p_{c,i}$ with lowest uncertainty $H_i(c)$. Then, the value $\tilde{w}_c = \arg\max_w p_c$ is considered as the mapped number of weeds in cell c, and the mean squared error (MSE) is computed with respect to the real value w_c for the whole field. Figure 2 shows how this error decreases as further observations are gathered from the field, taking as temporal reference the time $M\mathcal{T}_N$ necessary to gather M independent observations for each cell with N UAVs following the baseline **B**. When the communication is perfect, all UAVs share the same map, and any new observation contributes to reducing the uncertainty. The mapping error is initially rather high but decreases as UAVs discover and focus on interest areas. Notably, all the IG-based strategies scale well to the swarm size N and present the best performance when the greedy approach **G** is employed. In this case, the MSE gets better than the one of the baseline **B** for $M \approx 3$—when the error is about 0.15—and continues to decrease another order of magnitude with further observations. The baseline $\mathbf{B_{PF}}$ is instead less efficient in reducing the mapping error and outperforms **B** only when $M > 6$, hence requiring much longer than the best IG-based strategy **G**. The other heuristics perform slightly worse, meaning that additional randomness in selecting the next cell to visit does not increase performance. Specifically, the **R** and the $\mathbf{S}_5$ strategies have very similar profiles, and the worst performance is obtained with the $\mathbf{S}_1$ strategy.

The difference in performance among the proposed heuristics is also visible for the total completion times, shown in Fig. 3. Here, we measure the time $\mathcal{T}_C$ necessary to fully cover the entire field by passing over every cell at least once, scaled concerning the time $\mathcal{T}_N$ of the baseline **B**. The greedy strategy **G** performs better than any other strategy, with a coverage time that scales perfectly with group size and values that roughly correspond to the time required to have an MSE smaller than **B**. The other strategies generally have longer coverage times,

---

[1] We ignore here the initial relocation from a deployment station and also disregard the need to return to a predefined location.

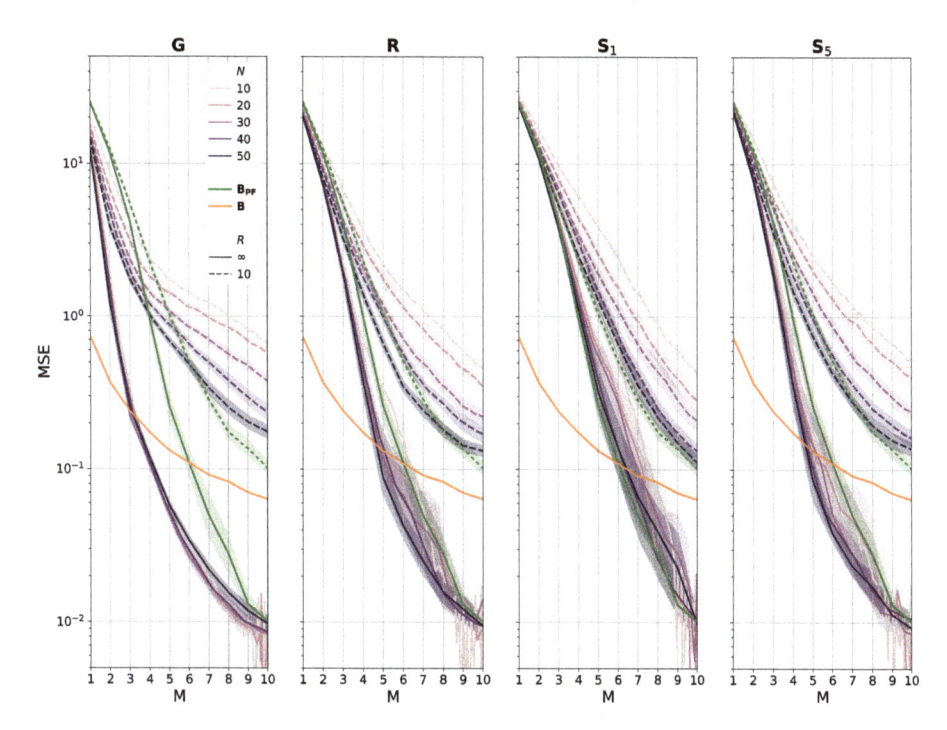

Fig. 2. MSE of the maps generated with different strategies (average and standard deviation), with respect to the optimal pre-planned monitoring baseline **B** and the PF-based RRW strategy $\mathbf{B_{PF}}$. Note that all plots are rescaled in time with respect to the time T_N required by N agents to fully cover the field once. As a consequence, the average MSEs from the baseline **B** for all values of N coincide. The baseline $\mathbf{B_{PF}}$ is shown for both $R = \infty$ and $R = 10$, and corresponds to the case with $\beta = 1$, $\sigma_a = 2$, $\sigma_r = 8$ and $N = 50$, which is the one with lowest MSE among all the tested parameters.

with slight improvements for larger group size N although never being on par with **G**.

The adaptive approach reduces the uncertainty below the optimal pre-planned strategy by mainly focusing on the areas where weed is present, avoiding monitoring those devoid of weeds. Indeed, Fig. 4 shows the Pearson's correlation coefficient between the number of weeds present in a cell and the number of independent observations gathered for it. It is possible to note that, as time goes by, the relationship between these two variables builds stronger, meaning that more time is spent over areas that require more observations to reduce the uncertainty substantially. In contrast, areas without weed are quickly abandoned.

The results obtained for a limited communication range tell a different story, however. First and foremost, the swarm does not reduce the error substantially below the baseline **B** (see dotted lines in Fig. 2). The monitoring and mapping performance is substantially worse due to an inefficient reduction of uncertainty as the observations from other UAVs are only shared locally. There is no mecha-

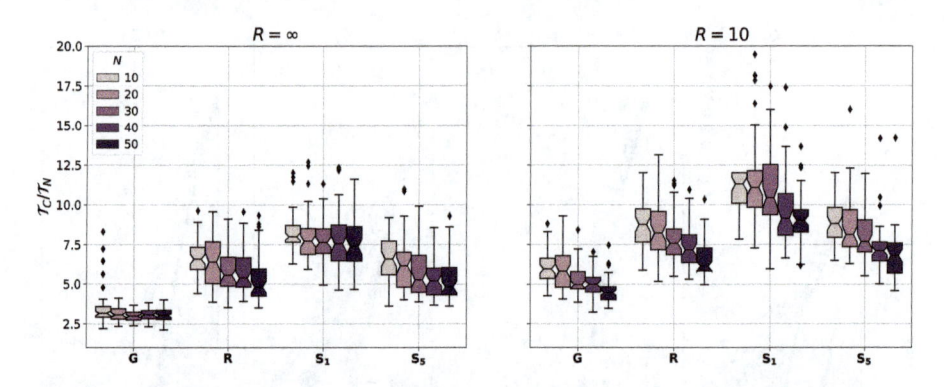

Fig. 3. Coverage time $\mathcal{T}_C$ relative to $\mathcal{T}_N$ for the different IG-based strategies, for $R = \infty$ (left) and $R = 10$ (right).

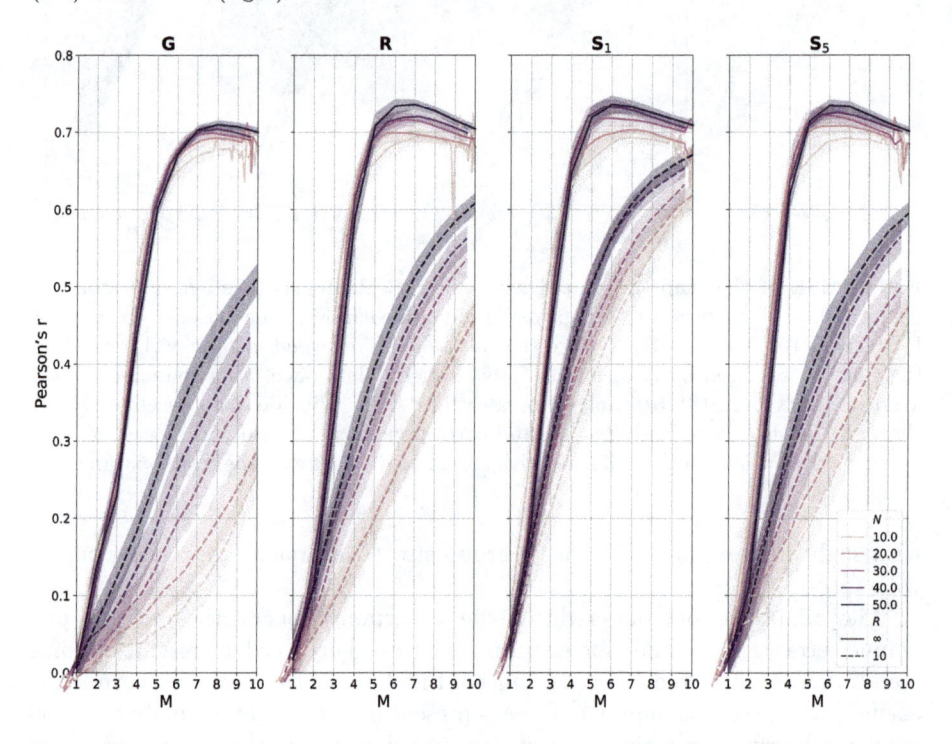

Fig. 4. Correlation between the number of weeds present and the number of observations performed, plotted over time for the different strategies.

nism to ensure that UAVs maintain the local maps aligned. The error gets lower as the group size increases due to the larger diffusion of messages thanks to the re-broadcasting communication protocol. Also, the coverage time is positively affected by group size, as shown in Fig. 3. Nevertheless, the density of agents is not sufficiently high to ensure that the communication topology is always connected, leading to a partial loss of communications. Consequently, different UAVs may map areas already sufficiently observed by others without any improvement for the collective. This also leads to a reduced correlation between the number of weeds in a cell and the number of observations, especially for low values of N as shown in Fig. 4.

5 Conclusions

This paper has demonstrated that a performance improvement is expected when adaptive approaches are employed, leading to a substantial reduction of uncertainty below what can be achieved through a blind acquisition of information using pre-planned trajectories. The mapping error is substantially reduced by investing a little more exploration time despite the low accuracy in the classification output of single images. Additionally, the best heuristic based on IG is rigorous. It bears no free parameters, being therefore ideal for deployment in a swarm robotics system notwithstanding the group size or the field dimensions, as no tuning is needed.

One observed limitation concerns the loss in performance for constrained communication, which jeopardises the benefits from an adaptive approach with a UAV swarm. This aspect is worth studying in detail, although it does not invalidate the ideal communication case results. In several scenarios—and precision agriculture in particular—one can reasonably assume that a good-enough communication channel can be set up in place, thanks to long-range radio communication or the upcoming 5G technology. The limited communication scenario may be addressed by either changing the communication protocol in a way to ensure better alignment among UAVs of the local world representations (e.g., by sharing not just the latest observation but the entire word representation), or by explicitly maintaining a high degree of connectivity within the swarm.

Another limitation of the proposed approach stands in the assumption of independence between observations, which likely does not hold in real-world scenarios, especially for slowly-changing environmental conditions as in crop fields. Indeed, multiple observations are likely to provide the same errors if these are related to specific features of the observed area (e.g., a volunteer potato hiding between two sugar beets), especially if observations are made from the same perspective a short time-frame. To mitigate this issue and increase the chances that repeated observations lead to substantial uncertainty reduction, a possible approach is to vary the observation position relative to the point of interest. Different perspectives can lead to more informative observations. Another complementary approach consists of using already available knowledge as a prior for the classification, exploiting specially-trained CNNs [17]. Combining these two possibilities can lead to a substantial reduction in uncertainty that can be modelled and exploited with the approach proposed in this paper. Future work will investigate this possibility while tackling real-world monitoring and mapping tasks.

Acknowledgements. This work has partially been funded by the Deutsche Forschungsgemeinschaft (DFG, German Research Foundation) under Germany's Excellence Strategy, EXC-2070 – 390732324 (PhenoRob). Vito Trianni acknowledges partial support from the project TAILOR (H2020-ICT-48 GA: 952215).

References

1. Albani, D., Manoni, T., Arik, A., Nardi, D., Trianni, V.: Field coverage for weed mapping: toward experiments with a UAV swarm. In: Compagnoni, A., Casey, W., Cai, Y., Mishra, B. (eds.) BICT 2019. LNICST, vol. 289, pp. 132–146. Springer, Cham (2019). https://doi.org/10.1007/978-3-030-24202-2_10
2. Albani, D., Manoni, T., Nardi, D., Trianni, V.: Dynamic UAV swarm deployment for non-uniform coverage. In: 17th International Conference on Autonomous Agents and MultiAgent Systems (AAMAS), pp. 523–531 (2018)
3. Albani, D., Nardi, D., Trianni, V.: Field coverage and weed mapping by UAV swarms. In: 2017 IEEE/RSJ International Conference on Intelligent Robots and Systems (IROS), pp. 4319–4325 (2017)
4. Bareiss, D., van den Berg, J.: Generalized reciprocal collision avoidance. Int. J. Robot. Res. **34**(12), 1501–1514 (2015)
5. Cabreira, T.M., Brisolara, L.B., Ferreira, P.R., Jr.: Survey on coverage path planning with unmanned aerial vehicles. Drones **3**(1), 1–38 (2019)
6. Das, J., et al.: Data-driven robotic sampling for marine ecosystem monitoring. Int. J. Robot. Res. **34**(12), 1435–1452 (2015)
7. Fawakherji, M., Youssef, A., Bloisi, D., Pretto, A., Nardi, D.: Crop and weeds classification for precision agriculture using context-independent pixel-wise segmentation. In: 2019 Third IEEE International Conference on Robotic Computing (IRC), pp. 146–152. IEEE (2019)
8. Galceran, E., Carreras, M.: A survey on coverage path planning for robotics. Robot. Auton. Syst. **61**(12), 1258–1276 (2013)
9. He, K., Zhang, X., Ren, S., Sun, J.: Deep residual learning for image recognition. In: IEEE Conference on Computer Vision and Pattern Recognition (CVPR), pp. 770–778 (2016)
10. Kapoutsis, A.C., Chatzichristofis, S.A., Kosmatopoulos, E.B.: DARP: divide areas algorithm for optimal multi-robot coverage path planning. J. Intell. Robot. Syst. **86**(3–4), 663–680 (2017)
11. King, A.: Technology: the future of agriculture. Nature **544**(7651), 21–23 (2017)
12. Kingma, D.P., Ba, J.: ADAM: a method for stochastic optimization. arXiv preprint arXiv:1412.6980 (2014)
13. Lin, T.-Y., et al.: Microsoft COCO: common objects in context. In: Fleet, D., Pajdla, T., Schiele, B., Tuytelaars, T. (eds.) ECCV 2014. LNCS, vol. 8693, pp. 740–755. Springer, Cham (2014). https://doi.org/10.1007/978-3-319-10602-1_48
14. Liu, J., Williams, R.K.: Monitoring over the long term: intermittent deployment and sensing strategies for multi-robot teams. In: Proceedings of the 2020 IEEE International Conference on Robotics and Automation (ICRA 2020), pp. 7733–7739 (2020)
15. Lottes, P., Khanna, R., Pfeifer, J., Siegwart, R., Stachniss, C.: UAV-based crop and weed classification for smart farming. In: 2017 IEEE International Conference on Robotics and Automation (ICRA), pp. 3024–3031 (2017)

16. Ma, K.-C., Ma, Z., Liu, L., Sukhatme, G.S.: Multi-robot informative and adaptive planning for persistent environmental monitoring. In: Groß, R., et al. (eds.) Distributed Autonomous Robotic Systems. SPAR, vol. 6, pp. 285–298. Springer, Cham (2018). https://doi.org/10.1007/978-3-319-73008-0_20
17. Magistri, F., Nardi, D., Trianni, V.: Using prior information to improve crop/weed classification by MAV swarms. In: 11th International Micro Air Vehicle Competition and Conference, pp. 67–75 (2019)
18. Milioto, A., Lottes, P., Stachniss, C.: Real-time blob-wise sugar beets vs weeds classification for monitoring fields using convolutional neural networks. ISPRS Ann. Photogrammetry Remote Sens. Spat. Inf. Sci. **4**, 41 (2017)
19. Nieuwenhuizen, A.T., Hofstee, J.W., van de Zande, J.C., Meuleman, J., van Henten, E.J.: Classification of sugar beet and volunteer potato reflection spectra with a neural network and statistical discriminant analysis to select discriminative wavelengths. Comput. Electron. Agric. **73**(2), 146–153 (2010)
20. Ognibene, D., Demiris, Y.: Towards active event recognition. In: IJCAI, vol. 13, pp. 2495–2501 (2013)
21. Ognibene, D., Mirante, L., Marchegiani, L.: Proactive intention recognition for joint human-robot search and rescue missions through Monte-Carlo planning in POMDP environments. In: Salichs, M.A., et al. (eds.) ICSR 2019. LNCS (LNAI), vol. 11876, pp. 332–343. Springer, Cham (2019). https://doi.org/10.1007/978-3-030-35888-4_31
22. Palazzolo, E., Stachniss, C.: Effective exploration for MAVs based on the expected information gain. Drones **2**(1), 9 (2018)
23. Ren, S., He, K., Girshick, R., Sun, J.: Faster R-CNN: towards real-time object detection with region proposal networks. In: Advances in Neural Information Processing Systems, pp. 91–99 (2015)
24. Rossello, N.B., Carpio, R.F., Gasparri, A., Garone, E.: Information-Driven Path Planning for UAV with Limited Autonomy in Large-scale Field Monitoring. arXiv (2020)
25. Smouse, P.E., Focardi, S., Moorcroft, P.R., Kie, J.G., Forester, J.D., Morales, J.M.: Stochastic modelling of animal movement. Philos. Trans. R. Soc. B: Biol. Sci. **365**(1550), 2201–2211 (2010)
26. Zhang, C., Kovacs, J.M.: The application of small unmanned aerial systems for precision agriculture: a review. Precision Agric. **13**(6), 693–712 (2012)

A Discrete Model of Collective Marching on Rings

Michael Amir[1(✉)], Noa Agmon[2], and Alfred M. Bruckstein[1]

[1] Technion - Israel Institute of Technology, Haifa, Israel
{ammicha3,freddy}@technion.ac.il
[2] Department of Computer Science, Bar-Ilan University, Ramat Gan, Israel
agmon@cs.biu.ac.il

Abstract. We study the collective motion of autonomous mobile agents on a ringlike environment. The agents' dynamics is inspired by known laboratory experiments on the dynamics of locust swarms. In these experiments, locusts placed at arbitrary locations and initial orientations on a ring-shaped arena are observed to eventually all march in the same direction. In this work we ask whether, and how fast, a similar phenomenon occurs in a stochastic swarm of simple agents whose goal is to maintain the same direction of motion for as long as possible. The agents are randomly initiated as marching either clockwise or counterclockwise on a wide ring-shaped region, which we model as k "narrow" concentric tracks on a cylinder. Collisions cause agents to change their direction of motion. To avoid this, agents may decide to switch tracks so as to merge with platoons of agents marching in their direction.

We prove that such agents must eventually converge to a local consensus about their direction of motion–all agents on each narrow track must eventually march in the same direction. We give asymptotic bounds for the expected amount of time it takes for such convergence or "stabilization" to occur, which depends on the number of agents, the length of the tracks, and the number of tracks. We show that when agents also have a small probability of "erratic", random track-jumping behaviour, a global consensus on the direction of motion across all tracks must eventually occur. Finally, we verify our theoretical findings in numerical simulations.

Keywords: Mobile robots · Swarms · Collective motion · Natural algorithms

1 Introduction

Birds, locusts, human crowds and swarm-robotic systems exhibit interesting collective motion patterns. The underlying autonomous agentic behaviours from which these patterns emerge have been an enduring topic of great interest [1–3]. In particular, a lot of research has centred around the analysis of formal mathematical models for swarm dynamics in its various manifestations [4–7]. Rigorous mathematical results are necessary for understanding swarms and for designing predictable and effective swarm-robotic systems. However, multi-agent swarms have a uniquely complex and "mesoscopic" nature [8], and relatively few standard techniques for the analysis of such systems have been established. Consequently, the analysis of new models of swarm dynamics is important for advancing our understanding of the subject.

In this work, we study the dynamics of "locust-like" agents moving on a discrete ringlike surface. The model we study is inspired by the following well-documented experiment [9]: place many locusts on a ringlike arena at random positions and orientations. They start to move around and bump into the walls and into each other, and as they do so, remarkably, over time, they begin to collectively march in the same direction; clockwise or counterclockwise (Fig. 1). Inspired by observing these experiments, we asked the following question: what are simple and reasonable myopic rules of behaviour that might lead to this phenomenon? Our goal is to study this question from an *algorithmic* perspective, by considering a model of discretized mobile agents that act upon a local algorithm. As with much of the literature on swarm dynamics [4,10,11], our goal is not to study a mathematical model of locusts in particular (the precise mechanisms underlying locusts' behaviours are very complex and subject to intense ongoing research, e.g. [2,9]), but to study the kinds of algorithmic local interactions that lead to collective marching and similar phenomena. As such, we are interested in simple "algorithmic" principles that yield collective marching or that approximate the biologically emerging behaviours. The resulting model is idealized and simple to describe, but the patterns of motion that emerge while the locusts progress towards stabilized collective marching are surprisingly complex.

Fig. 1. The collective clockwise marching of locusts in a ring arena (by Amir Ayali).

The starting point for this work is the following postulated "rationalization" of what a locust-like agent wants to do: it wants to keep moving in the same direction of motion (clockwise or counterclockwise) for as long as possible. We can therefore consider a model of locust-like agents that never change their heading unless they collide, heads-on, with agents marching in the opposite direction, and are forced to do so due to the pressure which is exerted on them. When possible, these agents prefer to *bypass* agents that are headed towards them, rather than collide with those agents. This is done by changing lanes; moving in an orthogonal manner between concentric narrow *tracks* which partition the ringlike arena. The formal description of this "rationalized" model is given in Sect. 2, and will be our subject of study.

Contribution. We describe and study a stochastic model of locust-like agents in a discretized ringlike arena which is modelled as multiple *tracks* that wrap around a cylinder. We show that our agents eventually reach a "local consensus" about the direction of marching, meaning all agents on the same track march in the same direction. We give asymptotic bounds for the amount of time this takes based on the number of agents and the size of the arena. Because of the idealized "precise" nature of our model, a global consensus where *all* locusts walk in the same direction (as in Fig. 1) is not guaranteed, since locusts in different tracks might never meet. However, we show that, when a small probability of "erratic", random behaviour is added to the model, such a global consensus must occur. We verify our claims via simulations and make some further empirical observations that may inspire further investigations into the model.

Despite being simple to describe, analyzing the model proved tricky in several respects. Our analysis strategy is to show that the model repeatedly passes between two phases: one in which it is "chaotic", such that locusts are arbitrarily moving about, and one in which it is "orderly", such that all locusts are in a kind of dense deadlock situation and collisions are frequent. We derive our asymptotic bounds from studying the well-behaved phase while bounding the amount of time the locusts can spend in the chaotic phase.

Related Work. The experiments inspiring our work are discussed in [2,9]. Mathematical models of the collective motion of natural organisms such as birds, locusts and ants, and the convergence of such systems of agents to stable formations, have been discussed in numerous works including [6,7,12]. Discrete mathematical models of transport on rings, such as *simple exclusion processes*, have also been investigated in the literature [13].

The central focus of this work regards consensus: do the agents eventually converge to the same direction of motion, and how long does it take? Similar questions are often asked in the field of opinion dynamics. Mathematically, if the agents' direction of motion (clockwise or counterclockwise) is considered an "opinion", we can compare our model to models in this field. When there are no empty locations at all in the environment, our model is fairly close to the *voter model* on a ring network with two distinct opinions, the main difference being that, unlike in the voter model, our agents' direction of motion determines which agents' opinions can influence them (an excellent survey on this topic is [14]). The comparison to the voter model breaks when we introduce empty locations and multiple ringlike tracks, at which point we must take into account the physical location of every agent when considering which agents can influence its opinion. Several works have explored models of opinion dynamics in a ring environment where the agents' physical location is taken into account [15,16]. Our model is distinct from these in several respects: first, in our model, an agents' internal state–its direction of motion–plays an active part in the algorithm that determines which locations an agent may move to. Second, we partition our ring topology into several narrow rings ("tracks") that agents may switch between, and an agents' decision to switch tracks is influenced by the presence of platoons of agents moving in its direction in the track that it wants to switch to. In other words, we model agents that actively attempt to "swarm" together with agents moving in their direction of motion.

Protocols for achieving consensus about a value, location or the collective direction of motion have also been investigated in swarm robotics and distributed algorithms [17–19]. However, in this work, we are not searching for a protocol that is designed to efficiently bring about consensus; we are investigating a protocol that is inspired by natural phenomena and want to see *whether* it leads to consensus and how long this takes on average.

2 Model and Definitions

We postulate here a model for locust-inspired marching in a wide ringlike arena, which is divided into narrow concentric rings. For simplicity, we map the arena to the surface of a discretized cylinder of height k partitioned into k narrow rings of length n, which are called *tracks*. For example, the environment of Fig. 2 corresponds to $k = 3, n = 8$ (3 tracks of length 8). The coordinate (x, y) refers to the xth location on the yth track (which can also be seen as the xth location of a ring of length n wrapped around the cylinder at height y). Since we are on a cylinder, we have that $\forall x, (x + n, y) \equiv (x, y)$.

A swarm of m identical agents, or "locusts", which we label $A_1, \ldots, A_m$, are dispersed at arbitrary locations on the cylinder and move autonomously at discrete time steps $t = 0, 1, \ldots$. A given location (x, y) on the cylinder can contain at most one locust. Each locust A_i is initiated with either a "clockwise" or "counterclockwise" *heading*, which determines their present direction of motion. We define $b(A_i) = 1$ when A_i has clockwise heading, and $b(A_i) = -1$ when A_i has counterclockwise heading.

The locusts move synchronously at discrete time steps $t = 0, 1, \ldots$. At every time step, locusts try to move in their direction of motion: if a locust A is at (x, y), it will try to move to $(x + b(A), y)$. A clockwise movement corresponds to adding 1 to x, and a counterclockwise movement corresponds to subtracting 1. The locusts have physical dimension, so if the location a locust attempts to move to already contains another locust at the beginning of the time step, the locust instead stays put. If A_i and A_j both attempt to move to the same location, one is chosen at random to move to the location and the other stays put.

Locusts that are adjacent exert pressure on each other to change their heading: if A_i has a clockwise heading and A_j has a counterclockwise heading, and they lie on the coordinates (x, y) and $(x + 1, y)$ respectively, then at the end of the current time step, one locust (chosen uniformly at random) will flip its heading to the other locust's heading. Such an event is called a **conflict** between A_i and A_j. A conflict is *won* by the locust that successfully converts the other locust to their heading.

Let A be a locust at (x, y). If the locust A has clockwise heading, then the *front* of A is the first locust after A in the clockwise direction, and the *back* of A is the first locust in the counterclockwise direction. Vice-versa when A has counterclockwise heading. Formally, the *front* of A is the location of the form $(x + b(A)i, y)$ which minimizes i subject to the constraints that $(x + b(A)i, y)$ contains a locust and $i \geq 1$. The *back* of A is the location of the form $(x - b(A)i, y)$ which minimizes i subject to analogous constraints. The locusts in the front and back of A are denoted $A^\rightarrow$ and $A^\leftarrow$ respectively, and are called A's *neighbours*; these are the locusts that are directly in front of and behind A. Note that when a track has two or less locusts, $A^\rightarrow = A^\leftarrow$, and when a track has one locust, $A = A^\rightarrow = A^\leftarrow$.

Besides moving in the direction of their heading within their track, a locust A at (x, y) can switch tracks, moving vertically from (x, y) to $(x, y+1)$ or $(x, y-1)$ (unless this would cause it to go above track k or below track 1). Such vertical movements occur *after* the horizontal movements of locusts along the tracks. Locusts are incentivized to move vertically when this enables them to avoid changing their heading (inertia). Specifically, A may move to the location $E = (x, y \pm 1)$ at time t when:

1. At the beginning of time t, A and $A^\rightarrow$ are not adjacent and $b(A) \neq b(A^\rightarrow)$.
2. After A moves to E, the new $A^\leftarrow$ and $A^\rightarrow$ have heading $b(A)$.
3. No locust will attempt to move horizontally to E at time $t+1$.

Condition (1) states that there is an imminent conflict between A and $A^\rightarrow$ which is bound to occur. Condition (2) guarantees that, by changing tracks to avoid this conflict, A is not immediately advancing towards another conflict; A's new neighbours will have the same heading as A. Condition (3) guarantees the location A wants to move to on the new track isn't the current destination of another locust already on that track. Together, these conditions mean that locusts only change tracks if this results in avoiding collisions and in "swarming" together with other agents marching in the same direction. A locust is allowed to switch tracks only when it senses that (1), (2) and (3) are fulfilled.

Besides these conditions, we make no assumptions about *when* locusts move vertically. In other words, locusts do not always need to change tracks when they are allowed to by rules (1)–(3); they may do so arbitrarily, say with some probability q or according to any internal scheduler or algorithm. We do not determine in any sense the times when locusts move tracks–but only determine the preconditions required for such movements; our results in the following sections remain true regardless. This makes our results general in the sense that they hold for many different track-switching "swarming" rules, so long as those rules do not break the conditions (1)–(3). Figure 2 illustrates one time step of the model, split into horizontal and vertical movement phases.

In order to slightly simplify our analysis of the model, *we assume that every track has at least 2 locusts at all times*, although our results remain true without this assumption. Finally, everywhere in this work, *the beginning* of a time step refers to the configuration of the swarm at that time step before any locusts moved, and *the end* of a time step refers to the configuration at that time step after all locust movements are complete.

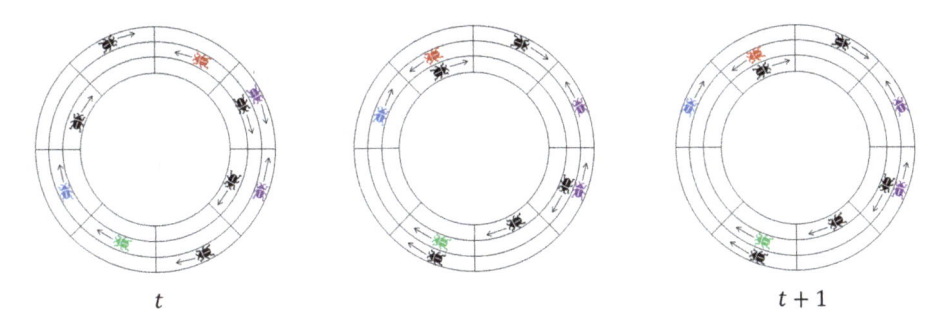

Fig. 2. The outcome of one time step of the model. The leftmost figure is the configuration at the start of time t (with $k = 3, n = 8$). The rightmost is the beginning of time $t + 1$. The middle figure illustrates the outcome of conflicts and horizontal movements specifically, before any vertical movements. The *front* and *back* of the blue locust are the red and green locusts respectively. The purple locusts conflict with each other. Since conditions (1)–(3) are fulfilled, the blue locust may switch tracks, and it does so.

3 Stabilization Analysis

We will mainly be interested studying the stability of the headings of the locusts over time. Does the model reach a point where the locusts stop changing their heading? If so, are their headings all identical? How soon does this occur?

In the case of a single track ($k = 1$), we shall see that the locusts all eventually stabilize with identical heading, and bound the expected time for this to happen in terms of m and n. In the multi-track case, we shall see that the locusts stabilize and agree on a heading *locally* (i.e., all locusts *on the same track* eventually have identical heading and thereafter never change their heading), and bound the expected time to stabilization in terms of m, n, k. In the multi-track case, we show further that a small probability of "erratic" track-switching behaviour induces *global* consensus: all locusts across all tracks converge to the same heading.

Due to space constraints, some of the proofs and technical diagrams in this section have been cut and may be found in the extended version of this paper, available at arXiv [20]. The extended version also contains further results.

3.1 Locusts on Narrow Ringlike Arenas ($k = 1$)

We start by studying the case $k = 1$, that is, we study a swarm of m locusts marching on a single track of length n. Throughout this section, we assume this is the case, except in Definition 2, which we will use in later sections.

For the rest of this section, let us call the swarm *non-stable* at time t if there are two locusts A_i and A_j such that $b(A_i) \neq b(A_j)$; otherwise, the swarm is *stable*. We wish to bound the number of time steps it takes for the system to become stable, T_{stable}. The primary goal of the rest of this section is to prove the following Theorem, which tells us that the expected time to stabilization grows quadratically in the number of locusts m, and linearly in the track length n.

Theorem 1. *For any configuration of m locusts on a ring with a single track, $\mathbb{E}[T_{stable}] \leq m^2 + 2(n-m) = \mathcal{O}(m^2 + n - m)$. Additionally, there are initial locust configurations for which $\mathbb{E}[T_{stable}] = \Omega(m^2 + n - m)$.*

In particular, Theorem 1 tells us that all locusts must have identical heading within finite expected time. This fact in isolation (without the time bounds in the statement of the theorem) is relatively straightforward to prove, by noting that the evolution of the locusts' headings and locations can be modelled as a finite Markov chain, and the only absorbing classes in this Markov chain are ones in which all locusts have the same heading (see [21]).

We give next a definition of *segments*, which are sets of agents all on the same track that share a direction of motion. This will allow us to partition the swarm into segments, such that every locust belongs to a unique segment (see Figs. 3a, 3b). Although this section focuses on the case of a single track (and claims in this section are made under the assumption that there is only a single track), the definition is general, as we will use it in subsequent sections. Note that segments are only well-defined for tracks where not all locusts have the same heading.

Definition 2. *At time t, let $\mathcal{K}$ be one of the k tracks, such that not all locusts in $\mathcal{K}$ have identical heading. Any locust A in $\mathcal{K}$ for which $b(A^{\leftarrow}) \neq b(A)$ is called a **segment tail** at time t. When A is a segment tail, define the sequence of locusts $B_0 = A$ and $B_{i+1} = B_i^{\rightarrow}$. Let B_q be the first locust in this sequence for which $b(B_q) \neq b(B_0)$. The set $\{B_0, B_1, \ldots B_{q-1}\}$ is called the **segment** of the agents $B_0, \ldots B_{q-1}$ at time t. The locust B_{q-1} is called a **segment head**.*

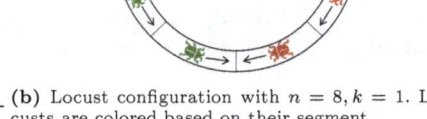

(a) Locust configuration with $n = 8, k = 3$. Locusts are colored based on their segment.

(b) Locust configuration with $n = 8, k = 1$. Locusts are colored based on their segment.

Fig. 3. An illustration of locust segments in two arbitrary locust configurations.

Only locusts which are segment heads at the beginning of a time step can change their heading at that time step. When the heads of two segments are adjacent to each other, the resulting conflict causes one to change its heading, leave its previous segment, and instead become part of the other segment. If the head of a segment is also the tail of a segment, the segment is eliminated when it changes heading. Segment tails can also cease being tails when two segments of the same heading merge (due to the segment of opposite heading that separated them being eliminated), in which case the number of segments decreases. No other action by an agent can change the number of segments. Hence, the number of segments and segment tails can only go down, and no agent ever becomes a segment tail due to changing its heading. Therefore there must exist a locust, determined probabilistically based on the evolution of the swarm, which remains a segment tail at all times $t < T_{stable}$ and never changes its heading. Denote one such locust A_W (if there is more than one, choose arbitrarily).

Definition 3. *The segment of A_W at the beginning of time t is called the **winning segment** at time t, denoted $SW(t)$. The head of $SW(t)$ is denoted $H_W(t)$. If at time t_0 the swarm is stable, we let $SW(t_0)$ equal the entire swarm.*

Lemma 4. *The expected number of time steps $t < T_{stable}$ in which $|SW(t)|$ changes is bounded by m^2.*

Lemma 5. *The expected number of time steps $t < T_{stable}$ in which $|SW(t)|$ does not change is bounded by $2(n-m)$.*

The arguments used to prove Lemma 4 and 5 are somewhat technical, and are detailed in [20]. Using these Lemmas, the proof of Theorem 1 now follows.

Proof. Lemma 5 says that before time T_{stable}, $|SW(t)|$ does not change in at most $2(n-m)$ time steps in expectation, and Lemma 4 tells us that the expected number of changes to $|SW(t)|$ before time T_{stable} is at most m^2. Hence, for any configuration of m agents on a ring of track length n, $\mathbb{E}[T_{stable}] \leq m^2 + 2(n-m)$.

Let us now show a locust configuration for which $\mathbb{E}[T_{stable}] = \Omega(m^2 + n)$, to asymptotically match the upper bound we found. Consider a ring with $k = 1$, m divisible by 2, and an initial locust configuration where locusts are found at coordinates $(0,1), (1,1), \ldots (m/2, 1)$ with clockwise heading and at $(-1,1), (-2,1), \ldots (-m/2 - 1, 1)$ with counterclockwise heading, and the rest of the ring is empty. This is a ring with exactly two segments, each of size $m/2$. Since after every conflict, the segment sizes are offset by 1 in either direction, the expected number of conflicts between the heads of the segments that is necessary for stabilization is equal to the expected number of steps a random walk with absorbing boundaries at $m/2$ and $-m/2$ takes to end, which is $m^2/4$. Since the heads of the segments start at distance $n-m$ from each other, it takes $\Omega(n-m)$ steps for them to reach each other. Hence the expected time for this ring to stabilize is $\Omega(m^2 + n - m)$. □

3.2 Locusts on Wide Ringlike Arenas ($k > 1$)

Let us now investigate the case where $k > 1$, that is, m locusts are marching on a cylinder of height k partitioned into k tracks of length n. The first question we should ask is whether, just as in the case of the locusts in a $k = 1$ environment, there is some time T where all locusts have identical heading. The answer is "not necessarily": consider for example the case $k = 2$ where on the $k = 1$ track, all locusts march clockwise, and on the $k = 2$ track, all locusts march counterclockwise. By the rules for switching tracks (see Sect. 2), no locusts will ever switch tracks in this configuration, hence the locusts will perpetually have opposing headings. As we shall prove in this section, on the cylinder, swarms stabilize *locally*–meaning that eventually, all locusts *on the same track* have identical heading, but this heading is not always shared globally.

We say that the yth track is stable if all locusts whose location is $(\cdot, y)$ have identical heading. Once a track becomes stable, it remains this way forever, as by the model, the only locusts that may move into the track must have the same heading as its locusts. Let T_{stable} be the first time when every all the k tracks are stable. Our goal will be to prove the following asymptotic bounds on T_{stable}:

Theorem 6. $\mathbb{E}[T_{stable}] = \mathcal{O}(\min(\log(k)n^2, mn + m^2))$.

The bound $\mathcal{O}(mn + m^2)$ is tighter when there is a relatively small number of locusts; $m \lesssim \sqrt{\log(k)}n$. The bound $\mathcal{O}(\log(k)n^2)$ is better otherwise.

Recalling Definition 2, each locust belongs to some segment. Each track has its own segments, and these may sometimes change size due to conflicts, or agents moving vertically from their current segment to a segment on another track. In this section, we will treat segments as having persistent identities, similar to SW in the previous section. This is encapsulated in Definition 7:

Definition 7. *Let S be a segment whose tail is A. Then $S(t)$ refers to the segment whose tail is A at the beginning of time t. If A is not a segment tail at time t, then we will say $S(t) = \emptyset$ (this can happen once A changes its heading, moves to another track, or due to another segment merging with $S(t)$ which might cause $b(A^{\leftarrow})$ to equal $b(A)$, thus making A no longer the tail).*

If S is a segment, define S_1 to be the segment tail and $S_{i+1} = S_i^{\rightarrow}$.

Let us give a few examples of the notation in Definition 7. Suppose at time t_1 we have some segment S. Then the tail of S is S_1, and the head is $S_{|S|}$. $S(t)$ is the segment whose tail is S_1 at time t, hence $S(t_1) = S$. Finally, $S(t)_{|S(t)|}$ is the head of the segment $S(t)$.

In the $k > 1$ setting, locusts can frequently move between tracks, which complicates our study of T_{stable}. One crucial fact however, is that the number of segments on any individual track is non-increasing. This is because, first, as in the previous section, locusts moving and conflicting on the same track can never create new segments. Second, by the locust model, locusts can only move into another track when this places them between two locusts that already belong to some (clockwise or counterclockwise) segment. That being said, locusts moving in and out of a given track makes it much harder to measure how close a track is to stability by the method we used in the previous section. Hence, in the following definitions of *compact* and *deadlocked* locust sets, our goal is to identify configurations of locusts which locusts cannot enter into, which makes their analysis easier. Subsequently, we will show that segments must enter into deadlock "relatively often", which is what will enable the analysis.

Definition 8. *Let X be any set of locusts $\{X_1, X_2, \ldots X_N\}$ such that $X_{i+1} = \overrightarrow{X_i}$ (X is not necessarily a segment). X is called **compact** if either:*

1. *Every locust in X has clockwise heading and for every locust $X_i \in X$, $i < N$, there is a locust $X_j \in X$ at most two clockwise steps from X_i, **or***
2. *Every locust in X has counterclockwise heading and for every locust $X_i \in X$, $i < N$, there is a locust $X_j \in X$ at most two counterclockwise steps from X_i*

Definition 9. *Let $X = \{X_1, X_2, \ldots X_N\}$ and $Y = \{Y_1, Y_2, \ldots Y_M\}$ be two compact sets, such that the locusts of X have clockwise heading and the locusts of Y have counterclockwise heading. X and Y are **in deadlock** if X_N and Y_M are adjacent. (See Fig. 3b for an example of two deadlocked segments).*

A compact set of locusts X is essentially a platoon of locusts all on the same track and headed in the same direction, jammed together with at most one empty space between each consecutive pair. As long as X remains compact, no new locusts can enter the track between any two locusts of X, because the model states that locusts do not move vertically into empty locations to which a locust is attempting to move horizontally, and the locusts in a compact set are always attempting to move horizontally to the empty location in front of them.

Lemma 10. *Let P and Q be the only segments on track K at time t_0, s.t. P's locusts head clockwise. Let d be the clockwise distance from P_1 to Q_1. At time $t_0 + 3d$, $P(t_0 + 3d)$ and $Q(t_0 + 3d)$ are in deadlock, or the track is stable.*

Lemma 11. *Let $seg(t)$ denote the set of segments in all tracks at time t. At time $t + 3n$, either every segment is in deadlock with some other segment, or $|seg(t+3n)| < |seg(t)|$.*

Theorem 12. $\mathbb{E}[T_{stable}] \leq \frac{3}{4}mn + \frac{\pi^2}{24}m^2 = \mathcal{O}(mn + m^2)$

Proof. Let $|seg(t)|$ denote the number of segments at time t. $\mathbb{E}[T_{stable}]$ can be computed as the sum of times $\mathbb{E}[T_2 + T_4 + \ldots + T_{|seg(0)|}]$, where T_i is the expected time until the number of segments drops below i, if it is currently i (we increment the index by 2 since segments are necessarily eliminated in pairs).

Let us estimate $E[T_{2i}]$. Suppose that at time t, the number of segments is $2i$. Then after $3n$ steps at most, either the number of segments has decreased, or all segments are in deadlock. There are in total i pairs of segments in deadlock, and as there are m locusts, there must be a pair $P(t+3n), Q(t+3n)$ that contains at most m/i locusts. By the properties of deadlocked segments, $P(t+3n), Q(t+3n)$ remain in deadlock until either P or Q is eliminated. We can compute precisely how long this takes, since at every time step after time $t + 3n$, the heads of P and Q conflict, resulting in one of the segments increasing in size and the other decreasing. Hence, the expected time it takes P or Q to be eliminated is precisely the expected time it takes a symmetric random walk starting at 0 to reach either $|P(t+3n)|$ or $-|Q(t+3n)|$, which is $|P(t+3n)| \cdot |Q(t+3n)| \leq (\frac{m}{2i})^2$. Hence, $E[T_{2i}] \leq 3n + (\frac{m}{2i})^2$. Consequently:

$$\mathbb{E}[T_2 + T_4 + \ldots + T_{|seg(0)|}] \leq 3n \cdot \frac{|seg(0)|}{2} + \sum_{i=1}^{\infty}\left(\frac{m}{2i}\right)^2 \leq \frac{3}{4}nm + \frac{\pi^2}{24}m \quad (1)$$

Where we use $|seg(0)| \leq m/2$ and the identity $\sum_{i=1}^{\infty}(\frac{1}{i})^2 = \frac{\pi^2}{6}$. □

We prove next that $E[T_{stable}] = \mathcal{O}(\log(k)n^2)$. We require first a lemma:

Lemma 13. *Consider k independent random walks with absorbing barriers at 0 and $2n$, i.e., random walks that end once they reach 0 or $2n$. The expected time until **all** k walks end is $\mathcal{O}(n^2 \log(k))$.*

Theorem 14. $\mathbb{E}[T_{stable}] = \mathcal{O}(\log(k) \cdot n^2)$

Proof. (Sketch.) Let $seg_i(t)$ denote the number of segments in track i at time t, and define $\mathcal{M}_t = \max_{1 \leq i \leq k} seg_i(t)$. Let us bound the expected time until $\mathcal{M}_t$ to decreases. Define the set $K(t)$ to be all tracks having $|\mathcal{M}_t|$ segments at time t. Then $\mathcal{M}_t$ decreases at the first time $t' > t$ when all tracks in $K(t)$ have had their number of segments decrease. We may bound this as follows: by slightly generalizing Lemma 11, if $\mathcal{M}_t$ doesn't decrease after $3n$ time steps (i.e., $\mathcal{M}_t = \mathcal{M}_{t+3n}$), all tracks in $K(t+3n)$ now have all their segments in deadlock. The number of deadlocked segment pairs at every track in $K(t+3n)$ is $\mathcal{M}_t/2$, so in every such track there is such a pair with at most $2n/\mathcal{M}_t$ locusts. By Lemma 13, using similar reasoning to Theorem 12, these pairs of deadlocked segments all resolve into single segments after at most $c \cdot \log(k)\left(\frac{2n}{\mathcal{M}_t}\right)^2$ expected time for some constant c. Hence, the number of expected time steps for $\mathcal{M}_t$ to decrease is bounded above by $3n + c\log(k)\left(\frac{2n}{\mathcal{M}_t}\right)^2$.

T_{stable} is the first time when $\mathcal{M}_t = 0$. Let us assume n is even for simplicity (the computation will hold regardless, up to rounding). Then the maximum possible number of segments at time $t = 0$ is n, and $\mathcal{M}_t$ decreases in jumps of 2 or more (since segments can only be eliminated in pairs). Hence, T_{stable} is bounded by the amount of time it takes $\mathcal{M}_t$ to decrease at most $n/2$ times. By linearity of expectation and the previous paragraph, this can be bounded by summing $3n + c\log(k)\left(\frac{2n}{\mathcal{M}_t}\right)^2$ over $\mathcal{M}_t = n, n-2, n-4, \ldots 2$:

$$\mathbb{E}[T_{stable}] \leq \frac{n}{2} \cdot 3n + c\log(k)\left(\frac{2n}{n}\right)^2 + c\log(k)\left(\frac{2n}{n-2}\right)^2 + \ldots + c\log(k)\left(\frac{2n}{2}\right)^2 \leq$$
$$\leq \frac{3}{2}n^2 + 4c\log(k)n^2\sum_{i=1}^{\infty}\left(\frac{1}{2i}\right)^2 = \frac{3}{2}n^2 + \frac{\pi^2}{6}c\log(k)n^2 = \mathcal{O}(\log(k)n^2) \quad (2)$$

As claimed. □

Theorem 6 follows from Theorems 12 and 14, by taking the minimum. □

Erratic Track Switching and Global Consensus. Theorem 6 shows that, in finite expected time, all locusts on a given track have identical heading. This is a stable *local* consensus, in the sense that two different tracks may have locusts marching in opposite directions forever. We might ask what modifications to the model would force a *global* consensus, i.e., make it so that stabilization occurs only when all locusts across *all* tracks have identical heading. There is in fact a simple change that would force this to occur: let us assume that any locust, with some probability $p \in (0,1)$, exhibits "erratic" track switching behaviour: meaning that they disregard the vertical movement conditions (1)–(3) of the model (see Sect. 2) and attempt to move vertically to an adjacent empty space on the track above or below them regardless. This behaviour is "erratic" in the sense that such vertical movements do not help locusts keep moving in the same direction, and may in fact force them into conflict with other locusts.

The next Theorem shows that the existence of erratic behaviour forces a global consensus of locust headings. The goal is to prove that there is some finite time after which all locusts must have the same heading. Note that the bound we find for this time is crude, and is not intended to approximate T_{stable}. We study the question of how p affects T_{stable} empirically in the next section.

Theorem 15. *Assuming there is at least one empty space (i.e., $m < nk$), and the probability of erratic track switching is $0 < p < 1$, the locusts all have identical heading in finite expected time.*

4 Simulation and Empirical Evaluation

We explore some questions about the expected value of T_{stable} through numerical simulations. One thing we could not study in our formal analysis (since we analyzed only the *worst-case* possible behaviour) are the helpful effects of track switching on T_{stable}. Recall that our model allows locusts to switch tracks if this would enable them to avoid a conflict and join a track where *locally*, locusts are marching in their same direction. As we shall see numerically, this often speeds up the time to consensus. This justifies track-switching behaviour as a mechanism that, despite being highly local, enables the locusts to come to local consensus about the direction of motion sooner.

In Fig. 4, (a) and (b), we measure T_{stable} as it varies with n and k. We simulate two different locust configurations: a "dense" configuration, and a "sparse" configuration. In the dense configuration, 50% of locations are initiated with a locust, with the locations chosen at random. In the sparse configuration, 10% of locations are initiated with a locust. The locusts are initiated with random heading. We measure the effect of track switching on T_{stable}: the opaque lines measure T_{stable} when locusts switch tracks as often as they can (while still obeying the rules of the model), and the dotted lines measure T_{stable} when locusts never switch tracks. For every value of n, k, we ran the simulation 1000 to 3000 times and averaged T_{stable} over all simulations. As we can see, in the sparse configuration, track-switching has a significantly positive effect on time to stabilization. For example, with $k = 30$, $n = 30$, T_{stable} is approximately 13.5 when locusts switch tracks as soon as they can, and approximately 25 when they never switch tracks–nearly double. In the dense configuration, we see that enabling locusts to move tracks has little to no effect, since the locust model rarely allows them to do so due to the tracks being overcrowded.

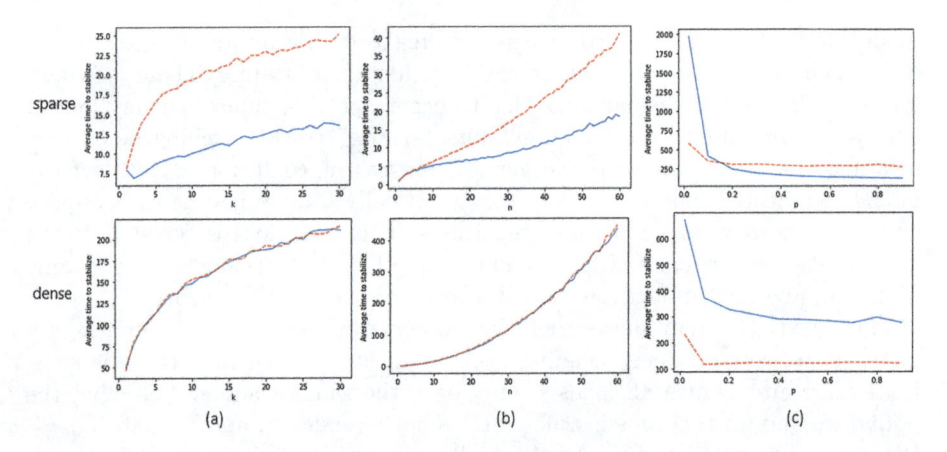

Fig. 4. Simulations of the locust model. The y axis is T_{stable}. Column (a) measures T_{stable} for $k = 1...30$, with n fixed at 30. Column (b) measures T_{stable} for $n = 1...60$, with k fixed at 5. Column (c) measures T_{stable} with $n = 30, k = 5$, and p (erratic behaviour probability) going from 0 to 1. The top row measures T_{stable} for a sparse locust configuration, and the bottom row does so for a dense configuration. The dashed line estimates T_{stable} when locusts never switch tracks non-erratically; the blue line estimates T_{stable} when locusts switch tracks as often as the model rules allow.

In column (c) of Fig. 4, we measure how the probability p of erratic behaviour affects T_{stable}. As we proved in the previous section, whenever $p > 0$, stabilization requires *global* rather than local consensus. Hence, we cannot directly compare the T_{stable} of these graphs with columns (a) and (b), where T_{stable} measures the time to local consensus. We see that $\mathbb{E}[T_{stable}]$ approaches ∞ as p goes to 0, as expected, since when $p = 0$, global stability might never occur in some initial configurations. $\mathbb{E}[T_{stable}]$ decreases sharply as p goes to some critical point around 0.1, and at a slower rate afterwards. It is interesting that low probability of erratic behaviour affects $\mathbb{E}[T_{stable}]$ significantly more in the *sparse* configuration, where for $p = 0.02$, if locusts also switch tracks whenever the model allows them, $\mathbb{E}[T_{stable}]$ was measured as being approximately 1974, as opposed to 669 in the dense configuration. We make also the curious observation that, while non-erratic track switching accelerates local consensus, for some track-switching behaviours, it will in fact decelerate the attainment of global consensus. This is seen by the fact that frequent non-erratic track-switching was helpful in Columns (a) and (b) of Fig. 4, but mostly increased time to stability in Column (c). We further discuss this phenomenon in [20].

5 Concluding Remarks

We studied collective motion in a model of discrete locust-inspired swarms, and bounded the expected time to stabilization in terms of the number of agents m, the number of tracks k, and the length of the tracks n. We showed that when the swarm stabilizes, there must be a local consensus about the direction of motion. We also showed that, when the model is extended to allow a small probability of erratic behaviour to perturb the system, global consensus occurs.

A direct continuation of our work would be to find upper bounds on time to stabilization when there is some probability of erratic behaviour. Furthermore, our empirical simulations suggest several curious phenomena related to erratic behaviour: first, there seems to be a clash between "erratic" and non-erratic, "rational" track-switching, as when locusts switch tracks non-erratically in order to avoid collisions, this seems to accelerate the attainment of local consensus, but mostly hinder the attainment of global consensus. Second, increasing the probability of erratic track-switching p behaviour was helpful in accelerating global consensus up to a point, but in simulations, its impact seemed to fall off past a small critical value of p. In future work, it would be interesting to investigate these aspects of the model.

Although our dynamics model is inspired by experiments on locusts, it can be understood in more abstract terms as a model that describes a specific situation where many agents that wish to maintain a direction of motion are confined to a small space where they exert pressure on each other. It is natural to ask what kinds of collective dynamics, if any, we should expect when this small space has a different topology; rather than a ringlike arena, we might consider, e.g., a square arena. We believe that rich models of swarm dynamics can be discovered through observing natural organisms exert pressure on each other in such environments.

Acknowledgements. This research was partially supported by the Israel Science Foundation grant no. 2306/18. The authors would like to thank Prof. Amir Ayali (Tel Aviv University) for bringing our attention to the locust experiments and for graciously letting us use the image in Fig. 1, Prof. Ofer Zeitouni (Weizmann Institute of Science) for helpful discussions, and the anonymous reviewers for constructive comments.

References

1. Altshuler, Y., Pentland, A., Bruckstein, A.M.: Introduction to swarm search. In: Altshuler, Y., Pentland, A., Bruckstein, A.M. (eds.) Swarms and Network Intelligence in Search. SCI, vol. 729, pp. 1–14. Springer, Cham (2018). https://doi.org/10.1007/978-3-319-63604-7_1
2. Ariel, G., Ayali, A.: Locust collective motion and its modeling. PLOS Comput. Biol. **11**(12), e1004522 (2015)
3. Fridman, N., Kaminka, G.A.: Modeling pedestrian crowd behavior based on a cognitive model of social comparison theory. Comput. Math. Organ. Theory **16**(4), 348–372 (2010)
4. Bruckstein, A.M.: Why the ant trails look so straight and nice. Math. Intelligencer **15**(2), 59–62 (1993)
5. Chaté, H., Ginelli, F., Grégoire, G., Peruani, F., Raynaud, F.: Modeling collective motion: variations on the Vicsek model. Eur. Phys. J. B **64**(3), 451–456 (2008)
6. Czirók, A., Vicsek, M., Vicsek, T.: Collective motion of organisms in three dimensions. Physica A: Stat. Mech. Appl. **264**(1–2), 299–304 (1999)

7. Ried, K., Müller, T., Briegel, H.J.: Modelling collective motion based on the principle of agency: general framework and the case of marching locusts. PloS One **14**(2), e0212044 (2019)
8. Chazelle, B.: Toward a theory of Markov influence systems and their renormalization. arXiv preprint arXiv:1802.01208 (2018)
9. Amichay, G., Ariel, G., Ayali, A.: The effect of changing topography on the coordinated marching of locust nymphs. PeerJ **4**, e2742 (2016)
10. Chazelle, B.: Natural algorithms and influence systems. Commun. ACM **55**(12), 101–110 (2012)
11. Amir, M., Bruckstein, A.M.: Probabilistic pursuits on graphs. Theor. Comput. Sci. **795**, 459–477 (2019)
12. Shiraishi, M., Aizawa, Y.: Collective patterns of swarm dynamics and the Lyapunov analysis of individual behaviors. J. Phys. Soc. Jpn. **84**(5), 054002 (2015)
13. Kriecherbauer, T., Krug, J.: A pedestrian's view on interacting particle systems, KPZ universality and random matrices. J. Phys. A: Math. Theor. **43**(40), 403001 (2010)
14. Dong, Y., Zhan, M., Kou, G., Ding, Z., Liang, H.: A survey on the fusion process in opinion dynamics. Inf. Fusion **43**, 57–65 (2018)
15. Chandra, A.K., Basu, A.: Diffusion controlled model of opinion dynamics. Rep. Adv. Phys. Sci. **1**(01), 1740008 (2017)
16. Hegarty, P., Martinsson, A., Wedin, E.: The Hegselmann-Krause dynamics on the circle converge. J. Differ. Equations Appl. **22**(11), 1720–1731 (2016)
17. Barel, A., Manor, R., Bruckstein, A.M.: Come together: multi-agent geometric consensus. arXiv preprint arXiv:1902.01455 (2017)
18. Manor, R., Bruckstein, A.M.: Chase your farthest neighbour. In: Groß, R., et al. (eds.) Distributed Autonomous Robotic Systems. SPAR, vol. 6, pp. 103–116. Springer, Cham (2018). https://doi.org/10.1007/978-3-319-73008-0_8
19. Olfati-Saber, R., Fax, J.A., Murray, R.M.: Consensus and cooperation in networked multi-agent systems. Proc. IEEE **95**(1), 215–233 (2007)
20. Amir, M., Agmon, N., Bruckstein, A.M.: A discrete model of collective marching on rings (2020). arXiv:2012.04980
21. Grinstead, C.M., Snell, J.L.: Introduction to Probability. American Mathematical Society (2012)

Map Learning via Adaptive Region-Based Sampling in Multi-robot Systems

Gianni A. Di Caro[✉] and Abdul Wahab Ziaullah Yousaf

Carnegie Mellon University in Qatar, Doha, Qatar
gdicaro@cmu.edu, abdulwaz@andrew.cmu.edu

Abstract. We present a novel approach for informative path planning in multi-robot systems for mapping under time and communication constraints. The approach is based on modeling the mapping task as a regression problem using Gaussian processes (GP) and adaptively directing the robots towards the regions that are most informative for GP learning. The methodology is based on a multi-stage process where either a robot or a group of robots search for the best convex region where to sample new data, identify the most informative sampling locations in the region, and compute an optimized path through them. The process is iterated over time adapting to newly gathered evidence and is performed collaboratively. Techniques from Monte Carlo, sequential Bayesian inference, and orienteering optimization models are combined in an integrated strategy. Fully distributed and leader-follower architectures are designed to implement the multi-stage strategy and have been evaluated in simulation, showing up to 69% of improvement over a baseline strategy.

Keywords: Informative path planning · Mapping in MRS · GP learning

1 Introduction

We consider scenarios where a *mobile multi-robot system* (MRS) is used for building a *spatial map of a physical value of interest* over a *potentially large and unknown region* in a given *time budget*. We are motivated by the use of robot teams in aquatic environments, where robots have sensors for measuring the quantity of interest, such as depth, temperature, salinity, pH, contaminants, etc. In these scenarios the use of an MRS can allow to optimize task completion by exploiting inherent *parallelism, spatial distribution, and redundancy of resources.*

The use of MRS can bring important advantages but also presents major challenges [14]. In particular, *given a limited mission time*, a central challenge consists in planning the robot paths to maximize the information which is jointly gathered along the paths, where robots take measurements *at specific locations*. Here, we focus on these problems of *informative path planning* (IPP) [1,16].

We assume that measures and fluctuations of the target data map values show some spatial correlation (e.g., water's pH doesn't locally change abruptly)

and might slowly evolve over time, naturally defining a *spatial stochastic process* which we treat as *stationary for the duration of the mission*. Under these assumptions, the process is modeled as a *Gaussian Process* (GP) [17]. Therefore, the problem of estimating the ground-truth map is represented as regression learning problem via GP model. Robots are therefore data sampling agents in a *Bayesian inference scheme*, where the current GP model acts as a prior which is incrementally updated according to gathered measurements.

Unfortunately, due to spatial correlations, the utility of a sequence of measures (e.g., sampled over a path) is not constant but depends and is correlated on all previous sequences, showing a *diminishing return* property, related to submodularity [9]. This sets IPP as an *NP-hard* problem even in the (simplest) case of single robot for static sensor placement aimed to gathering measures for map reconstruction [9]. Moreover, given that the region to map is initially unknown, the diminishing return property requires planning to be *adaptive* to keep directing robots to the locations where observation data are expected to be more valuable accounting for incrementally gathered new evidence.

In order to address both realistic and general use cases, we focus on scenarios where an optimized map building has to be performed within a *restricted time budget*, and over a *large region* where a *global communication infrastructure is not in place*. These types of scenarios ask for tackling IPP challenges in a MRS using approaches which are optimized in terms of final map accuracy as well as *distributed* and *scalable* for both communications and computations.

In this work we present a *novel strategy for online, distributed IPP in MRSs for mapping tasks under time and communication limitations*. Moreover, we present and study different *distributed architectures* for the multi-robot implementation of the strategy. The core concept of the strategy approach is the iterative and adaptive definition of *region of interests* (ROIs) of high expected Bayesian utility where a robot or a group of robots travels to and gathers measurements for map building. The expected utility of a ROI is predicted based on current GP model. Inside a ROI, a subset of locations of high predicted utility is selected, defining a traveling and sampling *plan* for each robot in the ROI. *Monte Carlo methods, sequential Bayesian optimization,* and *orienteering models* [19] are the main ingredients we use to tackle IPP problems. ROI/plans are adaptively and iteratively redefined by the robots based on a rolling time schedule or newly gathered evidence, implementing a *closed-loop process for GP learning*.

In the following sections, we describe the strategy approach and the proposed architectures for individual and groups of robots. The methods have been comparatively evaluated in a custom simulation environment where *real bathymetry data* are passed to the robots through a noisy sensor. Results show good scalability and significant improvement in accuracy compared to a baseline approach.

Paper contributions can be summarized as follows:

- A novel decentralized multi-robot algorithmic framework for IPP in map construction tasks under multiple constraints, with an emphasis on limited time and communications. Framework's strategy performs iterative and adaptive

multi-stage computation of convex regions of maximal expected informativeness, that are used by individual or group of robots for data sampling.
- Original integration of greedy, sequential Bayesian optimization, and Monte Carlo sampling and estimation to identify the regions that are expected to be the most informative ones. We extend the common notion of point-based sampling to region-based sampling, and then find an optimized path through good sampling locations respecting time constraints.
- Two complementary distributed architectures for implementing the framework: a fully distributed one, which is computationally light and favors robot spatial distribution, and a leader-follower one, which groups the robots, supporting tight communications and data sharing for effective coordination.

2 Related Work

There is a vast literature tackling IPP problems for single robots, such as [11] which can be considered as a reference state-of-the-art work. The authors consider GP-based map learning for a non-stationary environmental monitoring scenario. Our approach shares some common traits with this work, including use of mutual information, definition of full sampling paths, replanning based on a GP predictability criterion. However, our focus is on the use of MRSs, such that we address how to let the robots establish synergies, we propose different architectures to account for limitations in networking and risks of collisions, and we emphasize the role of limited time budget. How to transfer solutions in [11] to a cooperative MRS under similar constraints is not immediate, and this concern applies to other single-robot work, such that below we focus on IPP for MRSs.

In [10] multi-robot construction of a communication map is considered using a leader-follower architecture, where the leader robot computes the locations to be visited by single or multiple followers. Both single and multiple-leader teams are considered. A rendezvous phase is utilized to consolidate the data for multiple-leader case. The approach has been tested in ROS simulations and real experiments and compared against a random baseline.

In [5] a robotic swarm is considered where each robot explicitly penalizes its utility function at the location of the waypoint of other robots present in the vicinity. This minimizes the risk of collisions and favors spreading of robots. The method is aimed for source localization as well as mapping of the environment.

In [8] anytime algorithms combining sampling-based planning with branch and bound techniques are proposed that achieve efficient and scalable information gathering in continuous spaces with motion constraints. The algorithms are asymptotically optimal and have been validated in water monitoring tasks.

An MRS moving in a polygonal environment is considered in [4] under continuous connectivity constraints. The measures acquired by the robots are sent to a centralized planner that computes connected paths. An integer programming formulation is adopted to get computationally-efficient and collision free solutions. Approach has been validated in simulations using 10 robots.

All these different works with MRS can be seen as individual contributions toward the solution of a class of problems which is inherently very challenging. In each work, different assumptions are made and different methods are employed. Validations with large robots teams are still missing as well as a general solution framework to use in practice. In our work we are precisely going towards the direction of defining a possibly more general solution strategy for a broad application problem domain (data mapping) and realistic constraints on time and communications. We propose a solution strategy which can be efficiently implemented in practice and it allows to choose among different distributed architectural designs that can be selected according to the properties of the problem environment and of the robots. Moreover, we are running validation studies with relatively large(r) number of robots, precisely to encourage the development of a truly scalable and efficiently parallel solution.

3 Problem Description: Multi-robot IPP for Map Learning Under Time and Communication Limitations (IPP-MRS)

We address a generic scenario where an MRS with n robots is employed for constructing a probabilistic map of an unknown (potentially large) environment $\mathcal{R}$ given a total mission time T_{max} and a scalar measure of interest. Robots are equipped with appropriate *sensors for taking measurements*. A robot does not sample data according to a given rate while moving, but instead it *takes measurements at specific given locations* (sampling by location features). At a sampling location x^i a robot doesn't move but takes and averages multiple sensor readings. In the experiments, we assume that robots' sensors are affected by *unbiased Gaussian noise* and average 10 measures at each location, taking 30s.

Randomness and spatial distribution of the mapping data are modeled as a stationary Gaussian Process (GP). Data spatial correlation is captured via an appropriate choice of a *kernel function*, which we select as the Matérn kernel [17]. Data sampled by robots are used to incrementally learn a non-parametric GP model. The *informativeness* of sampling at a location x^i for improving accuracy of GP learning is expressed as a measure of *information gain*, while informativeness of a *path* is usually quantified using measures of *mutual information*, precisely accounting for diminishing returns [9]. Here we build on this way of proceeding, using the notion of *utility* to quantify single point informativeness.

The IPP problem that we tackle consists in finding the *jointly most informative sampling paths for the MRS* under limited time and potential restrictions on communications such that the target GP process is everywhere reconstructed with high accuracy. The goal is to find a sequence Φ^t of *plans: directed paths of discrete sampling locations for each robot*. The time parameter t reflects the iterative updating of plans according to gathered evidence and GP updating. We will refer to the problem as described in this section as *IPP-MRS*.

Note that IPP problems can be represented either in the continuous or the *discrete domain*. We consider the latter, where data sampling is only done at discrete locations. In this case the IPP can be represented on a graph, where each node provides a utility. Treating the expected utility along a path as purely additive (i.e., not accounting for diminishing returns), an IPP can be modeled as an *orienteering* optimization problem for a single robot, and as a *team orienteering* problem for multiple robots [19], where both problem classes are NP-hard.

In general, *prior observation data* (a startup GP) might be available or not. We assume to start from scratch such that an initial phase of the mission (not accounting for the time budget T_{max}) is devoted to build a *sparse prior* by sending out robots sampling at uniformly selected random locations. We therefore start with a set $\mathcal{D}^m = [X^m, \boldsymbol{y}^m]$, of observations for the GP, where $X^m = \bigcup_{k=1}^{k=m} \boldsymbol{x}^k$ is the union of m-prior observation locations $\boldsymbol{x}^k$ indexed by k-th location, and $\boldsymbol{y}^m = \bigcup_{k=1}^{k=m} y^k$ is the union of recorded values y^k at those locations.

Since, the region $\mathcal{R}$ can be remote or hardly accessible, a global communication infrastructure cannot be in general assumed. Thus, robots shall rely on *ad hoc, multi-hop communications*. This implies the additional constraint that robots shall be able to maintain, on average, a pairwise communication distance to ensure at least *delay-tolerant multi-hop communication* of coordination controls and observations. In the following we will elaborate on these aspects, considering architectures that can provide different communication guarantees.

A side, yet practically useful constraint to consider, is that paths shall be mutually disjoint while performing sampling actions, to *minimize collision risks*. Again, we will discuss different ways of addressing this constraint.

Note that the absence of a global communication infrastructure/cloud access, as well as the NP-hardness of IPP problems, rules out the use of a centralized system for efficiently computing joint robots plans during mission deployment. In other words, computations must happen *online* and be *distributed*.

4 Tackling IPP-MRS with Different Architectures

As briefly discussed in the Introduction, we tackle IPP-MRS problems with a general strategy combining Monte Carlo, sequential Bayesian optimization, and orienteering models. The strategy can be implemented adopting either leader-follower or fully distributed architectures. In the former case, the robot team is organized in groups, where in each group computations are effectively distributed between a leader robot and follower robots. The groups acts as a *collaborative unit*, optimizing data sampling in the elliptic ROI selected by the leader. In the fully distributed case *each robot acts as an individual unit* and performs all computations by its own, independently from other robots but possibly accounting for their plans. The latter depends on whether robots communicate or not. The fully distributed model is described in the next section, while the leader-follower model is in Sect. 6. Three limit cases of architectures are illustrated in Fig. 1.

It doesn't matter the architecture, the process for selecting and sampling in the ROI(s) happens iteratively, following a rolling horizon approach. Time budget T_{max} is partitioned in *equally sized time intervals* of length T_{pmax}. During each interval T_{pmax}, each robot gathers measure in its ROI and updates the GP. Adopting such an *interval-based rolling horizon*, group- or individual plans are only computed for the next T_{pmax} time units. Either at the end of the interval, or

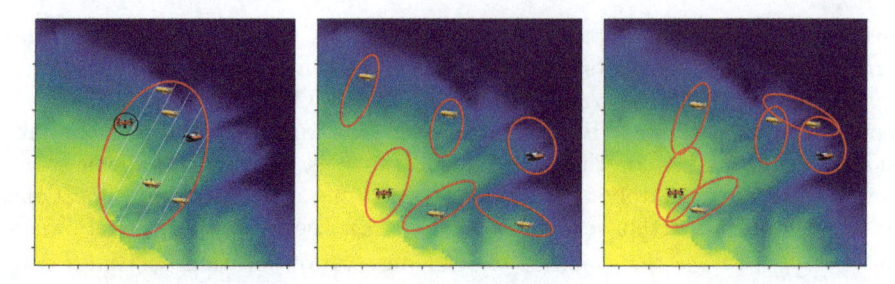

Fig. 1. Illustration of different architectures, 6-robots team. In the 2D heatmap colors indicate water depths. (Left) Team is organized in one Leader-follower group, leader is the black circled robot. Each robot samples in a different portion of the ROI. (Middle) Team is organized in a fully distributed architecture, where each robot independently selects its own ROI. Robots communications avoid ROI overlapping, minimizing risks of diminishing returns and collisions. (Right) Same as in Middle, but robots do not communicate, such that ROIs can overlap.

when the measures gathered in the ROI aren't anymore significantly informative for GP learning, a new ROI is identified (by leader or by individual robots, depending on the architecture). Robots move there for continuing sampling. Process is iterated until time budget T_{max} is fully used.

5 Fully Distributed Adaptive Selection of ROI and Paths

In the fully distributed architecture, each individual robot adaptively searches for a ROI that encompasses the following criteria. The ROI must: (i) not feature the presence of another robot; (ii) have the highest utility, in comparison to other regions in the reach; (iii) be traversable in the finite time interval T_{pmax}; (iv) be in *reachable distance*, defined as a fraction ϕ of current time budget T_{pmax}, to prevent traveling far and exhausting the budget.

Once the ROI is identified, the robot moves to it to gather measures there. To optimize usage of time resources, the robot finds a *feasible path* that minimizes the travel to and within the ROI and traverses key informative locations inside it. Either at the end of the time interval, or when the measures gathered in the ROI indicates that available information has been already gathered, a new ROI is selected. This process is explained step-by-step in the following sub-sections.

5.1 Identification of the ROI with the Largest Utility

The main goal is to identify a ROI, a sub-region $r \subset \mathcal{R}$ which is expected to provide the *largest utility should we decide to explore that area via sampling*. Moreover, the ROI should not overlap with the ROI of other robots, to reduce risks of diminishing returns and collisions. Mutual information about ROIs is obtained through notification *messages* that the robots *locally broadcast periodically*. In a message a robot includes the parameters identifying its current ROI and can include or not sampled data (as reflected in the models FDC and FDNC of Sect. 7). Regions mapping to ROIs of other robots are not removed from the search but are penalized, discounting the expected utility of their points.

To optimize the calculations involved in ROI selection and to ease navigation, r is chosen as an *ellipse*, that acts as local *containment region* for the robot(s). Ellipses define convex regions that are fully described by 5 parameters only and impose no major restrictions to size and orientation of a ROI.

Each robot independently generates several ellipses (e.g., hundreds) that satisfy the criteria (i–iv) above. Each ellipse is sized for letting the robot, in the given time budget T_{pmax}, navigate (at a constant speed) the ellipse from one end to the other and take measurements.

The utility for ROI selection is a measure of *predicted information gain* from the ROI. However, the only information the robot has for making predictions about the most promising region is the current GP. At this aim, we utilize Monte Carlo to generate multiple ellipses at random inside the region $\mathcal{R}$ and to evaluate its overall expected utility $\mathcal{U}$, eventually selecting the one with the largest utility according to current GP. This is formalized as an optimization problem:

$$(\mathbf{c}^*, a_l^*, a_w^*, \theta^*) = \mathrm{argmax}\, \mathcal{U}(\mathbf{c}, a_l, a_w, \theta, n) \tag{1}$$

subject to:

$$\frac{\|\mathbf{c}^* - \mathbf{c}_1\|}{v} \leq \phi T_{pmax} \tag{2}$$

$$\frac{2\sqrt{(a_l^2 + a_w^2)} + 2a_w}{v} = T_{pmax} \tag{3}$$

The objective function (1) seeks the parameters for the ellipse, where $\mathbf{c}^*$ is the center and (a_l^*, a_w^*) are major-minor axis, such that a utility function $\mathcal{U}$ which is defined over the ellipse and depends on the size n of the robotic team, returns maximum value. The utility function quantifies an estimate of the expected information that can be extracted from the region. Constraint (2) ensures that the travelling time from robot's current position c_l to the center c^* of the ellipse is within some threshold limit ϕ of the travel horizon T_{pmax}. This constraint avoids selecting an ellipse that would lead the robot traveling too far from current position before sampling. Constraint (3) is for sizing the ellipse according to finite horizon time budget T_{pmax}. Sizing is such that a planar graph formed by a continuous line from all four vertices of the ellipse can be traversed by a robot travelling at a constant velocity v within the budget T_{pmax}.

5.2 Defining Predicted Utility Values

To quantify the expected utility $\mathcal{U}$ in (1) we have to rely on the information from the current GP model, which consists of the m recorded observations so far, $GP(\mathbf{x}^{1:m})$. In particular, using GP's predicted mean and variance, we can assign a *point-based utility* to individual locations, which would tell how much improvement it can be expected if a robot would sample the physical value of interest at that location. In the literature, there are various metrics for improvement and various measures that appropriately balances between exploration and exploitation [18]. Here, we use an information-theoretic utility, the *mutual information* $MI(x)$ [3], which returns the predicted information that can be obtained if we sample at point x based on the knowledge of the whole region we have so far, in the form of a GP. We use mutual information to extend the idea of point-based utility to a *region-based utility*, which is what we need in (1).

In principle, a continuous elliptical region is described by the infinite points x inscribed in its boundary, where a utility would be defined for each point x. In order to compute the utility of the region we would then need to compute its integral measure over the ellipse. Since we don't have a closed form expression for MI over the ellipse, we have to rely to a numeric approximation, using a *Monte Carlo approach*. More precisely, we randomly generate N sample locations $\xi^k, k = 1, \ldots, N$ inside the elliptical containment region and we assign a mutual information measure to each sample. The mutual information is estimated by fitting the GP to all the m recorded observations so far, such that the mutual information measure at the ξ^i-th location is $MI(\xi^i \mid GP(x^{1:m}))$. This results in the following utility for the N locations in the ellipse:

$$\mathcal{U}_{ellipse}(\mathbf{c}, a_l, a_w, \theta) = \frac{\sum_{i=0}^{N} MI(\xi^i \mid GP(x^{1:m}))}{N \; \sigma(\sum_{i=0}^{N} MI(\xi^i \mid GP(x^{1:m})))} \qquad (4)$$

$$\xi_1^k = a \cos(t)\cos(\theta) - b \sin(t)\sin(\theta) + c_1 \qquad (5)$$

$$\xi_2^k = a \sin(t)\cos(\theta) + b \cos(t)\sin(\theta) + c_2 \qquad (6)$$

$$0 < a \le a_l, 0 < b \le a_w \qquad (7)$$

$$0 \le t, \theta \le 2\pi \qquad (8)$$

The *utility score* (4) is the ratio between average mutual information and variance for the N points in the region. A high score implies large average mutual information and low variance in the region, providing an indication of stable expected utility sampling from the region and its a way to balance exploration and exploitation. The score is in the form of an *inverse coefficient of variation* (CV), that avoids numerical issue of direct CV for vanishing utilities. A location point ξ^k inside the ellipse is sampled by uniform sampling of t, θ, a, b in the their domains as defined in (7–8) and plugging sampled values in (5–6), that are parametric equations of a 2D ellipse (note that a and b are sampled in $\sqrt{U(0,1)}$ and then scaled by a_l and a_w, respectively, to ensure a uniform sampling).

5.3 Identify Candidate Sampling Locations Inside Selected ROI

Once a ROI has been selected among all the generated ones, the robot knows the region to visit for sampling. However, where to precisely gather measures respecting the time budget T_{pmax} still has to be defined. At this aim, the robot begins by defining s candidate points inside the ellipse, that are expected to provide a large utility and that at the same time are minimally correlated with each other to not incur in the diminishing return issue when actual data sampling is performed. The robot generates the s points using first a form of the *Krigging Believer* (KB) algorithm [6] and then reusing *cached data from previous Monte Carlo generation* with an additional *distance constraint*.

The KB algorithm is used to generate points by "simulating" a sequential sampling experiment through *greedy Bayesian optimization*. In the KB algorithm, a new data point is inserted into the GP dataset using the value suggested by the point-based mutual information utility, instead of the real observation. To keep the dataset $D^{1:m} = [X^{1:m}, y^{1:m}]$ with the actual observations unaltered, a duplicate dataset $\overline{D^{1:m+s}} = [X^{1:m+s}, y^{1:m+s}]$ is created with the m original observations and the s observations obtained sequentially by Bayesian optimization. In the simulation, each time a candidate point is returned, instead of physically going to that location, the observation y^k is returned by evaluating the GP, and adding the result to the duplicate dataset. The GP is then updated with the duplicate dataset and in turn used by the utility function for new location suggestion. The process is repeated until a batch of s-samples is obtained. The effect of this approach is that expected variance at the suggested location gets reduced, morphing the utility function of actual sequential sampling.

The KB algorithm is a *greedy* algorithm and is known to be prone to getting stuck at local optima [6]. If this happens and the size of returned set is $\hat{s} < s$, the remaining points $s - \hat{s}$ are gathered by reusing cached locations previously sampled using Monte Carlo. Rejection sampling is applied to ensure that selected points are at a certain distance d away from $\hat{s}$ and from each other. To define d we have used the *kernel lengthscale* of the copy GP and a custom defined *threshold distance* d_{thr}, selecting the minimum between the two to set d.

5.4 Compute a Resourceful and Informative Path Using Orienteering

Once the sets of locations and utilities are computed, the robot defines and solve an *orienteering problem* [19] over its region. The goal is to find a *path* going trough a subset of the candidate sampling points. The subset is expected to *jointly maximize the predicted rewards* (which corresponds to the utility 4) while at the same time the traveling time satisfies the *time budget constraint* T_{pmax}.

Unfortunately, orienteering problems are *NP-hard*. Trying to solve optimal path in an online setting with several candidate points would be not resourceful. Therefore, we adopt a *heuristic* solution approach, based on a modified version of the ant-colony metaheuristic [13]. The fact that we work with an approximation is consistent with the fact that the scenario features many sources of uncertainty so we utilize adaptive replanning to cope with these.

5.5 Executing the Plan, Termination and Replanning

Once a path is computed, the robot travels to the assigned region, where it samples physical measures and update its model of the environment in the form of GP. Data are locally shared with other robots, such that a *shared GP representation* is maintained over time. Given that the region and the points inside were selected based on the GP($x^{1:m}$) learned from the m points sampled so far, it is fair to expect that major deviations will arise between predicted and actual utilities. If this happens, a replanning should be triggered, to stop gathering data in the current region and define a new, more promising, region for sampling.

At this aim, we utilize an *online replanning criterion* which looks at the sampled data and checks against predicted results from the GP to determine if the information quality of the region is consistent within a threshold ε. In practice, if the average information gain from the ROI gets too small compared to the gain from the initial sample within it, future samples might deliver only minimal amount of new information. If this is the case, a termination signal is triggered. The robot stops sampling and compute a new ROI to visit, re-iterating the steps. This is summarized in the pseudo-code, where x_{m+1} is the first point sampled in the ellipse, and $\mathcal{U}(x_{m+1})$ is its measured utility based on GP($x^{1:m}$).

1: **procedure** ONLINE REPLANNING STRATEGY
2:     Given: GP($x^{1:m+1}$), $\mathcal{U}_0 = \mathcal{U}(x^{m+1})$, sample index $k = 1$
3:     **while** There still are points to sample in the ROI according to plan **do**
4:         x^{m+k} is new sampling point;
5:         **if** $\left(\mathcal{U}_0 - \frac{1}{k}\sum_{i=m+1}^{m+k}\mathcal{U}(x_i)\right) < \varepsilon$ **then**
6:             GP($x^{1:m+k}$) ← UpdateGP(x^{m+k});
7:             $k \leftarrow k + 1$;
8:         **else**
9:             Terminate sampling at current ROI and replan;

6 Leader-Follower Architecture

The steps described in the previous Subsect. 5.1–5.5 are the main strategy for tackling IPP-MRS problems for an individual robot. In the case of a leader-follower architecture, the process is the essentially same but computations are partly centralized and partly distributed. A *leader* robot performs the calculations described in Subsect. 5.1–5.2, while in parallel the *followers* (all robots) individually perform the computations described in Subsect. 5.3–5.5.

Robots starts as a *connected group*. A leader robot, randomly selected, is in charge of selecting a ROI for the group using the same equations as for a single robot. However, the size of the sampled ROIs are scaled to group size. Once a ROI has been selected, the leader splits it in more or less equally sized sub-regions and assigns each sub-region to a follower (see Fig. 1, Left). In parallel, each follower selects candidate points and computes a path in the assigned sub-region, solving an orienteering problem, precisely as in the case of a single robot. Once all robots have computed their paths, the group moves to the ROI staying connected, and robots start sampling in their sub-regions.

Note that within the containment region, given that it has been suitably sized, robots are expected to get *local connectivity* for sharing GP data and coordinating a replanning. Moreover, navigation is collision-free by region construction. Replanning can be triggered by any robot in the group, which becomes leader, according to the conditions described in Sect. 5.5. The replanning is communicated to rest of the group and robots move to the new ROI.

7 Computational Experiments

We have validated our approach in a custom simulation environment in Python. Experiments have addressed *accuracy in map estimation* among the different architectures, *scalability* of performance, and comparison to a non-informed baseline approach where robots move according to a *random waypoint model* [2] (RWP), similarly to what has been done in other works [10,15].

In particular, we consider the cases of a *fully distributed architecture* (FDC) where the robots can enjoy a fully connected communication network, a *fully distributed architecture where the robots do not communicate* (FDNC), and a *leader-follower* (LF) architecture with one single group of robots. In practice, we address limit scenarios for communications and coordination. In the FDC case, robots are fully supported by a network such that they can act independently but can locally coordinate at the same time. The FDNC is an extreme swarm approach, where robots act totally independently from each other. LF is in between, the organization is such that robots can exploit multi-hop connectivity and therefore full GP sharing and coordination, but they have to pay the "price" of sticking somehow together. Making an study over these "limit" cases we can get an informed a guess about what would be the behavior for in-between cases.

As mentioned in the Introduction, we are motivated by surveying marine environments, so the test scenarios consists of *estimation of bathymetry maps* using small robot boats, the Lutra Prop. To ensure that computational experiments are close physical experiments we considered a realistic time budget of $T_{max} = 4h$ for the mission, with the boats moving at a constant velocity $v = 1m/s$, and relatively large areas to survey, which are scaled to 6000× 6000 m^2 and discretized into a grid of 10000× 10000 sampling points. To serve as a ground-truth, multiple bathymetry data have been taken from the GEBCO database (www.gebco.net). Robot's sensor is affected by unbiased Gaussian noise equal to 5% of the mean.

A *prior model* for a GP is formed from initial sparse random sampling of the scenario dataset, with a total of 10 data points, D_{10} (e.g., similar to [10,12]).

7.1 Accuracy in Map Estimation

For quantifying accuracy in map estimation, the *root mean square error* (RMSE) of the learned GP against the ground-truth has been used as main quality metric. Robot teams of size $n = 5$ have been used for the comparative analysis among the different architectures and the RWP baseline. For each dataset, we have executed the algorithms for 10 times and averaged the values. Results are shown in the table below. Coordinates of the mapping regions are reported in the table, which shows improvement percentage of RMSE of our approaches vs. RWP.

We can observe that both FDC and LF provide a consistent improvement over the RWP baseline, with FDC significantly outperforming all the other approaches. Somehow surprisingly, also FDNC does relatively well, but not as good as FDC and LF. This is however, reasonable, since in FDNC robots are blind to each other (apart for collision avoidance). For the first three datasets the improvement of FDC over RWP is substantial (almost 70%), which fully supports the quality of our multi-staged optimization strategy. For the Philippines sea, where only a minimal improvement over the baseline is observed, the ground-truth map is quite flat, with long correlation lengths, such that uniform random sampling can be as good as utility-directed sampling. For Pacific Ocean D, the ground-truth features a quite uniformly distributed local variability that needs uniform sampling, that FNDC cannot ensure, resulting worse than RWP.

Dataset	lon0	lon1	lat0	lat1	FDC%	FDNC%	LF%
Genoa	8.306	9.314	43.085	44.022	69.73	64.28	33.38
Oman Coast	58.389	23.66	58.914	24.197	41.56	2.34	21.11
North Pacific Ocean A	168.347	172.856	39.831	43.647	60.49	14.60	32.54
North Pacific Ocean B	145.814	152.731	31.177	35.993	17.32	5.83	0.13
North Pacific Ocean C	165.352	170.806	22.956	26.514	35.43	24.02	10.97
North Pacific Ocean D	156.477	159.997	30.952	33.943	25.72	−6.99	9.60
South Pacific Ocean	129.681	134.743	17.289	22.218	9.74	14.31	7.22
Philippine Sea	129.681	134.743	17.289	22.218	8.88	1.41	5.63

Parameters: $n = 5$, $T_{pmax} = 4$ h, $\phi = 45\%$, $v = 1$m/s, $\varepsilon = 0.5$; percentages refer to improvement vs. RWP baseline; **FDC** is fully distributed with communication; **FDNC** is fully distributed without communication; **LF** is leader-follower.

Interestingly, FDC outperforms LF. This is likely due to the diminishing return effect for the robots in the group being sampling about the same (yet large) region. However, it must also be pointed out that FDC is an "ideal" case, where communications are nearly global, which is hardly the case in real scenarios. Nevertheless, the relatively good performance of FDNC is quite comforting, since it says that, most likely, a fully distributed approach with local and intermittent communications will have a performance consistently better than baseline and approaching that of FDC with good communications support.

7.2 Scalability in Performance and in Computations

To test the significance of using multiple robots to minimize mission time, we compute the *scalability performance* of the **FDC** approach with increasing number of robots. We considered realistic team sizes of up to 20 robots. The scaling criteria is based on the time it takes for a team size to reach target RMSE of 200 in Genoa dataset which represents roughly a 7% point-by-point error considering the depth values in the dataset. Ideally, the more the robots, the more parallel sampling actions and the faster the team would reach the target RMSE value.

Scalability results are shown in Fig. 2. An increasing team size results in reduction in mission completion time, with a rapid decrease from about 3.51 h

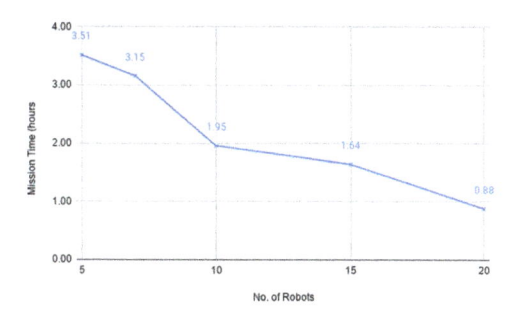

Fig. 2. Mission times vs. team size to reach a target RMSE of 200 in the Genoa dataset for the FDC architecture.

(mission time) for team of 5 robots, to a few minutes for a team of 20 robots. However, completion time goes to zero only asymptotically due to submodularity [7], such that adding robots would not significantly decrease mission time.

8 Conclusions

We have presented a novel multi-staged strategy for tackling IPP-MRS problems for map construction under limited time budgets as defined in Sect. 3. The proposed approach is based on modeling the mapping problem after a GP, and learning the GP by directing robots to sample in regions expected to provide large utilities. The developed strategy is a form of distributed adaptive data sampling that relies on concepts from Monte Carlo sampling and estimation, sequential Bayesian optimization, orienteering models. Contrary to many other approaches in literature, we focus on region-based utilities, rater than point-based ones. We have also presented different decentralized architectures (fully distributed and leader-follower) for implementing the strategy in a collaborative way, aiming to design systems which can be effective in GP learning, scale well in performance and computations can profit of available networking.

Effectiveness of the approaches for map learning and performance scaling have been evaluated in simulation, showing very promising results for both fully distributed and leader-follower implementations. In terms of map accuracy our approaches have clearly outperformed a baseline approach where the robots follow an RWP strategy for moving and sampling. In terms of scaling the performance with increasing the number of robots (up to 20) results have been extremely satisfactory. In future work, we will perform extensive testing in a ROS environment both in simulation with larger teams, and with real robots. Real robots' experiments will be particularly important to test the effects of multi-hop opportunistic communications on coordination and data sharing in the different architectures.

Acknowledgments. This work was made possible by NPRP Grant 10-0213-170458 from the Qatar National Research Fund (a member of Qatar Foundation). The findings herein are solely the responsibility of the authors.

References

1. Arora, S., Scherer, S.: Randomized algorithm for informative path planning with budget constraints. In: IEEE ICRA, pp. 4997–5004 (2017)
2. Bettstetter, C., Resta, G., Santi, P.: The node distribution of the random waypoint mobility model for wireless ad hoc networks. IEEE Trans. Mobile Comput. **2**(3), 257–269 (2003)
3. Contal, E., Perchet, V., Vayatis, N.: Gaussian process optimization with mutual information. In: International Conference on Machine Learning (ICML), pp. 253–261 (2014)
4. Dutta, A., Ghosh, A., Kreidl, O.P.: Multi-robot informative path planning with continuous connectivity constraints. In: IEEE ICRA, pp. 3245–3251 (2019)
5. Ghassemi, P., Chowdhury, S.: Informative path planning with local penalization for decentralized and asynchronous swarm robotic search. In: International Symposium on Multi-Robot and Multi-Agent Systems (MRS), pp. 188–194. IEEE (2019)
6. Ginsbourger, D., Le Riche, R., Carraro, L.: Kriging is well-suited to parallelize optimization. In: Tenne, Y., Goh, C.-K. (eds.) Computational Intelligence in Expensive Optimization Problems. ALO, vol. 2, pp. 131–162. Springer, Heidelberg (2010). https://doi.org/10.1007/978-3-642-10701-6_6
7. Hidaka, S., Oizumi, M.: Fast and exact search for the partition with minimal information loss. PloS One **13**(9), e0201126 (2018)
8. Hollinger, G.A., Sukhatme, G.S.: Sampling-based robotic information gathering algorithms. Int. J. Robot. Res. **33**(9), 1271–1287 (2014)
9. Krause, A., Guestrin, C.: Submodularity and its applications in optimized information gathering. ACM Trans. Intell. Syst. Technol. **2**(4), 1–20 (2011)
10. Li, A.Q., et al.: Multi-robot online sensing strategies for the construction of communication maps. Auton. Robot. **44**(3), 299–319 (2020). https://doi.org/10.1007/s10514-019-09862-3
11. Ma, K.C., Liu, L., Heidarsson, H.K., Sukhatme, G.S.: Data-driven learning and planning for environmental sampling. J. Field Robot. **35**(5), 643–661 (2018)
12. Mishra, R., Chitre, M., Swarup, S.: Online informative path planning using sparse Gaussian processes. In: OCEANS-MTS/IEEE OCEANS, pp. 1–5 (2018)
13. Montemanni, R., Gambardella, L.: An ant colony system for team orienteering problems with time windows. Found. Comput. Decis. Sci. **34**(4), 287 (2009)
14. Nieto-Granda, C., Rogers, J.G., Christensen, H.I.: Coordination strategies for multi-robot exploration and mapping. Int. J. Robot. Res. (IJRR) **33**(4), 519–533 (2014)
15. Popović, M., et al.: An informative path planning framework for UAV-based terrain monitoring. Auton. Robots **44**, 1–23 (2020)
16. Popović, M.: An informative path planning framework for UAV-based terrain monitoring. Auton. Robot. **44**(6), 889–911 (2020)
17. Rasmussen, C.E., Williams, C.K.I.: Gaussian Processes for ML. MIT Press, Cambridge (2006)
18. Shahriari, B., Swersky, K., Wang, Z., Adams, R.P., De Freitas, N.: Taking the human out of the loop: a review of Bayesian optimization. Proc. IEEE **104**(1), 148–175 (2015)
19. Vansteenwegen, P., Gunawan, A.: Orienteering Problems. Springer, Heidelberg (2019). https://doi.org/10.1007/978-3-030-29746-6

Collective Transport via Sequential Caging

Vivek Shankar Vardharajan[1]([✉]), Karthik Soma[2], and Giovanni Beltrame[1]

[1] Polytechnique Montreal, Montreal, QC H3T 1J4, Canada
vivek-shankar.varadharajan@polymtl.ca
[2] National Institute of Technology, Tiruchirapalli, India

Abstract. We propose a decentralized algorithm to collaboratively transport arbitrarily shaped objects using a swarm of robots. Our approach starts with a task allocation phase that sequentially distributes locations around the object to be transported starting from a seed robot that makes first contact with the object. Our approach does not require previous knowledge of the shape of the object to ensure caging. To push the object to a goal location, we estimate the robots required to apply force on the object based on the angular difference between the target and the object. During transport, the robots follow a sequence of intermediate goal locations specifying the required pose of the object at that location. We evaluate our approach in a physics-based simulator with up to 100 robots, using three generic paths. Experiments using a group of KheperaIV robots demonstrate the effectiveness of our approach in a real setting.

Keywords: Collaborative transport · Task allocation · Caging · Robot swarms

1 Introduction

Several insect species exhibit an incredible level of coordination to lift and carry heavy objects to their nest, whether for building materials or food. Paratrechina longicornis can collectively transport heavy food from a source location to its nest purely through local interaction with neighboring ants [10]. These ants are capable of carrying an object of arbitrary shape and ten times heavier than their bodies by collaborating in an effective manner. Designing approaches to realize collaborative transport using a group of robots can find its application in warehouse management [19] and collaborative construction of structures [14]. In this work, we take inspiration from these natural insect species to design an approach to collaboratively transport heavy objects (i.e. that cannot be carried by a single robot) of arbitrary shape using a swarm of robots. The main challenges of this type of collective transport are: 1. the effective placement of robots around the transported object, 2. the effective application of force around the object to avoid tugs-of-war, and 3. adapting to the objects center-of-mass movement

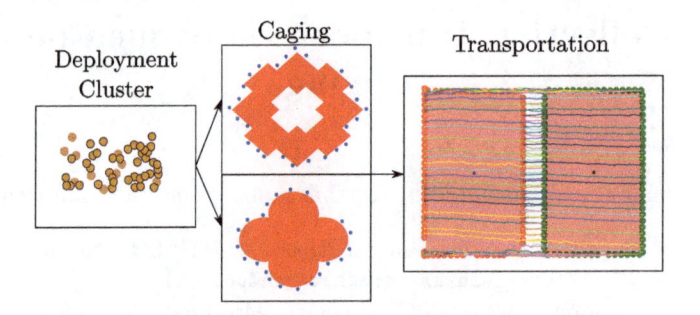

Fig. 1. Illustration of the process of object transpiration, with the robots starting from a deployment cluster, caging the object and transporting the object.

and the alignment of forces between neighbors. Our approach starts with a task allocation phase, where the robots are sequentially deployed around the object, a process known as caging [23]. Completing the task allocation phase, the robots transport the object towards a target location as shown in Fig. 1.

Collaborative transport is a well-studied topic: some approaches use ground robots [3,7,14,24] for cooperative manipulation either with explicit communication [8] or force based coordination [9], while others use quadcopters carrying a cable-suspended [5] or rigidly attached [27] object that is heavier than one robot's maximum payload.

Many of the approaches discussed earlier do not explicitly consider robot's interaction with other robots to continuously maintain cage formation, while including an obstacle avoidance mechanism within their control framework. This latter is particularly limiting when the perimeter of the pushed object is small, as it might prevent the robots to get close to the object to apply an effective force. In our approach, we design a sequential placement of robots to avoid this scenario and maintain the initial formation continuously while moving.

Approaches like [2] assume that the position and shape of the object is either continuously known or periodically updated using a central system. Measuring the position and shape of the object in a real world scenarios might be difficult and would limit the use of the transport system to some indoor applications. Furthermore, some approaches either assume that the robots are readily placed around the object to be manipulated [20] and design control strategies for the manipulation of the object. A few approaches provide emphasis on caging or design a control policy for a specific type of caging [13]. In this paper, we design a complete system that allows the robots to start from a deployment cluster and take up positions around the object to be transported, in a way that is similar to [7]. Our approach periodically estimates the centroid of the object using only the relative positional information shared by neighboring robots, avoiding the need to have continuous external measurements. Our task allocation allows heterogeneous robots with varying capabilities (e.g. path planning), as well as provides a way to addressing robot failures [22]. We provide sufficient conditions for the convergence of our caging, and show that our approach terminates for convex objects.

2 Related Work

The concept of caging was first introduced by Rimon and Blake [18] for a finger gripper. Caging is a concept of trapping the object to be manipulated by a gripping actuator, in our work we use a group of robots to act as a single entity to grip the object using form closure as in [23]. A simple form of caging and leader-follower based strategies [26] are employed to push the object by sensing the resultant forces. The main constraint of these works is that the robots cannot follow paths with sharp turns.

Pereira et al. [13] introduce a new type of caging called object closure, in which the object to be transported is loosely caged until the configuration satisfies certain conditions in an imaginary *closure configuration space*. Each robot in the team has to estimate the orientation and position of the object to use this approach. Wan et al. [23] propose robust caging to minimize the number of robots to form closure using translation and rotation constraints. Wan et al. also extended their work for polygons, balls and cylinders to be transported on a slope [24]. This approach requires continuous positional updates from an external system and uses a central system to compute the minimum number of robots required.

An approach to caging L-shaped objects is proposed in [7]: the robots switch between different behavioral states to approach the object and achieve *potential caging*. This approach requires the robots to know certain properties of the object beforehand, such as the minimum and maximum diameter of the caged object. A caging strategy for polygonal convex objects is proposed in [6], the approach uses a sliding mode fuzzy controller to traverse predefined paths. A leader robot coordinates the transportation using the relative position of all the other robots.

Gerkey et al. [11] propose a strategy for pushing an object by assigning "pushers" and "watchers". Watchers monitor the position of the object and other robots, while pushers perform the pushing task. The approach provides fault tolerance to robot failures and relies heavily on the performance of the watchers. Chen et al. [3] propose an approach in which the robots are placed around an object that occludes the transportation destination: the robots that do not see the destination are the ones that push the object. The intuition behind this placement is that the location that is most effective for pushing the object is the occluded region, i.e. the opposite side of the direction of movement. Our strategy is similar to [3], where we estimate the angular difference between the object and the goal using the proximity sensor, and only the robots that are below a threshold apply a force to the object. Our approach places robots all around the object to adjust and follow a sequence of changing target location, as opposed to only placing robots in the occluded region [3].

The work in [25] combines reinforcement learning and evolutionary algorithms to coordinate 3 types of agents to learn to push an object. "Vision robots" estimate the positions of the object and other robots, "evolutionary learning agents" generate plans for the "worker robots", and these latter execute the plan to push the object. Alkilabi et al. [1] use an evolutionary algorithm to tune a recurrent neural network controller that allows a group of e-puck robots to col-

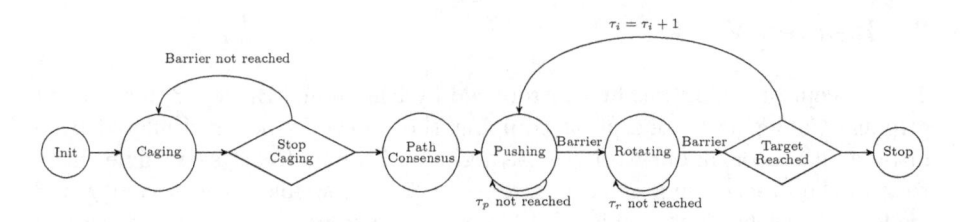

Fig. 2. High level state diagram

laboratively transport a heavy object. The robots use an optical flow sensor to determine and achieve an alignment of forces. The authors demonstrate that the approach works well for different object sizes and shapes, however, the proper functioning of the algorithm relies heavily on the performance of the optical flow sensors.

3 Methodology

Figure 2 shows the high level state machine used in our collaborative transport behavior. At initialization, the robots enter into a sequence of task allocation rounds allowing the robots to take up positions around the object to be transported with a desired inter-robot distance. Once caging is complete, the robots that are a part of the object cage agree on the desired object path and start pushing and rotating the object. The path is represented as a sequence of points, and all the robots in the cage use a barrier (i.e. they wait for consensus) to go through each intermediate point.

3.1 Problem Formulation

Let $C_o(t) \in \mathbb{R}^2$ be the centroid of an arbitrary shaped object at time t, $x_i(t) \in \mathbb{R}^2$ the position of robot i at time t, and $X(t) = \{x_1(t), x_2(t), ..., x_n(t)\}$ the set containing the positions of the robots at time t. Let S_o be a closed, convex set representing the perimeter of the object to be transported. Given a sequence of target locations $\mathbb{T} = \{\tau_1, \tau_2, ..., \tau_n\}$ the task of the robots is to take up locations around the perimeter of the object S_o with a desired inter-robot distance $I_d \in \mathbb{R}$ and drive the centroid of the object C_o from a known initial state $C_o(0)$ to a final state $C_o(t_f)$ at some time t_f, passing through the target locations in $\mathbb{T}$.

We assume that the robots can perceive the goal in the environment and know an estimate of the initial centroid location $C_o(0)$ of the object to be transported. The mass of the object is assumed to be proportional to the size with a minimum density for the object. We also assume that the line connecting two subsequent targets τ_{x-1} and τ_x does not go through obstacles and the perimeter of the object S_o to be transported is greater than $3 * I_d$. We consider a point mass model for the robots ($\dot{x}_i(t) = u_i$) and assume that the robots are fully controllable with u_i. The robots are assumed to be equipped with a range and bearing sensor to determine the relative positional information and communicate with neighboring robots within a small fixed communication range d_C. In our experiments, we used KheperaIV robots that are equipped with 8 proximity sensors at equal angles around the robot, and we assume similar capabilities in general.

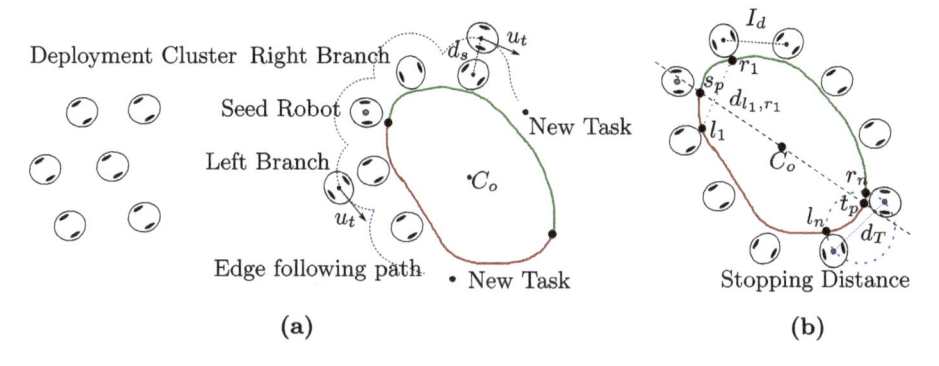

Fig. 3. Illustration of the process of task allocation based caging (a) illustrate the process of edge following to reach the new target (b) the stopping condition to terminate caging.

3.2 Task Allocation Based Caging

Consider a group of robots randomly distributed in a cluster and a known initial position of the object. The goal of the robots is to deploy to suitable location around the perimeter of the object C_o to guarantee object closure while respecting the required inter-robot distance I_d.

The caging process starts with the allocation of the first task (an estimation of the centroid of the object) to a seed robot (closest robot to the object, elected via bidding) as shown in Fig. 3 (a). The seed robot moves towards the center of the object until it detects the object with its on-board proximity sensors. As the seed robot touches the object, it creates two target locations (one to its left and one to its right, called *branches*). The robots bid for these new target locations and continue the process of spawning the new targets along their branch until the minimum distance between robots in the two branches is smaller than d_T.

Our task allocation algorithm performs the role of determining the appropriate target around the object for each robot, the *caging targets*. We consider a Single Allocation (SA) problem [4], where every robot is assigned a single task. The caging targets are sequentially available to the robots, i.e. a new target becomes available after a robot has reached its target. Note that these targets are considered to be approximate (created by establishing a local coordinate system like in [21]), hence they are refined by the robots using their proximity sensors and the position of the closest robot in the branch on reaching the assigned target. The approach of sequential caging is particularly appropriate for scenarios where the shape of the object is initially or continuously unknown, the robots sequentially assign robots to the closure and enclose the transported object.

We use a bidding algorithm [22] (described in thesupplementary material[1]): the robots locally compute bids for a task and recover the lowest bid of the team from a distributed, shared tuple space [16]. The robots update the tuple space if the local bid is lower, with conflicts resolved using the procedure outlined in [22]). After a predetermined allocation time (T_a), the lowest bid in the tuple space is declared as the winner. T_a has to be selected considering the communication topology and delays to avoid premature allocation as detailed in [22].

To reach the assigned target, the robots edge-follow the neighbors in their target branch. The control inputs (u_i) are generated using range and bearing information from neighbors: the robots find their closest neighbor and create a neighbor vector x_n using the range and bearing information. The control inputs are then $u_i = (\perp x_n) + (||x_n|| - I_d) * x_n$; the first term makes sure the robot orbits the neighbor and the second term applies a pulling or pushing force to keep the robot at distance I_d. On reaching the target (detected using the proximity sensors), the robot measure the distance to the closest neighbor of the branch and apply a distance correction to keep the inter-robot caging distance to be I_d. The robot creates an obstacle vector using the proximity values: $x_o = \frac{\sum_{i \in P} p_i}{|P|}$, with $P = \{p_0, ..., p_7\}, p_i \in \mathbb{R}^2$ being the set containing the individual proximity readings as vectors. The inter-robot distance correction control inputs are generated using: $u_i = \perp x_o$. When there are not enough robots to complete the caging, the robots can adapt by increasing I_d and applying inter-robot distance correction control until the termination condition is met.

Proposition 1. Consider two sets L and R denoting the left and right branch respectively, L contains all the attachment points of the left branch robots to the object, and R contains the attachment points on the right branch. The caging terminates if $\exists p \in L, q \in R$ such that $d_{pq} \leq d_T$ while $d_{lr} > d_T \quad \forall l \in L - \{p\} \land \forall r \in R - \{q\}$.

Referring to Fig. 3(b), consider two closed, convex sets, $L_o = \{l_1, ..., l_n\}$ containing all the points on the curve $l_1 l_n$ and $R_o = \{r_1, ..., r_n\}$ containing all the points on the curve $r_1 r_n$. Set L_o and R_o are ordered and the distance between the two constitutive points satisfy $d(l_n, ln - 1) > d_T$.

3.3 Behaviors for Pushing and Rotating

Once two robots around the perimeter of the object satisfy the termination condition and a consensus is reached on the path, the robots initiate the target following routine. The path is represented as a sequence of desired object centroid target locations $\mathbb{T}$ and each entry $\tau_i = (\tau_p, \tau_r)$, with $\tau_p \in \mathbb{R}^2$ and $\tau_r \in [0, 2\pi]$. τ_p is the local target location and τ_r is the desired object orientation along the z-axis (yaw) at that target location. The main intuition behind having these local targets is to use a geometric path planner. One of the robot in the swarm with the ability to compute a path to the user defined target τ_n compute the

---

[1] https://mistlab.ca/papers/CollectiveTransport/.

path and share it as a sequence of states(targets) using virtual stigmergy [16]. The robots sequentially traverse the targets in $\mathbb{T}$, on reaching a τ_p, the robots rotate the object to τ_r. Each robot in caging computes u_{fp} as in Eq. 1 to exert a forward force and push the object. Similarly, for rotating the object the robots apply u_{fr} as in Eq. 2.

$$u_{fp} = u_t + u_f + u_{cp}, \tag{1}$$
$$u_{fr} = u_r + u_f + u_{cr} \tag{2}$$

where, u_t and u_r are a force to move the object towards the target by pushing, and a torque to rotate the object to the desired angle, respectively. u_f, as shown in Eq. 3, is the contribution that makes sure the robots stay in the same formation. u_{cp} (Eq. 4) and u_{cr} (Eq. 5) are contributions that ensure the robots stay in contact with the object during pushing and rotation.

Maintaining Formation. The robot formation from the caging operation tends to get distorted as a result of its application of pushing force on the object to move it towards the target. The robots in the cage apply a force to stay in this formation throughout the transportation task: they store a set $N_f = \{(d_i, \theta_i) | d_i \leq k * I_d, \forall i \in N\}$ that contains the range and bearing measurements of their neighbors, with k being a design parameter. The control input u_f to maintain formation is:

$$u_f = \sum_{\forall i \in N_f} \frac{K_f(d_i - d_{cur})}{d_i} \begin{bmatrix} d_i \cos \theta_i - \cos \frac{\theta_i - \theta_{cur}}{\theta_i} \\ d_i \sin \theta_i - \sin \frac{\theta_i - \theta_{cur}}{\theta_i} \end{bmatrix} \tag{3}$$

The first term in the Eq. 3 is the desired inter-agent distance correction, while the second term applies the desired orientation correction. This formulation is inspired from the commonly used edge potential to preserve a lattice structure among the robots [12]. We apply this edge potential among adjacent robots in a cage to preserve the formation and the desired inter-agent distance.

Maintaining Contact with the Object. The robots in the formation need to determine if they need to apply a force and stay in contact with the object. During pushing, the robots apply a control input to stay in contact with the object, determining its effectiveness in pushing as in Eq. 4. The effectiveness of a robot's pushing depends on the position of the robot with respect to the object and the target.

The angle θ_p (a parameter) determines if the robot is an effective pusher: if the angle between the object x_o and the target τ_p is greater or equal to θ_p, pushing is considered effective, and the robots apply an input u_{cp} to maintain contact. Similarly, the robots apply u_{cr} to maintain contact during rotations:

$$u_{cp} = \begin{cases} [0,0]^T, & \angle(x_o, \tau_p) \geq \theta_p \\ \frac{K_{cp} x_o}{||x_o||}, & \angle(x_o, \tau_p) < \theta_p \end{cases} \tag{4}$$

$$u_{cr} = \frac{K_{cr} x_o}{||x_o||} \tag{5}$$

where x_o is the proximity vector that determines the current object location in robots coordinate system, and θ_p, K_{cp} and K_{cr} are design parameters. θ_p is a design parameter that defines the effective pushing perimeter around the object, as shown in Fig. 4.

Applying Forces. The robots have to exert force in the right direction to move and rotate the object according to the targets in $\mathbb{T}$. The robots must apply force in the desired angular window around the perimeter of the object to avoid tugs-of-war. The control inputs u_t and u_r make sure the robots exert the force in the right direction:

$$u_t = \begin{cases} [0,0]^T, & ||\tau_{p_l} - x_i|| \leq d_{tol} \\ \frac{K_t[\tau_{p_l} - x_i]}{||\tau_{p_l} - x_i||}, & ||\tau_{p_l} - x_i|| > d_{tol} \end{cases} \quad (6)$$

$$u_r = \begin{cases} \begin{bmatrix} 0 & -1 \\ 1 & 0 \end{bmatrix} \frac{K_r(x_i - C_o)}{||x_i - C_o||}, & \angle(\tau_p, C_o) < \theta_r \\ [0,0]^T, & \angle(\tau_p, C_o) \geq \theta_r \end{cases} \quad (7)$$

where, d_{tol} is a design parameter that defines the distance tolerance, τ_{p_l} is the local target computed by the robot using the centroid position and its position along the perimeter of the object. On reaching an intermediate target τ_i the robots share their approximate position with respect to a common coordinate system computed as in [21] in a distributed, shared tuple space [16] with all the other robots. The robots retrieve this positional information and compute the centroid of the object C_o, which is then used during the rotation of the object.

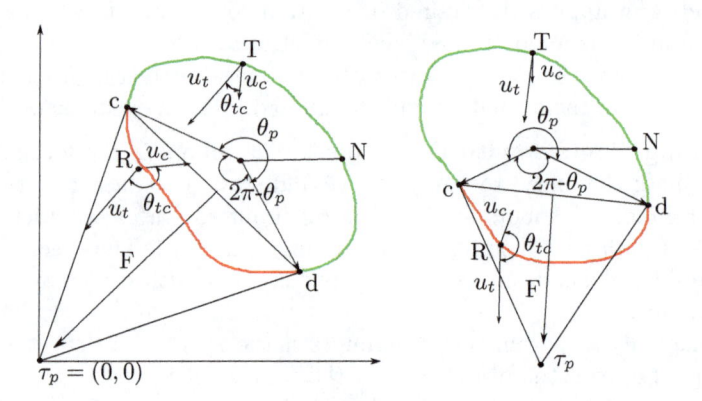

Fig. 4. Illustration of the resultant force and the angle of effective robots, the effective pushing positions on the perimeter of the object is shown in green.

As in [3], we can prove that the object reaches the goal as $t \to \infty$, with the difference being that the robots exerting force are not based on the occluded perimeter of the object, but are instead the robots satisfying $\angle(x_o, \tau_p) < \theta_p$.

Theorem 1. *The distance between the centroid of the object C_o and the target location τ_p strictly decreases if the velocity of the transported object is governed by the translation dynamics equation of the object $\dot{v}_o = kF$. For $t \to \infty$, the center of the object C_o reaches τ_p, where $\dot{v}_o$ is the derivative of the object velocity and kF is the fraction of the force that is transfered to the object from the robots.*

Proof. Figure 4 shows the resultant F transferred from the robots to the object and the effective angular window (along the curve cTd) on the perimeter of the object to exert force. Consider all the robots along the curve cTd are applying a force using a control input determined by the unit vector u_t. The overall force transferred to the object is $F = (c_x - c_y) - (d_x - d_y)$, which is the tangent vector $(d - c)$ rotated by $(\frac{\pi}{2})$ [3].

Consider the squared distance between the target $\tau_p = [0, 0]^T$ and centroid C_o at time t to be $d_g(t) = ||\tau_p - C_o||^2$, taking the time derivative gives $\dot{d}_g = 2k * C_o * F$, substituting F with the resultant force gives $\dot{d}_g = k * ((C_{o_y} C_x - C_{o_x} c_y) - (C_{o_y} d_x - C_{o_x} d_y))$. The distance $d_g(t) \geq 0, \forall t > 0$, when the center of the object lies outside the desired goal τ_p and since $C_o * F < 0$ is strictly decreasing because of the force applied by the robots, we get $\lim_{t\to\infty} \dot{d}_g = 0$. In other words, the center C_o will eventually reach the goal τ_p.

4 Experiments

We performed a set of experiments in a physics-based simulator (ARGoS3 [17]) with a KheperaIV robot model under various conditions to study the performance of our approach. We implemented our behavioral state machine for the robots using Buzz [15]. We set the number of robots $N_r \in \{25, 50, 100\}$ and adapt the size $S \in \{[2, 2], [3.6, 6], [7.2, 12]\}m$ and mass $M \in \{5.56, 30.024, 120.096\}kg$ of a cuboid object according to the number of robots. The mass of the object is calculated assuming a constant density hollow material. In another set of experiments, we used three irregular objects: cloud, box rotation, and clover. We set the design parameters of the algorithm to the values shown in Table 1. We choose the gain parameters for maintaining formation(k_f), contact with object(k_{cp}) and force application (k_t) based on several rounds of trail-and-error simulations. The tolerance parameters d_{tol} and *Orient. tol.* are chosen to fit our non-holonomic robots: a large part of the error shown in Fig. 7 is due to the non-holonomic nature of our robots. We evaluate the various performance metrics over three benchmark paths: a straight line, a zigzag, and straight line with two 90° rotations (straight_rot in Fig. 5). All the paths consists of 9 waypoints (WPs) and straight_rot has its rotations at WP 3 and 6. Each experiment is repeated 30 times with random initial conditions.

Results. We assess the performance of the algorithm observing the time taken to cage the object and push it along the benchmark paths, plotted in Fig. 6. The time to cage the object increases with the perimeter of the object: the median times to cage are 247 s, 779 s and 2753 s for 25, 50 and 100 for robots, respectively transporting objects of size $\{2, 2\}$, $\{3.6, 6\}$, $\{7.2, 12\}$. The 3 irregular shapes took

Table 1. Experimental parameters

Caging		Pushing		Rotating	
d_s	0.35 m	θ_p	115°	K_{cr}	450
I_d	0.45 m	K_{cp}	40(< 115°), 20(≥ 115°)	Orient. tol.	5.72°
d_{tol}	0.05 m	d_{tol}	0.1m	K_f	400
d_T	$1.85 I_d$	K_f	40	K_r	600
K_t	30	K_t	60	Barrier	90%
Prox. thres.	0.7	Barrier	90%		

Fig. 5. Trajectory taken by the centroid of the object vs the desired path in the three benchmarking paths.

around 300 s to cage when using 30 robots. The time taken to push the object is approximately 100 s for the straight path (regardless of the object size) and about 160 s for the zigzag with 25 and 50 robots.

When using 100 robots for the straight line and the zigzag, the system was slightly faster, which could be explained by the higher cumulative force exerted by the robots. The time taken following a straight path with rotations increases sub-linearly with the number of robots with median times being 135 s, 155 s and 200 s for 25, 50 and 100 robots, respectively. Figure 7 shows the centroid estimation error, position error and orientation error on each row from left to right. The

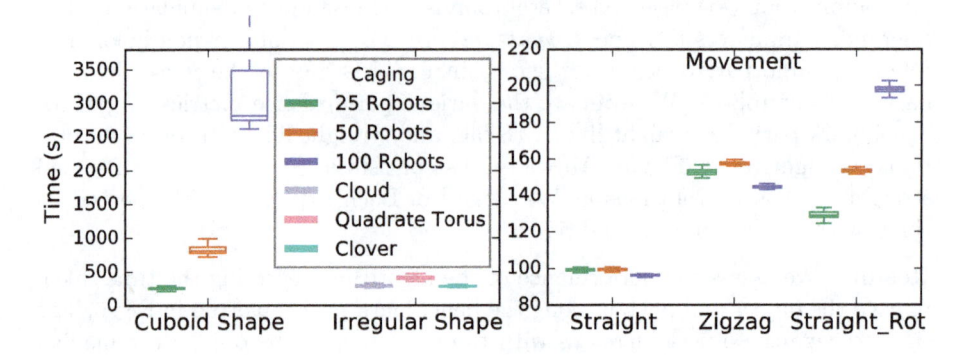

Fig. 6. Time to complete caging (left) and push an object along 3 paths (right).

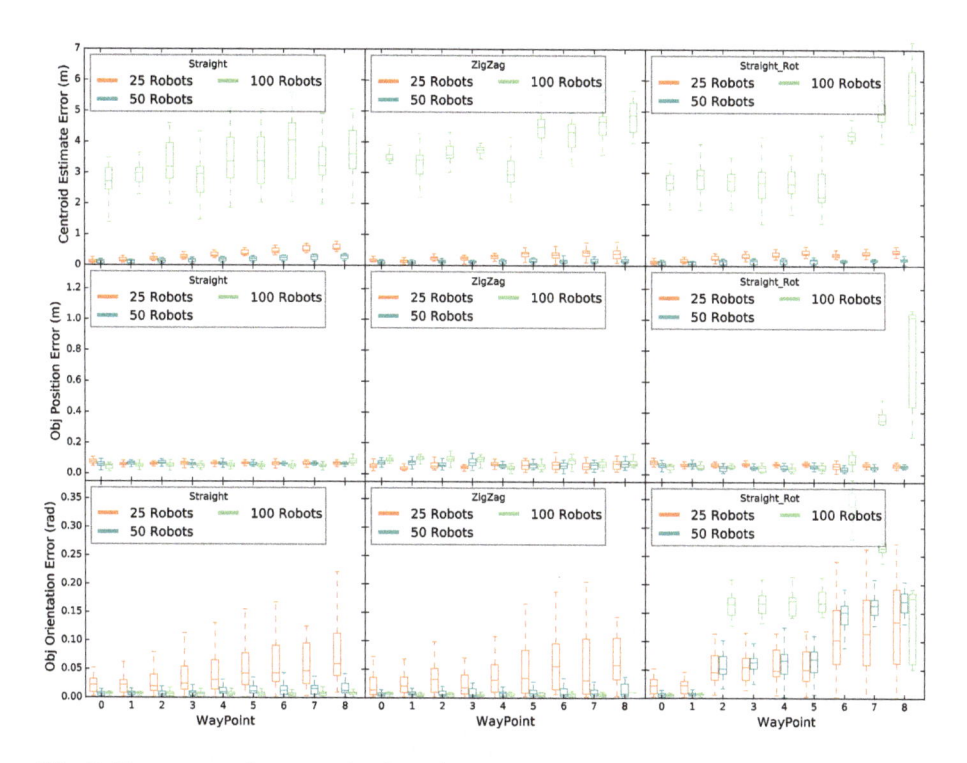

Fig. 7. From top to bottom: the first plot shows the average centroid estimation error, the second shows the object centroid position error, and the third shows the object orientation error; from left to right, the figure shows the results for our 3 benchmark paths.

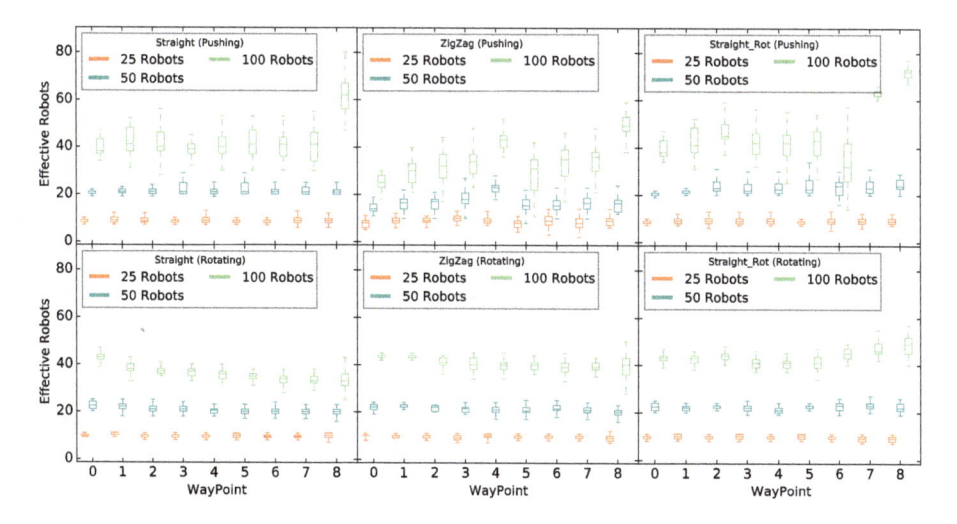

Fig. 8. Number of robots that were effective in pushing and rotating at different waypoints of the three benchmark paths.

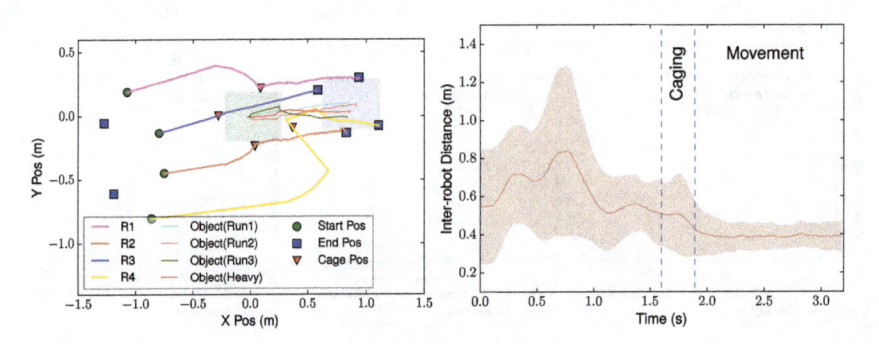

Fig. 9. Trajectories taken by the robots (left) and the inter-robot distance between the two adjacent robot in the cage (right).

centroid estimation error increases as the robots progress along the path, which can be explained by the distortion in formation as the robots progress towards the final target. The centroid estimation error for 100 robots is relatively large and shows some variability, which could be largely influenced by the communication topology during the centroid estimation process, as detailed in Sect. 3. The object position error computed using the difference between the desired position and the ground truth position appears stable around 0.1m, which is within our design tolerance (d_{tol}). In the $straight_{rot}$ case with 100 robots, the position error drastically increases around the final WPs, likely due to the drift induced by the second rotation. The orientation error accumulates slowly for the other paths, likely because the pushing force applied towards the target induces a small torque. Without global positioning, the error accumulates at every rotation.

Figure 8 shows the number of effective robots for pushing and rotating the transported object, computed using 4. The number of effective pushers appears to increase slowly as the robots progress towards the final target in all cases, which could be due to distortions of the caging formation. The number of effective rotators stays constant for most of the cases, but increases during rotations, due to the robots either getting closer to the corners or the mid point of the object. This could be caused by the large error in the estimation of the centroid resulting in a generation of biased control input to rotate the object.

Robot Experiments. We perform a small set of experiments using a group of 6 KheperaIV robots. The robots use a hub to compute and transmit the range and bearing information from a motion capture system, for more details on the experimental setup, we refer the reader to [22]. We performed two sets of experiments with robots transporting a foam box of size (0.285, 0.435) m: without any payload, and with 4 kg of LiPo battery on the object. Figure 9 shows the trajectories followed by the robots and the inter-robot distance during the 3 runs without any payload and one run with the payload. It can be observed that the robots were able to consistently reach the target (0, 0.9) by following the 3 WPs placed at every 0.3 m. The inter-robot distance between the two adjacent robots in a cage approximates well the desired $I_d = 0.45$ m at caging and it is maintained consistently during pushing in all runs with a maximum standard deviation of 0.1m.

5 Conclusions

We propose a decentralized algorithm to cage an arbitrary-shaped object and transport it along a desired path consisting of a set of object poses. The robots periodically estimate the centroid of the object based on the positional information shared by the robots caging the object, and use this information to transport the object. We study the performance of our algorithm using a large set of simulation experiments with up to 100 robots traversing 3 benchmark paths and a small set of experiments on KheperaIV robots. As future work, we plan to implement a path planner to provide the object path in a cluttered environment.

References

1. Alkilabi, M.H.M., Narayan, A., Tuci, E.: Cooperative object transport with a swarm of e-puck robots: robustness and scalability of evolved collective strategies. Swarm Intell. **11**(3–4), 185–209 (2017). https://doi.org/10.1007/s11721-017-0135-8
2. Melo, R.S., Macharet, D.G., Campos, M.F.M.: Collaborative object transportation using heterogeneous robots. In: Santos Osório, F., Sales Gonçalves, R. (eds.) LARS/SBR -2016. CCIS, vol. 619, pp. 172–191. Springer, Cham (2016). https://doi.org/10.1007/978-3-319-47247-8_11
3. Chen, J., Gauci, M., Li, W., Kolling, A., Groß, R.: Occlusion-based cooperative transport with a swarm of miniature mobile robots. IEEE Trans. Rob. **31**(2), 307–321 (2015)
4. Choi, H., Brunet, L., How, J.P.: Consensus-based decentralized auctions for robust task allocation. IEEE Trans. Rob. **25**(4), 912–926 (2009)
5. Cotsakis, R., St-Onge, D., Beltrame, G.: Decentralized collaborative transport of fabrics using micro-UAVs. In: 2019 International Conference on Robotics and Automation (ICRA), pp. 7734–7740. IEEE (2019)
6. Dai, Y., Kim, Y.G., Wee, S.G., Lee, D.H., Lee, S.G.: Symmetric caging formation for convex polygonal object transportation by multiple mobile robots based on fuzzy sliding mode control. ISA Trans. **60**, 321–332 (2016)
7. Fink, J., Michael, N., Kumar, V.: Composition of vector fields for multi-robot manipulation via caging. In: Robotics: Science and Systems, vol. 3 (2007)
8. Franchi, A., Petitti, A., Rizzo, A.: Distributed estimation of state and parameters in multiagent cooperative load manipulation. IEEE Trans. Control Netw. Syst. **6**(2), 690–701 (2019)
9. Gabellieri, C., Tognon, M., Sanalitro, D., Pallottino, L., Franchi, A.: A study on force-based collaboration in swarms. Swarm Intell. **14**(1), 57–82 (2019). https://doi.org/10.1007/s11721-019-00178-7
10. Gelblum, A., Pinkoviezky, I., Fonio, E., Ghosh, A., Gov, N., Feinerman, O.: Ant groups optimally amplify the effect of transiently informed individuals. Nat. Commun. **6**, 1–9 (2015)

11. Gerkey, B.P., Mataric, M.J.: Pusher-watcher: an approach to fault-tolerant tightly-coupled robot coordination. In: International Conference on Robotics and Automation (ICRA), vol. 1, pp. 464–469. IEEE (2002)
12. Mesbahi, M., Egerstedt, M.: Graph Theoretic Methods in Multiagent Networks, vol. 33. Princeton University Press (2010)
13. Pereira, G.A., Campos, M.F., Kumar, V.: Decentralized algorithms for multi-robot manipulation via caging. Int. J. Robot. Res. **23**(7–8), 783–795 (2004)
14. Petersen, K.H., Napp, N., Stuart-Smith, R., Rus, D., Kovac, M.: A review of collective robotic construction. Sci. Robot. **4**(28), eaau8479 (2019)
15. Pinciroli, C., Beltrame, G.: Buzz: an extensible programming language for heterogeneous swarm robotics. In: International Conference on Intelligent Robots and Systems, pp. 3794–3800 (2016)
16. Pinciroli, C., Lee-Brown, A., Beltrame, G.: A tuple space for data sharing in robot swarms. In: Proceedings of the 9th EAI International Conference on Bio-inspired Information and Communications Technologies, BICT'15, pp. 287–294 (2016)
17. Pinciroli, C., et al.: ARGoS: a modular, parallel, multi-engine simulator for multi-robot systems. Swarm Intell. **6**(4), 271–295 (2012). https://doi.org/10.1007/s11721-012-0072-5
18. Rimon, E., Blake, A.: Caging planar bodies by one-parameter two-fingered gripping systems. Int. J. Robot. Res. **18**(3), 299–318 (1999)
19. Rosenfeld, A., Noa, A., Maksimov, O., Kraus, S.: Human-multi-robot team collaboration for efficient warehouse operation. In: Autonomous Robots and Multirobot Systems (ARMS) (2016)
20. Rubenstein, M., Cabrera, A., Werfel, J., Habibi, G., McLurkin, J., Nagpal, R.: Collective transport of complex objects by simple robots. In: Proceedings of the 2013 International Conference on Autonomous Agents and Multi-agent Systems, pp. 47–54 (2013)
21. Rubenstein, M., Cornejo, A., Nagpal, R.: Programmable self-assembly in a thousand-robot swarm. Science **345**(6198), 795–799 (2014)
22. Varadharajan, V.S., St-Onge, D., Adams, B., Beltrame, G.: Swarm relays: distributed self-healing ground-and-air connectivity chains. IEEE Robot. Autom. Lett. **5**(4), 5347–5354 (2020)
23. Wan, W., Fukui, R., Shimosaka, M., Sato, T., Kuniyoshi, Y.: Cooperative manipulation with least number of robots via robust caging. In: IEEE/ASME International Conference on Advanced Intelligent Mechatronics, AIM, pp. 896–903 (2012)
24. Wan, W., Shi, B., Wang, Z., Fukui, R.: Multirobot object transport via robust caging. IEEE Trans. Syst. Man Cybern. Syst. **50**(1), 270–280 (2020)
25. Wang, Y., de Silva, C.W.: Cooperative transportation by multiple robots with machine learning. In: 2006 IEEE International Conference on Evolutionary Computation, pp. 3050–3056. IEEE (2006)
26. Wang, Z., Hirata, Y., Kosuge, K.: Control a rigid caging formation for cooperative object transportation by multiple mobile robots. In: IEEE International Conference on Robotics and Automation, Proceedings, ICRA'04, vol. 2, pp. 1580–1585. IEEE (2004)
27. Wang, Z., Singh, S., Pavone, M., Schwager, M.: Cooperative object transport in 3D with multiple quadrotors using no peer communication. In: Proceedings - IEEE International Conference on Robotics and Automation, pp. 1064–1071 (2018)

ReactiveBuild: Environment-Adaptive Self-Assembly of Amorphous Structures

Petras Swissler[✉] and Michael Rubenstein

Northwestern University, Evanston, IL 60208, USA
pswissler@u.northwestern.edu, rubenstein@northwestern.edu

Abstract. *ReactiveBuild* is an algorithm that enables robot swarms to build a variety of robust, amorphous, and environment-adaptive structures without pre-planning. Robots form structures by climbing their peers until either reaching a point closest to a goal location or until a neighboring robot recruits it for structural reinforcement. This contrasts with typical approaches to robotic self-assembly which generally seek to form some a priori, latticed shape. This paper demonstrates a simulated swarm of *FireAnt3D* robots using *ReactiveBuild* to form towers, chains, cantilevers, and bridges in three-dimensional environments.

Keywords: Self-assembly · Swarm robotics · Bio-inspired · Climbing robot

1 Introduction

Self-assembly is a set of behaviors that enable organisms such as cells and social insects to join together and form structures [1–3]. Self-assembled structures enhance swarm capability by enabling organisms to adapt to their environment. These benefits have driven research into the field of robotic self-assembly [4].

Of particular interest are the structures self-assembled by certain ant species, including towers [5], bridges [6], and chains [7]. In each case, the insects form these structures by *self-climbing* (climbing over their peers) and are thought to grow the structures based on environmental conditions and insect-insect interactions. These structures are therefore not only functional, but also adaptive to dynamic conditions. In contrast to this environment-reactive approach, most robotic self-assembly algorithms seek to form a priori, prescribed shapes [8–11], limiting the adaptability of self-assembled structures.

One exception to this is [12], in which self-assembly and disassembly result from robots reacting to local traffic, taking inspiration from [6]. Another exception is [13], in which robots use local force information to build cantilevers an order of magnitude longer than those built without using local force information; local forces also appear to guide self-assembly in nature [2,5]. In both examples robotic self-assembly followed simple, decentralized rules.

The organization of self-assembled structures also differs between those of insects and those of most robotic self-assembly platforms. Insects form and reinforce self-assembled structures by grabbing each other at seemingly arbitrary

Fig. 1. Examples in this paper use FireAnt3D, which consists of three spheres and can walk on a floor, wall, and ceiling of its peers [22].

locations using their mandibles and legs, resulting in *amorphous* (not constrained to a lattice) structures [14,15]. Contrasting this, most examples of robotic self-assembly form discretized, latticed structures [16–18]. Although such discretization simplifies the localization of individual agents within a structure [10,11], environment-adaptive structures do not appear to require such localization. A latticed approach is problematic for structures starting from more than one location; misalignments are likely without careful alignment of all starting locations (difficult when operating in unknown, arbitrary environments). *Amorphous* structures can also better conform to arbitrary terrain.

Several robots capable of *amorphous* self-assembly have been developed. In most, operation is limited to a 2D plane [12,19–21], and several require grippers to be properly oriented prior to attachment, limiting spontaneity [12,20]. To our knowledge, only FireAnt3D [22] has demonstrated 3D *self-climbing* while also forming the strong connections necessary to self-assemble robust structures (see Fig. 1). Unfortunately, work on algorithms for robust self-assembly of *amorphous* structures is also sparse [12] and has been limited to 2D.

This paper presents *ReactiveBuild*, a novel algorithm for the self-assembly of *amorphous*, environment-adaptive 3D structures. We validate *ReactiveBuild* using simulated FireAnt3D robots [22] to self-assemble four types of environment-adaptive structures: towers, chains, cantilevers, and bridges. Structures arise via a single set of environment-adaptive behaviors, are not latticed, and control stresses between individual robots to maintain structural integrity throughout construction.

2 Algorithm Description

ReactiveBuild enables robots to self-assemble environment-adaptive structures without requiring these structures to conform to a lattice. Any robotic platform with the following capabilities can execute *ReactiveBuild*:

- *Self-climb* and to climb the environment.
- Send messages to contacting robots.
- Sense local forces and the direction to some goal position.

These capabilities match what we expect to be possible with updated versions of FireAnt3D. As discussed below, local communication only requires the sending and receiving of small integers; no other communication is required.

Algorithm 1: ReactiveBuild: Moving Robots

initialize: *role* ← *moving*, *next_move_direction* ← Sense(*direction to goal*)
while *role == moving* **do**
$\quad$ StepTowards(*next_move_direction*)
$\quad$ *last_move_direction* ← *next_move_direction*
$\quad$ *next_move_direction* ← Sense(*direction to goal*)
$\quad$ **if** *next_move_direction == last_move_direction* **then**
$\quad\quad$ // Robot would reverse
$\quad\quad$ *role* ← *structural*
$\quad$ **end**
$\quad$ *comm_in* ← communication from other robots
$\quad$ **if** max(*comm_in*) ≥ 1 **then**
$\quad\quad$ // Robot has been recruited
$\quad\quad$ *role* ← *structural*
$\quad$ **end**
end

Algorithm 2: ReactiveBuild: Structural Robots

while *true* **do**
$\quad$ Sense(*sensed_force*)
$\quad$ *recruit_value* ← min(floor($B^{(sensed\_force/F-1)}$), J)
$\quad$ **for** *all contact zones* **do**
$\quad\quad$ *in_neighbor* ← max(*comm_in(this_zone)*) − 1
$\quad\quad$ *in_others* ← max(*comm_in(other_zones)*) − 2
$\quad\quad$ *comm_out(this_zone)* ← max(*recruit_value, in_neighbor, in_others*)
$\quad$ **end**
end

Robots running *ReactiveBuild* take on one of two roles: a *moving robot*, or a *structural robot*. Robots start as *moving robots* (Algorithm 1) and move towards some goal location (e.g. moving towards a light). A *moving robot* joins the structure as a *structural robot* under one of two circumstances: first, the next step towards the goal would be a backwards step (i.e. it has already passed the closest point to the goal); second, if a *structural robot* recruits it into the structure. This transition to being a *structural robot* is permanent.

After joining the structure, *structural robots* (Algorithm 2) sense local forces and output recruitment values based on the force it senses as well as the recruitment values of neighboring *structural robots*. The calculation of *recruit_value* depends on three factors. F is the threshold force; when *sensed_force* exceeds F, the robot begins to recruit structural reinforcement. J is the maximum *recruit_value*; this controls how far away from an area of high stress a robot can recruit structural reinforcement (i.e. the extent of its jurisdiction). B controls the exponential growth of *recruit_value* after *sensed_force* exceeds F.

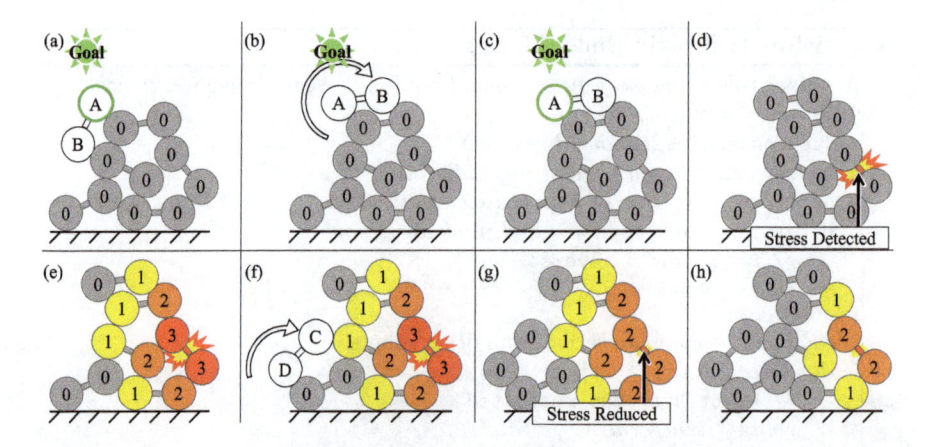

Fig. 2. An illustration of a small swarm of 2D robots executing *ReactiveBuild*. Numbers represent the *recruit_value* output by each sphere.

The algorithm assumes that robots can spatially differentiate incoming communications using discrete contact zones (e.g. the individual spheres of FireAnt3D), though *ReactiveBuild* still functions using only a single contact zone (removing spatial differentiation). Recruitment values decrement by one for each *hop distance* the message travels, similar to [23]. Both inter-robot contacts and intra-robot contact zones are a *hop distance* of 1 apart.

To simplify the algorithm we assume that there is only one *moving robot* at a time; this paper does not explore ways in which moving robots may interact, such as the traffic effects explored in [12]. Figure 2 illustrates the function of *ReactiveBuild* using a 2D robot similar to [21].

(a) A *moving robot* climbs the tower and must now decide in which direction to move. It determines that sphere A is closest direction to the goal.
(b) The robot therefore steps in the direction of sphere A.
(c) The robot finds once again finds that sphere A is closest to the goal. Because a step in this direction would result in backwards motion, the robot stops and joins the structure as a *structural robot*. If the goal were further to the right and sphere B was the closest to the goal, the robot would have continued moving towards the goal since this would result in forwards motion.
(d) The highlighted *structural robot* detects an increase in structural stress.
(e) This robot uses the observed stress to calculate a *recruit_value* of 3, which it then communicates to neighbor robots. Robots propagate this recruitment signal through the structure, decrementing it by 1 for each *hop*.
(f) A new *moving robot* begins to climb the structure and, eventually, sphere C contacts a sphere with a non-zero recruitment signal. The robot therefore stops and joins the structure as a *structural robot*.
(g) The addition of the new robot to the structure reduces structural stresses. The lowered stresses result in a decrease in *recruit_value* from 3 to 2.
(h) The decrease in *recruit_value* cascades through the structure, shrinking the recruitment region.

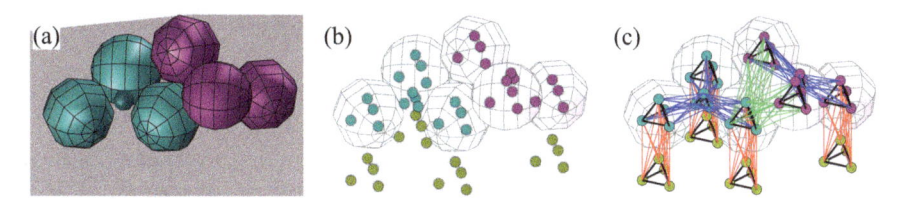

Fig. 3. Robot configurations (a) are converted into nodes arranged in tetrahedrons (b). Linear truss elements (shown as lines) connect nodes to create the FEM (c).

3 Simulator Overview

A bespoke simulator was built to validate *ReactiveBuild* using a virtual swarm of FireAnt3D robots. The simulated robots move towards a specified goal location using the locomotive method described in [22]. The simulation adds one robot at a time to the environment, waiting until the robot finishes climbing and joins the structure before adding another.

To accurately and efficiently calculate connection stresses and force sensor measurements, the simulation formulates and solves a finite element model (FEM) [24] after a robot joins the structure. To do this, it first creates four nodes arranged in a tetrahedron at the location of each robot sphere and center, as well as at the location of each environmental contact. Each node has three translational degrees of freedom with the exception of those associated with environmental contacts, which are fixed. A 0.25-unit gravitational load is applied to each robot sphere node (per-sphere weight is 1 unit).

Linear truss elements (limited to pure tension or compression) are then created between nodes. First, each tetrahedron of nodes is fully connected. Elements are then created to fully connect the nodes of each sphere to its associated robot center nodes, representing the robot structure. Finally, connections are modeled by fully connecting the nodes of contacting spheres, or between sphere and environmental contact nodes. Figure 3 illustrates this process. The code used to formulate and run this model is adapted from [25].

The simulation solves the FEM in one step, assuming small deflections. To simulate robot force sensors, axial and shear forces, as well as bending and torsion moments are calculated from element forces. A robot's measured force, *sensed_force*, is the sum of axial forces and bending moments averaged across all sphere-center connections; robots can only measure axial forces and bending moments. Connection stresses are calculated using all forces and moments associated with the connection and assuming a circular contact with radius 0.5. Because the focus of this paper is not the direct application of *ReactiveBuild* on real-world robots, element stiffnesses were selected based on qualitative review of the simulated robots. For completeness, the chosen stiffnesses are (in simulation units): $5e10$ for in-sphere elements, $1e10$ for sphere connection elements, and $2e9$ for robot structure elements.

4 Algorithm Validation

In this section a swarm of virtual FireAnt3D robots running *ReactiveBuild* spontaneously forms structures resulting from environmental conditions, all while controlling stresses within these structures. Based on the position of the goal location in the environment, robots form four general types of structure: tower, chain, cantilever, and bridge. The results in this section are observational, with limited, high-level interpretation.

100 structures of 100 robots each are simulated at varying combinations of threshold force F, exponent base B, and maximum *recruit_value* J to understand their effect on structure formation. We chose values of F, B, and J based on their exhibition of different construction behaviors, specifically:

- The impact of F using $F = [1, 2.5, 5, 25]$, with $B = 3$ and $J = 5$.
- The impact of B using $B = [1.5, 3, 6, 10]$ at $F = [2.5, 5]$, with $J = 5$.
- The impact of J using $J = [1, 2, 5, 1000]$ at $F = [2.5, 5]$, with $B = 3$.

All distances given in this section are in sphere radii and are measured to sphere centers. Values of F do not correspond to any particular real-world values but are related to the weight of each sphere. The decision to limit experiments to 100 structures of 100 robots each was arbitrary and was not due to any simulator constraint or specific statistical reason. A simulation video is available at [26].

4.1 Tower

Robots build towers by executing *ReactiveBuild* and targeting a goal 65 units above an environment consisting of a flat plane, as shown in Fig. 4. Robots are added one at a time, starting at random positions and orientations outside the tower, on the plane. A tower should be as tall as possible while also being strong enough to withstand its own weight.

As F increases towers become taller, skinnier, and experience a higher peak connection stress (see Fig. 5). It was found that F is directly proportional to the peak stress present in the final tower ($R^2 = 0.999$), showing that F directly

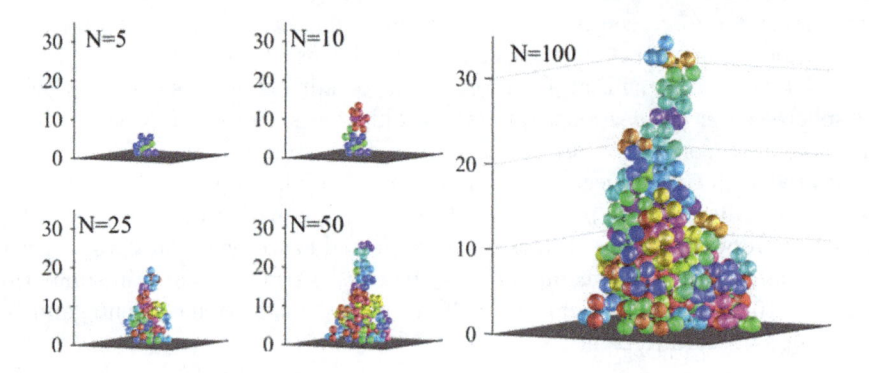

Fig. 4. Tower growth using $F = 10$, $B = 3$, $J = 5$.

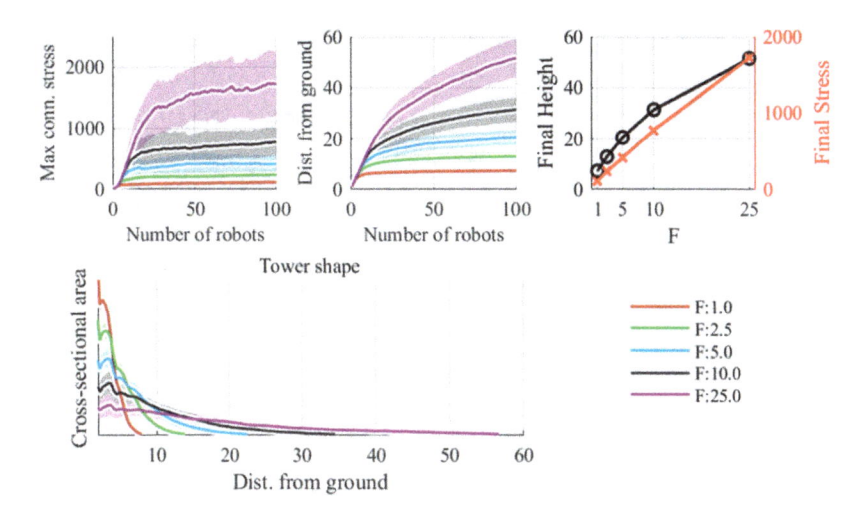

Fig. 5. Tower growth at five values of F; $B = 3$, $J = 5$. Lines represent the mean across 100 runs; shaded areas represent a region ± 1 standard deviation.

influences the maximum stresses present in the tower. We also found that the final height is proportional to the square root of F ($R^2 = 1.000$). After the structure matured (after $N = 25$), stress grew linearly with N ($R^2 = 0.870$), and height grew proportional to the square root of N ($R^2 = 0.983$).

Structure shape is evaluated by summing the number of spheres within distance 1 of a given height across all final towers. The number of spheres (analogous to the cross-sectional area) is proportional to the square of the distance from the top (mean $R^2 = 0.992$ across all tested F). This indicates that the weight of the tower above any given location is proportional to the cube of distance from the top, and by extension the weight supported per unit area is linearly proportional to distance from the top. Therefore, these are not towers of constant stress, as ants appear to build [5]. This is not surprising: such towers require tighter and tighter packing of agents in the lower levels of the tower, and the towers formed by the simulated FireAnt3D robots have relatively uniform packing.

In separate trials, we found that both stress and height for the final structure are proportional to the inverse of B (across both F, $R^2 = 0.993$ and 0.976). B did not obviously change the shape of the structure. Similarly, towers became taller and experienced higher stresses as J decreased, with minimal differences between $J = 5$ and $J = 1000$. This suggests that *ReactiveBuild* controls structural stresses such that recruitment values beyond 5 are rarely used.

4.2 Chain

A chain consists of robots linked together, dangling from an edge to reach the lowest possible point while reinforcing against gravitationally induced stresses. To demonstrate *ReactiveBuild* achieving such a chain, simulated robots target a goal 600 units below the edge of the plane. As seen in Fig. 6, the environment

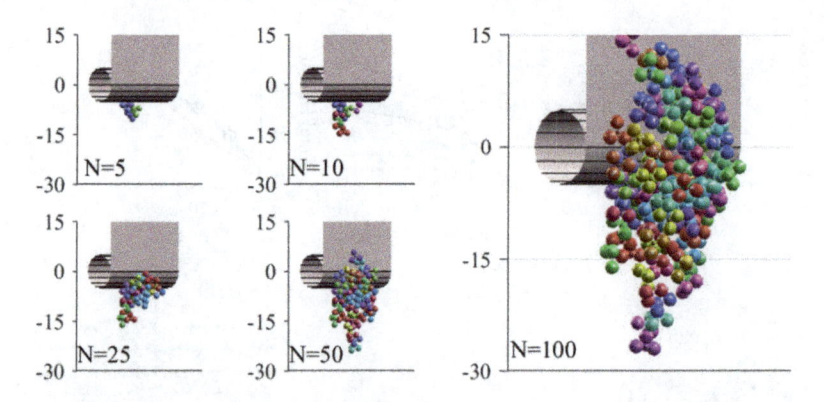

Fig. 6. Chain growth using $F = 10$, $B = 3$, $J = 5$.

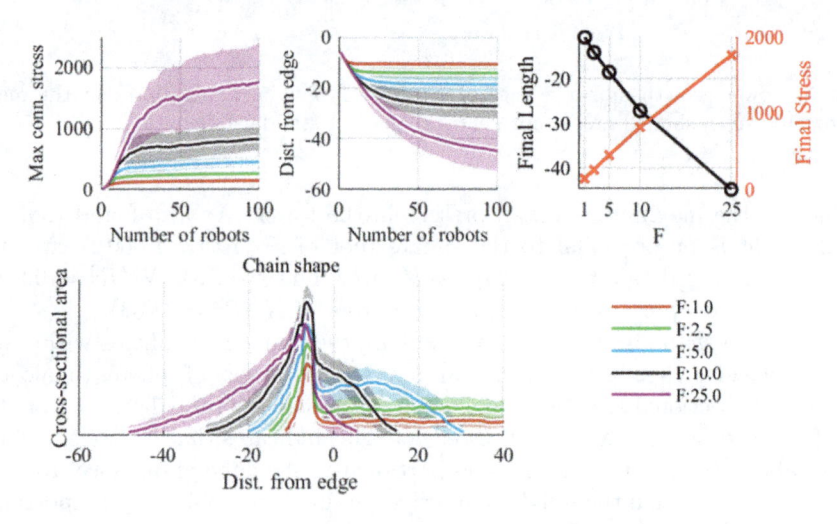

Fig. 7. Chain growth at five values of F; $B = 3$, $J = 5$. Lines represent the mean across 100 runs; shaded areas represent a region ± 1 standard deviation.

terminates in a cylinder at the edge of the plane. Robots are added to the chain one at a time, at random distances and orientations behind the furthest-back above the highest robot.

The results for the chain are similar to the tower, as seen in Fig. 7; F is linearly proportional to the stress in the final structure ($R^2 = 0.998$). When ignoring the portion near the edge, the resulting structure has a similar shape to that of the tower, but is proportional to the distance from the tip raised to the 1.75 power (mean $R^2 = 0.992$ across all values of F). Another difference is that the final height is proportional to F raised to the 0.75 power ($R^2 = 0.999$). These differences may result from the chain having flatter shape relative to the tower, due to environmental geometry. Differences are also seen in the growth of the structure, as both stress and length are proportional to $log(N)$ after maturation (after $N = 25$) ($R^2 = 0.970$ and 0.901).

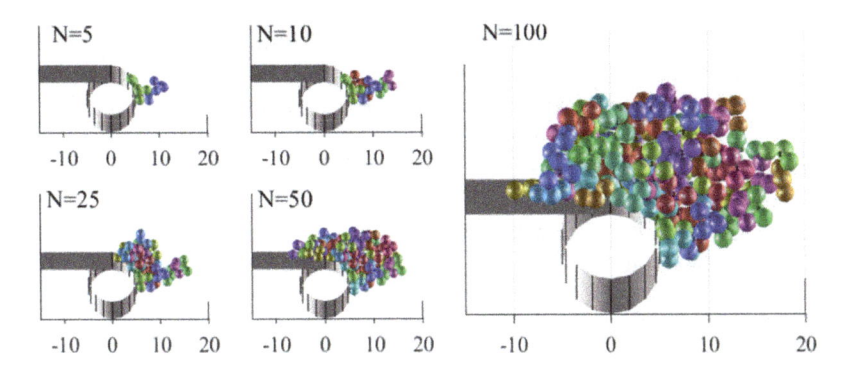

Fig. 8. Cantilever growth using $F = 10$, $B = 3$, $J = 5$.

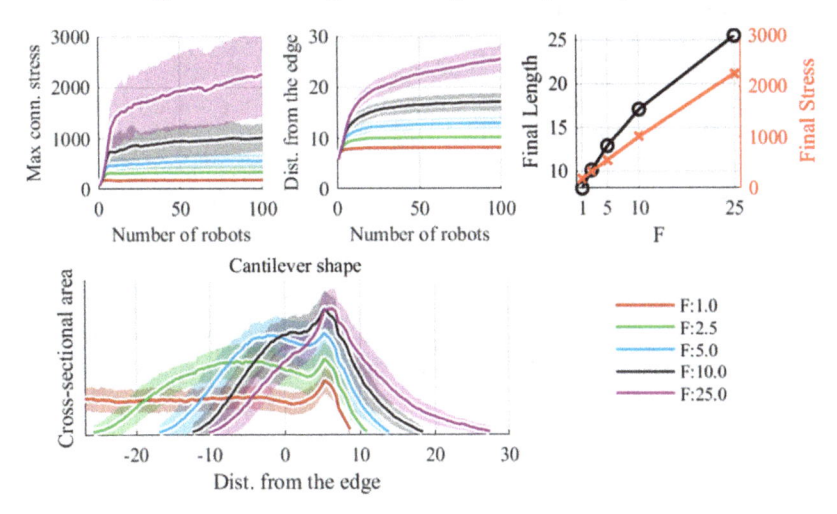

Fig. 9. Cantilever growth at five values of F; $B = 3$, $J = 5$. Lines represent the mean across 100 runs; shaded areas represent a region ±1 standard deviation.

Trends for varying B match those from the tower: both stress and length are inversely proportional to B. Similarly, smaller values of J correspond to higher stresses and longer chains, with minimal difference between $J = 5$ and $J = 1000$.

4.3 Cantilever

A cantilever is a structure that extends horizontally out from an edge; this structure has particularly challenging loading conditions. To demonstrate *ReactiveBuild* achieving a cantilever, simulated robots target a goal 45 units outward horizontally and 10 units upward vertically from the edge of a horizontal plane. As seen in Fig. 8, the environment terminates in a cylinder at the edge of the plane. Robots are added to the cantilever one at a time, at random distances and orientations behind the furthest-back robot.

As with the chain, a support structure emerges behind the edge. The support structure begins forming early in the construction of the cantilever and eventually becomes the primary area of construction. The support structure begins as reinforcement to an area of high stress resulting from the lengthening of the cantilever; this reinforcement eventually grows into a small tower at that location, which then requires its own support. As seen in Fig. 9, this diminishes further lengthening of the cantilever after about $N = [5, 17, 20]$ for $F = [1, 2.5, 5]$; the cantilevers constructed using $F = 10$ and $F = 25$ continue to lengthen for the duration of these trials. Tests with varying B and J showed similar effects as with the tower and chain and did not affect the formation of the support structure.

4.4 Bridge

The final structure is a bridge built simultaneously over a gap between two cantilever-style environments (see Fig. 10). Robots are added one at a time, alternating between the left and right side of the environment. All robots share the same set of behaviors, with the exception that the robots on the left target a goal location on the right, and vice versa; there is no coordination between sides before meeting in the middle. Robots are added to the structure until one robot successfully crosses the bridge to the other side or until $N = 200$. Bridges are considered to be unsuccessful if $N \geq 100$. Table 1 lists the parameter

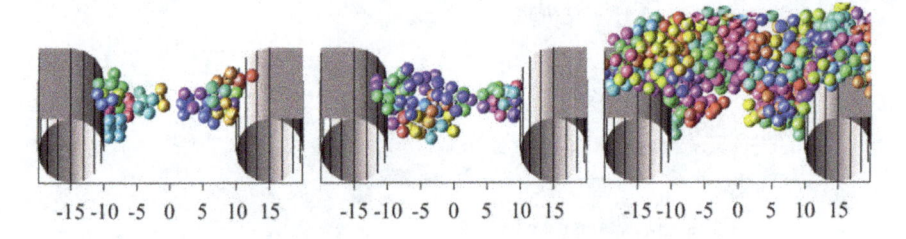

Fig. 10. Three separate bridges grown using $F = 10$, $B = 3$, and $J = 5$. There are three general types of bridges: (left) two separated cantilevers, (center) two cantilevers with a narrow connection, (right) *unified support structure*.

Table 1. Table showing the number of robots to build a bridge across a gap. Medians and variances are based on a Weibull fit of simulation results to account for censored data. Success Rate (SR) is the proportion of bridges completed in under 100 robots.

| Parameters ||| Gap Width |||||||
| F | B | J | 20 ||| 25 ||| 30 |||
			Med.	Var	SR	Med.	Var	SR	Med.	Var	SR
5	3	5	9.5	6.3	99%	27	20	22%	-	-	0%
10	3	5	6.6	1.9	100%	17.8	14.1	100%	49.0	36.5	67%
25	3	5	6.5	1.8	100%	11.7	3.6	100%	18.6	10.7	99%
10	3	2	6.6	1.9	100%	15.9	8.4	100%	49.0	31.5	87%
10	3	10	6.6	1.9	100%	16.2	9.9	98%	47.1	31.6	67%
10	1.5	5	6.6	1.9	100%	14.8	6.4	100%	40.4	29.1	95%
10	6	5	6.7	2.2	100%	19.7	17.3	97%	44	37	24%

configurations for each of the trials run, with each trial consisting of 100 bridges built. As a consequence of environment geometry, robots would sometimes move under the environment cylinders, failing to reach the other side; these robots count towards the number of robots necessary to build a successful bridge.

Higher F resulted in more reliable crossing of wider gaps. Lower B also increased reliability. Notably, for a 30-unit gap, runs using $F = 10$, $B = 1.5$ had a similar success rate as runs using $F = 25$, $B = 3$, but did so with a higher median number of robots (i.e. with a greater occurrence of the *unified support structure* type of bridge), which may be desirable in certain circumstances. Lower J also corresponded to higher success rates.

5 Conclusion

Through the simulation of FireAnt3D robots, this paper demonstrated three primary benefits of amorphous and environment-adaptive self-assembly. First, robots were able to adapt the structure to the geometry of the environment. Second, robots were able to climb about the structure without consideration for proper alignment to neighbors or the environment. Finally, robots were able to build multi-origin structures without need for long-range communication or alignment.

In assessing *ReactiveBuild* algorithm parameters, we found that, in most situations, the effects of B and J were indistinguishable from an adjusted value of F, but for a multi-origin bridge small B and J appear to be advantageous by localizing reinforcement to areas of highest stress. It therefore seems ideal to default to the use of small values of B and J, scaling F based on robot capability. Because these parameters were able to control connection stresses it will be possible to use *ReactiveBuild* for specific self-assembling robots.

There are several readily-apparent ways to extend the work presented in this paper. First and most importantly would be to adapt the algorithm and simulation to support multiple active robots at once. Second, to explore the impact of environment geometry on self-assembled structures. Finally, another interesting extension would be to evaluate the performance of *ReactiveBuild* on other robotic systems, both lattice-based and amorphous.

ReactiveBuild approaches self-assembly in a way distinct from traditional methods. To our knowledge, this is the first demonstration of a *amorphous*, environment-adaptive robotic self-assembly algorithm in 3D. The simulations performed in this paper validate *ReactiveBuid*, the FireAnt3D concept demonstrated in [22], as well as the benefits of *amorphous* self-assembly in general. We hope that the work outlined in this paper encourages others to explore the self-assembly of *amorphous* and environment-adaptive structures.

Acknowledgement. We thank Orion Kafka and Newell Moser for their help in strengthening the portions of this paper related to solid mechanics, and Thomas Bochynek for his help in strengthening the portions of this paper related to biology.

References

1. Anderson, C., Theraulaz, G., Deneubourg, J.: Self-assemblages in insect societies. Insectes Sociaux **49**(2), 99–110 (2002)
2. Peleg, O., Peters, J., Salcedo, M., Mahadevan, L.: Collective mechanical adaptation of honeybee swarms. Nat. Phys. **14**(12), 1193–1198 (2018)
3. Gumbiner, B.: Cell adhesion: the molecular basis of tissue architecture and morphogenesis. Cell **84**(3), 345–357 (1996)
4. Yim, M., et al.: Modular self-reconfigurable robot systems [grand challenges of robotics]. IEEE Robot. Autom. Mag. **14**(1), 43–52 (2007)
5. Phoneko, S., Mlot, N., Monaenkova, D., Hu, D.L., Tovey, C.: Fire ants perpetually rebuild sinking towers. R. Soc. Open Sci. **4**(7), 170475 (2017)
6. Reid, C., et al.: Army ants dynamically adjust living bridges in response to a cost-benefit trade-off. Proc. Natl. Acad. Sci. **112**(49), 15113–15118 (2015)
7. Lioni, A., Sauwens, C., Theraulaz, G., Deneubourg, J.: Chain formation in Oecophylla longinoda. J. Insect Behav. **14**(5), 679–696 (2001). https://doi.org/10.1023/A:1012283403138
8. Werfel, J., Petersen, K., Nagpal, R.: Designing collective behavior in a termite-inspired robot construction team. Science **343**(6172), 754–758 (2014)
9. Gauci, M., Nagpal, R., Rubenstein, M.: Programmable self-disassembly for shape formation in large-scale robot collectives. In: Groß, R., et al. (eds.) Distributed Autonomous Robotic Systems. SPAR, vol. 6, pp. 573–586. Springer, Cham (2018). https://doi.org/10.1007/978-3-319-73008-0_40
10. Tucci, T., et al.: A distributed self-assembly planning algorithm for modular robots. In: AAMAS (2018)
11. Stoy, K., Nagpal, R.: Self-repair through scale independent self-reconfiguration. In: International Conference on Intelligent Robots and Systems, pp. 2062–2067. IEEE (2004)
12. Malley, M., et al.: Eciton robotica: design and algorithms for an adaptive self-assembling soft robot collective. In: ICRA (2020)
13. Melenbrink, N., et al.: Using local force measurements to guide construction by distributed climbing robots. In: IROS (2017)
14. Mlot, N., et al.: Fire ants self-assemble into waterproof rafts to survive floods. Proc. Natl. Acad. Sci. **108**(19), 7669–7673 (2011)
15. Foster, P.C., et al.: Fire ants actively control spacing and orientation within self-assemblages. J. Exp. Biol. **217**(12), 2089–2100 (2014)
16. Jorgensen, M.W., et al.: Modular ATRON: modules for a self-reconfigurable robot. In: International Conference on Intelligent Robots and Systems, pp. 2068–2073. IEEE (2004)
17. Neubert, J., et al.: A robotic module for stochastic fluidic assembly of 3D self-reconfiguring structures. In: ICRA (2010)
18. Romanishin, J.W., et al.: 3D M-Blocks: self-reconfiguring robots capable of locomotion via pivoting in three dimensions. In: ICRA (2015)
19. Shimizu, M., et al.: Adaptive reconfiguration of a modular robot through heterogeneous inter-module connections. In: ICRA (2008)
20. Mondada, F., et al.: SWARM-BOT: a new distributed robotic concept. Auton. Robot. **17**(2–3), 193–221 (2004). https://doi.org/10.1023/B:AURO.0000033972.50769.1c
21. Swissler, P., Rubenstein, M.: FireAnt: a modular robot with full-body continuous docks. In: International Conference on Robotics and Automation, pp. 6812–6817. IEEE (2018)

22. Swissler, P., et al.: FireAnt3D: a 3D self-climbing robot towards non-latticed robotic self-assembly. In: International Conference on Intelligent Robots and Systems. IEEE (2020)
23. Espinosa, G., et al.: Using hardware specialization and hierarchy to simplify robotic swarms. In: International Conference on Robotics and Automation, pp. 7667–7673. IEEE (2018)
24. Fish, J., et al.: A First Course in Finite Elements. Wiley, West Sussex (2007)
25. Ananthasayanam, B.: 3D truss analysis/user interface in FEM, MATLAB Central File Exchange. https://www.mathworks.com/matlabcentral/fileexchange/6832-3d-truss-analysis-user-interface-in-fem. Accessed 16 Oct 2020
26. Swissler, P.: DARS_2021_Simulation_Videos. https://www.youtube.com/watch?v=YLXcj7RptPw. Accessed 11 Mar 2021

Processes for a Colony Solving the Best-of-N Problem Using a Bipartite Graph Representation

Puneet Jain and Michael A. Goodrich

Department of Computer Science, Brigham Young University, Provo, UT, USA
puneetj@byu.edu, mike@cs.byu.edu

Abstract. Agent-based simulations and differential equation models have been used to analyze distributed solutions to the best-of-N problem. This paper shows that the best-of-N problem can be also solved using a graph-based formalism that abstractly represents (a) agents and solutions as vertices, (b) individual agent states as graph edges, and (c) agent state dynamics as edge creation (attachment) or deletion (detachment) between agent and solution. The paper identifies multiple candidate attachment and detachment processes from the literature, and then presents a comparative study of how well various processes perform on the best-of-N problem. Results not only identify promising attachment and detachment processes but also identify model parameters that provide probable convergence to the best solution. Finally, processes are identified that may be suitable for the best-M-of-N problem.

1 Introduction

Swarms and *colonies* are important ways to organize bio-inspired agents [3,18,20,38,41,44]. Varying Brambilla et al.'s swarm taxonomy [4], spatial swarms are characterized by persistently colocated agents. Hub-based colonies, by contrast, include agents colocated at a hub and spatially distributed agents with few inter-agent interactions. Hub-based colonies are important because they (a) can potentially include more agents than swarms [32, ch. 1] and (b) provide distributed solutions to the best-of-N and best-M-of-N problems [24,41,45–47].

Two approaches are often used to design and analyze hub-based colonies: *agent-based* (AB) models and *differential equation* (DE) models. AB models typically use state machines to generate agent behavior, which are often used in empirically oriented experiments to explore how various settings affect colony behavior [11,46]. DE models include mean-field models [6,7], evolutionary models [44, ch. 4], and certain kinds of probabilistic models [44, ch. 3]. DE models are often used to find theoretical properties such as stable attractors, bifurcations, and steady state distributions [26,28,34]. Combining AB and DE models can link micro-level agent behaviors to macro-level swarm phenomena [10].

For spatial swarms, *graph-based models* have proven useful for evaluating how group size, communication networks, and misinformation can affect the swarm.

Both theoretical and empirical results can be derived for graph-based models, complementing the results provided by differential equation models [31,40]. In contrast, graph-based approaches are rarely applied to hub-based colonies.

This paper details a dynamic bipartite graph-based model for a best-of-N problem. A mathematical formalism is outlined and empirical demonstrations are provided based on two fundamental processes: the probability that an agent will "attach" to a site (e.g., assess it, recruit to it) and "detach" from a site (e.g., return to the hub to rest, be recruited to assess another site). Different attachment and detachment processes are compared, including both homogeneous and heterogeneous behaviors. The merits and drawbacks of various processes are discussed and applied to the best-M-of-N problem.

2 Related Literature

A graph-based model of a hub-based colony was proposed in [10]. Unlike [10], which represented agent states as graph *vertices*, this paper represents agent states as graph *edges* between an agent and various points of interest in the world. Preliminary work in [19] describes a bipartite graph formalism with nominal "attachment" and "detachment" processes but fails to discuss why the attachment/detachment processes work or how they compare to other processes.

Preferential attachment [2,35,36] motivates the degree-based attachment process proposed below. Bipartite graph models, similar to the one used in this paper, have been used with preferential attachment to model virus spread [37]. Ant colony optimization with bipartite graphs has also been used for assigning cells to switches and for reconstructing newspaper articles [16,42]. Preferential attachment has been used in bio-inspired applications to model animal decisions to join or leave a group [44, ch. 2], similar to the processes in the present paper in which agents cluster in a group around important sites.

A survey of best-of-N problems is presented in [46], and examples of the problem in dynamic environments are in [33,39]. Best-of-N problems include selecting the best candidate for a job [8], selecting the best nest [41], finding the best foraging site [20], and finding the top nodes in social networks [25,48].

3 Bipartite Graph Formalism

Best-of-N problems are usually characterized by two entities: *agents* and *solutions*. Agents actively explore the world, communicate with other agents, and participate in distributed decision making. Since many best-of-N solutions correspond to decisions about physical locations, we refer to solutions as *sites*.

Agents in both agent-based and differential equation models typically adopt one of several possible states. For example, honeybees finding a new nest can be in states that *explore* the world, *assess* a possible nest site, *dance* to advertise a possible nest site, *observe* other honeybees in the nest, *rest*, or *commit* to a nest site [9,34,41]. Similarly, consensus decision in various ant species include scout, forager, recruiter, and passive states [14,20,27]. Transitions between agent states depend on what other agents do as well as what is observed in the environment [14,20,41]. Agents, sites, and state transitions can be modeled as a bipartite graph.

Bipartite Graph Formalism. Let $G = (V, E)$ be the bipartite graph constructed by partitioning a set of vertices, V, into *agent* vertices and *site* vertices; $V = V_{\text{agent}} \cup V_{\text{site}}$ and $V_{\text{agent}} \cap V_{\text{site}} = \emptyset$. Since G is bipartite, the edge set consists only of edges connecting an agent vertex to a site vertex, $E = \{(a,s) | a \in V_{\text{agent}} \text{ and } s \in V_{\text{site}}\}$. Without loss of generality, each site is assigned a quality in the range $0 \leq \text{qual}(s) \leq 1$.

An edge between agent a and site s represents a subset of possible *site-oriented* states. For honeybee and ant colonies, site-oriented states include assess, dance, recruit, pipe, or commit states. An agent without an edge indicates that the agent is in a *site-agnostic* state; for honeybee and ant colonies, site-agnostic states include rest, observe, or explore states. Thus, the presence or absence of edges between an agent and a site are abstract representations of possible agent states. Agents in hub-based colonies typically assess, promote, or take action for only one site at a time, whence $\forall a \in V_{\text{agents}}, \deg(a) \leq 1$.

Graph Dynamics. There are multiple ways the $|V_{\text{agents}}|$ agents can be connected to the $|V_{\text{sites}}|$ sites. A specific graph is called a *configuration* and is denoted by x, and G_t denotes a time-indexed random variable of possible configurations. Graph dynamics model how agents transition between site-oriented and site-agnostic states. A transition from a site-agnostic state to a site-oriented state is represented by adding an edge, and vice versa by deleting an edge.

Adding (removing) a graph edge between an agent and a site is called *attachment* (*detachment*). The probabilities of attachment or detachment abstractly encode the nondeterminism in transitions between site-oriented and site-agnostic states. These probabilities induce a random process that maps one graph to another, $G_t \to G_{t+1}$. G_t and G_{t+1} can differ in at most two edges because attachment and detachment are independent and only affect one edge at a time. Time steps abstractly represent individual agent transitions between a site-oriented and a site-agnostic state rather than real-time estimates of colony behavior.

4 Attachment and Detachment Probabilities

This section uses patterns from biology to motivate attachment and detachment processes. Let A and R denote Attachment and Detachment random variables, respectively[1].

4.1 Attachment Probability

First, an agent a is *selected* with uniform probability $\frac{1}{|V_{\text{agents}}|}$. If $\deg(a) > 0$, the agent is already attached to a site, so no edge is added. Second, an edge (a, s) is potentially *added* using an attachment process below.

---

[1] R denotes edge removal/detachment so that D can be used to denote degree.

Motivation for Attachment Processes. Biological models of the best-of-N problem suggest that agents are more likely to connect to popular sites.

- The probability that an observing honeybee will be recruited to a site grows with the number of honeybees dancing for the site [41]
- The amount of attraction pheromone deposited on a trail grows with the number of ants following the trail [12]
- "[T]he number of nestmates [an ant] encounter[s] ... [acts] as a stimulus to switch ... to recruitment by carrying ... nestmates to a new nest" [14]
- Encountering more returning foragers stimulate more ants to forage [20].

Agents can vary in their ideal group size, which means that popularity-based aggregation can be modulated by *individual* preferences.

- The probability of a *Holocnemus pluchei* spider staying or leaving a shared web depends on how large and well fed the spider is [23].
- The group sizes of large mammalian herbivores might be affected by individual preferences [13,17].
- Group formation models assume that individuals have sociality thresholds [5].
- Localized interactions can cause fish to differ in whether they join a shoal [21].
- Some ants prefer risky exploration to joining a majority [22].

Baseline Attachment. Given agent a, a site s is chosen with uniform probability $\frac{1}{|V_{\text{sites}}|}$ and an edge (a, s) is formed. This baseline provides insight into clustering driven solely by the detachment process.

Nominal Attachment. Popularity-based networks are well-modeled by *preferential attachment* [2][44, ch. 2]. Thus, nominal attachment probabilistically favors sites with higher degree.

Select Degree. There can be several different sites with the same degree. Let $\text{Deg}(x_t) = \{d : \exists s \in V_{\text{sites}} \text{ for which } \deg(s) = d\}$, be the set of unique degrees in configuration x_t. The monotonic function $f(d, G_t) = d + 1/|\text{Deg}(x_t)|$ encodes two factors that govern the probability that a degree is chosen: (1) higher degree sites should induce more agents to attach, and (2) sites with zero degrees should have a non-zero probability of attachment so that new sites can be "discovered". The probability of selecting a degree is obtained by sampling from the set $\deg(G)$ obtained by normalizing $f(\cdot)$, yielding $P_{D|G_t}(d|x_t) = f(d)/[\sum_{d' \in \text{Deg}(x_t)} f(d')]$, where D denotes the degree random variable.

Select Site. Given the degree, a site is chosen with uniform probability from sites with that degree, $P_{S|D}(s|d, x_t) = 1/|\{s' : \deg(s') = d\}|$. The probability of sampling site s given the graph is derived using the chain rule and marginalizing,

$$P_{S|G_t}(s|x_t) = \sum_d P_{S|D,G_t}(s|d, x_t) P_{D|G_t}(d|x_t) = \frac{P_{D|G_t}(\deg(s)|x_t)}{|\{s' : \deg(s') = \deg(s)\}|}. \quad (1)$$

Add Edge. Let E denote the "add edge" random variable. No edge is added if the randomly selected agent is already attached. An edge is added to an unattached agent using the product of Eq. (1) and the uniform probability of selecting the agent, yielding $P_{E|G_t}[(a,s)|x_t] = \frac{P_{D|G_t}(\deg(s)|x_t)}{|V_{\text{agents}}|\cdot|\{s':\deg(s')=\deg(s)\}|}$.

Saturated Degree Attachment. A variation of preferential attachment attaches agents to sites with probability (a) linearly proportional to the site's degree up to a specified saturation degree, d^{sat}, and (b) zero if $\deg(s) > d^{\text{sat}}$. This saturated attachment process uses the *Select Site* and *Add Edge* probabilities from the nominal attachment process, but uses the *Select Degree* probability,

$$P_{D|G_t}(d|x_t) = \begin{cases} \frac{f(d)}{\sum_{d' \in \text{Deg}(x_t), d' \leq d^{sat}} f(d')}, & d \leq d^{\text{sat}} \\ 0, & \text{otherwise} \end{cases} \qquad (2)$$

Preferred Degree Attachment. Variations of preferential attachment also allow individual agents to prefer different site degrees. Let d_i^* denote the preferred degree for agent a_i, and let $\Delta_{ij} = |d_i^* - \deg(s_j)|$ denote the difference between agent a_i's preferred degree and the degree of site s_j. When agents are allowed a degree preference, all agents may prefer the same degree (*homogeneous*) in which case $\forall i\ d_i^* = d^*$, or agents may prefer different degrees (*heterogeneous*). Section 5 specifies parameters for homogeneous and heterogeneous degree preferences.

Select Site. The monotonically decreasing function $f(\Delta_{ij}) = 1/(1+\Delta_{ij}^2)$ favors sites with degree close to the preferred degree. Normalizing yields

$$P_{S|G_t}(s_j|x_t) = \frac{f(\Delta_{ij})}{\sum_j f(\Delta_{ij})}. \qquad (3)$$

Add Edge. As before, no edge is added if the agent is already attached. An edge is added to an unattached agent using the product of Eq. (3) and the uniform agent selection probability, yielding $P_{E|G_t}[(a_i, s_j)|x_t] = \frac{P_{S|G_t}(s_j|x_t)}{|V_{\text{agents}}|}$.

4.2 Detachment Probability

Edge $(a, s) \in E$ is first *selected* with uniform probability $1/|E|$. Second, the edge is *removed* using a detachment process from below.

Motivation for Detachment Processes. Biological models suggest that individual agents persist longer in states that favor higher quality sites.

- The length of a waggle dance and the number of assessment runs made by a honeybee is higher for high quality sites than low quality sites [41]
- Ants deposit more pheromone on returning from high quality foraging sites, creating trails that persist longer [12].

Quality-based detachment processes can exhibit individual preferences.

- Ants can have different grain size preference when building a nest [1].
- Birds can have different sugar preferences for nectar [29].

Baseline Detachment. An edge (a, s) randomly selected from the edge set is removed from the edge set with fixed probability $1/20$. This value was subjectively set so that edges are probably not removed immediately after creation.

Nominal Detachment. Recall that $\text{qual}(s) \in [0, 1]$. The probability of removing (a, s) decreases with site quality, $P_{R|G_t}[(a, s)|x_t] = (1 - \text{qual}(s))/|E|$.

Detachment with Quality Preferences. Individual agents may have preferences for different site qualities. Thus, we differentiate between the *objective* quality of a site, $\text{qual}(s_j)$ and the variations in the *subjective* site quality q_i^*, for agent a_i. The difference between the objective and subjective site qualities, $\delta_{ij} = \text{qual}(s_j) - q_i^*$, abstracts the differences in how agents assess site quality.

Homogeneous detachment experiments, where $\forall i \; q_i^* = q^* = 1$, are omitted because they are equivalent to nominal attachment.

Heterogeneous preferences mean that individual agents subjectively prefer sites of different qualities.

The edge removal probability is defined as

$$P_{R|G_t}[(a_i, s_j)|x_t] = \begin{cases} \frac{1-\delta_{ij}}{|E|} & \text{if } \delta_{ij} \geq 0 \\ 0.25 & \text{otherwise} \end{cases}. \tag{4}$$

The probability of removing an edge is minimum when the objective quality is close to the subjective quality. When objective quality is less than the subjective preference, Eq. (4) assigns a uniform probability of detachment. Parameters for heterogeneous quality preferences are given in Sect. 5.

5 Methods

Twenty Agents and Five Sites. Pilot experiments indicated that 20 agents and 5 sites reasonably represent a range of larger colonies.

Twenty Trials and Runs. A single run consists of using one set of parameters from the cases in Table 1 and running the graph based simulation for 700 time steps. A trial is 20 runs with the same parameters. An experiment (or each case) consists of 20 trials with the same parameters. Multiple trials generate mean and interquartile range estimates for success probability. The initial graph configuration had no agent-site connections.

Quality Distributions. The probability of converging to a successful configuration depends on the objective site qualities. Linear, exponential, and sublinear distributions of quality represent many resource allocation problems [15], problems finding the most influential nodes in a social network [30], and problems like foraging [43], respectively. These distributions are subjectively chosen as $\mathbf{q}_{\text{lin}} = \theta$, $\mathbf{q}_{\text{exp}} = e^{5(\theta-1)}$, and $\mathbf{q}_{\text{sub}} = (\theta^{1/2})/2$, respectively, where $\theta = [0.2, 0.3, 0.5, 0.75, 0.9]$ is a parameter vector.

Experiment Conditions. Five attachment conditions (random, nominal, homogeneous, heterogeneous, and saturation), three detachment conditions (random, nominal, and heterogeneous), and three site quality distributions (linear, exponential, sublinear) were used. Note that random detachment ignores site quality. All combinations were considered, but results are shown only for interesting conditions. For homogeneous attachment $d^* = \frac{|V_{\text{agents}}|}{2} = 10$, for heterogeneous attachment the d_i^*'s were independently sampled from $\mathcal{N}(10, 2)$, for saturation $d^{sat} = 10$, and for heterogeneous detachment the q_i^*'s were independently sampled from $\mathcal{N}(1, 0.05)$. The same set of d_i^*, q_i^* and d^{sat} were used for each sample run within each trial set.

6 Results

Results are shown in Table 1. The first column is a reference number denoted *case*. Lines between rows and the daggers indicate comparison groups. The second column specifies the attachment, detachment, and quality conditions. The third column shows the sample mean and interquartile range for the probability of finding a successful configuration as a function of time. The plurality rule in these figures breaks ties in favor of the best site. A *successful configuration* is one in which more agents are connected to the site with highest objective quality than any other site. In all the figures, the brief initial high success rate is an artifact of ranking sites when no agents are attached.

The fourth column (Ratio of Sites) shows a stacked bar graph that indicates how often each site had the highest number of agents attached to it at the end of each set and trial. The top bar indicates the site with highest objective quality, the second bar indicates the site with second highest quality, and so on. The width of the bar is the percentage of time that site has the largest number of agents attached at the end of each simulation run. When there are multiple sites that tie for the most numbers of edges, the tie is broken in favor of the lowest quality site; this is done to present a conservative representation of ranking that balances the tie-breaker used in the average success plots. Case 9 shows two bars which show the percentage of time sites are ranked first (left stack) or second (right stack). The key to the stacked bar graphs is given in the first row. Objective site qualities are ranked $q(s_4) > q(s_3) > \ldots > q(s_0)$.

The fifth column shows typical graph configuration at the end of a simulation run, when showing the example is helpful. The monochromatic vertices in the upper left represent agents, and the colormapped vertices represent sites.

Baseline: Cases 1–2. Random attachment and detachment results (not shown in the table) are typified by the following: Each site is equally likely to be chosen as the best solution which means that the highest quality site is chosen about 1/5 of the time, the site chosen as best changes often over time, and a typical configuration has very few edges. Nominal attachment and random detachment in Case 1 show a similar pattern, where each site is equally likely to be chosen as the best-of-N solution and the highest quality site is chosen 1/5 of the time. Unlike random attachment/detachment, the site that is chosen under nominal

Table 1. Results for different attachment (Att), detachment (Det), and quality distributions (Qual). Success probability over time, ratio of sites ranked as first, and example final graph configurations are shown.

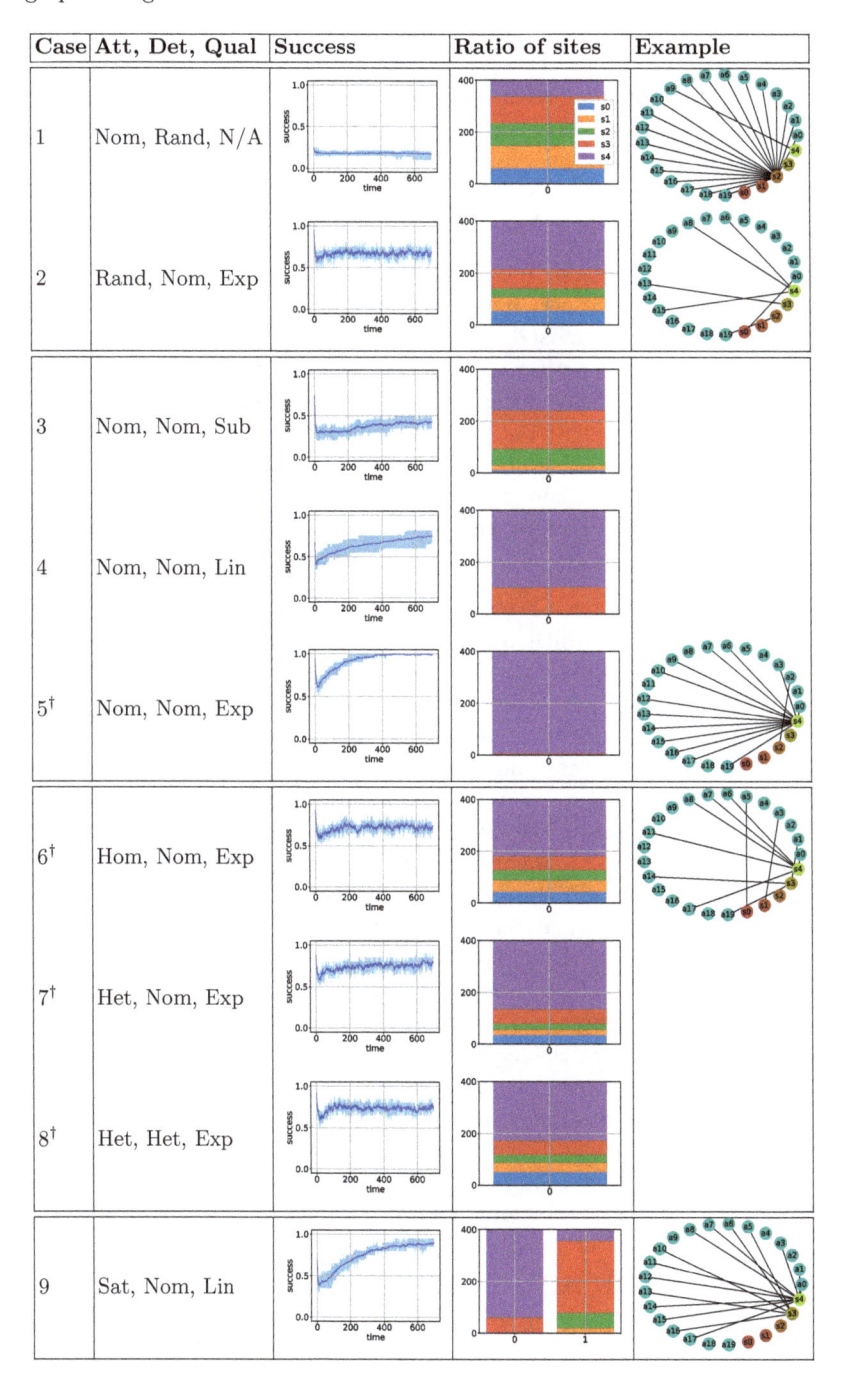

Case	Att, Det, Qual	Success	Ratio of sites	Example
1	Nom, Rand, N/A			
2	Rand, Nom, Exp			
3	Nom, Nom, Sub			
4	Nom, Nom, Lin			
5[†]	Nom, Nom, Exp			
6[†]	Hom, Nom, Exp			
7[†]	Het, Nom, Exp			
8[†]	Het, Het, Exp			
9	Sat, Nom, Lin			

attachment rarely changes after the first 100 time steps. The example configuration shows that popularity-based attachment causes many agents to cluster around a common site regardless of the site's quality.

Random attachment and nominal detachment in Case 2 show the following: the highest quality site is chosen more than 50% of the time because agents persist at the highest quality site longer than at other sites. However, clusters are small, as illustrated by the example configuration, making it difficult for the best site to "hold onto" agents for a long time. Linear and sublinear quality distributions decrease the success probability compared to the exponential quality distribution because there is less difference between site qualities.

Effect of Quality: Cases 3–5. Nominal (popularity-based) attachment and nominal (quality-based) detachment exhibit the following: The first or second highest quality sites are most likely chosen as the best-of-N solution, and the chosen site rarely changes after 100 iterations. The typical configuration shows that many agents cluster at the highest quality site. Comparing Cases 3–5 reveals that the type of quality distribution affects success. When the distance between site qualities is low (linear/sublinear), quality-based persistence is less effective, making other sites more likely to be chosen. Importantly, the red band in the stacked bar chart indicates that the second highest quality site is often chosen when two sites have nearly equal quality.

Effect of Heterogeneity: Cases $5^\dagger$-$8^\dagger$. Comparing Case 5, which has nominal attachment, to Case 6, which has homogeneous attachment, appears to show that homogeneity decreases the success probability. It is not homogeneity per se that is responsible for the differences in these cases since all agents in the nominal attachment condition behave the same. Rather, the difference is that the probability of attachment under homogeneous attachment increases as degree increases until it reached $d^* = |V_{\text{agents}}|/2 = 10$, but when more than 10 agents are attached to a site the probability of a new agent attaching to it decreases. By contrast, the nominal attachment probability grows linearly with degree. This results in lower average success rates under homogeneous attachment because it is easier for the probabilistic dynamics to induce switches between configurations that are constrained in their popularity.

Case 7 shows that heterogeneous agents are more likely than homogeneous agents to select the highest quality site. This is because there is a chance that agents will prefer sites with degree $d_i^* > |V_{\text{agents}}|/2 = 10$. This allows more agents to be attracted to more popular sites provided that agents who prefer less popular sites attach first and then stay attached due to the persistence of high quality sites. Case 8, which uses both heterogeneous attachment and detachment, shows that the highest quality site is selected less frequently than Case 7 (note the larger blue, orange, and green bands in the bar graph). Agents that subjectively prefer sites with lower objective quality persist longer at those sites while popularity-based attachment recruits other agents to the site.

Best-M-of-N: Cases 3, 4, 8, 9. Case 9 indicates that saturating the effect of popularity causes agents to attach to other sites. As illustrated in the example configuration and the stacked bar chart, agents are likely to form a second cluster around the second highest quality site. In effect, popularity-based clustering is divided across multiple possible sites, allowing quality-based persistence to divide agents among the two best sites. Indeed, two stacked bar charts indicate that the two objectively highest quality sites are almost always selected as the best-two-of-N solutions. Note that an exponential distribution decreases the probability of selecting the second highest quality site because relative persistence at that site is lower, and a sublinear distribution increases the relative persistence at the second highest quality site. Cases 3,4,8,9 all exhibit a division of agent clusters around multiple sites. These suggest that algorithms that successfully solve the best-M-of-N will perform best when the algorithm includes some form of saturation, degree preference, and heterogeneity. Moreover, best-M-of-N might be easier to solve when distances between site qualities are small.

7 Conclusion and Future Work

A graph-based abstraction of a hub-based colony can plausibly be used to solve the best-of-N problem. Moreover, degree-based preferential attachment combined with quality-based detachment appear sufficient for solving the problem. A colony is more likely to successfully solve the best-M-of-N problem when popularity-based saturation and agent heterogeneity are added to the algorithm.

Future work includes (a) modeling how configurations will evolve in real-time, (b) exploring the effects of noisy estimates of popularity and quality on the likelihood of finding the best solution, (c) finding useful blends of quality, popularity, and other environment information (e.g., site distance) in both attachment and detachment processes, (d) modeling existing agent-based models so that abstract analyses of these models can be performed, and (e) developing formal analysis tools for the graph random process to establish theoretical colony properties.

Acknowledgements. This paper was partially supported by a grant from the US Office of Naval Research under grant number N00014-18-1-2503. All opinions, findings, and results are the responsibility of the authors and not the sponsoring organization.

References

1. Aleksiev, A.S., Longdon, B., Christmas, M.J., Sendova-Franks, A.B., Franks, N.R.: Individual choice of building material for nest construction by worker ants and the collective outcome for their colony. Anim. Behav. **74**(3), 559–566 (2007)
2. Barabási, A.L., Albert, R.: Emergence of scaling in random networks. Science **286**(5439), 509–512 (1999)
3. Binitha, S., Sathya, S.S., et al.: A survey of bio inspired optimization algorithms. Int. J. Soft Comput. Eng. **2**(2), 137–151 (2012)
4. Brambilla, M., Ferrante, E., Birattari, M., Dorigo, M.: Swarm robotics: a review from the swarm engineering perspective. Swarm Intell. **7**(1), 1–41 (2013). https://doi.org/10.1007/s11721-012-0075-2

5. Brown, J.L.: Optimal group size in territorial animals. J. Theor. Biol. **95**(4), 793–810 (1982)
6. Bussemaker, H.J., Deutsch, A., Geigant, E.: Mean-field analysis of a dynamical phase transition in a cellular automaton model for collective motion. Phys. Rev. Lett. **78**(26), 5018 (1997)
7. Carrillo, J.A., Choi, Y.-P., Hauray, M.: The derivation of swarming models: mean-field limit and Wasserstein distances. In: Muntean, A., Toschi, F. (eds.) Collective Dynamics from Bacteria to Crowds. CICMS, vol. 553, pp. 1–46. Springer, Vienna (2014). https://doi.org/10.1007/978-3-7091-1785-9_1
8. Chow, Y., Moriguti, S., Robbins, H., Samuels, S.: Optimal selection based on relative rank the ("secretary problem"). Israel J. Math. **2**(2), 81–90 (1964). https://doi.org/10.1007/BF02759948
9. Cody, J.R., Adams, J.A.: An evaluation of quorum sensing mechanisms in collective value-sensitive site selection. In: 2017 International Symposium on Multi-Robot and Multi-Agent Systems (MRS), pp. 40–47. IEEE, Los Angeles (2017)
10. Coppola, M., Guo, J., Gill, E., De Croon, G.C.: The PageRank algorithm as a method to optimize swarm behavior through local analysis. Swarm Intell. **13**(3–4), 277–319 (2019). https://doi.org/10.1007/s11721-019-00172-z
11. Dorigo, M., Birattari, M., Blum, C., Christensen, A.L., Reina, A., Trianni, V.: Ant Colony Optimization and Swarm Intelligence. LNCS, vol. 11172. Springer, Heidelberg (2018). https://doi.org/10.1007/978-3-540-87527-711th International Workshop, ANTS 2018. Proceedings
12. Dorigo, M., Bonabeau, E., Theraulaz, G.: Ant algorithms and stigmergy. Futur. Gener. Comput. Syst. **16**(8), 851–871 (2000)
13. Estes, R.D.: Social organization of the African bovidae. Behav. Ungulates Relat. Manag. **1**, 166–205 (1974)
14. Franks, N.R., Dornhaus, A., Best, C.S., Jones, E.L.: Decision making by small and large house-hunting ant colonies: one size fits all. Anim. Behav. **72**(3), 611–616 (2006)
15. Fu, T.P., Liu, Y.S., Chen, J.H.: Improved genetic and ant colony optimization algorithm for regional air defense WTA problem. In: First International Conference on Innovative Computing, Information and Control-Volume I (ICICIC'06), vol. 1, pp. 226–229. IEEE (2006)
16. Gao, L., Wang, Y., Tang, Z., Lin, X.: Newspaper article reconstruction using ant colony optimization and bipartite graph. Appl. Soft Comput. **13**(6), 3033–3046 (2013)
17. Gerard, J.F., Bideau, E., Maublanc, M.L., Loisel, P., Marchal, C.: Herd size in large herbivores: encoded in the individual or emergent? Biol. Bull. **202**(3), 275–282 (2002)
18. Ghaffari, M., Musco, C., Radeva, T., Lynch, N.: Distributed house-hunting in ant colonies. In: Proceedings of the 2015 ACM Symposium on Principles of Distributed Computing, pp. 57–66. ACM (2015)
19. Goodrich, M., Jain, P.: Swarm Intelligence. In: 12th International Conference, ANTS 2020, Extended Abstracts. Springer (2020)
20. Gordon, D.M.: Ant Encounters: Interaction Networks and Colony Behavior, vol. 1. Princeton University Press, Princeton (2010)
21. Hoare, D.J., Couzin, I.D., Godin, J.G., Krause, J.: Context-dependent group size choice in fish. Anim. Behav. **67**(1), 155–164 (2004)
22. Imirzian, N., Zhang, Y., Kurze, C., Loreto, R.G., Chen, D.Z., Hughes, D.P.: Automated tracking and analysis of ant trajectories shows variation in forager exploration. Sci. Rep. **9**(1), 1–10 (2019)

23. Jakob, E.M.: Individual decisions and group dynamics: why pholcid spiders join and leave groups. Anim. Behav. **68**(1), 9–20 (2004)
24. Kempe, D., Kleinberg, J., Tardos, É.: Maximizing the spread of influence through a social network. In: Proceedings of the ninth ACM SIGKDD International Conference on Knowledge Discovery and Data mining, pp. 137–146 (2003)
25. Kimura, M., Saito, K., Nakano, R., Motoda, H.: Extracting influential nodes on a social network for information diffusion. Data Min. Knowl. Disc. **20**(1), 70–97 (2010). https://doi.org/10.1007/s10618-009-0150-5
26. Laomettachit, T., Termsaithong, T., Sae-Tang, A., Duangphakdee, O.: Decision-making in honeybee swarms based on quality and distance information of candidate nest sites. J. Theor. Biol. **364**, 21–30 (2015)
27. Lee, C., Lawry, J., Winfield, A.: Negative updating combined with opinion pooling in the best-of-n problem in swarm robotics. In: Dorigo, M., Birattari, M., Blum, C., Christensen, A.L., Reina, A., Trianni, V. (eds.) ANTS 2018. LNCS, vol. 11172, pp. 97–108. Springer, Cham (2018). https://doi.org/10.1007/978-3-030-00533-7_8
28. Leonard, N.E.: Multi-agent system dynamics: bifurcation and behavior of animal groups. Annu. Rev. Control. **38**(2), 171–183 (2014)
29. Lotz, C.N., Schondube, J.E.: Sugar preferences in nectar-and fruit-eating birds: behavioral patterns and physiological causes 1. Biotropica: J. Biol. Conserv. **38**(1), 3–15 (2006)
30. Lusher, D., Koskinen, J., Robins, G.: Exponential Random Graph Models for Social Networks: Theory, Methods, and Applications. Cambridge University Press, Cambridge (2013)
31. Mesbahi, M., Egerstedt, M.: Graph Theoretic Methods in Multiagent Networks. Princeton University Press, Princeton (2010)
32. Moffett, M.W.: The Human Swarm: How Our Societies Arise, Thrive, and Fall. Basic Books, New York (2019)
33. Nedić, A., Olshevsky, A., Uribe, C.A.: Nonasymptotic convergence rates for cooperative learning over time-varying directed graphs. In: 2015 American Control Conference (ACC), pp. 5884–5889. IEEE (2015)
34. Nevai, A.L., Passino, K.M., Srinivasan, P.: Stability of choice in the honey bee nest-site selection process. J. Theor. Biol. **263**(1), 93–107 (2010)
35. Newman, M.E.: Clustering and preferential attachment in growing networks. Phys. Rev. E **64**(2), 025102 (2001)
36. Newman, M.E.: Properties of highly clustered networks. Phys. Rev. E **68**(2), 026121 (2003)
37. Omic, J., Kooij, R., Van Mieghem, P.: Virus spread in complete bi-partite graphs. In: 2nd International ICST Conference on Bio-Inspired Models of Network, Information, and Computing Systems (2008)
38. Passino, K.M.: Biomimicry of bacterial foraging for distributed optimization and control. IEEE Control Syst. Mag. **22**(3), 52–67 (2002)
39. Prasetyo, J., De Masi, G., Ranjan, P., Ferrante, E.: The best-of-n problem with dynamic site qualities: achieving adaptability with stubborn individuals. In: Dorigo, M., Birattari, M., Blum, C., Christensen, A.L., Reina, A., Trianni, V. (eds.) ANTS 2018. LNCS, vol. 11172, pp. 239–251. Springer, Cham (2018). https://doi.org/10.1007/978-3-030-00533-7_19
40. Ren, W., Beard, R.W., Atkins, E.M.: A survey of consensus problems in multi-agent coordination. In: Proceedings of the 2005, American Control Conference, pp. 1859–1864. IEEE (2005)

41. Seeley, T.D., Buhrman, S.C.: Nest-site selection in honey bees: how well do swarms implement the "best-of-N" decision rule? Behav. Ecol. Sociobiol. **49**(5), 416–427 (2001). https://doi.org/10.1007/s002650000299
42. Shyu, S.J., Lin, B.M., Hsiao, T.S.: Ant colony optimization for the cell assignment problem in PCS networks. Comput. Oper. Res. **33**(6), 1713–1740 (2006)
43. Sinervo, B.: Optimal foraging theory: constraints and cognitive processes. University of Southern California Santa Cruz, pp. 105–130 (1997). https://printfu.org/foraging+animals
44. Sumpter, D.J.: Collective Animal Behavior. Princeton University Press, Princeton (2010)
45. Sumpter, D.J., Beekman, M.: From nonlinearity to optimality: pheromone trail foraging by ants. Anim. Behav. **66**(2), 273–280 (2003)
46. Valentini, G., Ferrante, E., Dorigo, M.: The best-of-N problem in robot swarms: formalization, state of the art, and novel perspectives. Front. Robot. AI **4**, 9 (2017)
47. Wilson, J.G.: Optimal choice and assignment of the best m of n randomly arriving items. Stochast. Process. Appl. **39**(2), 325–343 (1991)
48. Zhang, Y., Zhou, J., Cheng, J.: Preference-based top-K influential nodes mining in social networks. In: 2011 IEEE 10th International Conference on Trust, Security and Privacy in Computing and Communications, pp. 1512–1518. IEEE, Changsha (2011)

Decentralized Navigation in 3D Space of a Robotic Swarm with Heterogeneous Abilities

Shota Tanaka[✉], Takahiro Endo, and Fumitoshi Matsuno

Kyoto University, Kyoto, Japan
`tanaka.shota.64x@st.kyoto-u.ac.jp`

Abstract. This paper proposes a decentralized method for navigation of multiple robots, each of which has different abilities, by a single leader robot in 3D space, especially focusing on the connectivity maintenance. We assume a swarm of robots whose sensing ranges, maximum speeds and maximum accelerations are different. For such robots, we propose a control method for maintaining the whole connectivity by each agent's keeping local connectivity in a decentralized way. We also mathematically prove that the proposed method can enable multiple robots to navigate in 3D space while keeping the connectivity. Finally, numerical simulation results are presented to confirm the effectiveness of the proposed method.

Keywords: Swarm control · Navigation control · Heterogeneous agents

1 Introduction

The robotic swarm consists of multiple robots and requires their cooperation to achieve specific tasks. It is pointed out that swarm robots are useful due to their "robustness", "flexibility" and "scalability" [1,2]. "Robustness" is a property that prevents some agents' failure from depriving the original function of the whole swarm, and it is important when using the robotic swarm in a harsh environment. "Flexibility" is a property that allows a robotic swarm to adapt to various situations and tasks. For example, a large object that is too large for a single robot to carry could be transported by a multiple robots' cooperation. "Scalability" is a property that the system still works well even if the number of agents increases, when each agent determines their action by only using local information (called "decentralized control"). In decentralized control, even though each agent acts according to their simple control law, the whole swarm exhibits good performance due to each agent's interaction. For these reasons, robotic swarms are expected to be applied to various situations.

Robotic swarms are used in such tasks as exploration [3], transportation [4], and surveillance [5]. They have a basic task in common: "moving to the destination while maintaining the swarm structure," and to realize this task by decentralized control, each agent needs to get information about the surroundings by sensing or communication and to determine their action based on it.

Considering that agents have a physical limit to the sensing/communication distance, the distance between agents should be kept to such an extent that each agent can obtain the necessary information.

There have been some researches on connectivity maintenance control, but many of them have an assumption that agents have the same capabilities [6,7]. On the other hand, the robotic swarm consisting of heterogeneous agents is expected to adapt to a wider variety of tasks as each of them makes use of their abilities and cooperate with each other. In [8], the authors state that we have to incorporate heterogeneity into swarm robot systems in order to apply them to the actual environment. They proposed a swarm system composed of hand-bots that can climb up or down walls and work as a manipulator, foot-bots that navigate and carry the hand-bots, and eye-bot to scan the environment and send information. In [9], a task example for heterogeneous robotic swarms are shown, in which drones with different abilities monitor several targets which moves differently.

There are some researches on "heterogeneous" (meaning "agents have different capabilities" in this context) swarms [10–12]. In [10], a decentralized control method is proposed with which agents with different sensing ranges estimate connectivity from the graph Laplacian of the connected graph only using local information, and move in such a way as to keep connectivity. However, the research has an assumption that each agent's sensing range could be enlarged if needed, and does not consider agents' physical constraints such as the maximum speed. On the other hand, [11] proposed a control method for mobile robots with different sensing range, maximum speed and maximum acceleration. In this method, a single leader navigates the other agents while maintaining connectivity, without agents' communicating or getting information about their surroundings, and each agent moves based only on their relative position to neighboring agents. [12] is the expansion of this research, to which the authors have added the constraint on agents' viewing angle, and the stability analysis is proved in [13]. However, [11,12] handles moving robots in a 2D plane, and navigations method for heterogeneous robots in 3D space have not been proposed.

When we use the method [11] to 3D space, we have two main problems. The first problem is in the vector resolution of the control input. As [11], where the input vector is decomposed using circular coordinates, let us consider the vector resolution of the control input using the spherical coordinates. The singular point in circular coordinates is the point where an agent and its target are in the same position. Here, this situation cannot physically occur. However, in spherical coordinates, the singular points are points where an agent comes right above or below its target, and this situation is likely enough to occur. At this singularity, the basis vectors are no longer determined uniquely, which is a big obstacle to maintaining the continuity of the agent's input vector.

The second problem is about decentralized control. In [11], the leader's speed constraint is a constant and can be uniquely determined from each follower's ability, which is known in advance by the leader. However, when trying to apply the same method to our study, the leader's constraint will include the variables that only followers could obtain. Specifically, the leader would have to know the angle $\theta_i(t)$ (introduced in Chap. 3.3), which the leader cannot obtain without assuming the unlimited sensing range. We will have to use a centralized control system where the leader can obtain each follower's measured value with communication. Moreover, another problem also occurs that the leader's speed constraint is no longer determined uniquely and depends on time.

In this paper, based on the study [11], we propose a control method that can solve the problems above. In other words, we propose a decentralized control method in which a single leader robot navigates the other robots, each of which has different capabilities, while maintaining connectivity in 3D space. We assume that each agent has a different sensing range, maximum speed, and maximum acceleration and does not use the communication with other agents.

2 Problem Setting

2.1 Modeling of Agents

We consider a leader-follower navigation problem in 3D space $D \in \mathbb{R}^3$ with no obstacles. The robotic swarm consists of a single leader and n followers, and followers are numbered $1, 2, \ldots, n$ while the leader is numbered $n + 1$. These numbers are set only for the convenience of description, and agents do not recognize them in the actual situation. We define sets of indices of all agents as $\mathcal{A} = \{1, 2, \ldots, n + 1\}$, and of all followers as $\mathcal{F} = \{1, 2, \ldots, n\}$.

We define a position vector and control input for agent $i \in \mathcal{A}$ at time t as $\boldsymbol{x}_i(t)$ and $\boldsymbol{u}_i(t)$ respectively, and its equation of motion is described as $\dot{\boldsymbol{x}}_i(t) = \boldsymbol{u}_i(t)$.

We assume that each follower $i \in \mathcal{F}$ has the following physical constraints about their input vector and its time derivative:

$$\|\boldsymbol{u}_i(t)\| \leq U_i, \ \|\dot{\boldsymbol{u}}_i(t)\| \leq A_i, \ \boldsymbol{u}_i(t): \text{Continuous}. \tag{1}$$

Here, we define $\dot{\boldsymbol{u}}_i(t)$ as the larger of left and right derivative of $\boldsymbol{u}_i(t)$ at time t, so we do not need to think of acceleration constraints. If we consider the flying robot as the agent, the actual flying robot has a limit to speed and acceleration, so we assume that they cannot do such movements that exceed the limit. In addition to that, the input vector should be continuous for controlling the actual robots. The leader agent does not have those constraints above, other than the next section's speed constraint.

We assume that each follower $i \in \mathcal{F}$ can measure with an on-board sensor its relative displacement $\boldsymbol{x}_{ij}(t)$ to any agent $j \in \mathcal{A}$ in the neighborhood area $S_i(t)$, where $\boldsymbol{x}_{ij}(t) := \boldsymbol{x}_j(t) - \boldsymbol{x}_i(t)$ and $S_i(t) = \{\boldsymbol{X}(t) \in D \mid \|\boldsymbol{x}_i(t) - \boldsymbol{X}(t)\| \leq \rho_i, \ \rho_i > 0\}$. Here, ρ_i is the sensing range of agent i and is not necessarily uniform for all agents. We also assume that followers can always get the vertically upward direction $\boldsymbol{e}_z$ on the inertial frame. In addition, we assume that as long as an agent in $S_i(t)$ is staying there, each follower can distinguish it from the other agents inside $S_i(t)$.

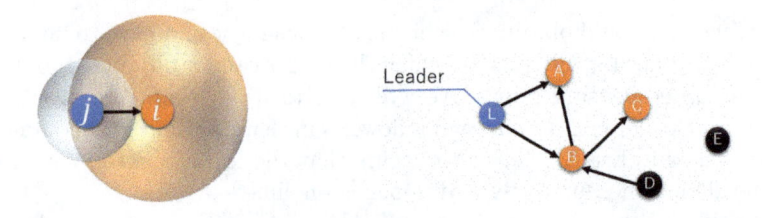

Fig. 1. ij semi-connection.

Fig. 2. Connectivity expression as a directed graph.

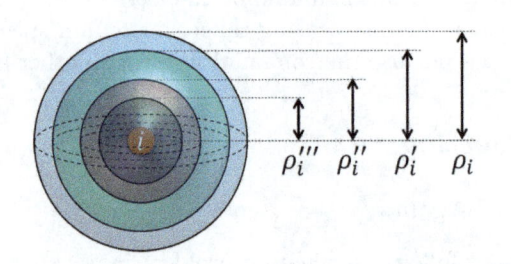

Fig. 3. Sensing radius ρ_i and design parameters ρ'_i, ρ''_i, ρ'''_i for agent i.

2.2 Connectivity of Agents

We use graph theory to describe connectivity between agents [11].

We define a group of nodes as $\mathcal{N} := \{N_1, N_2, \ldots, N_{n+1}\}$, and a group of directed edges between nodes at time t as $\mathcal{E}(t)$. A directed graph can be described $\mathcal{G}(t) = \mathcal{N} \times \mathcal{E}(t)$, and we describe a directional branch from N_j to N_i as $E_{ji} \in \mathcal{E}(t)$. If E_{ji} exists, we call N_j a parent of N_i, and call N_i a child of N_j. If a path can be tracked along directional branches (like E_{kj}, E_{ji}), it is called a directed path. If N_{n+1} has no parent and every node in $\mathcal{N}$ other than N_{n+1} has only one parent, $\mathcal{G}(t)$ is called a spanning tree with a root N_{n+1} We define the following leader semi-connectivity [11].

Definition 1 (Leader Semi-Connectivity). *If E_{ji} exists, $i \in \mathcal{A}$ and $j \in \mathcal{A}$ are ij semi-connected (Fig. 1). If there is a directed path from N_{n+1} to N_i, follower $i \in \mathcal{F}$ is leader semi-connected. A swarm of agents is leader semi-connected if all followers are leader semi-connected. In other words, the directed graph $\mathcal{G}(t)$ is leader semi-connected if $\mathcal{G}(t)$ has a spanning tree with a root N_{n+1}.*

Figure 2 is an example of directed paths for a robot swarm with five followers. Here, followers A, B, C are leader semi-connected while D, E are not.

Next, we add an assumption about the initial position at time $t = 0$. Assumption 1 is an assumption for the initial position of each followers.

Assumption 1 (Initial Placement). *We define a positive constants ρ'_i, ρ''_i, ρ'''_i satisfying $0 < \rho'''_i < \rho''_i < \rho'_i < \rho_i$, where ρ'_i and ρ''_i are criteria for switching the control input and ρ'''_i is equivalent to the lower limit for the distance between* agent i and the agent which agent i is following (Fig. 3). For follower $i \in \mathcal{F}$, we set regions $S_i'(t)$ and $S_i''(t)$ as follows:

$$S_i'(t) := \{\boldsymbol{X}(t) \in D \mid \|\boldsymbol{x}_i(t) - \boldsymbol{X}(t)\| \leq \rho_i'\},$$
$$S_i''(t) := \{\boldsymbol{X}(t) \in D \mid \|\boldsymbol{x}_i(t) - \boldsymbol{X}(t)\| \leq \rho_i''\}.$$

and define their boundary as $\partial S_i'(t), \partial S_i''(t)$, respectively.

We set $B_i(t) := S_i'(t) \setminus \{\partial S_i'(t) \cup S_i''(t)\}$ and assume that all followers have at least one other agent in $B_i(t)$, do not have any other agents in $S_i'''(t)$, and are static ($\boldsymbol{u}_i(0) = \boldsymbol{0}$), at the time $t = 0$. We also assume that at least one follower includes the leader in $B_i(t)$ at $t = 0$.

2.3 Control Objective

In this paper, we propose a decentralized control method and leader's constraint to keep the state where, for any $t \geq 0$, each agent satisfies the physical constraints (1), and every follower is leader semi-connected. The leader is assumed to obtain the followers' properties ($\rho_i, \rho_i', \rho_i'', \rho_i''', U_i, A_i$) off-line in advance, and to move within the speed constraint shown below. Each follower can only use information about itself and its relative position to other agents within its sensing range.

3 Proposed Method

We show a method for creating and maintaining a spanning tree structure of $\mathcal{G}(t)$. In this method, each follower chooses one other agent as a target, and makes efforts to keep semi-connectivity with it, which leads to leader semi-connectivity of all agents. We also give them some degrees of freedom for changing the swarm shape. The proposed method consists of three parts: leader's speed constraint, selection of a target agent, and control inputs for maintaining connectivity.

3.1 Leader's Speed Constraint

In order to maintain leader semi-connectivity, we constrain the leader's speed $\|\boldsymbol{u}_{n+1}(t)\|$ as $\|\boldsymbol{u}_{n+1}(t)\| \leq U_{n+1}$, where U_{n+1} is determined by:

$$U_{n+1} \leq \min_{i \in \mathcal{F}} U_i, \quad U_{n+1} \leq \min_{i \in \mathcal{F}} \sqrt{A_i h_i / \{\sqrt{5}(2 + \sqrt{3})\}} \qquad (2)$$

where $h_i := \min\{\rho_i - \rho_i', \rho_i'' - \rho_i'''\}$. The first constraint is for obeying followers' speed constraints and is derived from the proof of Theorem 2. The second one is for acceleration constraints and arises from the proof of Theorem 4.

3.2 Target Selection Process

Each follower $i \in \mathcal{F}$ chooses as its target the first agent which has passed $\partial S_i'(t)$ or $\partial S_i''(t)$ at $t > 0$. If multiple agents become the target candidates simultaneously, the target will be selected from them randomly. From Assumption 1, no followers start moving earlier than the leader, so the spanning tree of $\mathcal{G}(t)$ is created through this process. We define t_i as the time when agent i has determined its target, and the target does not change after that.

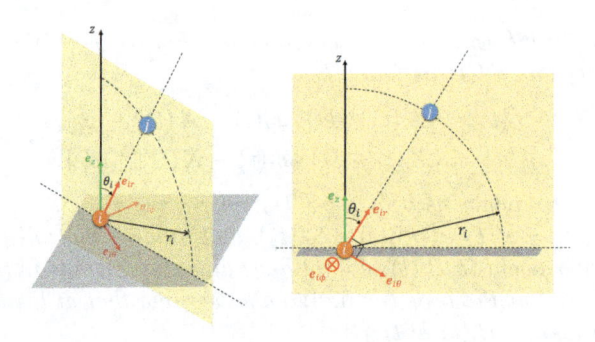

Fig. 4. Local basis vectors $e_{ir}(t), e_{i\theta}(t), e_{i\phi}(t)$ of agent i.

3.3 Control Input

We express the control input $u_i(t)$ to follower $i \in \mathcal{F}$ as follows (Fig. 4):

$$u_i(t) = u_{ir}(t)e_{ir}(t) + u_{i\theta}(t)e_{i\theta}(t) + u_{i\phi}(t)e_{i\phi}(t). \tag{3}$$

Here, $e_{ir}(t)$ is defined as $e_{ir}(t) := x_{ij}(t)/r_i(t)$ for $r_i(t) > 0$, where $r_i(t) := \|x_{ij}(t)\|$ and agent j is the target of agent i. We also define $\theta_i(t)$ as the angle measured from e_z to $e_{ir}(t)$, which ranges from 0 to π. In addition, $e_{i\theta}(t)$ is an unit vector which is on the same plane as $e_{ir}(t)$ and e_z and is orthogonal to $e_{ir}(t)$. Also, $e_{i\theta}(t)$ faces toward the direction in which $\theta_i(t)$ increases. Further, $e_{i\phi}(t)$ is an unit vector which faces such direction that $e_{ir}(t), e_{i\theta}(t), e_{i\phi}(t)$ make a right-handed coordinate system. Note that $e_{i\theta}(t)$ and $e_{i\phi}(t)$ cannot be uniquely determined when $\sin\theta_i(t) = 0$. If $r_i(t) = 0$, we set $e_{ir}(t) = e_{i\theta}(t) = e_{i\phi}(t) = \mathbf{0}$.

We set $u_i(t) = \mathbf{0}$ for $t < t_i$, and design the control input for $t \geq t_i$ as follows:

1. The case of $\sin\theta_i(t) \neq 0$
   (a) $\rho_i''' \leq r_i(t) \leq \rho_i''$:
   $$u_{ir}(t) = a_i'(r_i(t) - \rho_i''), \quad u_{i\theta}(t) = u_{i\phi}(t) = 0 \tag{4}$$

   (b) $\rho_i'' < r_i(t) < (\rho_i'' + \rho_i')/2$:
   $$\begin{aligned} u_{ir}(t) &= 0, \quad u_{i\theta}(t) = 2k_i l_i \sigma_i(t) U_i'(t)(r_i(t) - \rho_i'')\sin\theta_i(t)/(\rho_i' - \rho_i'') \\ u_{i\phi}(t) &= 2\sqrt{1-k_i^2} l_i \sigma_i(t) U_i'(t)(r_i(t) - \rho_i'')\sin\theta_i(t)/(\rho_i' - \rho_i'') \end{aligned} \tag{5}$$

   (c) $(\rho_i'' + \rho_i')/2 \leq r_i(t) < \rho_i'$:
   $$\begin{aligned} u_{ir}(t) &= 0, \quad u_{i\theta}(t) = 2k_i l_i \sigma_i(t) U_i'(t)(\rho_i' - r_i(t))\sin\theta_i(t)/(\rho_i' - \rho_i'') \\ u_{i\phi}(t) &= 2\sqrt{1-k_i^2} l_i \sigma_i(t) U_i'(t)(\rho_i' - r_i(t))\sin\theta_i(t)/(\rho_i' - \rho_i'') \end{aligned} \tag{6}$$

   (d) $\rho_i' \leq r_i(t) < \rho_i' + U_i'(t)/(2a_i)$:
   $$\begin{aligned} u_{ir}(t) &= a_i(r_i(t) - \rho_i'), \quad u_{i\theta}(t) = k_i\sigma_i(t)\sin\theta_i(t)u_{ir}(t) \\ u_{i\phi}(t) &= \sqrt{1-k_i^2}\sigma_i(t)\sin\theta_i(t)u_{ir}(t) \end{aligned} \tag{7}$$

(e) $\rho'_i + U'_i(t)/(2a_i) \leq r_i \leq \rho'_i + U'_i(t)/a_i$:

$$u_{ir}(t) = a_i(r_i(t) - \rho'_i), \quad u_{i\theta}(t) = k_i\sigma_i(t)\sin\theta_i(t)(U'_i(t) - u_{ir}(t))$$
$$u_{i\phi}(t) = \sqrt{1 - k_i^2}\sigma_i(t)\sin\theta_i(t)(U'_i(t) - u_{ir}(t)) \tag{8}$$

2. The case of $\sin\theta_i(t) = 0$

  Though $e_{i\theta}(t)$ and $e_{i\phi}(t)$ cannot be uniquely determined, we do not have to consider the direction of these basis vectors when we set $u_{i\theta}(t) = u_{i\phi}(t) = 0$. Then (3) gives $u_i(t) = u_{ir}(t)e_{ir}(t)$, and we set $u_{ir}(t)$ the same as (4)–(8).

  Here, $a_i := V_i/(\rho_i - \rho'_i)$, $a'_i := V_i/(\rho''_i - \rho'''_i)$, $k_i, \sigma_i(t) \in [-1, 1]$, $U'_i(t) := \max_{0 \leq \tau \leq t}|u_{ir}(\tau)|$. Further, l_i and V_i are defined as follows:

$$l_i := \min\left\{1, \frac{(\rho'_i - \rho''_i)A_i}{2\sqrt{9 + 2\sqrt{6}V_i^2}}\right\}, \quad V_i := \min\left\{U_i, \sqrt{\frac{A_i h_i}{\sqrt{5}(2 + \sqrt{3})}}\right\}. \tag{9}$$

In (9), V_i corresponds to the more stringent speed constraint imposed on the leader. $k_i \in [-1, 1]$ is an arbitrary constant which can be set for each agent, and determines the length of the input components in the direction of $e_{i\theta}$ and $e_{i\phi}$.

From the definition of $u_{ir}(t)$ when $r_i(t) \geq \rho'_i$, we obtain $r_i(t) = \rho'_i + u_{ir}(t)/a_i$. Since $a_i > 0$ and $u_{ir}(t) \leq U'_i(t)$, $r_i(t) \leq \rho'_i + U'_i(t)/a_i$ holds as long as ij semi-connectivity is kept. Then, since $r_i(t) > \rho'''_i$ also holds, we just need to consider the input for $\rho'''_i \leq r_i(t) \leq \rho'_i + U'_i(t)/a_i$. The parameter $\sigma_i(t) \in [-1, 1]$ represents how wide agents spread. The larger $|\sigma_i(t)|$ is, the wider they spread.

3.4 What Is Guaranteed by the Proposed Method

Under Assumption 1, when each agents moves according to the mentioned method, all followers are leader semi-connected for all $t \geq 0$, satisfy the physical constraints (1) for all $t \geq t_i$, and keep semi-connectivity to their target.

4 Mathematical Proof

Now, we prove that each follower keeps semi-connectivity with its target and satisfies its physical constraints (1) under the control method in Sect. 3. We assume that agent i's target is j and agent j's velocity vector is expressed as $u_j(t) = u_{jr}(t)e_{ir}(t) + u_{j\theta}(t)e_{i\theta}(t) + u_{j\phi}(t)e_{i\phi}(t)$ using agent i's basis vectors.

Theorem 1 (Semi-connectivity with the target). *Suppose Assumption 1 holds and $u_i(t)$ is set according to (3)–(8). If $\|u_j(t)\| \leq U_{n+1}$, then $\rho''_i - U_{n+1}/a'_i \leq r_i(t) \leq \rho'_i + U_{n+1}/a_i$.*

Proof: This can be proved in the same way as Theorem 1 in [12]. □

Theorem 2 (Maximum speed). *Suppose $u_i(t)$ is set according to (3)–(8). If $\|u_j(t)\| \leq U_{n+1}$, then $\|u_i(t)\| < U_{n+1}$.*

Proof: If $t < t_i$, the theorem clearly holds since $\boldsymbol{u}_i(t) = \boldsymbol{0}$. We discuss the case $t \geq t_i$. From Theorem 1, $\rho'' - U_{n+1}/a' \leq r \leq \rho' + U_{n+1}/a$ holds and, combining it with the control inputs (4)–(8), we can say that the velocity in the target direction u_{ir} satisfies $|u_{ir}| \leq U_{n+1}$.

1. The case of $\sin\theta = 0$ or $\rho'' - U_{n+1}/a' < r \leq \rho''$:
   Since $u_{i\theta} = u_{i\phi} = 0$, $\|\boldsymbol{u}_i\| = |u_{ir}| < U_{n+1}$ holds.
2. The case of $\sin\theta \neq 0$ and $\rho'' < r < \rho' + U'/a$:
   (a) $\rho'' < r < (\rho'' + \rho')/2$: From $0 < (r - \rho'')/(\rho' - \rho'') < 1/2$, $|l| \leq 1$, and $|\sigma| \leq 1$, the following holds:

$$\|\boldsymbol{u}_i\|^2 = u_{i\theta}^2 + u_{i\phi}^2 = 4l^2\sigma^2 U'^2 \left(\frac{r - \rho''}{\rho' - \rho''}\right)^2 \sin^2\theta < U'^2.$$

Because of $|u_{ir}| \leq U_{n+1}$ and $U'_i(t) := \max_{0 \leq \tau \leq t} |u_{ir}(\tau)|$, we obtain $U'_i(t) \leq U_{n+1}$. Therefore, $\|\boldsymbol{u}_i\| < U_{n+1}$ holds.

   (b) $(\rho'' + \rho')/2 \leq r < \rho'$: From $0 < (\rho' - r)(\rho' - \rho'') \leq 1/2$, $|l| \leq 1$, and $|\sigma| \leq 1$, the following holds:

$$\|\boldsymbol{u}_i\|^2 = u_{i\theta}^2 + u_{i\phi}^2 = 4l^2\sigma^2 U'^2 \left(\frac{\rho' - r}{\rho' - \rho''}\right)^2 \sin^2\theta \leq U'^2.$$

Using $U' < U_{n+1}$, we obtain $\|\boldsymbol{u}_i\| < U_{n+1}$.

   (c) $\rho' \leq r < \rho' + U'/(2a)$: From (7), $|u_{ir}| < U'/2$ holds. Using $U' < U_{n+1}$, we obtain $|u_{ir}| < U_{n+1}/2$. From $|\sigma| \leq 1$, the following holds:

$$\|\boldsymbol{u}_i\|^2 = u_{ir}^2 + u_{i\theta}^2 + u_{i\phi}^2 = (1 + \sigma^2 \sin^2\theta) u_{ir}^2 \leq 2u_{ir}^2 < U_{n+1}^2/2$$

Therefore, $\|\boldsymbol{u}_i\| < U_{n+1}$ holds.

   (d) $\rho' + U'/(2a) \leq r < \rho' + U'/a$: From (8), $U'/2 \leq u_{ir} < U'$ holds. From $|\sigma| \leq 1$, the following holds:

$$\|\boldsymbol{u}_i\|^2 = u_{ir}^2 + \sigma^2 \sin^2\theta (U' - u_{ir})^2 \leq u_{ir}^2 + (U' - u_{ir})^2$$
$$= 2\left(u_{ir} - \frac{U'}{2}\right)^2 + \frac{U'^2}{2} < U'^2$$

Combining it with $U' < U_{n+1}$, $\|\boldsymbol{u}_i\| < U_{n+1}$ holds.

Therefore, it is proved that if $\|\boldsymbol{u}_j(t)\| \leq U_{n+1}$, $\|\boldsymbol{u}_i(t)\| < U_{n+1}$ holds. □

Theorem 3 (Continuity of the control input). *Suppose Assumption 1 holds and $\boldsymbol{u}_i(t)$ is set according to (3)–(8). If $\|\boldsymbol{u}_j(t)\| \leq U_{n+1}$, then $\boldsymbol{u}_i(t)$ is continuous for any t.*

Proof: If $t < t_i$, $\boldsymbol{u}_i$ is always zero and the theorem clearly holds. We discuss the case $t \geq t_i$.

1. The case of $\sin\theta \neq 0$:
   Since $\|\boldsymbol{u}_j\| \leq U_{n+1}$, $|u_{jr}| \leq U_{n+1}$ and $|u_{ir}| \leq U_{n+1}$ holds. Hence $\dot{r} = u_{jr} - u_{ir}$ is bounded and then r is continuous for any t. Because $u_{ir}, u_{i\theta}, u_{i\phi}$ are continuous for r, these are also continuous for t.

   The basis vectors $\boldsymbol{e}_{ir}, \boldsymbol{e}_{i\theta}, \boldsymbol{e}_{i\phi}$ change according to $\dot\theta_i$ and $\dot\phi_i$, which are the relative velocity between agents i and j in the directions of $\boldsymbol{e}_{i\theta}$ and $\boldsymbol{e}_{i\phi}$ respectively. Here, $\dot\phi_i$ corresponds to the angular velocity around the Z-axis (See Fig. 4), and define the counterclockwise direction (viewed from $+Z$ direction) as positive.

   Using $r > 0$ and $\sin\theta \neq 0$, $\dot\theta$ and $\dot\phi$ is expressed as $\dot\theta = (u_{j\theta} - u_{i\theta})/r$, $\dot\phi = (u_{j\phi} - u_{i\phi})/(r\sin\theta)$. Since $\|\boldsymbol{u}_j\| \leq U_{n+1}$ and $\|\boldsymbol{u}_i\| \leq U_{n+1}$, they can be defined as bounded values. Therefore, the time derivative of basis vectors $\dot{\boldsymbol{e}}_{ir} = \dot\theta \boldsymbol{e}_{i\theta} + \dot\phi \sin\theta \boldsymbol{e}_{i\phi}$, $\dot{\boldsymbol{e}}_{i\theta} = -\dot\theta \boldsymbol{e}_{ir} + \dot\phi \cos\theta \boldsymbol{e}_{i\phi}$, $\dot{\boldsymbol{e}}_{i\phi} = -\dot\phi(\sin\theta \boldsymbol{e}_{ir} + \cos\theta \boldsymbol{e}_{i\theta})$ are also bounded values, so $\boldsymbol{e}_{ir}, \boldsymbol{e}_{i\theta}, \boldsymbol{e}_{i\phi}$ is continuous for any t.

   From the discussion above, $\dot{\boldsymbol{u}}_i = \dot{u}_{ir}\boldsymbol{e}_{ir} + u_{ir}\dot{\boldsymbol{e}}_{ir} + \dot{u}_{i\theta}\boldsymbol{e}_{i\theta} + u_{i\theta}\dot{\boldsymbol{e}}_{i\theta} + \dot{u}_{i\phi}\boldsymbol{e}_{i\phi} + u_{i\phi}\dot{\boldsymbol{e}}_{i\phi}$ can be defined as bounded values and $\boldsymbol{u}_i(t)$ is continuous for any t.

2. The case of $\sin\theta = 0$:
   Since $u_{i\theta} = 0$ and $u_{i\phi} = 0$, $\dot{\boldsymbol{u}}_i$ is expressed as $\dot{\boldsymbol{u}}_i = \dot{u}_{ir}\boldsymbol{e}_{ir} + u_{ir}\dot{\boldsymbol{e}}_{ir} + \dot{u}_{i\theta}\boldsymbol{e}_{i\theta} + \dot{u}_{i\phi}\boldsymbol{e}_{i\phi}$, where $\dot{u}_{i\theta}$ and $\dot{u}_{i\phi}$ derives from (4)–(8).

   Using $u_{i\theta} = 0$ and $\sin\theta = 0$, $\dot{\boldsymbol{e}}_{ir}$ is expressed as $\dot{\boldsymbol{e}}_{ir} = (u_{j\theta}/r)\boldsymbol{e}_{i\theta}$, hence we obtain $\dot{\boldsymbol{u}}_i = \dot{u}_{ir}\boldsymbol{e}_{ir} + \{(u_{ir}u_{j\theta})/r + \dot{u}_{i\theta}\}\boldsymbol{e}_{i\theta} + \dot{u}_{i\phi}\boldsymbol{e}_{i\phi}$.

   Here, u_{ir}, $u_{i\theta}$, and $u_{i\phi}$ are continuous and bounded for any t regardless of whether $\sin\theta$ goes to zero. Since $|u_{j\theta}| \leq U_{n+1}$, $\dot{\boldsymbol{u}}_i(t)$ can be defined as a bounded value regardless of the directions of $\boldsymbol{e}_{i\theta}$ and $\boldsymbol{e}_{i\phi}$, therefore $\boldsymbol{u}_i(t)$ is continuous for any t.

In conclusion, it is proved that $\boldsymbol{u}_i(t)$ is continuous for any t if $\|\boldsymbol{u}_j(t)\| \leq U_{n+1}$. □

Theorem 4 (Maximum Acceleration). *Suppose Assumption 1 holds, $\boldsymbol{u}_i(t)$ is set according to (3)–(8), and $\sigma(t)$ is constant. If $\|\boldsymbol{u}_j(t)\| \leq U_{n+1}$, then $\|\dot{\boldsymbol{u}}_i(t)\| < A_i$.*

Proof: If $t < t_i$, $\boldsymbol{u}_i$ is always zero and the theorem clearly holds. We discuss the case $t \geq t_i$. When $\sin\theta \neq 0$, the acceleration vector $\dot{\boldsymbol{u}}_i$ is expressed as $\dot{\boldsymbol{u}}_i = (\dot{u}_{ir} - \dot\theta u_{i\theta} - \dot\phi \sin\theta u_{i\phi})\boldsymbol{e}_{ir} + (\dot\theta u_{ir} + \dot{u}_{i\theta} - \dot\phi \cos\theta u_{i\phi})\boldsymbol{e}_{i\theta} + (\dot\phi \sin\theta u_{ir} + \dot\phi \cos\theta u_{i\theta} + \dot{u}_{i\phi})\boldsymbol{e}_{i\phi}$.

1. The case of $\sin\theta \neq 0$ and $\rho'' - U_{n+1}/a' < r(t) \leq \rho''$:

$$\|\dot{\boldsymbol{u}}_i\|^2 = \dot{u}_{ir}^2 + (\dot\theta^2 + \dot\phi^2 \sin^2\theta)u_{ir}^2 = a'^2 \dot{r}^2 + \frac{u_{j\theta}^2 + u_{j\phi}^2}{r^2}a'^2(r - \rho'')^2$$

$$\leq a'^2\{(u_{jr} - u_{ir})^2 + \frac{(r-\rho'')^2}{r^2}(U_{n+1}^2 - |u_{jr}|^2)\}.$$

Since $(u_{jr} - u_{ir})^2 \leq (|u_{jr}| + U_{n+1})^2$ and $(r-\rho'')^2/r^2 \leq (\rho''/\rho''' - 1)^2 := c (> 0)$, we get the following:

$$\|\dot{\boldsymbol{u}}_i\|^2 \leq a'^2 \{(|u_{jr}| + U_{n+1})^2 + c(U_{n+1}^2 - |u_{jr}|^2)\}$$
$$= a'^2 \{(1-c)|u_{jr}|^2 + 2U_{n+1}|u_{jr}| + (1+c)U_{n+1}^2\}.$$

For $0 \leq |u_{jr}| \leq U_{n+1}$, $\|\dot{u}_i\|^2$ gets the maximum value $4a'^2U_{n+1}^2$ at $|u_{jr}| = U_{n+1}$. Since $V_i < \sqrt{Ah/2}$, $U_{n+1} \leq V_i$, and $h \leq \rho'' - \rho'''$, it is proved that $\|\dot{u}_i\|^2 \leq 4a'^2U_{n+1}^2 < A^2$.

2. The case of $\sin\theta \neq 0$ and $\rho'' < r(t) < (\rho'' + \rho')/2$:

$$\dot{u}_{i\theta} = kl\sigma U' \frac{2}{\rho' - \rho''}\left\{u_{jr}\sin\theta + \frac{r - \rho''}{r}(u_{j\theta} - u_{i\theta})\cos\theta\right\}$$

$$\dot{u}_{i\phi} = \sqrt{1 - k^2}l\sigma U' \frac{2}{\rho' - \rho''}\left\{u_{jr}\sin\theta + \frac{r - \rho''}{r}(u_{j\theta} - u_{i\theta})\cos\theta\right\}.$$

Calculating $\|\dot{u}_i\|^2$ with the formulae above and evaluating it with $(r - \rho'')^2/r^2 < 1$, $|k| \leq 1$, and $|\sigma| \leq 1$, we get the following:

$$\|\dot{u}_i\|^2 < \left(\frac{2lU'}{\rho' - \rho''}\right)^2 \{u_{jr}^2 + (u_{j\theta} - u_{i\theta})^2(|u_{j\theta} - u_{i\theta}| + |u_{j\phi} - u_{i\phi}|)^2\}$$

$$\leq \left(\frac{2lU'}{\rho' - \rho''}\right)^2 \{(u_{jr}^2 + u_{j\theta}^2) + u_{i\theta}^2$$

$$+ 2|u_{j\theta}||u_{i\theta}| + (|u_{j\theta}| + |u_{j\phi}| + |u_{i\theta}| + |u_{i\phi}|)^2\}.$$

Since $\|u_j\| \leq U_{n+1}$ and $\|u_i\| \leq U_{n+1}$, we get $|u_{jr}| + |u_{j\theta}| + |u_{j\phi}| \leq \sqrt{3}U_{n+1}$ and $|u_{i\theta}| + |u_{i\phi}| < \sqrt{2}U_{n+1}$. Using them, we can further evaluate $\|\dot{u}_i\|^2$ as $\|\dot{u}_i\|^2 < (2lU'/(\rho' - \rho''))^2\{4 + (\sqrt{3} + \sqrt{2})^2\}U_{n+1}^2 = (9 + 2\sqrt{6})(2lU'U_{n+1}/(\rho' - \rho''))^2$. Using (9) and $U' < U_{n+1} \leq V_i$, $\|\dot{u}_i\|^2 < A^2$ holds.

For the other cases (including the case $\sin\theta = 0$), we can also prove $\|\dot{u}_i\|^2 < A^2$ in the same way of evaluation as the two cases above. □

From Theorem 1–4, it is shown that the followers can keep leader semi-connected and satisfy all of their physical constraints (1).

5 Simulation

We show the simulation results for the proposed navigation method. We set the number of followers as $n = 10$, sampling period as 0.001 s, and $\sigma_i(t)$ as constants. We set the follower parameters as shown in Table 1. Here, the leader's speed constraints derived from (1) give $U_{n+1} = 0.6000$ m/s.

In this simulation, we consider a situation where a leader's trajectory includes drastic turns, and confirm that followers satisfy their physical constraints. The leader moves at its max speed U_{n+1} [m/s]. The leader starts from $X = Y = Z = 0$ m, and change its direction at 20 s, 50 s, 120 s, 160 s, and stops at 190 s.

Each agent's trajectory is shown in Fig. 5(a). The black line corresponds to the leader's trajectory. Figure 5(b), 5(c), and 5(d) show the results of the simulation, each of which corresponds to the values of distance indicators $r_i(t)/\rho_i$, speed indicators $\|u_i(t)\|/U_{n+1}$, acceleration indicators $\|\dot{u}_i(t)\|/A_i$, and θ_i for each follower, respectively. From Fig. 5(b) we can say that all followers successfully keep

Table 1. Follower parameters

i	1	2	3	4	5	6	7	8	9	10
ρ_i [m]	9.0	6.0	8.0	8.0	7.0	7.0	8.0	8.0	7.0	9.0
ρ'_i [m]	7.0	4.0	5.5	5.5	5.0	5.5	6.0	5.5	5.0	7.5
ρ''_i [m]	4.0	3.0	4.0	4.5	3.0	3.0	4.0	3.5	3.0	4.0
ρ'''_i [m]	2.0	1.0	1.5	2.0	1.0	1.5	2.0	1.0	1.0	2.5
U_i [m/s]	0.6	0.7	0.8	0.6	0.7	0.8	0.6	0.7	0.8	0.6
A_i [m/s^2]	2.0	2.0	1.5	1.5	2.0	2.5	2.0	1.5	2.0	2.5
k_i	0.7	-0.9	0.5	-0.3	0.7	0.1	-0.5	-0.9	0.3	-0.7
σ_i	1	-1	-1	1	-1	-1	-1	1	1	-1

(a) Agents' path.

(b) Distance from the target.

(c) Followers' speed.

(d) Followers' acceleration.

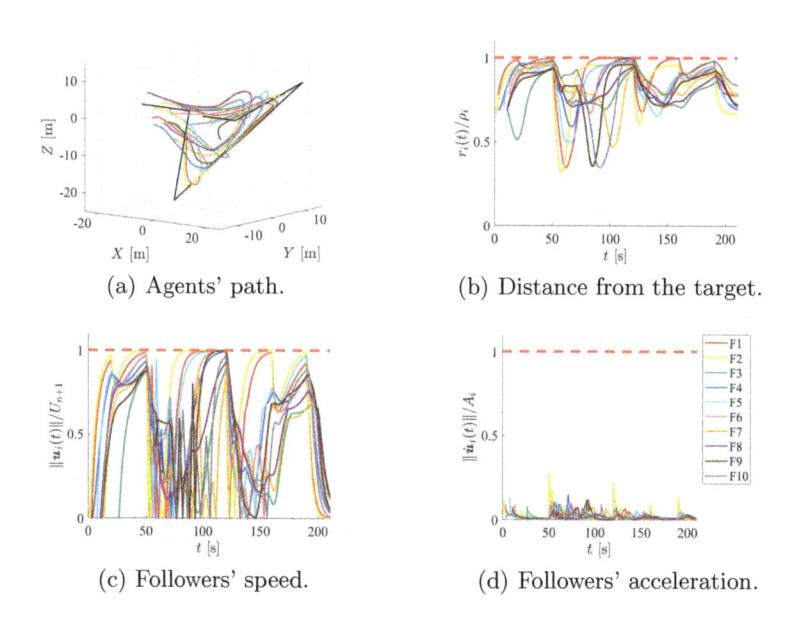

Fig. 5. Simulation results.

semi-connectivity with their target. In Fig. 5(d), some followers' acceleration values get larger values when the leader changes direction (especially at 50 s and 120 s) or stops moving, but Fig. 5(c) and 5(d) show that all followers still satisfy their physical constraints.

6 Conclusion

This paper proposed a control method for navigating robots with heterogeneous capabilities by a single leader in 3D space. In the proposed method, agents create a spanning tree by choosing the target agent to keep connectivity in a distributed way, and maintain the tree by keeping local connectivity with their target. This control method is decentralized, as each follower determines its action only using the local information. We have mathematically proved that leader semi-connectivity of the whole swarm is guaranteed for any leader's motion (under the leader's constraint) in 3D space with no obstacles, and confirmed the effectiveness of the proposed method by numerical simulation. For future work, we will try to show the proposed method can be applied to the actual environment by experiment. In addition, we hope that we could improve the control method for obstacle avoidance or inter-agent collision avoidance.

Acknowledgments. This work was supported in part by JST SICORP Grant Number JPMJSC18E4, Japan.

References

1. Brambilla, M., Ferrante, E., Birattari, M., Dorigo, M.: Swarm robotics: a review from the swarm engineering perspective. Swarm Intell. **7**(1), 1–41 (2013). https://doi.org/10.1007/s11721-012-0075-2
2. Navarro, I., Matia, F.: A survey of collective movement of mobile robots. Int. J. Adv. Robot. Syst. **10**, 1–9 (2013)
3. de Almeida, J.P.L.S., Nakashima, R.T., Neves-Jr, F., de Arruda, L.V.R.: Bio-inspired on-line path planner for cooperative exploration of unknown environment by a multi-robot system. Robot. Auton. Syst. **112**, 32–48 (2019)
4. Bechlioulis, C.P., Kyriakopoulos, K.J.: Collaborative multi-robot transportation in obstacle-cluttered environments via implicit communication. Front. Robot. AI **5**, 1–17 (2018)
5. Teruel, E., Aragues, R., Lopez-Nicolas, G.: A distributed robot swarm control for dynamic region coverage. Robot. Auton. Syst. **119**, 51–63 (2019)
6. Cardona, G.A., Calderon, J.M.: Robot swarm navigation and victim detection using rendezvous consensus in search and rescue operations. Appl. Sci. **9**(8), 1702 (2019)
7. Vilca, J., Adouane, L., Mezouar, Y.: Stable and flexible multi-vehicle navigation based on dynamic inter-target distance matrix. IEEE Trans. Intell. Transp. Syst. **20**(4), 1416–1431 (2019)
8. Dorigo, M., Floreano, D., Gambardella, L.M., Mondada, F., Nolfi, S., Baaboura, T.: Swarmanoid: a novel concept for the study of heterogeneous robotic swarms. IEEE Robot. Autom. Mag. **20**(4), 60–71 (2013)
9. Ferreira-Filho, E.B., Pimenta, L.C.A.: Abstraction based approach for segregation in heterogeneous robotic swarms. Robot. Auton. Syst. **122**, 103295 (2019)
10. Sabattini, L., Secchi, C., Chopra, N.: Decentralized estimation and control for preserving the strong connectivity of directed graphs. IEEE Trans. Cybern. **45**(10), 2273–2286 (2015)
11. Yoshimoto, M., Endo, T., Maeda, R., Matsuno, F.: Decentralized navigation method for a robotic swarm with nonhomogeneous abilities. Auton. Robot. **42**(8), 1583–1599 (2018). https://doi.org/10.1007/s10514-018-9774-x
12. Maeda, R., Endo, T., Matsuno, F.: Decentralized navigation for heterogeneous swarm robots with limited field of view. IEEE Robot. Autom. Lett. **2**(2), 904–911 (2017)
13. Endo, T., Maeda, R., Matsuno, F.: Stability analysis of swarm heterogeneous robots with limited field of view. Inform. Autom. **19**(5), 942–966 (2019)

Using Reinforcement Learning to Herd a Robotic Swarm to a Target Distribution

Zahi Kakish[1]([✉]), Karthik Elamvazhuthi[2], and Spring Berman[1]

[1] Arizona State University, Tempe, AZ 85281, USA
{zahi.kakish,spring.berman}@asu.edu
[2] University of California, Los Angeles, CA 90095, USA
karthikevaz@math.ucla.edu

Abstract. In this paper, we present a reinforcement learning approach to designing a control policy for a "leader" agent that herds a swarm of "follower" agents, via repulsive interactions, as quickly as possible to a target probability distribution over a strongly connected graph. The leader control policy is a function of the swarm distribution, which evolves over time according to a mean-field model in the form of an ordinary difference equation. The dependence of the policy on agent populations at each graph vertex, rather than on individual agent activity, simplifies the observations required by the leader and enables the control strategy to scale with the number of agents. Two Temporal-Difference learning algorithms, SARSA and Q-Learning, are used to generate the leader control policy based on the follower agent distribution and the leader's location on the graph. A simulation environment corresponding to a grid graph with 4 vertices was used to train and validate the control policies for follower agent populations ranging from 10 to 1000. Finally, the control policies trained on 100 simulated agents were used to successfully redistribute a physical swarm of 10 small robots to a target distribution among 4 spatial regions.

Keywords: Swarm robotics · Graph theory · Mean-field model · Reinforcement learning

1 Introduction

We present two Temporal-Difference learning algorithms [1] for generating a control policy that guides a mobile agent, referred to as a *leader*, to herd a swarm of autonomous *follower* agents to a target distribution among a small set of states. This leader-follower control approach can be used to redistribute a swarm of low-cost robots with limited capabilities and information using a single robot with sophisticated sensing, localization, computation, and planning capabilities, in scenarios where the leader lacks a model of the swarm dynamics. Such a control strategy is useful for many applications in swarm robotics, including exploration, environmental monitoring, inspection tasks, disaster response, and targeted drug delivery at the micro-nanoscale.

There has been a considerable amount of work on leader-follower multi-agent control schemes in which the leader has an attractive effect on the followers [2,3]. Several recent works have presented models for herding robotic swarms using leaders that have a repulsive effect on the swarm [4–6]. Using such models, analytical controllers for herding a swarm have been constructed for the case when there is a single leader [5,6] and multiple leaders [4]. The controllers designed in these works are not necessarily optimal for a given performance metric. To design optimal control policies for a herding model, the authors in [7] consider a reinforcement learning (RL) approach. While existing herding models are suitable for the objective of confining a swarm to a small region in space, many applications require a swarm to cover an area according to some target probability density. If the robots do not have spatial localization capabilities, then the controllers developed in [2–7] cannot be applied for such coverage problems. Moreover, these models are not suitable for herding large swarms using RL-based control approaches, since such approaches would not scale well with the number of robots. This loss of scalability is due to the fact that the models describe individual agents, which may not be necessary since robot identities are not important for many swarm applications.

In this paper, we consider a *mean-field* or *macroscopic* model that describes the swarm of follower agents as a probability distribution over a graph, which represents the configuration space of each agent. Previous work has utilized similar mean-field models to design a set of control policies that is implemented on each robot in a swarm in order to drive the entire swarm to a target distribution, e.g. for problems in spatial coverage and task allocation [8]. In this prior work, all the robots must be reprogrammed with a new set of control policies if the target distribution is changed. In contrast, our approach can achieve new target swarm distributions via redesign of the control policy of a single leader agent, while the control policies of the swarm agents remain fixed. The follower agents switch stochastically out of their current location on the graph whenever the leader is at their location; in this way, the leader has a "repulsive" effect on the followers. The transition rates out of each location are common to all the followers, and are therefore independent of the agents' identities. Using the mean-field model, herding objectives for the swarm are framed in terms of the distribution of the followers over the graph. The objective is to compute leader control policies that are functions of the agent distribution, rather than the individual agents' states, which makes the control policies scalable with the number of agents.

We apply RL-based approaches to the mean-field model to construct leader control policies that minimize the time required for the swarm of follower agents to converge to a user-defined target distribution. The RL-based control policies are not hindered by curse-of-dimensionality issues that arise in classical optimal control approaches. Additionally, RL-based approaches can more easily accommodate the stochastic nature of the follower agent transitions on the graph. There is prior work on RL-based control approaches for mean-field models of swarms in which each agent can localize itself in space and a state-dependent control policy can be assigned to each agent directly [9–11]. However, to our knowledge, there is no existing work on RL-based approaches applied to mean-field models for herding a swarm using a leader agent. Our approach provides an RL-based framework for designing scalable strategies to control swarms of resource-constrained robots using a single leader robot, and it can be extended to other types of swarm control objectives.

2 Methodology

2.1 Problem Statement

We first define some notation from graph theory and matrix analysis that we use to formally state our problem. We denote by $\mathcal{G} = (\mathcal{V}, \mathcal{E})$ a directed graph with a set of M vertices, $\mathcal{V} = \{1, ..., M\}$, and a set of $N_\mathcal{E}$ edges, $\mathcal{E} \subset \mathcal{V} \times \mathcal{V}$, where $e = (i, j) \in \mathcal{E}$ if there is an edge from vertex $i \in \mathcal{V}$ to vertex $j \in \mathcal{V}$. We define a *source map* $\sigma : \mathcal{E} \rightarrow \mathcal{V}$ and a *target map* $\tau : \mathcal{E} \rightarrow \mathcal{V}$ for which $\sigma(e) = i$ and $\tau(e) = j$ whenever $e = (i, j) \in \mathcal{E}$. Given a vector $X \in \mathbb{R}^M$, X_i refers to the ith coordinate value of X. For a matrix $A \in \mathbb{R}^{M \times N}$, A^{ij} refers to the element in the ith row and j^{th} column of A.

We consider a finite swarm of N follower agents and a single leader agent. The locations of the leader and followers evolve on a graph, $\mathcal{G} = (\mathcal{V}, \mathcal{E})$, where $\mathcal{V} = \{1, ..., M\}$ is a finite set of vertices and $\mathcal{E} = \{(i, j) \mid i, j \in \mathcal{V}\}$ is a set of edges that define the pairs of vertices between which agents can transition. The vertices in $\mathcal{V}$ represent a set of spatial locations obtained by partitioning the agents' environment. We will assume that the graph $\mathcal{G} = (\mathcal{V}, \mathcal{E})$ is strongly connected and that there is a self-edge $(i, i) \in \mathcal{E}$ at every vertex $i \in \mathcal{V}$. We assume that the leader agent can count the number of follower agents at each vertex in the graph. The follower agents at a location v only decide to move to an adjacent location if the leader agent is currently at location v and is in a particular behavioral state. In other words, the presence of the leader *repels* the followers at the leader's location. The leader agent does not have a model of the follower agents' behavior.

The leader agent performs a sequence of transitions from one location (vertex) to another. The leader's location at time $k \in \mathbb{Z}_+$ is denoted by $\ell_1(k) \in \mathcal{V}$. In addition to the spatial state $\ell_1(k)$, the leader has a behavioral state at each time k, defined as $\ell_2(k) \in \{0, 1\}$. The location of each follower agent $i \in \{1, ..., N\}$ is defined by a discrete-time Markov chain (DTMC) $X_i(k)$ that evolves on the state space $\mathcal{V}$ according to the conditional probabilities

$$\mathbb{P}(X_i(k+1) = \tau(e) \mid X_i(k) = \sigma(e)) = u_e(k) \quad (1)$$

For each $v \in \mathcal{V}$ and each $e \in \mathcal{E}$ such that $\sigma(e) = v \neq \tau(e)$, $u_e(k)$ is given by

$$u_e(k) = \begin{cases} \beta_e & \text{if } \ell_1(k) = \sigma(e) \text{ and } \ell_2(k) = 1, \\ 0 & \text{if } \ell_1(k) = \sigma(e) \text{ and } \ell_2(k) = 0, \\ 0 & \text{if } \ell_1(k) \neq \sigma(e), \end{cases} \quad (2)$$

where β_e are positive parameters such that $\sum_{\substack{e \in \mathcal{E} \\ v=\sigma(e) \neq \tau(e)}} \beta_e < 1$. Additionally, for each $v \in \mathcal{V}$, $u_{(v,v)}(k)$ is given by

$$u_{(v,v)}(k) = 1 - \sum_{\substack{e \in \mathcal{E} \\ v=\sigma(e) \neq \tau(e)}} u_e(k) \tag{3}$$

For each vertex $v \in \mathcal{V}$, we define a set of possible actions A_v taken by the leader when it is located at v:

$$A_v = \bigcup_{\substack{e \in \mathcal{E} \\ v=\sigma(e)}} \{e\} \times \{0,1\} \tag{4}$$

The leader transitions between states in $\mathcal{V} \times \{0,1\}$ according to the conditional probabilities

$$\mathbb{P}(\ell_1(k+1) = \tau(e), \ell_2(k+1) = d \mid \ell_1(k) = \sigma(e)) = 1 \tag{5}$$

if $p(k)$, the action taken by the leader at time k when it is at vertex v, is given by $p(k) = (e, d) \in A_v$.

The fraction, or *empirical distribution*, of follower agents that are at location $v \in \mathcal{V}$ at time k is given by $\frac{1}{N}\sum_{i=1}^{N} \chi_v(X_i(k))$, where $\chi_v(w) = 1$ if $w = v$ and 0 otherwise. Our goal is to learn a policy that navigates the leader between vertices using the actions $p(k)$ such that the follower agents are redistributed ("herded") from their initial empirical distribution $\frac{1}{N}\sum_{i=1}^{N} \chi_v(X_i(0))$ among the vertices to a desired empirical distribution $\frac{1}{N}\sum_{i=1}^{N} \chi_v(X_i(T))$ at some final time T, where T is as small as possible. Since the identities of the follower agents are not important, we aim to construct a control policy for the leader that is a function of the current empirical distribution $\frac{1}{N}\sum_{i=1}^{N} \chi_v(X_i(k))$, rather than the individual agent states $X_i(k)$. However, $\frac{1}{N}\sum_{i=1}^{N} \chi_v(X_i(k))$ is not a state variable of the DTMC. In order to treat $\frac{1}{N}\sum_{i=1}^{N} \chi_v(X_i(k))$ as the state, we consider the *mean-field limit* of this quantity as $N \to \infty$. Let $\mathcal{P}(\mathcal{V}) = \{Y \in \mathbb{R}_{\geq 0}^M; \sum_{v=1}^M Y_v = 1\}$ be the simplex of probability densities on $\mathcal{V}$. When $N \to \infty$, the empirical distribution $\frac{1}{N}\sum_{i=1}^{N} \chi_v(X_i(k))$ converges to a deterministic quantity $\hat{S}(k) \in \mathcal{P}(\mathcal{V})$, which evolves according to the following *mean-field model*, a system of difference equations that define the discrete-time Kolmogorov Forward Equation:

$$\hat{S}(k+1) = \sum_{e \in \mathcal{E}} u_e(k) B_e \hat{S}(k), \quad \hat{S}(0) = \hat{S}^0 \in \mathcal{P}(\mathcal{V}), \tag{6}$$

where B_e are matrices whose entries are given by

$$B_e^{ij} = \begin{cases} 1 & \text{if } i = \tau(e),\ j = \sigma(e), \\ 0 & \text{otherwise.} \end{cases}$$

The random variable $X_i(k)$ is related to the solution of the difference Equation (6) by the relation $\mathbb{P}(X_i(k) = v) = \hat{S}_v(k)$.

We formulate an optimization problem that minimizes the number of time steps k required for the follower agents to converge to $\hat{S}_{target}$, the target distribution. In this optimization problem, the reward function is defined as

$$R(k) = -1 \cdot \mathbb{E}||\hat{S}(k) - \hat{S}_{target}||^2. \tag{7}$$

Problem 1. Given a target follower agent distribution $\hat{S}_{target}$, devise a leader control policy $\pi : \mathcal{P}(\mathcal{V}) \times \mathcal{V} \to A$ that drives the follower agent distribution to $\hat{S}(T) = \hat{S}_{target}$, where the final time T is as small as possible, by minimizing the total reward $\sum_{k=1}^{T} R(k)$. The leader action at time k when it is at vertex $v \in \mathcal{V}$ is defined as $p(k) = \pi(\hat{S}(k), \ell_1(k)) \in A_v$ for all $k \in \{1, ..., T\}$, where $A = \cup_{v \in \mathcal{V}} A_v$.

2.2 Design of Leader Control Policies Using Temporal-Difference Methods

Two Temporal-Difference (TD) learning methods [1], *SARSA* and *Q-Learning*, were adapted to generate an optimal leader control policy. These methods' use of bootstrapping provides the flexibility needed to accommodate the stochastic nature of the follower agents' transitions between vertices. Additionally, TD methods are model-free approaches, which are suitable for our control objective since the leader does not have a model of the followers' behavior. We compare the two methods to identify their advantages and disadvantages when applied to our swarm herding problem. Our approach is based on the mean-field model (6) in the sense that the leader learns a control policy using its observations of the population fractions of followers at all vertices in the graph.

Sutton and Barto [1] provide a formulation of the two TD algorithms that we utilize. Let S denote the state of the environment, defined later in this section; A denote the action set of the leader, defined as the set A_v in Eq. (4); and $Q(S, A)$ denote the state-action value function. We define $\alpha \in [0, 1]$ and $\gamma \in [0, 1]$ as the learning rate and the discount factor, respectively. The policy used by the leader is determined by a state-action pair (S, A). R denotes the reward for the implemented policy's transition from the current to the next state-action pair and is defined in Eq. (7). In the SARSA algorithm, an on-policy method, the state-action value function is defined as:

$$Q(S, A) \leftarrow Q(S, A) + \alpha[R + \gamma Q(S', A') - Q(S, A)] \tag{8}$$

where the update is dependent on the current state-action pair (S, A) and the next state-action pair (S', A') determined by enacting the policy. In the Q-Learning algorithm, an off-policy method, the state-action value function is:

$$Q(S, A) \leftarrow Q(S, A) + \alpha[R + \gamma \max_{a} Q(S', a) - Q(S, A)] \tag{9}$$

Whereas the SARSA algorithm update (8) requires knowing the next action A' taken by the policy, the Q-learning update (9) does not require this information.

Both algorithms use a discretization of the observed state S and represent the state-action value function Q in tabular form as a multi-dimensional matrix, indexed by the leader actions and states. The state S is defined as a vector that contains a discretized form of the population fraction of follower agents at each location $v \in \mathcal{V}$ and the location $\ell_1(k) \in \mathcal{V}$ of the leader agent. The leader's spatial state $\ell_1(k)$ must be taken into account because the leader's possible actions depend on its current location on the graph. Since the population fractions of follower agents are continuous values, we convert them into discrete integer quantities serving as a discrete function approximation of the continuous fraction populations. Instead of defining F_v as the integer count of followers at location v, which could be very large, we reduce the dimensionality of the state space by discretizing the follower population fractions into D intervals and scaling them up to integers between 1 and D:

$$F_v = \text{round}\left(\frac{D}{N}\sum_{i=1}^{N} \chi_v(X_i(0))\right), \qquad (10)$$
$$\text{where } F_v \in [1,\ldots,D], \quad v \in \mathcal{V}.$$

For example, suppose $D = 10$. Then a follower population fraction of 0.24 at location v would have a corresponding state value $S_v = 2$. Using a larger value of D provides a finer classification of agent populations, but at the cost of increasing the size of the state S. Given these definitions, the state vector S is defined as:

$$S_{env} = [F_1,\ldots,F_M,\ell_1] \qquad (11)$$

The state vector S_{env} contains many states that are inapplicable to the learning process. For example, the state vector for a 2×2 grid graph with $D = 10$ has $10 \times 10 \times 10 \times 10 \times 4$ possible variations, but only $10 \times 10 \times 10 \times 4$ are applicable since they satisfy the constraint that the follower population fractions at all vertices must sum up to 1 (note that the sum $\sum_v F_v$ may differ slightly from 1 due to the rounding used in Equation (10).) The new state S_{env} is used as the state S in the state-action value functions (8) and (9).

The leader's control policy for both functions (8) and (9) is the following ϵ-greedy policy, where $X \in [0, 1]$ is a uniform random variable and ϵ is a threshold parameter that determines the degree of state exploration during training:

$$\pi(S_{env}) = \arg\max_{A} Q(S_{env}, A) \quad \text{if } X > \epsilon \qquad (12)$$

3 Simulation Results

An *OpenAI Gym* environment [12] was created in order to design, simulate, and visualize our leader-based herding control policies [13]. This open source virtual environment can be easily modified to simulate swarm controllers for different numbers of agents and graph vertices, making it a suitable environment for training leader control policies using our model-free approaches. The simulated controllers can then be implemented in physical robot experiments. Figure 1 shows the simulated environment for a scenario with 100 follower agents, represented by the blue × symbols, that are herded by a leader, shown as a red circle, over a 2 × 2 grid. The *OpenAI* environment does not store the individual positions of each follower agent within a grid cell; instead, each cell is associated with an agent count. The renderer disperses agents randomly within a cell based on the cell's current agent count. The agent count for a grid cell is updated whenever an agent enters or leaves the cell according to the DTMC (1), and the environment is re-rendered. Recording the agent counts in each cell rather than their individual positions significantly reduces memory allocation and computational time when training the leader control policy on scenarios with large numbers of agents.

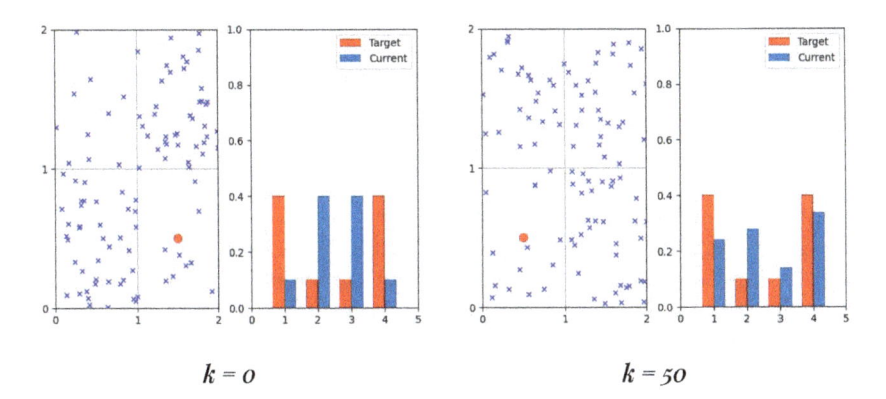

Fig. 1. Visual rendering of a simulated scenario in our *OpenAI* environment for iterations $k = 0$ and 50. The environment simulates a strongly connected 2 × 2 grid graph such as the one shown in Fig. 2. The leader (red circle) moves between grid cells in a horizontal or vertical direction. It may not move diagonally. Follower agents (blue × symbols) are randomly distributed in each cell. The borders of each cell are represented by the grid lines. The histogram to the right of each grid shows both the target (red) and current (blue) agent population fractions in each vertex at iteration k.

The graph $\mathcal{G}$ that models the environment in Fig. 1, with each vertex of $\mathcal{G}$ corresponding to a grid cell, was defined as the 2 × 2 graph in Fig. 2. In the graph, agents transition along edges in either a horizontal or vertical direction, or they can stay at the current vertex. The action set is thus defined as:

$$A = [\ Left,\ Right,\ Up,\ Down,\ Stay\ ] \tag{13}$$

Using the graph in Fig. 2, we trained and tested a leader control policy for follower agent populations of $N = 10$–100 at 10-agent increments. Both the SARSA and Q-Learning paradigms were applied and trained on 5000 episodes with 5000 iterations each. In every episode, the initial distribution $\hat{S}_{initial}$ and target distribution $\hat{S}_{target}$ of the follower agents were defined as:

$$\hat{S}_{initial} = \begin{bmatrix} 0.4 & 0.1 & 0.1 & 0.4 \end{bmatrix}^T \tag{14}$$

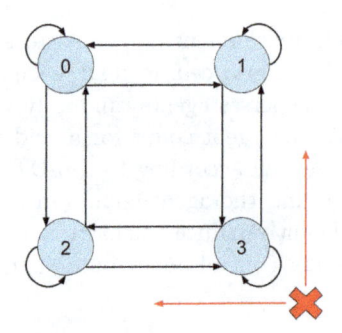

Fig. 2. The bidirected grid graph $\mathcal{G}$ used in our simulated scenario. The leader agent (red × symbol) is at vertex 3. The movement options for the leader are *Left* to vertex 2 or *Up* to vertex 1. The leader can also *Stay* at vertex 3, where its presence triggers follower agents at the vertex to probabilistically transition to vertex 1 or vertex 2.

$$\hat{S}_{target} = \begin{bmatrix} 0.1 & 0.4 & 0.4 & 0.1 \end{bmatrix}^T \tag{15}$$

The initial leader location, ℓ_1, was randomized to allow many possible permutations of states S_{env} for training. During training, an episode completes once the distribution of N follower agents reaches a specified terminal state. Instead of defining the terminal state as the exact target distribution $\hat{S}_{target}$, which becomes more difficult to reach as N increases due to the stochastic nature of the followers' transitions, we define this state as a distribution that is sufficiently close to $\hat{S}_{target}$. The learning rate and discount factor were set to $\alpha = 0.3$ and $\gamma = 0.9$, respectively. The follower agent transition rate β_e was defined as the same value β for all edges e in the graph and was set to $\beta = 0.025, 0.05,$ or 0.1. We use the mean squared error (MSE) to measure the difference between the current follower distribution and $\hat{S}_{target}$. The terminal state is reached when the MSE decreases below a threshold value μ. We trained our algorithms on threshold values of $\mu = 0.0005, 0.001, 0.0025,$ and 0.005.

After training the leader control policies on each follower agent population size N, the policies were tested on scenarios with the same environment and value of N. The policy for each scenario was run 1000 times to evaluate its performance. The policies were compared for terminal states that corresponded to the four different MSE threshold values μ, and were given 1000 iterations to converge within the prescribed MSE threshold of the target distribution (15) from the initial distribution (14).

Figure 3 compares the performance of leader control policies that were designed using each TD algorithm as a function of the tested values of μ. The leader control policies were trained on $N = 100$ follower agents, using the parameters $\beta = 0.05$ and $D = 10$ or 20, and tested in simulations with $N = 100$. The plots show that for both policies, the mean number of iterations required to converge to $\hat{S}_{target}$ decreases as the threshold μ increases for constant D, and at low values of μ, the mean number of iterations decreases when D is increased. In addition, as μ increases, the variance in the number of iterations decreases

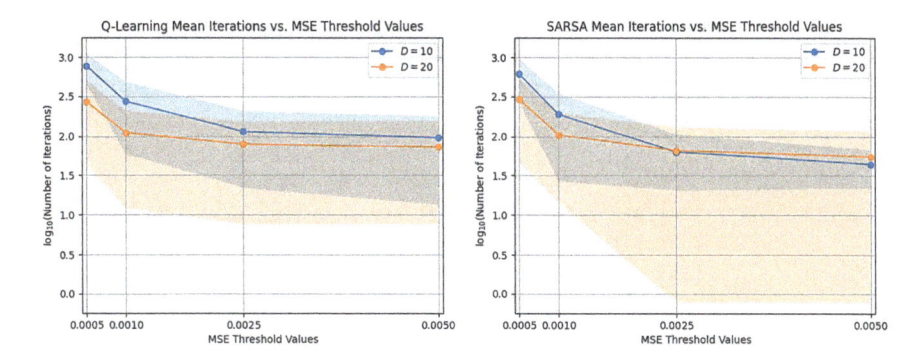

Fig. 3. Number of iterations until convergence to $\hat{S}_{target}$ (plotted on a log scale) versus the MSE threshold value μ for leader control policies that were learned using Q-Learning (*left*) and SARSA (*right*) with $\beta = 0.05$ and $N = 100$ follower agents. Each circle on the plots marks the mean number of iterations until convergence over 1000 test runs of a leader policy in the simulated grid graph environment in Fig. 2. The shaded regions indicate the range of ± 1 standard deviation about the mean numbers of iterations (blue for $D = 10$; orange for $D = 20$.)

(note the log scale of the y-axis in the plots) or remains approximately constant, except for the $D = 20$ case of SARSA.

Figure 4 compares the performance of leader control policies that were designed using each algorithm as a function of N, where the leader policies were tested in simulations with the same value of N that they were trained on. The other parameters used for training were $\mu = 0.0025$, $D = 20$, and $\beta = 0.025$, 0.05, or 0.1. The figures show that raising β from 0.05 to 0.1 does not significantly affect the mean number of iterations until convergence, while decreasing β from 0.05 to 0.025 results in a higher mean number of iterations. This effect is evident for both Q-Learning and SARSA trained leader control policies for $N > 50$. Both leader control policies result in similar numbers of iterations for convergence at each agent population size. Therefore, both the Q-Learning and SARSA training algorithms yield comparable performance for these scenarios.

The results in Fig. 4 show that as N increases above 50 agents, the mean number of iterations until convergence decreases slightly or remains approximately constant for all β values and for $\mu = 0.0025$. Moreover, from Fig. 3, we see that MSE threshold values $\mu < 0.0025$ for $N = 100$ result in a higher number of iterations than the $N = 100$ case in Fig. 4. This trend may be due to differences in the magnitude of the smallest possible change in MSE over an iteration k relative to the MSE threshold μ for different values of N. For example, for $N = 10$, a similarity in iteration counts for all four MSE thresholds μ can be attributed to the fact that the change in the MSE due to a transition of one agent, corresponding to a change in population fraction of $1/N = 1/10$, is much higher than the four MSE thresholds (i.e., $(1/10)^2 > 0.005$, 0.0025, 0.001, and 0.0005). Compare this to the iteration count for $N = 50$, which would have a corresponding change in MSE of $(1/50)^2$; this quantity is much smaller than

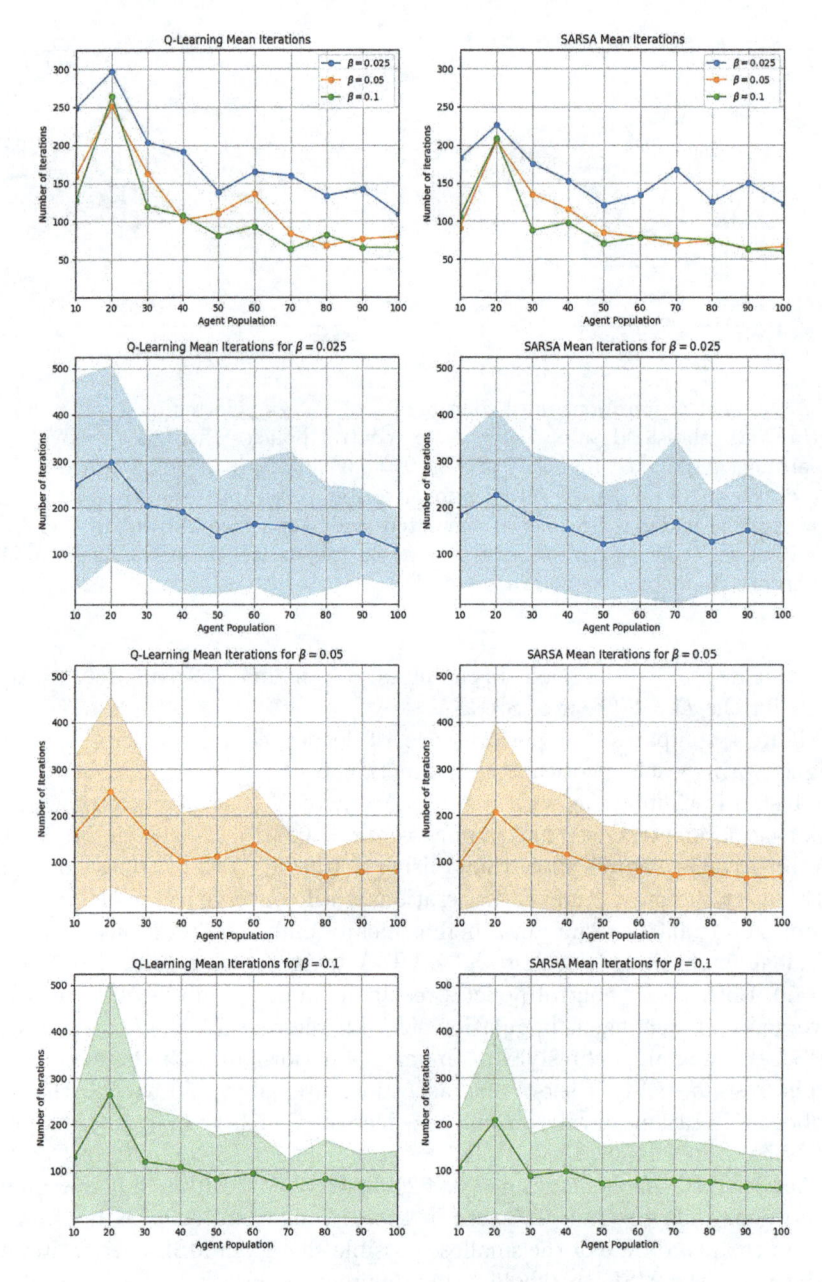

Fig. 4. Number of iterations until convergence to $\hat{S}_{target}$ versus number of follower agents N for leader control policies that were learned using Q-Learning (*left column*) and SARSA (*right column*) with $D = 20$ and $\mu = 0.0025$. Each circle on the plots marks the mean number of iterations until convergence over 1000 test runs of a leader policy in the simulated grid graph environment in Fig. 2 with the same value of N that the policy was trained on. The plot for each β value in the top two figures are reproduced individually in the three figures below them, along with shaded regions that indicate the range of ± 1 standard deviation about the mean numbers of iterations.

0.005 and 0.0025, but not much smaller than 0.001 and 0.0005. The iteration counts for $N = 100$ are much lower, since $(1/100)^2$ is much smaller than all four MSE thresholds.

Finally, Fig. 5 compares the performance of leader control policies that were designed using each algorithm as a function of N, where the leader policies were trained with $N = 10$, 100, or 1000 follower agents and tested in simulations with $N = 10$–100 (at 10-agent increments) and $N = 1000$ agents. This was done to evaluate the robustness of the policies trained on the three agents populations to changes in N. The other parameters used for training were $\mu = 0.0025$, $D = 20$, and $\beta = 0.025$, 0.05, or 0.1. As the plots in Fig. 5 show, policies trained on the smallest population, $N = 10$, yield an increased mean number of iterations until convergence when applied to populations $N > 10$. The reverse effect is observed, in general, for policies that are trained on higher values of N than they are tested on. An exception is the case where the policies are trained on $N = 100$ and 1000 and tested on $N = 10$, which produce much higher numbers of iterations than the policies that are both trained and tested on $N = 10$. This is likely a result of the large variance, and hence greater uncertainty, in the time evolution of such a small agent population. The lower amount of uncertainty in the time evolution of large swarms may make it easier for leader policies that are trained on large values of N to control the distribution of a given follower agent population than policies that are trained on smaller values of N. We thus hypothesize that training a leader agent with the mean-field model (6) instead of the DTMC model would lead to improved performance in terms of a lower training time, since the policy would only need to be trained on one value of N, and fewer iterations until convergence to the target distribution.

4 Experimental Results

We also conducted experiments to verify that our herding approach is effective in a real-world environment with physical constraints on robot dynamics and inter-robot spacing. Two of the leader control policies that were generated in the simulated environment were tested on a swarm of small differential-drive robots in the *Robotarium* [14], a remotely accessible swarm robotics testbed that provides an experimental platform for users to validate swarm algorithms and controllers. Experiments are set up in the Robotarium with MATLAB or Python scripts. The robots move to target locations on the testbed surface using a position controller and avoid collisions with one another through the use of barrier certificates [15], a modification to the robots' controllers that satisfy particular safety constraints. To implement this collision-avoidance strategy, the robots' positions and orientations in a global coordinate frame are measured from images taken from multiple *VICON* motion capture cameras.

A video recording of our experiments is shown in [16]. The environment was represented as a 2×2 grid, as in the simulations, and $N = 10$ robots were used as follower agents. The leader agent, shown as the blue circle, and the boundaries of the four grid cells were projected onto the surface of the testbed using an

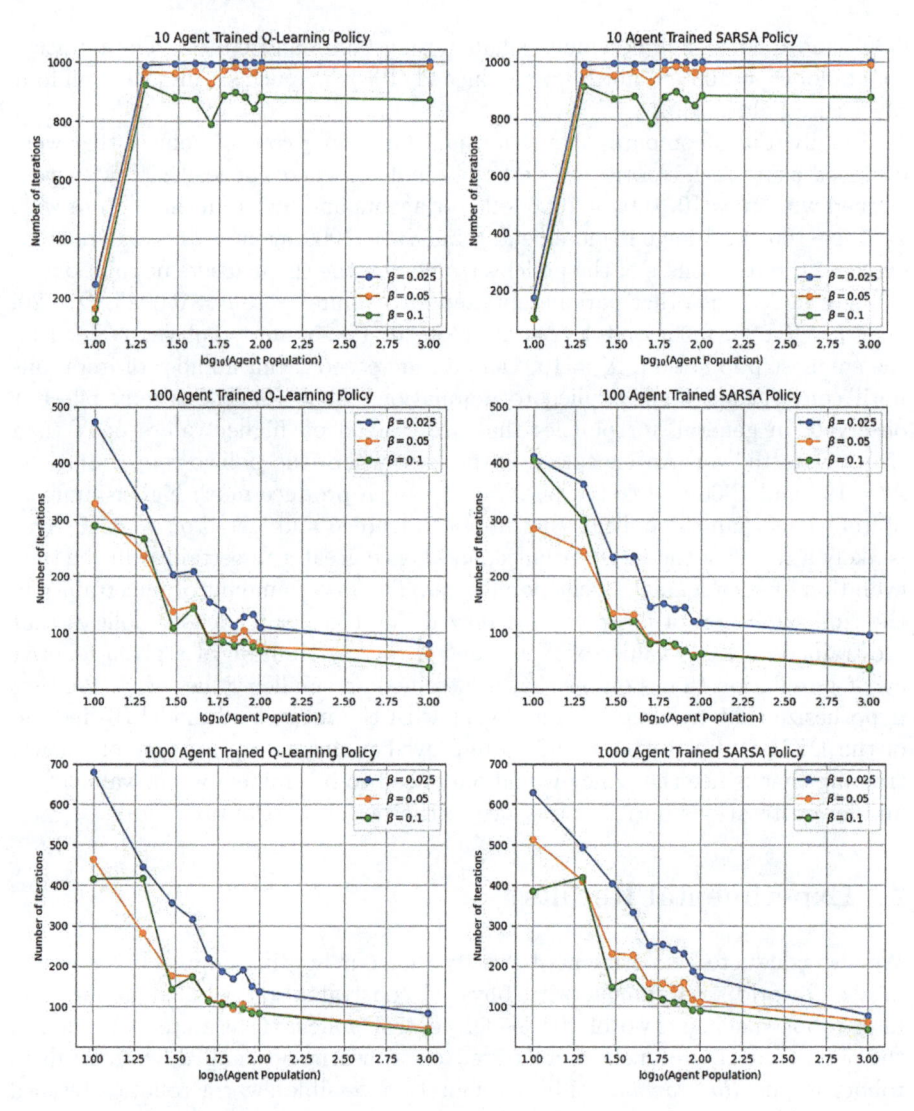

Fig. 5. Number of iterations until convergence to $\hat{S}_{target}$ versus number of follower agents N (plotted on a log scale) for leader control polices that were trained using Q-Learning (*left column*) and SARSA (*right column*) with $D = 20$; $\mu = 0.0025$; and $\beta = 0.025, 0.05$, or 0.1; and $N = 10, 100$, or 1000 agents. Each circle on the plots marks the mean number of iterations until convergence over 1000 test runs of a leader policy in the simulated grid graph environment in Fig. 2.

overhead projector. As in the simulations, at each iteration k, the leader moves from one grid cell to another depending on the action prescribed by its control policy. Both the SARSA and Q-Learning leader control policies trained with $N = 100$ follower agents, $D = 10$, $\mu = 0.0025$, and a $\beta = 0.1$ were implemented, and [16] shows the performance of both control policies. In the video, the leader is red if it is executing the *Stay* action and blue if it is executing any of the other actions in the set A (i.e., a movement action). The current iteration k and leader action are displayed at the top of the video frames. Actions that display ϵ next to them signify a random action as specified in (12). Each control policy was able to achieve the exact target distribution (15). The SARSA method took 59 iterations to reach this distribution, while the Q-Learning method took 23 iterations.

5 Conclusion and Future Work

We have presented two Temporal-Difference learning approaches to designing a leader-follower control policy for herding a swarm of agents as quickly as possible to a target distribution over a graph. We demonstrated the effectiveness of the leader control policy in simulations and physical robot experiments for a range of swarm sizes N, illustrating the scalability of the control policy with N, and investigated the effect of N on the convergence time to the target distribution. However, these approaches do not scale well with the graph size due to the computational limitations of tabular TD approaches and, in particular, our discretization of the system state into population fraction intervals. Our implementation requires a matrix with $D^M \times M \times |A|$ state-action values to train the leader control policy. For our 2×2 grid graph with $|A| = 5$ possible leader actions and $D = 20$ intervals, this is about $20^4 \times 4 \times 5$ values.

To address this issue, our future work focuses on designing leader control policies using the mean-field model rather than the DTMC model for training, as suggested at the end of Sect. 3. In this approach, the leader policies would be trained on follower agent population fractions that are computed from solutions of the discrete-time mean-field model (6), rather than from discrete numbers of agents that transition between locations according to a DTMC. The leader control policies can also be modified to use function approximators such as neural networks for our training algorithm, allowing for utilization of modern deep reinforcement learning techniques. Neural network function approximators provide a more practical approach than tabular methods to improve the scalability of the leader control policy with graph size, in addition to swarm size. In addition, the control policies could be implemented on a swarm robotic testbed in a *decentralized* manner, in which each follower robot avoids collisions with other robots based on its local sensor information.

Acknowledgment. This work was supported by the Arizona State University Global Security Initiative. Many thanks to Dr. Sean Wilson at the Georgia Tech Research Institute for running the robot experiments on the Robotarium.

References

1. Sutton, R.S., Barto, A.G.: Reinforcement Learning: An Introduction. MIT Press, Cambridge (2018)

2. Ji, M., Ferrari-Trecate, G., Egerstedt, M., Buffa, A.: Containment control in mobile networks. IEEE Trans. Autom. Control **53**(8), 1972–1975 (2008)
3. Mesbahi, M., Egerstedt, M.: Graph Theoretic Methods in Multiagent Networks. Princeton University Press, Princeton (2010)
4. Pierson, A., Schwager, M.: Controlling noncooperative herds with robotic herders. IEEE Trans. Rob. **34**(2), 517–525 (2017)
5. Elamvazhuthi, K., Wilson, S., Berman, S.: Confinement control of double integrators using partially periodic leader trajectories. In: American Control Conference, pp. 5537–5544 (2016)
6. Paranjape, A.A., Chung, S.-J., Kim, K., Shim, D.H.: Robotic herding of a flock of birds using an unmanned aerial vehicle. IEEE Trans. Robot. **34**(4), 901–915 (2018)
7. Go, C.K., Lao, B., Yoshimoto, J., Ikeda, K.: A reinforcement learning approach to the shepherding task using SARSA. In: International Joint Conference on Neural Networks, pp. 3833–3836 (2016)
8. Elamvazhuthi, K., Berman, S.: Mean-field models in swarm robotics: a survey. Bioinspiration Biomimetics **15**(1), 015001 (2019)
9. Šošić, A., Zoubir, A.M., Koeppl, H.: Reinforcement learning in a continuum of agents. Swarm Intell. **12**(1), 23–51 (2017). https://doi.org/10.1007/s11721-017-0142-9
10. Hüttenrauch, M., Adrian, S., Neumann, G.: Deep reinforcement learning for swarm systems. J. Mach. Learn. Res. **20**(54), 1–31 (2019)
11. Yang, Y., Luo, R., Li, M., Zhou, M., Zhang, W., Wang, J.: Mean field multi-agent reinforcement learning. In: International Conference on Machine Learning, pp. 5567–5576 (2018)
12. Brockman, G., et al.: OpenAI Gym. *arXiv preprint* arXiv:1606.01540 (2016)
13. Kakish, Z.: Herding OpenAI Gym Environment (2019). https://github.com/acslaboratory/gym-herding
14. Wilson, S., Glotfelter, P., Wang, L., Mayya, S., Notomista, G., Mote, M., Egerstedt, M.: The Robotarium: globally impactful opportunities, challenges, and lessons learned in remote-access, distributed control of multirobot systems. IEEE Control Syst. Mag. **40**(1), 26–44 (2020)
15. Wang, L., Ames, A.D., Egerstedt, M.: Safety barrier certificates for collisions-free multirobot systems. IEEE Trans. Rob. **33**(3), 661–674 (2017)
16. Kakish, Z., Elamvazhuthi, K., Berman, S.: Using reinforcement learning to herd a robotic swarm to a target distribution. Autonomous Collective Systems Laboratory YouTube Channel (2020). https://youtu.be/py3Pe24YDjE

Preservation of Giant Component Size After Robot Failure for Robustness of Multi-robot Network

Toru Murayama[1(✉)] and Lorenzo Sabattini[2]

[1] National Institute of Technology, Wakayama College,
Gobo Nada-cho Noshima 77, Wakayama, Japan
murayama@wakayama-nct.ac.jp
[2] University of Modena and Reggio Emilia, Reggio Emilia, Italy
lorenzo.sabattini@unimore.it

Abstract. This paper approaches a network topology control method for networked multi-robot systems. Although robustness of network connectivity against robot failures is a matter of concern for the multi-robot control, the robustification impedes the motion of robots because of limitations of the wireless communication. For mitigating the impediments, we focus our attention on the giant connected component size after a single robot fails, and aim to control such component size. A modified algebraic connectivity is introduced here as an indicator of the component size: a threshold for the algebraic connectivity is analyzed to preserve the component size. Theoretical properties and numerical examples are shown to demonstrate our control method.

Keywords: Fault tolerant · Network topology control · Multi-robot system

1 Introduction

A decentralized multi-robot system is expected to achieve collective behaviours based on inter-robot communications without a centralized controller. Since an appropriate wireless communication technology, such as a mobile ad-hoc network is provided on the robots to run such inter-robot communications, the network topology is an important factor for the multi-robot system to act as a team and it should be kept connected during their task performances in spite of robots' individual autonomous actions. For this reason, preservation of the connectivity of the whole network is a major research topics, and a lot of control methods to preserve the connectivity have been proposed in the literature [1,2]. Many of the connectivity preserving methods are based on the algebraic connectivity, since it is well-known that the second smallest eigenvalue of a Laplacian matrix (i.e., the algebraic connectivity) is greater than zero if and only if the network is connected. Studies related to connectivity preserving consensus, such as [3, 4], propose control laws which preserve all initial communication links without

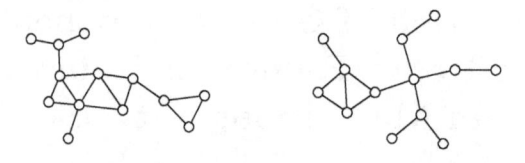

Fig. 1. Concept of component size robustness. A connected component with more than 9 nodes will remain in the left network after any one node is removed, while a component with only 5 nodes will remain in the right network. We claim that the left network is more robust than the right one from this perspective.

referencing the algebraic connectivity, thanks to a characteristics of consensus that allows to achieving the task without link disconnections. On diffusive tasks like a coverage [5,6], in contrast to cohesive tasks like consensus, some of the initial links may be disconnected for the task accomplishments and, therefore, each robots should carefully disconnect the links not to violate the connectivity.

Because some robot failures or malfunctions also can cause the network disconnection, and since the possibility of occurring robot malfunctions increases as the number of robots increases, the multi-robot system is demanded to maintain its performances as much as possible even if such robot failures occur. Some methods to robustify the network connectedness against a robot failure have been studied, such as [7–9], where the main concept is to let the network bi-connected, meaning that the network will remain connected if any one node is removed. Although the bi-connected multi-robot network stays connected even if any one robot fails, the bi-connectivity restricts the configuration space of the multi-robot system leading to limitations to the wireless communication.

Instead of the bi-connectivity, in [10], we introduced an alternative concept of the multi-robot network robustness which is related to the giant component size after any one robot failed (see Fig. 1). This is inspired by studies of complex networks [11] which discuss the giant component after random failures. Our concept relaxes the restriction arising from the bi-connectivity since it considers intermediate structures between tree graphs and bi-connected graphs. Thus, comparing to the bi-connectivity robustness, our concept allows the robots to move effectively on the diffusive tasks. Based on that concept, we constructed an algorithm to estimate the giant component size in [10] using the fact that an eigenvalue of a modified Laplacian matrix is related to the giant component size. We proposed a control method to improve the robustness based on a consensus control in [12] and a control method to preserve the giant component size in [13]. However, the proposed preservation method prohibits taking some network structures, since it was derived based on conservative conditions.

In this paper, we analyze some theoretical properties of the modified Laplacian eigenvalue to mitigate the conservativeness of the previous preservation methods. A distributed control law is formulated using the modified Laplacian eigenvalue which is redefined to fulfill the control objective, and an effective algorithm to find the parameters for the control is proposed, from the theoretical properties. The control method enables the multi-robot systems to relax the restriction of the bi-connectivity: performances of the diffusive tasks will be better than cases when the network is bi-connected. Numerical simulations demonstrate the validity of the proposed control method.

2 Definition and Formulation

In this study, we consider a multi-robot system consisting of N robots, and define an index set $\mathcal{V} = \{1, \ldots, N\}$ which is also regarded as a node set of a graph $\mathcal{G}$. The position of the robot i is denoted by $p_i \in \mathbf{R}^n$ where n is a dimension of the configuration space, and is controlled by a control input, as

$$\dot{p}_i = u_i^{\text{task}} + u_i^{\text{pre}}, \tag{1}$$

where $u_i^{\text{task}} \in \mathbf{R}^n$ denotes an input to achieve a cooperative task (like, e.g., a coverage control), and $u_i^{\text{pre}} \in \mathbf{R}^n$ denotes an input for preserving the component size defined below.

Define a weighted adjacency matrix $A \in \mathbf{R}^{N \times N}$, whose entry a_{ij} is given by

$$a_{ij} = \begin{cases} w(\|p_i - p_j\|), & \text{if } \|p_i - p_j\| \leq R, \\ 0, & \text{otherwise}, \end{cases} \tag{2}$$

where $R > 0$ denotes the upper-bound of the communication distance, and $w(r) \geq 0$ denotes a link weight function. In this study, we suppose the link weight function $w(r)$ is defined by a sigmoid function $w(r) = (1 - \tanh(\alpha r - \beta))/2$ with parameters α, β, and also suppose $\lim_{r \to R} w(r) < \varepsilon$ with a parameter $\varepsilon > 0$ defined later. Define the neighbor set of a node i by $\mathcal{N}_i = \{j \in \mathcal{V} : a_{ij} > 0\}$.

Consider a graph $\mathcal{G}_{-i}$ which is defined by deleting node i and its incident links from a graph $\mathcal{G}$. Connected components $\mathcal{C}_m^{(i)}$, $m \in \{1, \ldots, M^{(i)}\}$ of the graph $\mathcal{G}_{-i}$ are called *remaining connected components* of the graph $\mathcal{G}$ w.r.t. the node i, in this study. $M^{(i)}$ denotes a number of remaining connected components, and we say a node i is an articulation node if and only if $M^{(i)} > 1$. We describe the number of nodes in the remaining component $\mathcal{C}_m^{(i)}$ as $C_m^{(i)}$, and assume the sizes $C_1^{(i)}, \ldots, C_M^{(i)}$ are in ascending order. We can see the node i is an articulation node if and only if $C_M^{(i)} < N - 1$. We also define a set of the neighbors in the component $\mathcal{C}_m^{(i)}$ by $\mathcal{D}_m^{(i)} = \mathcal{C}_m^{(i)} \cap \mathcal{N}_i$, and the component weight $D_m^{(i)} = \sum_{j \in \mathcal{D}_m^{(i)}} a_{ij}$. The weight-sum set $\mathcal{W}_m^{(i)}$ is defined by $\mathcal{W}_m^{(i)} = \{\sum_{j \in S} a_{ij} | S \in 2^{\mathcal{D}_m^{(i)}} \setminus \{\emptyset, \mathcal{D}_m^{(i)}\}\}$, where $2^{\mathcal{D}_m^{(i)}}$ denotes the power set of the neighbor set $\mathcal{D}_m^{(i)}$.

To estimate the properties of graph $\mathcal{G}_{-i}$ in a distributed fashion, we introduce a perturbation parameter $\varepsilon > 0$ and we define the perturbed adjacency matrix $A^{(i)}(\varepsilon)$ whose jk-th entry is given by

$$a_{jk}^{(i)} = \begin{cases} \varepsilon a_{jk}, & \text{if } j = i \vee k = i, \\ a_{jk}, & \text{otherwise}. \end{cases} \tag{3}$$

The perturbed adjacency matrix $A^{(i)}$ is a modified version of an adjacency matrix such that only the i-th row and the i-th column are multiplied by the perturbation parameter ε. From the perturbed adjacency matrix $A^{(i)}(\varepsilon)$, we define a perturbed Laplacian matrix $L^{(i)}$ w.r.t. node i, by $L^{(i)}(\varepsilon) = D^{(i)}(\varepsilon) - A^{(i)}(\varepsilon)$ where $D^{(i)}$ denotes a degree matrix $D^{(i)} = \text{diag}(\sum_j a_{1j}^{(i)}, \sum_j a_{2j}^{(i)}, \ldots, \sum_j a_{Nj}^{(i)})$, and its second smallest eigenvalue $\lambda^{(i)}(\varepsilon)$. Additionally we define the eigenvector $v^{(i)}(\varepsilon)$ corresponding to the eigenvalue $\lambda^{(i)}(\varepsilon)$. Note that the eigenpair $(\lambda^{(i)}(\varepsilon), v^{(i)}(\varepsilon))$ can be computed by a distributed algorithm such as [14]. Thus we assume the eigenpairs $(\lambda^{(i)}(\varepsilon), v^{(i)}(\varepsilon))$ for all i are known by all the robots. Additionally, each robot knows its component sizes $C_m^{(i)}$ and the component weights $D_m^{(i)}$, exploiting the algorithm proposed in [10].

In the previous study [10], we analyzed that the value $\hat{\lambda}^{(i)} = \lim_{\varepsilon \to +0} \lambda^{(i)}(\varepsilon)/\varepsilon$, and we showed it is related to the size the of remaining connected components $C_m^{(i)}$, according to the following relations:

$$\hat{\lambda}^{(i)} = \frac{\sum_{j \in \mathcal{N}_i} a_{ij}(v_i^{(i)} - v_j^{(i)})^2}{\|v^{(i)}\|}, \tag{4}$$

$$\hat{\lambda}^{(i)} v_j^{(i)} = \frac{D_m^{(i)}}{C_m^{(i)}} \left(v_j^{(i)} - v_i^{(i)} \right), \ \forall j \in \mathcal{C}_m^{(i)}, \tag{5}$$

where $v^{(i)} = v^{(i)}(0)$. These relations imply the following facts:

1. $\|v_j^{(i)} - v_k^{(i)}\| = 0$ if nodes j and k are in the same connected component $(j, k \in \mathcal{C}_m^{(i)})$.
2. $\hat{\lambda}^{(i)}$ satisfies

$$\frac{D_{m^*}^{(i)}}{C_{m^*}^{(i)}} \leq \hat{\lambda}^{(i)} \leq \min_{m \neq m^*} \frac{D_m^{(i)}}{C_m^{(i)}} \ \text{where} \ m^* = \underset{m}{\text{argmin}} \ \frac{D_m^{(i)}}{C_m^{(i)}}. \tag{6}$$

The equation holds if and only if $v_i^{(i)} = 0$.

Moreover, from the relation (5), and the definition eigenvector corresponding to the second smallest Laplacian eigenvalue, namely $\sum_j v_j^{(i)} = 0$, we get the algebraic equation $f(\hat{\lambda}^{(i)}) = 0$ where

$$f(\hat{\lambda}^{(i)}) = \sum_{m=1}^{M^{(i)}} \frac{D_m^{(i)} C_m^{(i)}}{D_m^{(i)} - C_m^{(i)} \hat{\lambda}^{(i)}} + 1, \tag{7}$$

if $v_i^{(i)} \neq 0$. It is clear that the value $\hat{\lambda}^{(i)}$ is the minimal solution of the equation $f(\hat{\lambda}^{(i)}) = 0$, since the function $f(\hat{\lambda}^{(i)})$ is piecewise monotonic. We call the value $\hat{\lambda}^{(i)}$ the *perturbed algebraic connectivity* of node i.

Because the perturbed algebraic connectivity $\hat{\lambda}^{(i)}$ depends on pairs of the component size and the component weight $(C_m^{(i)}, D_m^{(i)})$, $m \in \{1, \ldots, M^{(i)}\}$, here we call $(C_m^{(i)}, D_m^{(i)})$ the 2-tuple of the remaining component $\mathcal{C}_m^{(i)}$. To describe cases when the component $\mathcal{C}_m^{(i)}$ is divided into two components, we define a *tearing* link set $\mathcal{T}^{(i)}(m,c,d)$, by which the 2-tuple $(C_m^{(i)}, D_m^{(i)})$ of the remaining component $\mathcal{C}_m^{(i)}$ is divided into two 2-tuples (c,d) and $(C_m^{(i)} - c, D_m^{(i)} - d)$. The graph generated by the tearing $\mathcal{T}^{(i)}(m,c,d)$ is denoted by $\mathcal{G}_{\mathcal{T}^{(i)}(m,c,d)}$.

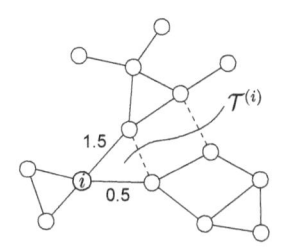

Fig. 2. Example of tearing. In this example, the tearing $\mathcal{T}^{(i)}$ divides the component $(C_M^{(i)}, D_M^{(i)}) = (11, 2.0)$ into two components $(6, 1.5)$ and $(5, 0.5)$.

The control objective of this study is to meet two requirements for all robots:

1. preserve the remaining component size $C_M^{(i)} \geq \bar{C}$,
2. and not prevent any tearing $\mathcal{T}^{(i)}(m,c,d)$ for all $m \neq M^{(i)}$,

where $\bar{C}$ denotes a component size threshold, assuming $\bar{C} > N/2$. Hence, we try to find a perturbed algebraic connectivity threshold $\bar{\lambda}^{(i)}(\mathcal{G}, \bar{C})$ such that

$$\hat{\lambda}^{(i)}(\mathcal{G}_{\mathcal{T}^{(i)}(M,C,D)}) \leq \bar{\lambda}^{(i)}, \ \forall C \in \{C_M^{(i)} - \bar{C}, \ldots, \bar{C}\}, D \in \mathcal{W}_M^{(i)}, \quad (8)$$

induced from the requirement 1 (preserving giant component size), and

$$\hat{\lambda}^{(i)}(\mathcal{G}_{\mathcal{T}^{(i)}(m,c,d)}) > \bar{\lambda}^{(i)}, \ \forall m \neq M^{(i)}, c \in \{1, \ldots, C_m^{(i)} - 1\}, d \in \mathcal{W}_m^{(i)}, \quad (9)$$

induced from the requirement 2 (not preventing any tearing in non-giant component). Since the inequalities (8)–(9) imply the following relation

$$\hat{\lambda}^{(i)}(\mathcal{G}_{\mathcal{T}^{(i)}(M,C,D)}) < \hat{\lambda}^{(i)}(\mathcal{G}_{\mathcal{T}^{(i)}(m,c,d)}), \quad (10)$$

then we analyze a condition to satisfy the relation (10) in Sect. 3. Then, we propose an algorithm to compute the threshold $\bar{\lambda}^{(i)}$ in Sect. 4, and a control method to satisfy $\bar{\lambda}^{(i)} < \hat{\lambda}^{(i)}$ for all time in Sect. 5. For simplicity of expressions, hereafter we omit the notation (i), e.g., $\hat{\lambda} = \hat{\lambda}^{(i)}$ and $\mathcal{T} = \mathcal{T}^{(i)}$, in obvious cases.

3 Modification for Perturbed Algebraic Connectivity

In this section, we propose a sufficient condition to satisfy the relation (10), that is preparatory to proposing the control method. Additionally, we redefine the perturbed link weight $a_{ij}^{(i)}(\varepsilon)$, with an alternative definition with respect to (3), to satisfy the relation (10).

Since we would like the robots to disconnect the communication links freely unless the giant component size C_M becomes less than the threshold $\bar{C}$, the lower-bound $\bar{\lambda}$ should not disturb the link disconnection in the non-giant components. We get a sufficient condition to fulfill this requirement as follows.

Lemma 1. *Consider two remaining components $\mathcal{C}_M$ and $\mathcal{C}_m$ such that $M = \arg\min_m D_m/C_m$ and $m \neq M$. Assume $|\mathcal{D}_M| > 1$, $|\mathcal{D}_m| > 1$, and the following is satisfied:*

$$D_M - a_M < \frac{a_m}{C_m - 1}, \text{ where } a_m = \min_{j \in \mathcal{D}_m} a_{ij}. \tag{11}$$

Then, the relation $\hat{\lambda}(\mathcal{G}_{\mathcal{T}(M)}) < \hat{\lambda}(\mathcal{G}_{\mathcal{T}(m)})$ holds for any tearing $\mathcal{T}(M) = \mathcal{T}(M, C, D)$ and $\mathcal{T}(m) = \mathcal{T}(m, c, d)$.

Proof. The value $\hat{\lambda}$ is represented as $\hat{\lambda} = \sum_{j \neq i} b_m (v_i - v_j)^2 / \|v\|$, where $b_m = D_m/C_m$ such that $j \in \mathcal{C}_m$. Thus, defining a sequence $b = \{b_1, \ldots, b_M\}$, we get

$$b(\mathcal{G}_{\mathcal{T}(M)}) \prec b(\mathcal{G}_{\mathcal{T}(m)}) \Rightarrow \hat{\lambda}(\mathcal{G}_{\mathcal{T}(M)}) < \hat{\lambda}(\mathcal{G}_{\mathcal{T}(m)}),$$

where the relation $x \prec y$ means that $x_m \leq y_m$ for all m and there exists at least one m satisfying $x_m < y_m$. Hereafter we prove the relation $b(\mathcal{G}_{\mathcal{T}(M)}) \prec b(\mathcal{G}_{\mathcal{T}(m)})$ is satisfied under assumption (11).

When the tearing $\mathcal{G} \setminus \mathcal{T}(M)$ is executed, the 2-tuple (C_M, D_M) is separated into $(\alpha_M C_M, \beta_M D_M)$ and $((1 - \alpha_M)C_M, (1 - \beta)D_M)$ where α_M, β_M are feasible values. Similarly, 2-tuples $(\alpha_m C_m, \beta_m D_m)$ and $((1 - \alpha_m)C_m, (1 - \beta)D_m)$ are generated by the tearing $\mathcal{G} \setminus \mathcal{T}(m)$. We assume $0 < \beta_M \leq \alpha_M < 1$ and $0 < \beta_m \leq \alpha_m < 1$ without loss of generality.

Consider a sequence of fractions $b_m = D_m/C_m$, ordered in ascending order. The sequence $b(\mathcal{G}_{\mathcal{T}(M)})$ of graph $\mathcal{G}_{\mathcal{T}(M)}$ will become

$$b(\mathcal{G}_{\mathcal{T}(M)}) = \left\{ \frac{\beta_M D_M}{\alpha_M C_M}, \frac{(1-\beta_M)D_M}{(1-\alpha_M)C_M}, \frac{D_m}{C_m}, \ldots \right\},$$

under assumption (11). Similarly, the sequence $b(\mathcal{G}_{\mathcal{T}(m)})$ will be

$$b(\mathcal{G}_{\mathcal{T}(m)}) = \left\{ \frac{D_M}{C_M}, \frac{\beta_m D_m}{\alpha_m C_m}, \frac{(1-\beta_m)D_m}{(1-\alpha_m)C_m}, \ldots \right\}.$$

Thus, the pairs of the fraction respectively satisfy

$$\frac{\beta_M D_M}{\alpha_M C_M} \leq \frac{D_M}{C_M}, \frac{(1-\beta_M)D_M}{(1-\alpha_M)C_M} < \frac{\beta_m D_m}{\alpha_m C_m}, \frac{D_m}{C_m} \leq \frac{(1-\beta_m)D_m}{(1-\alpha_m)C_m}, \ldots$$

we get $b(\mathcal{G}_{\mathcal{T}(M)}) \prec b(\mathcal{G}_{\mathcal{T}(m)})$ and the proof is completed. □

From Lemma 1, we can see that relation (9) is satisfied under the assumption (11) if we choose the threshold $\bar{\lambda}$ satisfying $\bar{\lambda} \leq \max_{C,D} \hat{\lambda}(\mathcal{G}_{T(M,C,D)})$. Here we redefine the elements $a_{jk}^{(i)}(\varepsilon)$ of the perturbed adjacency matrix $A^{(i)}(\varepsilon)$ to guarantee assumption (11), as

$$a_{jk}^{(i)}(\varepsilon) = \begin{cases} \varepsilon a_{jk}, & \text{if } (j=i \wedge k \in \mathcal{C}_M) \vee (k=i \wedge j \in \mathcal{C}_M), \\ \varepsilon(a_{jk} + C_M(D_M - a_M)), & \text{else if } (j=i \vee k=i) \wedge a_{jk} > 0, \\ a_{jk}, & \text{otherwise}, \end{cases} \quad (12)$$

where $a_M = \max_{j \in \mathcal{D}_M} a_{ij}$. Therefore, we can introduce the following result.

Theorem 1. *Assume the perturbed algebraic connectivity $\hat{\lambda}^{(i)} = \lim_{\varepsilon \to 0} \lambda^{(i)}(\varepsilon)/\varepsilon$ is derived from the perturbed adjacency matrix defined by (12), and the threshold $\bar{\lambda}^{(i)}$ is given by*

$$\bar{\lambda} = \max_{C,D} \hat{\lambda}(\mathcal{G}_{T(M,C,D)}), \ \text{s.t.} \ C \in \{C_M - \bar{C} + 1, \ldots, \bar{C} - 1\}, \ D \in \mathcal{W}_M. \quad (13)$$

Then, both the relations (8) and (9) are satisfied.

Proof. It is directly derived from Lemma 1 since

$$\max_{C,D} \hat{\lambda}^{(i)}(\mathcal{G}_{T^{(i)}(M,C,D)}) \leq \bar{\lambda}^{(i)} < \min_{c,d} \hat{\lambda}^{(i)}(\mathcal{G}_{T^{(i)}(m,c,d)}), \quad (14)$$

is satisfied. □

4 Threshold Computation

In this section, we discuss about an algorithm to compute the threshold (13). Since the computation (13) is a combinatorial optimization problem, it may take a lot of time to find the optimal solution by calculating all the combinations. Here we try to relax the computational complexity by solving the relaxed problem

$$\max_{C,D} \hat{\lambda}(\mathcal{G}_{T(M,C,D)}), \ \text{s.t.} \ C \in [C_M - \bar{C} + 1, \bar{C} - 1], \ D \in [a_M, D_M - a_M], \quad (15)$$

where $a_M = \min_{j \in \mathcal{D}_M} a_{ij}$. Since the optimal value of the relax problem (15) is greater than or equal to the one of the original problem (13), $\bar{\lambda}$ can be used as a threshold to preserve the giant component size $C_M \geq \bar{C}$.

The optimal solution of the relaxed problem (15) is shown as below.

Lemma 2. *Assume $C_M/2 < \bar{C}$ and $D_M - a_M < D_m/C_m$ for all $m \neq M$. If the solution d^* of the algebraic equation*

$$\sum_{m \neq M} \frac{D_m C_m}{(\bar{C}-1)(2(C_M - (\bar{C}-1))D_m - D_M C_m) - d^*(C_M - 2(\bar{C}-1))D_M C_m} \\ + \frac{1 - 2d^*}{(\bar{C}-1) - d^* C_M} + \frac{1}{2(\bar{C}-1)(C_M - (\bar{C}-1))} = 0, \quad (16)$$

satisfies $a_M \leq d^* D_M \leq a^*$ where $a^* = \min\left(D_M - a_M, (\bar{C} - 1)D_M/C_M\right)$, then the optimal value $\hat{\lambda}^*$ of the relaxed problem (15) is $\hat{\lambda}^* = \hat{\lambda}(\mathcal{G}_{T(M,\bar{C}-1,d^*D_M)})$. If $d^* D_M < a_M$ then $\hat{\lambda}^* = \hat{\lambda}(\mathcal{G}_{T(M,\bar{C}-1,a_M)})$. Otherwise $\hat{\lambda}^* = \hat{\lambda}(\mathcal{G}_{T(M,\bar{C}-1,a^*)})$.

Proof. Consider the implicit function $f(\hat{\lambda}(x), x)$ such that

$$f(\hat{\lambda}, x) = \frac{x_1 D_M x_2 C_M}{x_1 D_M - x_2 C_M \hat{\lambda}} + \frac{(1-x_1)D_M(1-x_2)C_M}{(1-x_1)D_M - (1-x_2)C_M \hat{\lambda}} + \sum_{m \neq M} \frac{D_m C_m}{D_m - C_m \hat{\lambda}} + 1, \tag{17}$$

where $x \in \mathbf{R}^2$, or more explicitly, $x_1 \in \mathcal{X}_1 = [a_M/D_M, 1 - a_M/D_M]$ and $x_2 \in \mathcal{X}_2 = [1 - (\bar{C}-1)/C_M, (\bar{C}-1)/C_M]$. From the symmetry of $\hat{\lambda}(x)$ w.r.t. $x = (0.5, 0.5)^T$, we assume $x_1 \leq x_2$ without loss of generality, and we define the feasible set $\mathcal{X} = \{x \in \mathcal{X}_1 \times \mathcal{X}_2 : x_1 \leq x_2\}$. Using the implicit function $f(\hat{\lambda}, x)$, the gradient $\partial \hat{\lambda}/\partial x$ is expressed as $\partial \hat{\lambda}/\partial x_i = -(\partial f/\partial x_i)/(\partial f/\partial \hat{\lambda})$ under $f(\hat{\lambda}, x) = 0$. Since the gradient

$$\frac{\partial f}{\partial \hat{\lambda}} = \frac{x_1 D_M (x_2 C_M)^2}{(x_1 D_M - x_2 C_M \hat{\lambda})^2} + \frac{(1-x_1)D_M((1-x_2)C_M)^2}{((1-x_1)D_M - (1-x_2)C_M \hat{\lambda})^2} + \sum_{m \neq M} \frac{D_m C_m^2}{(D_m - C_m \hat{\lambda})^2}, \tag{18}$$

is strictly positive, we can see $\operatorname{sgn}(\partial \hat{\lambda}/\partial x_i) = \operatorname{sgn}(-\partial f/\partial x_i)$. The gradient $\partial f/\partial x$ is expressed as

$$\begin{cases} \dfrac{\partial f}{\partial x_1} = -\dfrac{(x_2 - x_1)D_M^2 C_M^2 \hat{\lambda}((x_1 + x_2 - 2x_1 x_2)D_M - 2(1-x_2)x_2 C_M \hat{\lambda})}{(x_1 D_M - x_2 C_M \hat{\lambda})^2((1-x_1)D_M - (1-x_2)C\hat{\lambda})^2}, \\ \dfrac{\partial f}{\partial x_2} = -\dfrac{(x_2 - x_1)D_M^2 C_M^2 \hat{\lambda}((x_1 + x_2 - 2x_1 x_2)C_M \hat{\lambda} - 2(1-x_1)x_1 D_M)}{(x_1 D_M - x_2 C_M \hat{\lambda})^2((1-x_1)D_M - (1-x_2)C_M \hat{\lambda})^2}, \end{cases} \tag{19}$$

thus $\partial \hat{\lambda}/\partial x = (0,0)$ is satisfied if and only if $x_1 = x_2$. Since the set $x_1 = x_2$ is on the boundary of $\mathcal{X}$, the optimal solution $x^* = \arg\max \hat{\lambda}(x)$ is on the boundary $\partial \mathcal{X}$ of the feasible set $\mathcal{X}$.

Using the fact $f(\mu) > 0 \Rightarrow \hat{\lambda} < \mu$ and $f(\mu) < 0 \Rightarrow \hat{\lambda} > \mu$ where the parameter μ satisfies $x_1 D_M/x_2 C_M < \mu < (1-x_1)D_M/(1-x_2)C_M$, we can derive the fact:

$$\begin{cases} \operatorname{sgn}\left(f\left(\dfrac{(x_1 + x_2 - 2x_1 x_2)}{2x_2(1-x_2)}\dfrac{D_M}{C_M}\right)\right) = -\operatorname{sgn}\left(\dfrac{\partial f}{\partial x_1}\right) = \operatorname{sgn}\left(\dfrac{\partial \hat{\lambda}}{\partial x_1}\right), \\ \operatorname{sgn}\left(f\left(\dfrac{2x_1(1-x_1)}{(x_1 + x_2 - 2x_1 x_2)}\dfrac{D_M}{C_M}\right)\right) = \operatorname{sgn}\left(\dfrac{\partial f}{\partial x_2}\right) = -\operatorname{sgn}\left(\dfrac{\partial \hat{\lambda}}{\partial x_2}\right), \end{cases} \tag{20}$$

under the assumption $x_1 < x_2$.

By calculating the signs of

$$\begin{cases} f\left(\dfrac{x_1+x_2-2x_1x_2}{2x_2(1-x_2)}\dfrac{D_M}{C_M}\right) = \dfrac{(1-2x_1)2x_2(1-x_2)C_M}{x_2-x_1}+1 \\ \qquad + \displaystyle\sum_{m\neq M}\dfrac{2x_2(1-x_2)D_mC_mC_M}{2x_2(1-x_2)D_mC_M-(x_1+x_2-2x_1x_2)D_MC_m}, \\ f\left(\dfrac{2x_1(1-x_1)}{x_1+x_2-2x_1x_2}\dfrac{D_M}{C_M}\right) = \dfrac{(1-2x_2)(x_1+x_2-2x_1x_2)C_M}{x_2-x_1}+1 \\ \qquad + \displaystyle\sum_{m\neq M}\dfrac{(x_1+x_2-2x_1x_2)D_mC_mC_M}{(x_1+x_2-2x_1x_2)D_mC_M-2x_1(1-x_1)D_MC_m}, \end{cases} \quad (21)$$

we find that two values $\xi_1(x_2)$ and $\xi_2(x_1)$ exist such that

$$\operatorname{sgn}(x_1-\xi_1) = -\operatorname{sgn}\left(\dfrac{\partial\hat{\lambda}}{\partial x_1}\right), \quad \operatorname{sgn}(x_2-\xi_2) = \operatorname{sgn}\left(\dfrac{\partial\hat{\lambda}}{\partial x_2}\right). \quad (22)$$

This indicates that the optimal solution x^* is the closest point to the intersection of $\partial\hat{\lambda}/\partial x_1 = 0$ and $x_2 = (\bar{C}-1)/C_M$ (see Fig. 3). Substituting these conditions for the first equation of (20), we get the algebraic Eq. (16). □

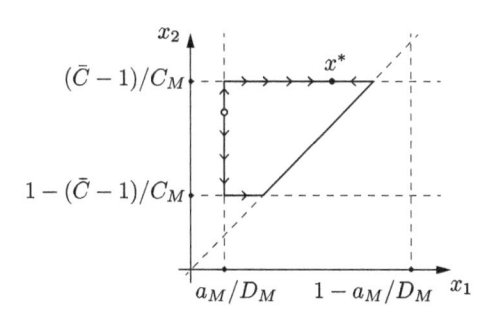

Fig. 3. Gradient of $\hat{\lambda}(x)$ on boundary of feasible space $\mathcal{X}$.

Using the optimal solution $(\bar{C}-1, d^*)$ of the relaxed problem (15) and the gradient shown in Fig. 3, we get the optimal solution $(\bar{C}-1, D^*)$ of the original problem (13), where D^* is either $D^* = \arg\min_{D\in\mathcal{W}_M}|d^*-D|$ s.t. $D \geq d^*$ or $D^* = \arg\min_{D\in\mathcal{W}_M}|d^*-D|$ s.t. $D \leq d^*$. The complexity of this computation (Algorithm 1) is $\mathcal{O}(|\mathcal{W}_M|)$ at most, while the number of the combination in the original problem (13) is $\mathcal{O}\left((2\bar{C}-C_M-1)|\mathcal{W}_M|\right)$. Therefore, we can reduce the computational complexity to find the threshold $\bar{\lambda}$.

5 Component Size Preservation Control

We propose a control law to guarantee the component size preservation condition $\hat{\lambda}^{(i)} > \bar{\lambda}^{(i)}$ in this section. The preservation condition will be maintained by

Algorithm 1. Compute threshold $\bar{\lambda}$

Require: 2-tuples (C_m, D_m), weight-sum set $\mathcal{W}_M$
Ensure: threshold $\bar{\lambda}$
  Find d^* satisfying the equation (16)
  $a^* \leftarrow \min\left(D_M - a_M, (\bar{C}-1)D_M/C_M\right)$
  if $a_M \leq d^* D_M \leq a^*$ **then**
    $D_+ \leftarrow \arg\min_{D \in \mathcal{W}_M} |d^* D_M - D|$ s.t. $D \geq d^* D_M$
    $D_- \leftarrow \arg\min_{D \in \mathcal{W}_M} |d^* D_M - D|$ s.t. $D \leq d^* D_M$
    return $\max\{\hat{\lambda}(\mathcal{G}_{T(M,\bar{C}-1,D_+)}), \hat{\lambda}(\mathcal{G}_{T(M,\bar{C}-1,D_-)})\}$
  else
    return $\max\{\hat{\lambda}(\mathcal{G}_{T(M,\bar{C}-1,a_M)}), \hat{\lambda}(\mathcal{G}_{T(M,\bar{C}-1,a^*)})\}$
  end if

an artificial potential function $U(\xi^{(1)}, \ldots, \xi^{(N)})$ such that U is monotonically decreasing and $\lim_{\xi^{(i)} \to 0+} U = +\infty$, where $\xi^{(i)} = \hat{\lambda}^{(i)} - \bar{\lambda}^{(i)}$. The effectiveness of the control method is shown by numerical simulations.

Considering an artificial potential function

$$U(p) = k_p \sum_{i \in \mathcal{V}} \coth(\hat{\lambda}^{(i)} - \bar{\lambda}^{(i)}), \tag{23}$$

where $p = (p_1^T, \ldots, p_N^T)^T \in \mathbf{R}^{nN}$ denotes state of the entire system, the preservation control law u_i^{pre} is defined by

$$\begin{aligned} u_i^{\text{pre}} &= -k_p \frac{\partial}{\partial p_i} U(p) \\ &= k_p \sum_{j \in \mathcal{V}} \operatorname{csch}^2(\hat{\lambda}^{(j)} - \bar{\lambda}^{(j)}) \left(\frac{\partial \hat{\lambda}^{(j)}}{\partial p_i} - \frac{\partial \bar{\lambda}^{(j)}}{\partial p_i} \right), \end{aligned} \tag{24}$$

where $k_p > 0$ denotes a control gain, and

$$\frac{\partial \hat{\lambda}^{(j)}}{\partial p_i} = \sum_{k \in \mathcal{N}_i} \left(v_i^{(j)} - v_k^{(j)} \right)^2 \frac{\partial a_{ik}^{(j)}}{\partial p_i}. \tag{25}$$

In this study, we assume $\partial \bar{\lambda}^{(j)}/\partial p_i$ is calculated by a numerical gradient, defined by

$$\frac{\partial \bar{\lambda}^{(j)}}{\partial p_{i,m}} \simeq \frac{\bar{\lambda}^{(j)}(p_{i,m} + \delta) - \bar{\lambda}^{(j)}(p_{i,m} - \delta)}{2\delta}, \tag{26}$$

where $\delta > 0$ denotes a sufficiently small spatial step, $m \in \{1, \ldots, n\}$ and $p_i = (p_{i,1}, \ldots, p_{i,n})^T$. We approximate the right-hand of (26) to be equal to zero if $j \notin \mathcal{N}_i \cup \{i\}$, since the 2-tuples $(C_m^{(j)}, D_m^{(j)})$ are unchanging unless the remaining component $C_m^{(j)}$ splits into some components or merges with other component. The perturbed algebraic connectivity $\hat{\lambda}^{(i)}$ is approximated as $\hat{\lambda}^{(i)} \simeq \lambda^{(i)}(\varepsilon)/\varepsilon$. Accordingly, the control law (24) can be executed in a distributed fashion.

(a) $N = 20$ (b) $N = 30$

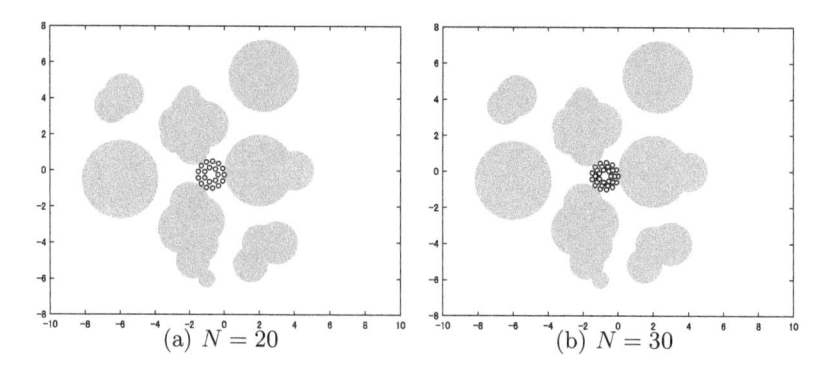

Fig. 4. Nominal initial positions in simulation. Initial positions $p(0) \in \mathbf{R}^{2N}$ are randomly generated based on these nominal positions.

5.1 Simulation Examples

Here we demonstrate the effectiveness of our control method using numerical simulations. The parameter are set as follows: the number of robots is $N \in \{20, 30\}$, the communication upper-bound is $R = 2.8$, the perturbation parameter is $\varepsilon = 0.003$, and the control gain is $k_p = 1$. The link weight function $w(r)$ in (2) is given by

$$w(r) = \frac{1 - \tanh(20(r/R - 0.73))}{2}. \tag{27}$$

Since $\lim_{r \to R} w(r) < \varepsilon$, the perturbed adjacency matrix $A^{(i)}(\varepsilon)$ approximately expresses the network structure such that both links incident to the node i and links with $a_{jk} < \varepsilon$ are removed.

Assuming a coverage task, to be performed with a connected network, we define an artificial potential function $U_1(p) = -\sum_{i \in \mathcal{V}} \sum_{j \in \mathcal{N}_i} \|p_i - p_j\|$ to increase distances among the robots. Obstacles to be avoided by the robots are set as illustrated by the gray circles in Fig. 4, and also we define an artificial potential for the obstacle avoidance as $U_2(p) = \sum_{i \in \mathcal{V}} \sum_{s \in \mathcal{O}} \|p_i - o_s\|^{-1}$, where $\mathcal{O}$ denotes an obstacle set and $o_s \in \mathbf{R}^2$ denotes the obstacle position. To assure the connectedness of the network, we employ the potential $U_3(p) = \coth(\lambda - 0.001)$. Then, the task control input u_i^{task} is given by

$$u_i^{\text{task}} = -\frac{\partial}{\partial p_i} (0.5 U_1 + 0.1 U_2 + U_3). \tag{28}$$

Initial positions of the robots $p_i(0)$ are randomly generated by Gaussian distribution $\mathcal{N}(\mu, \sigma^2)$ excepting non bi-connected configurations, where the expected value $\mu \in \mathbf{R}^2$ is illustrated by the white circles in Fig. 4, and the standard deviation $\sigma = 0.5$.

The component size threshold is set $\bar{C} \in \{11, \ldots, 19\}$ when $N = 20$ and $\bar{C} \in \{16, \ldots, 29\}$ when $N = 30$. We additionally ran simulations without the preservation control law ($u_i^{\text{pre}} = 0$) for comparison. Each simulation set was performed 10 times.

426   T. Murayama and L. Sabattini

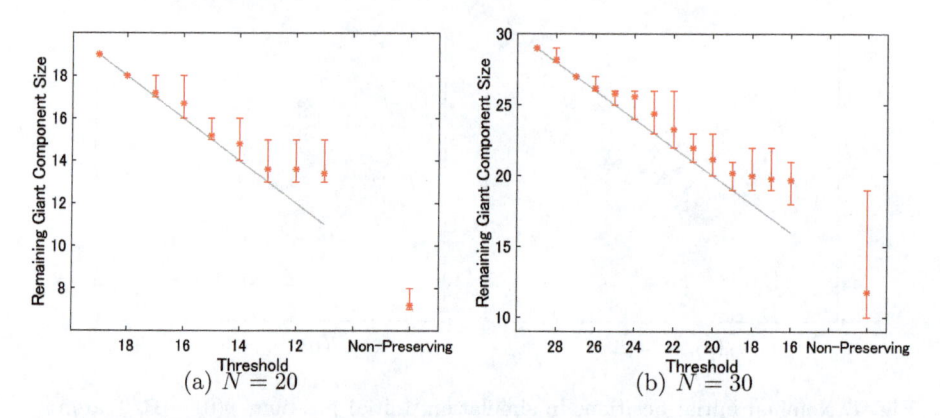

Fig. 5. Simulation results: remaining giant component size $\min_i C_M^{(i)}$ v.s. component size threshold $\bar{C}$. The star markers illustrate the average values of 10 trials, and the error bars show the min-max ranges.

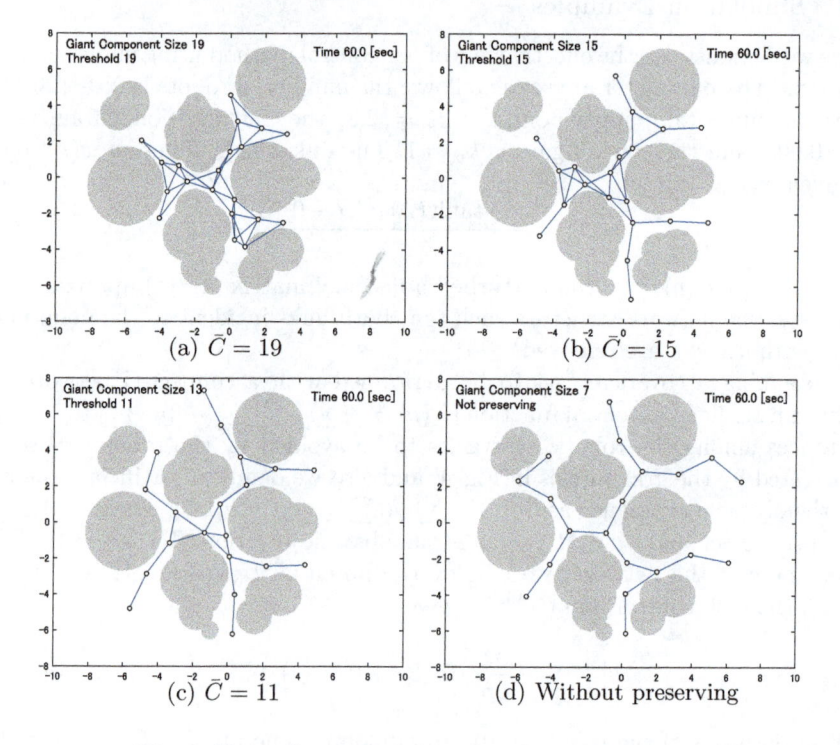

Fig. 6. Network structure snapshots in cases $N = 20$.

Statistical results of the 10 trials are shown in Fig. 5. We can observe that the remaining giant component size $C_M^{(i)}$ is greater than or equal to the threshold $\bar{C}$ for all cases, thanks to the component preservation control law proposed in this work. Most of the component sizes achieved without the preservation control

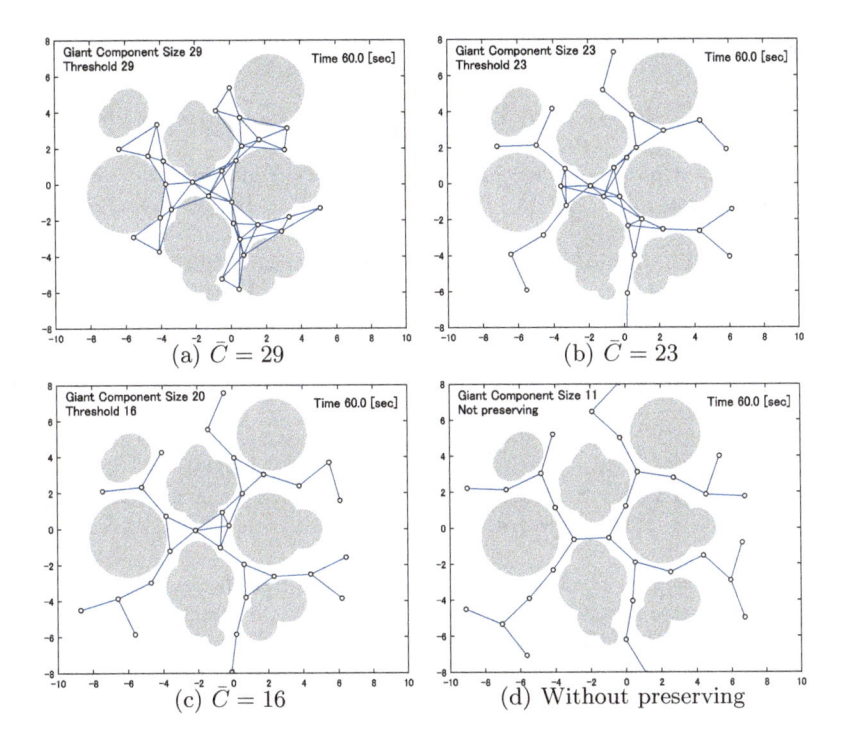

Fig. 7. Network structure snapshots in cases $N = 30$.

law go below the component sizes with the preservation. The network structure with the threshold $\bar{C} = N - 1$ stays bi-connected since it implies the network will be connected even if any one robot fails. These results are also confirmed in simulation snapshots Figs. 6–7.

6 Conclusion

In this paper, we proposed a control method to preserve the giant component size of the multi-robot network after one robot fails. We introduce the concept of perturbed algebraic connectivity which reflects the connectivity of the graph after a node has been removed, and we considered an artificial potential control to preserve the component size greater than a predefined threshold. To reduce conservativeness of the preservation, we analyzed a relation between link weights and the perturbed algebraic connectivity. Additionally we proposed an algorithm to find the perturbed algebraic connectivity threshold for the giant component size preservation.

Acknowledgements. This work was partially supported by JSPS KAKENHI Grant Number 20K19902.

References

1. Zavlanos, M.M., Pappas, G.J.: Distributed connectivity control of mobile networks. IEEE Trans. Robot. **24**(6), 1416–1428 (2008)
2. Sabattini, L., Chopra, N., Secchi, C.: Decentralized connectivity maintenance for cooperative control of mobile robotic systems. Int. J. Robot. Res. **32**(12), 1411–1423 (2013)
3. Dai, J., Zhu, S., Chen, C., Guan, X.: Connectivity-preserving consensus algorithms for multi-agent systems. In: Proceeding on 18th IFAC World Congress, pp. 5675–5680. IFAC, Milan (2011)
4. Hong, H., Yu, W., Fu, J., Yu, X.: Finite-time connectivity-preserving consensus for second-order nonlinear multiagent systems. IEEE Trans. Control Netw. Syst. **6**(1), 236–248 (2019)
5. Cortes, J., Martinez, S., Karatas, T., Bullo, F.: Coverage control for mobile sensing networks. IEEE Trans. Robot. Autom. **20**(2), 243–255 (2004)
6. Kantaros, Y., Thanou, M., Tzes, A.: Distributed coverage control for concave areas by a heterogeneous robot-swarm with visibility sensing constraints. Automatica **53**, 195–207 (2015)
7. Ghedini, C., Secchi, C., Ribeiro, C.H.C., Sabattini, L.: Improving robustness in multi-robot networks. In: Proceeding on 11th IFAC Symposium on Robot Control SYROCO, pp. 63–68. IFAC Salvador (2015)
8. Zareh, M., Sabattini, L., Secchi, C.: Enforcing biconnectivity in multi-robot systems. In: Proceedings on IEEE 55th Conference on Decision and Control CDC, pp. 1800–1805. IEEE, Las Vegas (2016)
9. Panerati, J., et al.: Robust connectivity maintenance for fallible robots. Auton. Robot. **43**(3), 769–787 (2018). https://doi.org/10.1007/s10514-018-9812-8
10. Murayama, T.: Articulation node importance estimation and its correctness for robustness of multi-robot network. In: Proceeding on 12th IFAC Symposium on Robot Control SYROCO, pp. 166–171. IFAC, Budapest (2018)
11. Barabasi, A.-L.: Network science, 1st edn. Cambridge University Press, Cambridge (2016)
12. Murayama, T., Sabattini, L.: Improvement of network fragility for multi-robot robustness. In: Proceeding on 1st IFAC Workshop on Robot Control WROCO, pp. 25–30. IFAC, Daejeon (2019)
13. Murayama, T., Sabattini, L.: Robustness of multi-robot systems controlling the size of the connected component after robot failure. In: Proceeding on 21st IFAC World Congress, pp. 3199–3205. IFAC, Berlin (2020)
14. Yang, P., Freeman, R.A., Gordon, G.J., Lynch, K.M., Srinivasa, S.S., Sukthankar, R.: Decentralized estimation and control of graph connectivity for mobile sensor networks. Automatica **46**(2), 390–396 (2010)

Swarm Localization Through Cooperative Landmark Identification

Sarah Brent[✉], Chengzhi Yuan, and Paolo Stegagno

ICRobots Lab, University of Rhode Island, Kingston, RI 02881, USA
{sbrent,cyuan,pstegagno}@uri.edu

Abstract. In this paper we propose a landmark-based map localization system for robotic swarms. The proposed system leverages the capabilities of a distributed landmark identification algorithm developed for robotic swarms presented in [1]. The output of the landmark identification consists of a vector of probabilities that each individual robot is looking at a particular landmark in the environment. In this work, this vector is used individually by each component of the swarm to feed the measurement update of a particle filter to estimate the robot location. The system was tested in simulation to validate its performance.

Keywords: Swarm · Localization · Landmark identification · Sensor fusion

1 Introduction

A robotic swarm generally consists of relatively inexpensive autonomous robots with limited computational, sensing and communication capabilities. In particular, robots may not have access to global information as their position, the position of their teammates, and commands of a centralized controller. Nevertheless, from the local behavior of the single robots and their interaction with the teammates, a global behavior should emerge to collectively perform some desired task [2]. However, many applications proposed in recent years require some degree of global knowledge, for example search and rescue [3], target search and tracking [4], information gathering and clean up of toxic spills [5], and even construction [6]. In these applications, the knowledge of global position would be beneficial or required for the task execution. Nevertheless, the availability to GPS may be limited by operational conditions as indoor environment.

Many works have addressed the problem of cooperative localization of a multi-robot system in a global frame of reference by using relative measurements between robots (e.g., [7]) and/or relative measurements of landmarks in known locations (e.g., [8]). Reliable relative measurements between robots may be difficult to achieve in practice in a swarm setup, unless dedicated hardware is mounted on the robots (e.g., [9]). Moreover, even if some measurement (e.g., position, bearing, distance) was available through a general purpose sensors as a camera, still the problem of associating the measurements to the specific id of the robot would require dedicated data association algorithms [10].

In this work, we propose a different vision-based approach that takes into account the challenges and limitations posed by a real robotic swarm. First, the robots are equipped with only a general purpose sensor as a camera. Secondly, we assume no relative measurements are available between robots. Although cameras would still allow for relative bearing measurements, due to the swarm setup reliably tagging the robots would not be feasible and we would therefore fall back into a data association problem. However, uniquely identifiable landmarks at known locations are available in the environment, although metric information as relative position or distance from the landmarks is not available. In this setup, each robot could independently perform a landmark based localization.

Single robot (or single sensor) landmark localization has been addressed deeply and broadly for many applications in robotics through either artificial or natural landmarks. An exhaustive review on this topic is outside the scope of this work, that is more focused on multi-robot landmark localization, and the following references should be taken as examples of current state of the art. In [11], authors employ a method that uses a single image from a camera and a minimum of three feature points to recover the camera's viewpoint. In [12], artificial landmarks on the ceiling in an indoor environment are used to localize a two degrees of freedom camera. In [13], authors create and maintain a sparse set of landmarks that are based on biologically motivated feature selection. In [14], the authors use a real-time camera feed from a drone and an AR tag to compute the position of a drone with respect to a point of origin. In [15], the author uses a 2D bar-code landmark for the self localization of mobile robots. In [16], authors propose a method for the global localization problem that uses two landmarks to localize the pose of the robot using bearing angle and distance of landmarks to calculate a possible area of the location of the robot and the particles. In [17], the authors present a landmark matching, triangulation, reconstruction and comparison algorithm that extracts natural landmarks to estimate the position of a robot. In [18], the authors propose a method for robot localization based on the shape and size changes of an object in the robot view as well as trigonometric concepts. In [19], authors deploy a localization method based on artificial as well as natural landmarks employing model based object recognition.

In general, many works proposed on this topic focus not only on the estimation of the robot position, but also on the selection and identification of the landmarks (e.g., [20]). Identification algorithms, however, can often provide wrong results, in particular when many different landmarks are present in the environment, and computational capabilities are limited. In a recent paper [1], we have proposed a system for cooperative identification of landmarks in a robotic swarm, in which the results of single-robot relatively shallow and low-accuracy Convolutional Neural Networks (CNNs) are shared among a robotic team to improve the overall accuracy. The system proposed in [1] is an extension of a previous paper [21] dealing with cooperative object recognition.

In this paper, each robot of the swarm will use the output of the system proposed in [1] to feed a particle filter and estimate its position. This approach is considerably different with respect to state-of-the-art landmark-based cooperative localization systems in literature. In several papers on cooperative localization (e.g., [22–25]) robots use each others as landmarks to improve odometric localization. Other authors [26–28] proposed to use relative position or distance measurements of the landmarks to compute estimates of the robots location via geometric considerations and sensor fusion techniques. In [29], each robot maintains its landmark-based pose estimate using an Unscented Kalman Filter in a common map. In a few papers, specific hardware is developed to use radio [30] or acoustic [9] landmarks and beacons.

None of the above mentioned papers is compatible to our assumptions, due to the use of relative and/or metric measurements, or additional hardware. Lidar is often used to address this type of problem, but there are several issues with using lidar. First, low-cost lidars usually have limited field of view. More expensive lidars can have up to several tens of meters of field of view but are not compatible with a swarm setup. Moreover, the use of lidars in localization usually requires a complete occupancy map of the environment, while we only assume that we know the coordinates of the landmarks in the world frame. From these considerations, we decided to use cameras as the only sensor for our localization algorithm, since for a camera a landmark will be recognizable from multiple distances and even with the interference of possible foreign smaller objects. To the best of our knowledge, this paper is the first to address explicitly the problem of vision-based localization with cooperative landmark identification in robotic swarms.

The rest of the paper is organized as follows. Section 2 will introduce the problem settings, including the robot model and sensor equipment, the communication graph as well as the formal definition of the localization problem of the swarm using cooperative place recognition. Section 3, will describe the methodology used including the system architecture, a recall of the cooperative landmark identification, as well as a description of the particle filter. In Sect. 4, a description of the simulation platform used to validate the localization system. Finally Sect. 5 will conclude the paper.

2 Problem Setting

Let $A = \{A_1, A_2, ..., A_n\}$ be a multi-robot system consisting of n agents. The generic robot A_i, $i = 1, \ldots, n$ moves in a 3D environment populated with a set $\Omega = \{\omega_1, \omega_2, ..., \omega_n\}$ of m objects ω_l, $l = 1, \ldots, m$. Its configuration $\mathbf{c}_i(k) = [\mathbf{c}_i(k) \ \phi_i(k)]^T$ at time step k in a world frame $W = O - XYZ$ is described by the position $\mathbf{q}_i(k) = [x_i(k) \ y_i(k)]^T \in \mathbb{R}^2$ and orientation $\phi_i(k) \in SO(2)$ of a frame attached to a representative point. A_i is modeled as a unicycle:

$$\begin{bmatrix} x_i(k) \\ y_i(k) \\ \phi_i(k) \end{bmatrix} = \begin{bmatrix} x_i(k-1) \\ y_i(k-1) \\ \phi_i(k-1) \end{bmatrix} + \begin{bmatrix} T\cos\phi_i(k) & 0 \\ T\sin\phi_i(k) & 0 \\ 0 & T \end{bmatrix} \begin{bmatrix} v_i(k) \\ \eta_i(k) \end{bmatrix}, \quad (1)$$

where $v_i(k)$ and $\eta_i(k)$ are the linear and angular velocities respectively, and T is the duration of the time step. In general, robots are not aware of their global positions in W, nor they have access to each other's relative position or bearing. However, we assume that the robots are able to communicate with each other within a certain range r. Hence, we define the communication graph as an ordered pair

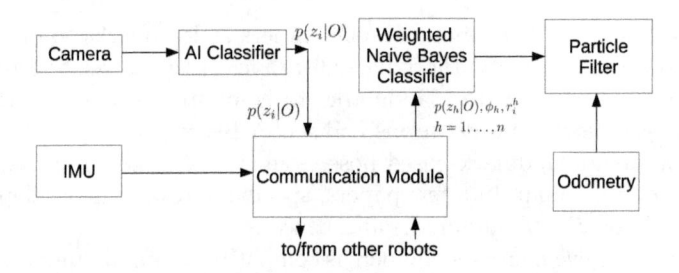

Fig. 1. Block scheme of the system running on robot A_i.

$G(k) = (\mathcal{N}, \mathcal{E}(k))$ consisting of nodes $\mathcal{N}$ (the robots) and edges $\mathcal{E}(k)$. Note that in general the communication graph is time variant. An edge $e = \{i, j\}$ is an unordered pair such that if $\{i, j\} \in \mathcal{E}(k)$, robots A_i and A_j can communicate at time step k. This implies that the underlying communication graph is undirected, i.e., if A_i communicates with A_j, then conversely A_j can communicate with A_i. We also will be operating under the assumption that the communication graph is connected, i.e., there is a path between any two nodes of the graph. However, it should be noted that if the communication network gets disconnected, the swarm will continue working as two or more separate subgroups and will perform the algorithm on the respective sub-networks. It is outside the scope of this paper to study the problem of controlling the swarm so that this assumption is verified. However, it can be achieved assuming that the robots move according to a connectivity maintenance swarming algorithm as the one proposed in [31].

Each A_i is equipped with and odometry module, that provide measurements $\mathbf{u}_i(k) = [\bar{v}_i(k)\ \bar{\eta}_i(k)]^T$ of the linear and angular velocity at all time steps. Each A_i is also equipped with a camera and gathers images $z_i(k)$ of an object $\omega^i(k) \in \Omega$, where the superscript i identifies the specific object observed by robot A_i at time step k. In general, different robots can potentially observe different objects in the environment at the same time. In addition, each robot A_i will be able to collect measurements $\bar{\phi}_i(k)$ of its own yaw angle $\phi_i(k)$ in the world frame W through a magnetometer. In the following, we will indicate with $Z(k) = \{z_i(k), i = 1, 2, ..., n\}$ the set of exteroceptive measurements collected by all the robots at time step k, and with $\Phi(k) = \{\bar{\phi}_i(k), i = 1, 2, ..., n\}$. Collectively, we will indicate with $Z_\Phi(k) = \{Z(k), \Phi(k)\}$ the set of all camera and yaw measurements.

Finally, we define n random variables $O^i(\omega, k), i = 1, ..., n$, one for each A_i, that represents the objects observed by $A_i, i = 1, ..., n$ at time step k:

$$O^i(\omega, k) = O^i(k) = l \Leftrightarrow \omega^i(k) = \omega_l \qquad (2)$$

Then, the probability $p(O^i(k) = l) = p(O^i(k))$ is the probability that $\omega^i(k) = \omega_l$.

The problem that we will address in this work is formally introduced as:

Problem 1. *The problem of localizing the agents in the world frame W is the problem of computing an estimate $\hat{\mathbf{c}}_i(k), i = 1, ..., n$ of their configurations $\mathbf{c}_i(k), i = 1, ..., n$ at each time step k given all exteroceptive and yaw measurements $Z_\Phi(s), s = 1, ..., k$ and all odometry measurements $\mathbf{u}_i(s), s = 1, ..., k$ at all time steps up to k.*

3 Methodology

3.1 System Architecture

The block scheme of the system running on each robot A_i is depicted in Fig. 1. The image collected by the camera, A_i's measurement $z_i(k)$, is passed through an AI classifier to determine which object A_i is observing. The output of the classifier is the m-vector of probabilities $P(z_i(k)|O^i(k))$ that A_i is observing ω_l. This information is then provided to the communication module that broadcast it to A_i's communication neighbors together with the measured yaw angle $\phi_i(k)$.

As all robots do the same, A_i's communication module also receives the probability vectors $P(z_h(k)|O^h(k))$ and the yaw angles $\phi_h(k)$, $h = 1, \ldots, n$, $h \neq i$ of all other robots in the swarm. The communication module also implements a gossiping algorithm so that each robot A_i in the team can receive the probability vectors and yaw angle measurements of all the robots in the swarm, even the ones that are not directly communicating with A_i itself. With an appropriate communication protocol, A_i also compute an estimate $r_i^h(k)$ of the communication distance between itself and the generic robot A_h, $h = 1, \ldots, n$, $h \neq i$. The communication distance is the number of communication steps that are needed for a message sent from robot A_h to reach robot A_i at time step k, and is equivalent to the graph length of the shortest path that connects nodes i and h in the communication graph $G(k)$.

The probability vector computed by the A_i's AI classifier, as well as the ones received by the other robots are passed to the Weighted Naive Bayesian Classifier (WNBC) together with the yaw angles $\phi_i(k)$, $\phi_h(k)$ and the estimated communication distances $r_i^h(k)$. This information is used by the WNBC to iteratively compute $P(O^i(k)|Z_\Phi(k))$. A_i uses this information computed locally to feed the measurement update of a particle filter that produces the estimate $\hat{\mathbf{c}}_i(k)$. The odometry measurements are used in the time update.

3.2 Cooperative Landmark Identification

This subsection recalls the concepts presented in [1] to perform the instantaneous cooperative landmark identification needed to feed the particle filters. Since the algorithm is instantaneous, for the sake of clarity in this section we will drop the time dependency represented by (k) in all variables. Each robot uses a standard single-view recognition algorithm, a convolution neural network (CNN) on the Tensorflow platform. CNN's are frequently used with image data for recognition purpose. To set up the individual CNN's, we have created a training and a testing dataset on the simulated world that we will be using to demonstrate our cooperative algorithm. In that world, we have deployed 17 buildings that are used as landmarks for the localization problem. Each dataset includes tens of images for each building collected from different points of view. A 5-layered CNN was learned using the training data set. The low number of layers in the relatively shallow CNN was mandated by the limited computational capabilities of the hardware this algorithm is meant for, the onboard computer of the robots. Then, we evaluated the single robot recognition capabilities with the testing dataset, resulting in a single robot correct recognition rate of 77%. The output of our CNN is a probability vector $p(z_i|O^i)$, which describes the probability that the observed object is of a certain known type.

Each A_i communicates its computed $p(z_i|O^i)$ over the network, together with its measured yaw angle ϕ_i. This means that the communication neighbors of R_i will receive its measured probabilities and yaw angle. However, every member of the team eventually needs to receive $p(z_i|O^i)$, $i = 1, \ldots, n$ to compute $P(\omega^i = \omega_l|Z_\Phi)$. Therefore, the robots enact a gossiping algorithm to spread the information among the team. This means that at a certain point, a generic robot A_j will send to its communication neighbors the data from A_i in a message that we denote as S_i^j, whose format is:

$$S_i^j = \left[p(z_i|O^i)^T \ r_i^j \ \phi_i \ i \right]^T, \tag{3}$$

where $p(z_i|O^i)$ and ϕ_i are the communicated data, and i is the indication of the owner of the measurements. r_i^j is an estimate of the communication distance between A_i and A_j, and is computed by using a counter that is increased every time that the message is rebounced by a robot to its communication neighbors.

The final goal of A_i is to compute the m-vector of probabilities $p(O^i|Z_\Phi)$. At this aim, a distributed Naive Bayes Classifier (NBC) was proposed in [21]. In [1], a weighting factor was introduced to take into account the possibility that robots may look at different objects at the same time. In the NBC, the yaw information is not used to compute $p(O^i|Z_\Phi) = p(O^i|Z)$, and the probability vectors computed by all robots converge to the same value $p(O^i|Z) = p(O^j|Z), \forall i, j = 1, \ldots, n$. To compute $p(O^i|Z)$, we begin by applying Bayes rule, and $p(O^i|Z)$ can be rewritten as:

$$p(O^i|Z) = \frac{p(O^i)p(Z|O^i)}{p(Z)} \tag{4}$$

By recursively applying the definition of conditional probability, the numerator of the right-hand side of Eq. (4) can be rewritten as:

$$\begin{aligned} p(O^i)p(Z|O^i) &= p(O^i)p(z_i, i = 1, \ldots, n|O^i) \\ &= p(O^i)p(z_1|O^i)p(z_2|O^i, z_1) \ldots p(z_n|O^i, z_1, \ldots, z_{n-1}) \end{aligned} \tag{5}$$

By assuming conditional independence of the measurements z_i, Eq. (5) can be simplified as:

$$p(O^i)p(Z|O^i) = p(O^i) \prod_{j=1}^{n} p(z_j|O^j) \tag{6}$$

Each A_i recursively computes Eq. (6) by maintaining at all communication steps b an estimate of $P(O^i|Z_i^b)$, where $Z_i^b = \{z_q, \forall q \in ID_i^b\}$ is the set of all measurements received by A_i up to the communication step b. Every time that A_i receives a new message S_h^j such that $h \notin ID_i^b$, it will update its current estimate incorporating the new measurements:

$$p(O^i|Z_i^b, z_h) = \frac{p(O^i)p(Z_i^b, z_h|O^i)}{p(Z_i^b, z_h)} = \frac{p(O^i)p(Z_i^b|O^i)p(z_h|O^i)}{p(Z_i^b)P(z_h)} \tag{7}$$

This algorithm relies on the assumption that A_i and A_h are collecting measurements of the same object ($w^i = w^h$). In the setting of this work, however, we have rejected this assumption as it is not compatible with a real world scenario. Therefore, we define the following random variable

$$R_h^i = \begin{cases} 1 & \text{if } w^i = w^h \\ 0 & \text{otherwise} \end{cases} \tag{8}$$

whose value is 1 if A_i and A_h are collecting measurements of the same object, and zero otherwise.

Introducing R_h^i, and considering that if $R_h^i = 0$ the measurement z_h carries no information on the object observed by robot A_i, we can write:

$$\begin{aligned} p(O^i|Z_i^b, z_h) &= p(O^i|Z_i^b, z_h, R_h^i)p(R_h^i) + p(O^i|Z_i^b, z_h, \bar{R}_h^i)p(\bar{R}_h^i) = \\ &= p(O^i|Z_i^b, z_h, R_h^i)p(R_h^i) + p(O^i|Z_i^b, \bar{R}_h^i)p(\bar{R}_h^i) = \\ &= p(O^i|Z_i^b, z_h)p(R_h^i) + p(O^i|Z_i^b)p(\bar{R}_h^i). \end{aligned} \tag{9}$$

Introducing Eq. (7) into (9):

$$\begin{aligned} p(O^i|Z_i^b, z_h) &= \frac{p(O^i)p(Z_i^b|O^i)p(z_h|O^i)P(R_h^i)}{p(Z_i^b)p(z_h)} + p(O^i|Z_i^b)p(\bar{R}_h^i) = \\ &= p(O^i|Z_i^b)\frac{p(z_h|O^i)p(R_h^i)}{p(z_h)} + p(O^i|Z_i^b)p(\bar{R}_h^i). \end{aligned} \tag{10}$$

In equation (10) the term $p(z_h)$ is a normalization factor α such that

$$\sum_l \frac{p(O^i = l|Z_i^b)p(z_h|O^i = l)}{p(z_h)} = 1, \tag{11}$$

therefore:

$$p(O^i|Z_i^b, z_h) = \alpha\, p(O^i|Z_i^b)p(z_h|O^i)P(R_h^i) + p(O^i|Z_i^b)p(\bar{R}_h^i). \tag{12}$$

Considering that

$$P(\bar{R}_h^i) = 1 - P(R_h^i) \tag{13}$$

the final step consists in computing the probability $P(R_h^i)$ that $w^i = w^h$. In general, $p(R_h^i)$ may depend on several factors and we do not have a standard way to compute it. In this work, we assumed that $p(R_h^i)$ depends on the distance and the relative orientation between A_i and A_h, and that these two factors are independent from each other. This is based on two considerations. First, the further apart the robots are, the less likely they are to be observing the same object. As the robots do not have direct access to their relative distance, they can use the estimated communication distance (which also provide an estimate of their cartesian distance) to compute the following:

$$p(R_h^i|r_i^h) = \frac{1}{r_i^{h\frac{r}{H}}} \tag{14}$$

where H is the average size of the objects in the world. Similarly, if the two robots look in different directions (i.e., have a relative orientation close to π), they are unlikely to be watching the same object. The relative orientation can be computed through the use of the yaw measurements ϕ_i, ϕ_h, therefore we have considered

$$p(R_h^i|\phi_i, \phi_h) = p(R_h^i|\phi_i - \phi_h) \qquad (15)$$

Finally, equations (14) and (15) can be combined into the following:

$$p(R_h^i) = p(R_h^i|\phi_i, \phi_h, c_i^h) = p(R_h^i|\phi_i, \phi_h)p(R_h^i|c_i^h) \qquad (16)$$

Note that by applying equation (12) instead of the NBC, each robot will compute its own personalized vector of probabilities that takes into account the relevance of other robot measurements to its individual identification process.

3.3 Particle Filter for Localization

Each A_i maintains a particle filter to estimate its configuration in the world frame W. Therefore, the state of each particle is an estimate $\hat{\mathbf{c}}_i^\delta(k) = [\mathbf{q}_k^\delta \; \phi_k^\delta]^T = [x_k^\delta \; y_k^\delta \; \phi_k^\delta]^T, p = 1, \ldots, \Delta$, including position and orientation of the robot in W, where Δ is the number of particles maintained. The filter consists in a time update, performed every time that a new odometry measurement $\mathbf{u}_i(k)$ is available, and a measurements update, performed every time that a new probability vector $P(O^i(k)|Z_\Phi(k))$ is computed. Note that the magnetometer readings $\bar{\phi}_i(k)$ where already used to compute $P(O^i(k)|Z_\Phi(k))$, so they will not be used in the particle filter to avoid reusing the same measurements twice.

3.3.1 Time Update

The time update of each particle follows the motion model of the unicycle (1):

$$\begin{aligned}
\phi_k^\delta &= \phi_{k-1}^\delta + T(\bar{\eta}_i(k) + \nu_\eta^\delta) \\
x_k^\delta &= x_{k-1}^\delta + T(\bar{v}_i(k) + \nu_v^\delta)\cos((\phi_k^\delta + \phi_{k-1}^\delta)/2) \\
y_k^\delta &= y_{k-1}^\delta + T(\bar{v}_i(k) + \nu_v^\delta)\sin((\phi_k^\delta + \phi_{k-1}^\delta)/2)
\end{aligned} \qquad (17)$$

where ν_η^δ and ν_v^δ are samples from the noise affecting the odometry measurements, assumed to be Gaussian with zero mean and known covariance.

3.3.2 Measurement Update

For the measurement update, we use a map of the environment in which the facade of each building ω_l observable by the robots is represented as a segment specified through the coordinates of a starting $\mathbf{a}_{ls}^W = [x_{ls}^W \; y_{ls}^W]^T$ and a final $\mathbf{a}_{lf}^W = [x_{lf}^W \; y_{lf}^W]^T$ point in W. This representation is compact, utilizes very low memory, and easy to implement as it requires very low information. It is also easily expandable to more complex environments.

For each particle δ, it is necessary to compute the expected measurement, i.e., what building the robot would see if it was in the location of that particle. First, the segments are rotated and translated in the frame attached to the robot:

$$\mathbf{a}_{l*}^{\delta} = R(\phi_k^{\delta})^T (\mathbf{a}_{l*}^W - \mathbf{q}_k^{\delta}), \; * = s, f, \; l = 1, \ldots, n, \tag{18}$$

where the superscript δ in $\mathbf{a}_{l*}^{\delta}$ indicates that the point is expressed in the frame δ identified by the particle, and $R(\phi_k^{\delta})$ is the 2D rotation matrix of an angle ϕ_k^{δ}.

The segment, now expressed in the frame δ, identifies a line in the form $ax + by + c = 0$. The coefficients are computed as:

$$\begin{aligned} a &= (y_{lf}^{\delta} - y_{ls}^{\delta}) \\ b &= -(x_{lf}^{\delta} - x_{ls}^{\delta}) \\ c &= -x_{ls}^{\delta}(y_{lf}^{\delta} - y_{ls}^{\delta}) + y_{ls}^{\delta}(x_{lf}^{\delta} - x_{ls}^{\delta}) \end{aligned} \tag{19}$$

To check if the particle is oriented in the direction of the building, we compute the intersection of the line $ax + by + c = 0$ with the line $y = 0$, obtaining $x = -c/a$.

Clearly, if the resulting $x < 0$, the particle is not oriented towards that particular building. Moreover, if the point $[-c/a \; 0]$ is between $\mathbf{a}_{ls}$ and $\mathbf{a}_{lf}$, the particle is oriented towards that l-th building.

At this point, there is still the possibility that the particle is behind the facade of the building. To check if this is the case, we compute a point $[x_n \; y_n]^T$ along the positive normal of the segment $[-(y_{lf}^{\delta} - y_{ls}^{\delta}) \; (x_{lf}^{\delta} - x_{ls}^{\delta})]^T$ and we check that the condition $c(ax_n + by_n + c) > 0$ is verified. If this is the case, the particle is oriented towards the l-th building from the correct direction, therefore we update its weight w_k^{δ} of a given particle by multiplying its current weight w_{k-1}^{δ} by $p(O^i(k) = \omega^l | Z_{\Phi}(k))$, that is, the probability that the robot is oriented towards the l-th building according to the measurements fused through the weighted Bayes classifier:

$$w_k^{\delta} = w_{k-1}^{\delta} * p(O^i(k) = \omega^l | Z_{\Phi}(k)) \tag{20}$$

Note that, being a probability, $0 \le p(O^i(k) = \omega^l | Z_{\Phi}(k)) \le 1$. Therefore, the weights of the particles will decrease a little if $p(O^i(k) = \omega^l | Z_{\Phi}(k)) \simeq 1$, and will decrease by several orders of magnitude if $p(O^i(k) = \omega^l | Z_{\Phi}(k)) \to 0$. However, if a particle is not associated to any building, as for example when the particle is outside the mapped area, following this algorithm it would eventually have the largest weight. To avoid this paradoxical situation, particles that are not associated to any building automatically receive very low weights. Moreover, since misidentifications are still possible, the resampling of the particles is not performed every measurements update, so that a single misidentification will not cause the removal of correct particles. The mean of the particles of the filter running on A_i is eventually selected as the estimate:

$$\hat{\mathbf{c}}_i(k) = \text{mean}_{\delta}(\hat{\mathbf{c}}_i^{\delta}(k)), \; i = 1, \ldots, n. \tag{21}$$

Fig. 2. Simulated city in Gazebo.

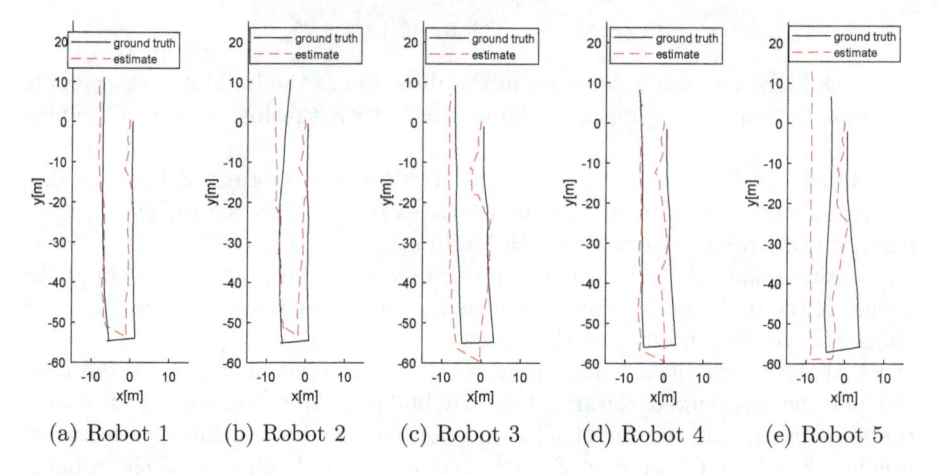

(a) Robot 1 (b) Robot 2 (c) Robot 3 (d) Robot 4 (e) Robot 5

Fig. 3. The estimated robot trajectories (red dashed lines) and the ground truth data (black solid lines) for all five robots in a typical experiment.

4 Simulations

We have tested the proposed algorithm in simulation using a complex environment in the Gazebo/ROS framework. In the simulated world, we have placed five small ($\sim$20 cm) robots and 17 unique buildings that act as landmarks. A view of the simulated world is provided in Fig. 2. In a typical simulation, the robots move in the world roughly in a line formation running the cooperative localization system. At the same time, ground truth data are collected.

The results of a typical experiment are presented in Figs. 3 and 4. Figure 3 reports the estimated robot trajectories (red dashed lines) and the ground truth data (black solid lines) for all five robots. The plots show how the estimates correctly track the actual paths of the robots. Figure 4 reports the particle distribution (red dots) for five time instants at 0%, 25%, 50%, 75% and 100% of

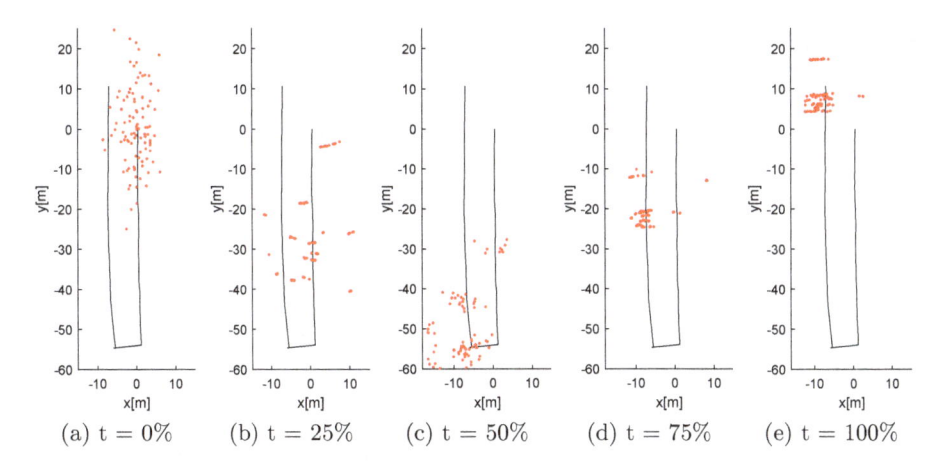

Fig. 4. The particle distribution (red dots) for five time instants at 0%, 25%, 50%, 75% and 100% of the simulation time for Robot 1. The black solid lines show the ground truth.

the simulation time for Robot 1. The black solid lines show the ground truth. From this plot, it is possible to see how the particles are initially scattered and slowly converges towards the ground truth.

5 Conclusions

In this paper, we have presented a landmark-based localization system that uses the output of a cooperative identification module to feed a set of independent particle filters. The proposed system achieves a collaborative localization setup without the need for knowledge, either in the form of an estimate or of a measurement, of relative positions and/or distance. Moreover, this result is obtained using only a general purpose sensor as a camera, without the need for additional hardware. These features makes the proposed system particularly suited for robotic swarms.

In the future, we plan to test the proposed localization system on real robots with larger data sets. This will also stress out the computational requirements of the algorithms. We foresee that larger maps may require adjustments in the measurement update steps to avoid checking all buildings in the maps, but rather selecting only a few candidate buildings in the proximity of each particle. Moreover, we will study more in depth the problem of particles not associated to any building, and we are planning to incorporate directly in the CNN's a case in which no landmarks are recognized.

Acknowledgments. This work was supported in part by the National Science Foundation under Grant CMMI 1952862.

References

1. Brent, S., Yuan, C., Stegagno, P.: Cooperative place recognition in robotic swarms. In: Proceedings of the 2021 Annual ACM Symposium on Applied Computing, ser. SAC 2021, Virtual Conference (2021)
2. Bayindir, L.: A review of swarm robotics tasks. Neurocomputing **172**, 08 (2015)
3. Bakhshipour, M., Ghadi, M.J., Namdari, F.: Swarm robotics search and rescue: a novel artificial intelligence-inspired optimization approach. Appl. Soft Comput. **57**, 708–726 (2017). http://www.sciencedirect.com/science/article/pii/S1568494617301072
4. Senanayake, M., Senthooran, I., Barca, J.C., Chung, H., Kamruzzaman, J., Murshed, M.: Search and tracking algorithms for swarms of robots: a survey. Robot. Auton. Syst. **75**, 422–434 (2016). http://www.sciencedirect.com/science/article/pii/S0921889015001876
5. Kakalis, N., Ventikos, Y.: Robotic swarm concept for efficient oil spill confrontation. J. Hazard. Mater. **154**, 880–887 (2008)
6. Kayser, M., et al.: Design of a multi-agent, fiber composite digital fabrication system. Sci. Robot. 3(22) (2018). https://robotics.sciencemag.org/content/3/22/eaau5630
7. Luft, L., Schubert, T., Roumeliotis, S.I., Burgard, W.: Recursive decentralized localization for multi-robot systems with asynchronous pairwise communication. Int. J. Robot. Res. **37**(10), 1152–1167 (2018). https://doi.org/10.1177/0278364918760698
8. Wang, B., Rathinam, S., Sharma, R.: Landmark placement for cooperative localization and routing of unmanned vehicles. In: International Conference on Unmanned Aircraft Systems (ICUAS) 2019, pp. 33–42 (2019)
9. Lin, Y., Vernaza, P., Ham, J., Lee, D.D.: Cooperative relative robot localization with audible acoustic sensing. In: IEEE/RSJ International Conference on Intelligent Robots and Systems 2005, pp. 3764–3769 (2005)
10. Franchi, A., Oriolo, G., Stegagno, P.: Mutual localization in multi-robot systems using anonymous relative measurements. Int. J. Robot. Res. **32**(11), 1302–1322 (2013). https://doi.org/10.1177/0278364913495425
11. Zhang, P., Milios, E.E., Gu, J.: Underwater robot localization using artificial visual landmarks. In: IEEE International Conference on Robotics and Biomimetics 2004, pp. 705–710 (2004)
12. Chen, D., Peng, Z., Ling, X.: A low-cost localization system based on artificial landmarks with two degree of freedom platform camera. In: 2014 IEEE International Conference on Robotics and Biomimetics, IEEE ROBIO 2014, pp. 625–630, 04 (2015)
13. Frintrop, S., Jensfelt, P.: Attentional landmarks and active gaze control for visual slam. Robot. IEEE Trans. **24**, 1054–1065 (2008)
14. Jayatilleke, L., Zhang, N.: Landmark-based localization for unmanned aerial vehicles. In: IEEE International Systems Conference (SysCon) 2013, pp. 448–451 (2013)
15. Kobayashi, H.: A new proposal for self-localization of mobile robot by self-contained 2D barcode landmark. In: Proceedings of SICE Annual Conference (SICE) 2012, pp. 2080–2083 (2012)
16. Han, S., Kim, J., Myung, H.: Landmark-based particle localization algorithm for mobile robots with a fish-eye vision system. IEEE/ASME Trans. Mechatron. **18**(6), 1745–1756 (2013)

17. Yuen, D.C.K., MacDonald, B.A.: Vision-based localization algorithm based on landmark matching, triangulation, reconstruction, and comparison. IEEE Trans. Rob. **21**(2), 217–226 (2005)
18. Bangash, S.A., Ghafoor, A.: Vision based mobile node localization using a landmark. In: The 5th International Conference on Automation, Robotics and Applications, pp. 255–259 (2011)
19. Jang, G., Kim, S., Lee, W., Kweon, I.: Robust self-localization of mobile robots using artificial and natural landmarks. In: Proceedings 2003 IEEE International Symposium on Computational Intelligence in Robotics and Automation. Computational Intelligence in Robotics and Automation for the New Millennium (Cat. No. 03EX694), vol. 1, pp. 412–417 (2003)
20. Sahdev, R., Tsotsos, J.K.: Indoor place recognition system for localization of mobile robots. In: 2016 13th Conference on Computer and Robot Vision (CRV), June 2016, pp. 53–60 (2016)
21. Stegagno, P., Massidda, C., Bülthoff, H.H.: Distributed target identification in robotic swarms. In: Proceedings of the 30th Annual ACM Symposium on Applied Computing, Salamanca, Spain, 2015, pp. 307–313 (2015). https://doi.org/10.1145/2695664.2695922
22. Kim, C., Yang, H., Kang, D., Lee, D.: 2-D cooperative localization with omnidirectional mobile robots. In: 2015 12th International Conference on Ubiquitous Robots and Ambient Intelligence (URAI), 2015, pp. 425–426 (2015)
23. Song, K., Tsai, C., Huang, C.: Multi-robot cooperative sensing and localization. In: IEEE International Conference on Automation and Logistics 2008, pp. 431–436 (2008)
24. Wang, L., Cai, X., Fan, W., Zhang, H.: An improved method for multi-robot cooperative localization based on relative bearing. In: IEEE International Conference on Robotics and Biomimetics 2009, pp. 1862–1867 (2008)
25. Li, Z., Acuna, R., Willert, V.: Cooperative localization by fusing pose estimates from static environmental and mobile fiducial features. In: 2018 Latin American Robotic Symposium, 2018 Brazilian Symposium on Robotics (SBR) and. Workshop on Robotics in Education (WRE) 2018, pp. 65–70 (2018)
26. Wu, P., Luo, Q., Kong, L.: Cooperative localization of network robot system based on improved MPF. In: IEEE International Conference on Information and Automation (ICIA) 2016, pp. 796–800 (2016)
27. Kang, H., Kawk, J., Kim, C., Jo, K.: Multiple robot formation keeping and cooperative localization by panoramic view. In: ICCAS-SICE 2009, pp. 3509–3513 (2009)
28. Jennings, C., Murray, D., Little, J.J.: Cooperative robot localization with vision-based mapping. In: Proceedings 1999 IEEE International Conference on Robotics and Automation (Cat. No.99CH36288C), vol. 4, pp. 2659–2665 (1999)
29. Xingxi, S., Bo, H., Tiesheng, W., Chunxia, Z.: Cooperative multi-robot localization based on distributed UKF. In: 2010 3rd International Conference on Computer Science and Information Technology, vol. 6, pp. 590–593 (2010)
30. Kim, C., Song, D., Xu, Y., Yi, J., Wu, X.: Cooperative search of multiple unknown transient radio sources using multiple paired mobile robots. IEEE Trans. Robot. **30**(5), 1161–1173 (2014)
31. Carpio, R.F., Di Giulio, L., Garone, E., Ulivi, G., Gasparri, A.: A distributed swarm aggregation algorithm for bar shaped multi-agent systems. In: 2018 IEEE/RSJ International Conference on Intelligent Robots and Systems (IROS), October 2018, pp. 4303–4308 (2018)

Author Index

A
Adams, Julie A., 214
Agmon, Noa, 16, 320
Albani, Dario, 306
Amini, Arash, 294
Amir, Michael, 320
Ayanian, Nora, 31

B
Begemann, Marian Johannes, 190
Beltrame, Giovanni, 95, 349
Berman, Spring, 401
Bourgeois, Julien, 70, 108
Brent, Sarah, 429
Bruckstein, Alfred M., 320

C
Cao, Yanjun, 95
Carbone, Carlos, 306
Charles, Christian, 190
Chaudhuri, Diptanil, 256
Chen, Austin K., 282
Chen, Jiahe, 269
Choset, Howie, 227
Cowley, Aidan, 95
Crosscombe, Michael, 82

D
Desai, Arjav, 45
Di Caro, Gianni A., 335
Diehl, Grace, 214
Digani, Valerio, 58
Dreier, Sven, 190
Dutta, Ayan, 163

E
Elamvazhuthi, Karthik, 401
Endo, Takahiro, 389

G
Goodrich, Michael A., 376

H
Hamann, Heiko, 190
Hannawald, Max Ferdinand, 190
Harvey, David, 120
Hauert, Sabine, 120
Hirayama, Michiaki, 148
Hogg, Elliott, 120

J
Jain, Puneet, 376
Jimbo, Tomohiko, 202

K
Kaiser, Tanja Katharina, 190
Kakish, Zahi, 401
Kamezaki, Mitsuhiro, 148
Kaufmann, Marcel, 95
Kootstra, Gert, 306
Kress-Gazit, Hadas, 269
Kumar, Vijay, 282

L
Lang, Christine, 190
Lawry, Jonathan, 82
Li, Xueting, 242
Liu, Guangyi, 294
Liu, Yifang, 269

M

Macharet, Douglas G., 282
Magistri, Federico, 306
Martinoli, Alcherio, 134, 148
Marwitz, Florian Andreas, 190
Matsuno, Fumitoshi, 389
Menashe, Nofar, 16
Michael, Nathan, 45
Motee, Nader, 294
Murayama, Toru, 415

N

Napp, Nils, 269
Nardi, Daniele, 306
Naz, André, 108
Niwa, Takahiro, 202

O

O'Kane, Jason M., 163, 256
Ognibene, Dimitri, 306

P

Pacheck, Adam, 269
Panerati, Jacopo, 95
Pappas, George J., 282
Patrick Kreidl, O., 163
Petersen, Kirstin, 269
Petzold, Julian, 190
Piranda, Benoît, 70, 108

Q

Quraishi, Anwar, 134

R

Rahmani, Hazhar, 256
Ramtoula, Benjamin, 95
Rao, Ananya, 227

Richards, Arthur, 120
Rubenstein, Michael, 1, 363

S

Sabattini, Lorenzo, 58, 95, 415
Sartoretti, Guillaume, 227
Shell, Dylan A., 256
Shibata, Kazuki, 202
Shishika, Daigo, 282
Soma, Karthik, 349
Stachniss, Cyrill, 306
Stegagno, Paolo, 176, 429
St-Onge, David, 95
Strawn, Kegan, 31
Swissler, Petras, 363
Sycara, Katia, 242

T

Takáč, Martin, 294
Tanaka, Shota, 389
Thalamy, Pierre, 108
Thivanka Perera, R. A., 176
Trianni, Vito, 306

V

Vardharajan, Vivek Shankar, 349

W

Wang, Hanlin, 1
Wasik, Alicja, 148

Y

Yi, Sha, 242
Yuan, Chengzhi, 176, 429

Z

Ziaullah Yousaf, Abdul Wahab, 335

CPSIA information can be obtained
at www.ICGtesting.com
Printed in the USA
LVHW081100090123
736757LV00003B/20